OPERATIONS RESEARCH

SECOND EDITION

R. PANNEERSELVAM

Professor and Head
Department of Management Studies
School of Management
Pondicherry University

PHI Learning Private Limited

New Delhi-110001

2011

Rs. 295.00

OPERATIONS RESEARCH, Second Edition
R. Panneerselvam

ISBN-978-81-203-2928-7

The export rights of this book are vested solely with the publisher.

Seventeenth Printing (Second Edition) ... ... **August, 2011**

Published by Asoke K. Ghosh, PHI Learning Private Limited, M-97, Connaught Circus, New Delhi-110001 and Printed by Syndicate Binders, A-20, Hosiery Complex, Noida, Phase-II Extension, Noida-201305 (N.C.R. Delhi).

To

My Grandparents

CONTENTS

Preface   *xi*

1. OVERVIEW OF OPERATIONS RESEARCH   1–8

1.1 Introduction *1*
1.2 Concept of a Model *2*
    1.2.1 Steps of Modelling *3*
1.3 Important Topics of Operations Research *3*
1.4 Scope of Operations Research *4*
1.5 Operations Research—A Tool for Decision Support System 6
1.6 Operations Research—A Productivity Improvement Tool *7*
    1.6.1 Increased Output for the Same Input *7*
    1.6.2 Decreased Input for the Same Output *7*
    1.6.3 Increase in the Output is more than the Increase in the Input *7*
    1.6.4 Decrease in the Input is more than the Decrease in the Output *8*
    1.6.5 Increase in the Output with Decrease in the Input *8*

Questions *8*

2. LINEAR PROGRAMMING   9–70

2.1 Introduction *9*
2.2 Concept of Linear Programming Model *9*
    2.2.1 Product Mix Problem *9*
    2.2.2 Assumptions in Linear Programming *12*
    2.2.3 Properties of Linear Programming Solution *12*
2.3 Development of LP Models *13*
2.4 Graphical Method *21*
2.5 Linear Programming Methods *25*
    2.5.1 Simplex Method *25*
    2.5.2 Big *M* Method *30*
    2.5.3 Dual Simplex Method *34*
    2.5.4 Two-phase Method *38*
2.6 Special Cases of Linear Programming *40*
    2.6.1 Identification of Special Cases from Simplex Table *46*
2.7 Duality *47*
    2.7.1 Formulation of Dual Problem *47*
    2.7.2 Application of Duality *53*
2.8 Sensitivity Analysis *56*
    2.8.1 Changes in the Right-hand Side Constants of Constraints *57*

2.8.2 Changes in the Objective Function Coefficients *59*
2.8.3 Adding a New Constraint *61*
2.8.4 Adding a New Variable *63*
Questions 65

3. TRANSPORTATION PROBLEM 71–126

3.1 Introduction *71*
3.2 Mathematical Model for Transportation Problem *72*
3.3 Types of Transportation Problem *74*
    3.3.1 Balanced Transportation Problem *74*
    3.3.2 Unbalanced Transportation Problem *74*
3.4 Methods to Solve Transportation Problem *76*
    3.4.1 Finding the Initial Basic Solution *77*
    3.4.2 Optimizing the Basic Feasible Solution Applying U–V Method *78*
3.5 Transshipment Model *106*
    3.5.1 Transshipment Problem with Sources and Destinations Acting as Transient Nodes *106*
    3.5.2 Transportation Problem with some Transient Nodes between Sources and Destinations *111*
3.6 Modelling the Transportation Problem with Quantity Discounts *115*
    3.6.1 Model for AQDS *117*
    3.6.2 Model for IQDS *118*
Questions 122

4. ASSIGNMENT PROBLEM 127–153

4.1 Introduction *127*
4.2 Zero-One Programming Model for Assignment Problem *128*
4.3 Types of Assignment Problem *130*
4.4 Hungerian Method *130*
4.5 Branch-and-Bound Technique for Assignment Problem *147*
Questions 151

5. NETWORK TECHNIQUES 154–195

5.1 Introduction *154*
5.2 Shortest-path Model *154*
    5.2.1 Systematic Method *154*
    5.2.2 Dijkstra's Algorithm *163*
    5.2.3 Floyd's Algorithm *171*
5.3 Minimum Spanning Tree Problem *174*
    5.3.1 PRIM Algorithm *175*
    5.3.2 Kruskal's Algorithm *181*
5.4 Maximal Flow Problem *186*
    5.4.1 Linear Programming Modelling of Maximal Flow Problem *186*
    5.4.2 Maximal Flow Problem (MFP) Algorithm *188*
Questions 193

6. INTEGER PROGRAMMING
196–229

6.1   Introduction   *196*
6.2   Integer Programming Formulations   *197*
6.3   The Cutting-plane Algorithm   *201*
6.4   Branch-and-Bound Technique   *209*
6.5   Zero-One Implicit Enumeration Algorithm   *219*
     6.5.1   Generalized 0-1 Programming Problem   *221*
     6.5.2   Zero-One Implicit Enumeration Technique   *221*

Questions   *228*

7. INVENTORY CONTROL
230–274

7.1   Introduction   *230*
7.2   Models of Inventory   *230*
     7.2.1   Purchase Model with Instantaneous Replenishment and without Shortages   *231*
     7.2.2   Manufacturing Model without Shortages   *236*
     7.2.3   Purchase Model with Instantaneous Replenishment and with Shortages   *238*
     7.2.4   Manufacturing Model with Shortages   *241*
7.3   Operation of Inventory System   *245*
7.4   Quantity Discount   *247*
7.5   Implementation of Purchase Inventory Model   *250*
     7.5.1   Fixed Order Quantity System (Q System)   *250*
     7.5.2   Periodic Review System (P System)   *251*
7.6   Multiple-item Model with Shortage Limitation   *254*
7.7   Purchase Model of Inventory for Multi-item with Inventory Carrying Cost Constraint   *256*
7.8   EOQ Model for Multi-item Joint Replenishment   *258*
     7.8.1   Purchase Model of Inventory for Multi-item Joint Replenishment without Shortages   *259*
     7.8.2   Manufacturing Model of Inventory with Multi-item Joint Replenishment without Shortages   *262*
7.9   EOQ for the Purchase Model of Inventory for Multi-item Joint Replenishment with Space Constraint   *265*
7.10   Determination of Stock Level of Perishable Items under Probabilistic Condition   *269*

Questions   *272*

8. DYNAMIC PROGRAMMING
275–297

8.1   Introduction   *275*
8.2   Application of Dynamic Programming   *276*
     8.2.1   Capital Budgeting Problem   *276*
     8.2.2   Reliability Improvement Problem   *278*
     8.2.3   Stage-coach Problem (Shortest-path Problem)   *281*
     8.2.4   Cargo Leading Problem   *284*
     8.2.5   Minimizing Total Tardiness in Single Machine Scheduling Problem   *286*
     8.2.6   Optimal Subdividing Problem   *290*

8.2.7 Solution of Linear Programming Problem through Dynamic Programming *292*

Questions 295

9. QUEUEING THEORY — *298–354*

9.1 Introduction *298*
9.2 Terminologies of Queueing System *299*
9.3 Empirical Queueing Models *300*
    9.3.1 $(M/M/1) : (GD/\infty/\infty)$ Model *301*
    9.3.2 $(M/M/C) : (GD/\infty/\infty)$ Model *305*
    9.3.3 $(M/M/1) : (GD/N/\infty)$ Model *309*
    9.3.4 $(M/M/C) : (GD/N/\infty)$ Model (for $C \leq N$) *313*
    9.3.5 $(M/M/C) : (GD/N/N)$ Model (for $C < N$) *318*
    9.3.6 $(M/M/1) : (GD/N/N)$ Model (for $N > 1$) *322*
9.4 Simulation *325*
    9.4.1 Need for Simulation *325*
    9.4.2 Types of Simulation *326*
    9.4.3 Major Steps of Simulation *327*
    9.4.4 Simulation using High-level Languages *327*
    9.4.5 General Purpose Simulation System (GPSS) *337*

Questions 352

10. PROJECT MANAGEMENT — *355–408*

10.1 Introduction *355*
10.2 Phases of Project Management *358*
10.3 Guidelines for Network Construction *359*
10.4 Critical Path Method (CPM) *359*
10.5 Gantt Chart (Time Chart) *365*
10.6 Project Evaluation and Review Technique (PERT) *368*
10.7 Crashing of Project Network *375*
    10.7.1 General Guidelines for Network Crashing *376*
    10.7.2 Crashing of Project Network with Cost Trade-off *377*
10.8 Project Scheduling with Constrained Resources *390*
    10.8.1 Resource Levelling Technique *390*
    10.8.2 Resource Allocation Technique *398*

Questions 404

11. DECISION THEORY — *409–423*

11.1 Introduction *409*
11.2 Decision under Certainty (Deterministic Decision) *409*
11.3 Decision under Risk *409*
    11.3.1 Expected Value Criterion *410*
    11.3.2 Expected Value Combined with Variance Criterion *411*
11.4 Decision under Uncertainty *411*
    11.4.1 Laplace Criterion *412*
    11.4.2 Maximin Criterion *413*

Contents • ix

    11.4.3 Minimax Criterion *414*

    11.4.4 Savage Minimax Regret Criterion *414*

    11.4.5 Hurwicz Criterion *416*

11.5 Decision Tree *417*

Questions *421*

12. GAME THEORY   424–470

12.1 Introduction *424*

    12.1.1 Terminologies of Game Theory *424*

12.2 Game with Pure Strategies *426*

12.3 Game with Mixed Strategies *428*

12.4 Dominance Property *430*

12.5 Graphical Method for $2 \times n$ or $m \times 2$ Games *440*

12.6 Linear Programming Approach for Game Theory *453*

Questions *467*

13. REPLACEMENT AND MAINTENANCE ANALYSIS   471–493

13.1 Introduction *471*

13.2 Types of Maintenance *471*

13.3 Types of Replacement Problem *472*

13.4 Determination of Economic Life of an Asset *472*

    13.4.1 Basics of Interest Formulae *473*

    13.4.2 Examples of Determination of Economic Life of an Asset *475*

13.5 Simple Probabilistic Model for Items which Completely Fail *485*

Questions *492*

14. PRODUCTION SCHEDULING   494–537

14.1 Introduction *494*

14.2 Single-machine Scheduling *494*

    14.2.1 Measures of Performance *495*

    14.2.2 Shortest Processing Time (SPT) Rule to Minimize Mean Flow Time *496*

    14.2.3 Weighted Shortest Processing Time (WSPT) Rule to Minimize Weighted Mean Flow Time *497*

    14.2.4 Earliest Due Date (EDD) Rule to Minimize Maximum Lateness *498*

    14.2.5 Model to Minimize Total Tardiness *499*

    14.2.6 Introduction to Branch-and-Bound Technique to Minimize Mean Tardiness *502*

    14.2.7 Model to Minimize Sum of Weighted Number of Early and Tardy Jobs *512*

14.3 Flow Shop Scheduling *515*

    14.3.1 Johnson's Algorithm for n Jobs and Two Machines Problem *517*

    14.3.2 Extension of Johnson's Algorithm for n Jobs and Three Machines Problem *519*

    14.3.3 Branch-and-Bound Method for n Jobs and m Machines *521*

14.4 Job Shop Scheduling *532*
    14.4.1 Two Jobs and *m* Machines Job Shop Scheduling *533*
Questions 535

15. GOAL PROGRAMMING   538–548

15.1 Introduction *538*
15.2 Simplex Method for Solving Goal Programming *541*
Questions 547

16. PARAMETRIC LINEAR PROGRAMMING   549–561

16.1 Introduction *549*
16.2 Changes in Objective Function Coefficients (C_j Values) *550*
16.3 Changes in Right-hand Side Constants (B_i Values) of Constraints *554*
16.4 Introduction to Changes in Resource Requirements Vector(s), P_j *559*
Questions 560

17. NONLINEAR PROGRAMMING   562–590

17.1 Introduction *562*
17.2 Lagrangean Method *562*
17.3 Kuhn–Tucker Conditions *569*
17.4 Quadratic Programming *572*
17.5 Separable Programming *581*
17.6 Chance-constrained Programming or Stochastic Programming *585*
Questions 589

Appendix   *591*

Suggested Further Reading   *593–594*

Answers to Questions   *595–604*

Index   *605–608*

PREFACE

An organizational system consists of various subsystems. The most ideal approach to optimize the performance of a system is to consider different subsystems as an integrated single unit. In some reality, integrating all the subsystems as a single unit will make the problem-solving process more complex, because of its size and different constraints. Under such situation, it is inevitable to optimize the performance of each subsystem. Operations research consists of topics to achieve each of these objectives depending on the reality.

Based on the feedback from academicians, I have revised this book in the following lines.

- Inclusion of quantity discount models for transportation problem.
- Inclusion of more worked-out examples in many chapters. This will help the students to have enhanced understanding of the concepts and techniques, which are discussed in different chapters.
- Inclusion of additional topics in dynamic programming and inventory control.
- Inclusion of chapter-end questions for the additional topics, which are included in this edition.

The quantity discount in transportation problem can be classified into all quantity discount scheme (AQDS) and incremental quantity discount scheme (IQDS). A mathematical model and a numerical illustration for each of these two quantity discount schemes are presented at the end of the chapter on transportation problem.

The chapter on dynamic programming contains an additional topic on minimizing total tardiness in single machine scheduling problem. Here, the single machine scheduling problem is mapped in such a way that the dynamic programming technique is applied to it for minimizing the total tardiness.

Under inventory control, the following topics have been included:

- Multiple-item model with storage limitation.
- Purchase model of inventory for multi-item with inventory carrying cost constrains.
- EOQ model for multi-item joint replenishment without shortages for purchase model of inventory as well as for manufacturing model of inventory.
- EOQ for the purchase model of inventory for multi-item joint replenishment with space constraint.

The methods/models in each of the additional topics are illustrated with suitable worked-out examples.

I express my profound gratitude and appreciation to academic colleagues who gave valuable feedback about the first edition of this text which helped me to improve its content.

My heartfelt thanks are due to the editorial and production teams of Prentice-Hall of India for their meticulous processing of the manuscript of this second edition.

Any suggestions to further improve the contents of this edition would be warmly appreciated.

R. PANNEERSELVAM

OVERVIEW OF OPERATIONS RESEARCH

1

1.1 INTRODUCTION

Operations research is a scientific approach to problem solving for executive decision making which requires the formulation of mathematical, economic and statistical models for decision and control problems to deal with situations arising out of risk and uncertainty. In fact, decision and control problems in any organization are more often related to certain daily operations such as inventory control, production scheduling, manpower planning and distribution, and maintenance.

According to Operations Research Society of America (ORSA), it is a tool which is concerned with the design and operation of the man-machine system scientifically, usually under conditions requiring the optimum allocation of limited resources.

As per the Operations Research Society of Great Britain, operations research is the application of the scientific methods to complex problems arising in the direction and management of large systems of men, machines, materials and money in industry, business and government.

The origin and development of operations research can be studied under the following classification:

1. Pre-World War II developments
2. Developments during World War II
3. Post-World War II developments
4. Computer era
5. Inclusion of uncertainty models.

Pre-World War II developments

Many of the techniques of today's operations research have been actually developed and used even before the term 'operations research' was coined. Some of the techniques are: inventory control, queueing theory, and statistical quality control.

In 1915, Ford Harris developed a simple EOQ (economic order quantity) model to optimize the total cost of inventory system, which was eventually analyzed in 1934 by R.H. Wilson. Around the same time (1916), A.K. Erlang, a Danish telephone engineer, was responsible for many of the early theoretical developments in the area of queueing theory.

In the early 1900s, routine quality checks conducted by inspectors were not found to be satisfactory for some companies. The problem was analyzed in the inspection engineering department of Western Electric's Bell Laboratory by Shewhart who ultimately designed control charts in 1924. These are called as the first Shewhart control charts. During the period 1925–26, the Western Electric Company defined various terminologies associated with acceptance sampling of

1

quality control that was used as a tool for controlling attributes of raw materials/components/ finished products. The terminologies include consumer's risk, producer's risk, probability of acceptance, operating characteristics (OC) curve, lot tolerance percent defective (LTPD), double sampling plan, type I error, type II error and so on. In 1925, Dodge introduced the basic concept of sampling inspection. Ten years later, Pearson developed the British Standard Institution Number 600, entitled 'Application of statistical method to international standardization and quality control'. In 1939, H. Roming presented his work on variable sampling plan in his Ph.D. dissertation.

Developments during World War II

During the World War II, the effective utilization of scarce resources was the top-most concern of the military in Britain. So, in Britain, scientists from different fields were jointly directed to do research on military operations for improving its effectiveness with the limited resources. Later on, this scientific and interdisciplinary approach became an important problem-solving aspect of operations research methodologies.

Post-World War II developments

After the World War II, the industries in America and Britain concentrated in applying the operations research methodologies to industrial problems for maximizing the profitability with limited resources.

In 1947, Dantzig, developed **simplex method** to solve linear programming problem. Thereafter the Operations Research Society of America, and the Institute of Management Science were founded in 1952 and 1953, respectively.

Computer era

Many of the operations research techniques involve complex computations and hence they take longer time for providing solutions to real life problems. The developments of high speed digital computers made it possible to successfully apply some of the operations research techniques to large size problems. The developments of recent interactive computers make the job of solving large size problems even more simple because of human intervention towards sensitivity analysis.

Inclusion of uncertainty models

The use of probability theory and statistics to tackle undeterministic situations made the operations research techniques more realistic.

1.2   CONCEPT OF A MODEL

Model is an abstraction of reality. Some examples of models are road map of a city to trace the shortest route from a given source to a given destination, three-dimensional view of a factory to plan the material movements in its shop floor, electrical network to compute the current flow in a particular arc, and linear equation to forecast the demand of a product.

An operations research model is defined as an idealized (simplified) representation of a real-life system. Operations research uses a number of models to obtain solutions of various realistic problems.

1.2.1 Steps of Modelling

The steps of modelling are listed below with reference to Figure 1.1:

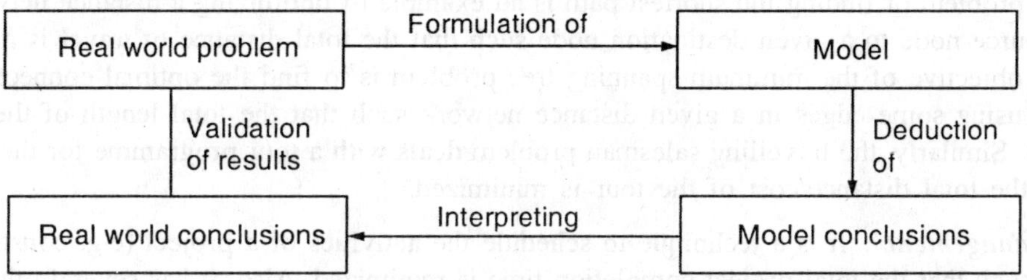

Figure 1.1 Schematic representation of modelling steps.

Step 1: Identify the management decision problem of real world.

Step 2: Formulate a model for the real world problem which in turn has the following steps:

    2.1 Identify the parameters and variables which are involved in the management decision problem. Define them verbally and then introduce symbols to represent each one of them.

    2.2 Select the variables that appear to be most influential so that the model may be kept as simple as possible. Classify the variables into controllable and non-controllable variables.

    2.3 State verbal relationship among the variables based upon known principles, specially, gathered data and intuition. Make assumptions of predictions concerning the behaviour of the non-controllable variables.

    2.4 Construct a model by combining all real world relationships into a system of symbolic relationships.

Step 3: Perform symbolic manipulations such as solving a system of equations, iterating set of steps or making statistical analysis to optimize certain measure of performance and draw model conclusions.

Step 4: Interpret the model conclusions in terms of real world problem's characteristics.

Step 5: Test and validate the results.

Step 6: Implement the results.

Step 7: Revise the model as and when necessary.

1.3 IMPORTANT TOPICS OF OPERATIONS RESEARCH

Operations research (OR) mainly discusses the following optimization techniques:

Linear programming. It involves linear objective function with a set of linear constraints.

Integer programming. It is an extension of linear programming with only integer values for the decision variables of the problem.

Distance related network techniques. These techniques are applied to deal with, transportation problem, shortest path problem, minimum spanning tree problem and travelling salesman problem.

Business organizations with multiple manufacturing units, sales and distribution outlets and warehouses need to select transportation routes and optimal shipping schedule of a commodity from a set of sources to a set of destinations such that the total cost of shipment is minimized.

The problem of finding the shortest path is an example of optimizing a distance network from a given source node to a given destination node such that the total distance of travel is minimized.

The objective of the minimum spanning tree problem is to find the optimal connection of all the nodes using some edges in a given distance network such that the total length of the edges is minimized. Similarly, the travelling salesman problem deals with a tour programme for the salesman such that the total distance/cost of the tour is minimized.

Project management.   It is a technique to schedule the activities of a project (e.g. construction of a bridge) such that the total project completion time is minimized. Also, it has several other control techniques.

Inventory control.   It is a technique to optimally plan and procure/produce raw materials/semi-finished products/finished products such that the total cost of the inventory system is minimized.

Dynamic programming.   It is systematic complete enumeration technique to solve a problem optimally in an emerging area by integrating the solutions of its subproblems.

Queueing theory.   It is a technique based on probability to study the waiting behaviour of some real life queueing systems.

Replacement analysis.   It is a technique to determine the economic life of an asset as compared to the minimum total cost. Also, it discusses the method of selecting the best option between individual replacement policy and group replacement policy such that the total cost of the system is minimized.

Game theory.   It is a technique to deal with uncertainty situations related to management decisions, such as bidding for tenders.

Goal programming.   The idea here is to convert the multiple objectives into a single goal, i.e. to reach a compromise solution for multiobjective models.

Simulation.   It is a technique to deal with probabilistic situation where empirical/mathematical models fail to provide solutions to real life problems.

Scheduling.   It is the process of preparing calendar for executing a set of jobs mostly in shop floors.

Nonlinear programming.   It is an extended version of linear programming problems with nonlinear objective function and linear constraints or with nonlinear objective function and nonlinear constraints which will enable analysts to incorporate realistic assumptions while solving problems.

1.4   SCOPE OF OPERATIONS RESEARCH

The techniques of operations research can be applied to several real world problems. A sample set of applications in different sectors is presented below:

Defence applications

Many of defence operations involve scientific decision making. The methods, viz. network techniques for shortest path problems, scheduling algorithms for vehicle routing, and allocation techniques for shipping food grains and ammunition, can be applied in defence establishments.

Industrial applications

Industrial management includes managing four functional areas: production, marketing, personnel and finance. In addition to these functions, there are other support services that contribute toward carrying out the business of any industrial enterprise.

A sample set of techniques which are used in the production function is presented below:

- Linear programming for aggregate planning
- Integer programming for shop floor production scheduling
- Network based techniques for line balancing and project management
- Inventory control techniques for planning and procuring raw materials
- Replacement analysis for equipment replacement decision
- Queueing theory for designing in-process buffer stock.

A sample set of techniques which are used in the marketing function is presented below:

- Linear programming for product mix problem
- Game theory for order bidding decision
- Distance network related techniques for shipping finished goods.

A sample set of techniques which are used in the personnel function is presented below:

- Linear programming for manpower planning
- Queueing theory for determining the size of maintenance crews
- Scheduling techniques for manpower scheduling.

A sample set of techniques which are used in the finance function is presented below:

- Integer programming for capital budgeting
- Linear programming for break-even analysis
- Integer programming or dynamic programming for portfolio selection.

Applications in public system

Some of the operations research techniques can be applied to plan, design and operate different public systems like, government and government institutions/departments, postal system, banks, highways, railways, airways, hospitals, and educational institutions.

In some government offices, techniques like goal programming for policy decisions, and integer programming for budgeting can be used; while in postal systems, techniques which are related to vehicle scheduling, manpower planning, formulating transfer policy, etc. can be used.

Banks would generally apply techniques like, queueing theory for determining the number of counters, and portfolio models for effective deployment of funds subject to government regulations.

For managing highways, the techniques of integer programming for project selection and vendor selection, simulation for traffic system design, etc. can be used. But the Railways need to apply techniques like, linear programming/integer programming for cargo loading, scheduling techniques for railway traffic control, queueing theory to determine the number of platforms.

In airways, simulating the air traffic, runway design, linear/integer programming for cargo loading, are the main areas where OR techniques will be immensely useful.

A sample set of techniques which are used in hospitals are:

- Queueing theory for out-patient system design.
- Linear programming/integer programming for scheduling the duties of nurses and doctors.
- Inventory control for procurement of medicines.
- Algorithms and models for managing the operation theatre.

1.5 OPERATIONS RESEARCH—A TOOL FOR DECISION SUPPORT SYSTEM

In the process of managing various subsystems of the organization, executives at different levels of the organization have to take several management decisions. These decisions are classified into strategic decisions, tactical decisions and operational decisions. The strategic decisions are taken at the top management level. Under this category, the definition of goals and policies, and selection of the decisions are based on organizational objectives. The tactical decisions are taken at middle management level, and these include acquisition of resources, plant location, new products establishments and monitoring of budgets. The operational decisions, which are taken at the bottom level of management, include effective and efficient use of existing facilities and resources to carry out activities within budget constraints. If we closely examine the relationship among the number of decisions taken at different levels, it would vary from high (at the bottom level management) to low (at the top level management). This concept of relative frequency of the number of decisions taken at different levels is presented in Figure 1.2. In this figure, the

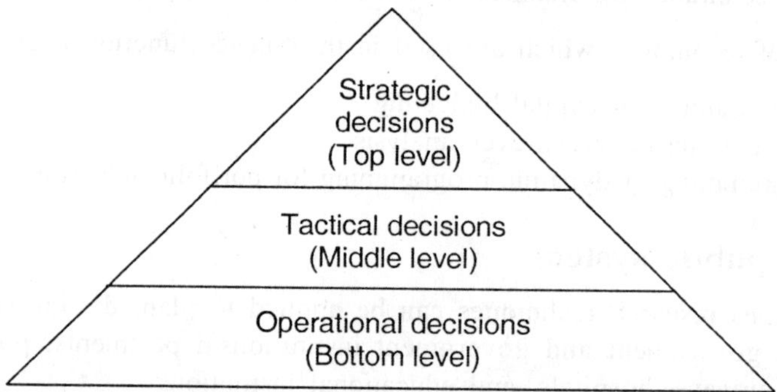

Figure 1.2 Relative frequency of decisions at different levels of management.

approximate area in each level represents the proportion of number of decisions taken in relation to other levels. The degree of structuredness of decisions taken at various levels varies in the decreasing order from the bottom to the top-level management. If the decisions are highly structured, managers can use 'Management Information Systems' for taking decisions.

Under semi-structured decision environment, decision support systems are used by managers. Decision support systems are aimed at assisting managers in their decision making process and not to replace them. Generally, decision support system consists of different models of operations research, which provide guidelines to the managers for taking effective decisions.

1.6 OPERATIONS RESEARCH—A PRODUCTIVITY IMPROVEMENT TOOL

Productivity is a relationship between the output (products/services) and the input (resources consumed in providing them) of a business system. Mathematically,

$$\text{Productivity} = \frac{\text{Output}}{\text{Input}}$$

For the survival of any organization, this productivity ratio must be at least 1. If it is more than 1, the organization is in a comfortable position. All organizations aim at improving this ratio as much as possible. So, the objective of the system is to identify ways and means of improving its productivity to the highest level possible.

There are several strategies for improving the productivity. These are discussed in the following sections:

1.6.1 Increased Output for the Same Input

In this strategy, the output is increased while keeping the input constant. Let us assume that in a steel plant, the layout of existing facilities is not proper. By slightly changing the location of billet-making section, say bringing it more closer to the furnace which produces hot metal, the scale formation at the top of ladles will be reduced to a greater extent. The molten metal is usually carried in ladles to the billet-making section. In the long run, this would give more yield in terms of tonnes of billet produced. In this exercise, there is no extra cost involved. The only task is the relocation of the billet-making facility by bringing it more closer to the furnace which involves insignificant cost. So, this is an example where the output is increased without any increase in the input.

1.6.2 Decreased Input for the Same Output

In this strategy, the input is decreased to produce the same output. Let us assume that there exists a substitute raw material to manufacture a product and it is available at a low price. If we can identify such material and use it for manufacturing the product, then certainly it will reduce the input cost. In this exercise, the job of the purchase department is to identify an alternate substitute material. The process of identification does not involve any extra cost. Thus, the productivity ratio will increase because of the decreased input by way of using the cheaper raw material to produce the same output.

1.6.3 Increase in the Output is more than the Increase in the Input

Consider the example of introducing a new product into the existing product mix of an organization. Let us make an assumption that the existing facilities are not fully utilized. So, the R&D wing of the company has identified a new product which has a very good market and which can be manufactured with the surplus facilities of the organization. If the manufacturing of the new product is taken up, it would lead to the following advantages:

- There will be an increase in the revenue of the organization by way of selling the new product in addition to the existing product mix.
- The increase in the material cost and operation and maintenance cost of machineries will be proportionately less in producing the new product because of utilization of the excess capacity of the existing facilities.

If we closely examine these two increases, the proportionate increase in the revenue will be more than the proportionate increase in the input cost. Hence, there will be a net increase in the productivity ratio.

1.6.4 Decrease in the Input is more than the Decrease in the Output

Let us consider the reverse case of the previous example, i.e. dropping an uneconomical product from the existing product mix.

This will result into the following:

- There will be a decrease in the revenue of the organization because of dropping a product from the existing product mix.
- There will be a decrease in the material cost, and operation and maintenance cost of machineries because of dropping an existing product from the product mix.

If we closely examine these two decreases, the proportionate decrease in the input cost will be more than the proportionate decrease in the revenue. Hence, there will be a net increase in the productivity ratio.

1.6.5 Increase in the Output with Decrease in the Input

Let us assume that there are advanced automated technologies like robot, automated guided vehicle system (AGVS), etc., which are available in the market. These can be employed in the organization for streamlining the operational procedures. If we employ these modern tools, the following will result:

- There will be a drastic reduction in the operation cost. Initially the cost on equipment would be very high. But in the long run, the reduction in the operation cost would break-even the high initial investment and offer more savings on the input.
- These advanced facilities would help in achieving higher productivity, which in turn will yield more revenue.

In this example, there is an increase in the revenue while there is a decrease in the input in the long run. Hence the productivity ratio will increase at a faster rate.

Many of the above productivity improvement strategies involve the usage of mathematical models and algorithms. With the advent of computer technology, these models, when applied to different productivity related situations, will yield solutions within reasonable time. Thus, operations research is considered to be a tool for the improvement of productivity in an organization.

QUESTIONS

1. Define operations research.
2. Discuss the origin and development of operations research with a suitable classification.
3. Define model. Discuss the steps of modelling.
4. Discuss the scope of operations research.
5. "Operations research is a tool for Decision Support System." Justify.
6. What are the ways of improving productivity? Also, explain the role of operations research in improving productivity.

LINEAR PROGRAMMING $\quad$ 2

2.1 INTRODUCTION

Linear programming is a mathematical programming technique to optimize performance (e.g. profit or cost) under a set of resource constraints (e.g. machine-hours, man-hours, money, materials, etc.) as specified by an organization. With the advent of highly efficient computer codes, the usefulness of this technique is maximized even though the computations in a linear programming model are too elaborate.

A sample list of applications of the linear programming problem is presented below:

1. Product mix problem
2. Diet planning problem
3. Cargo loading problem
4. Capital budgeting problem
5. Manpower planning problem.

These are explained with illustrations at a later stage in this chapter.

2.2 CONCEPT OF LINEAR PROGRAMMING MODEL

The model of any linear programming problem will contain: objective function, set of constraints and non-negativity restrictions. Each of the components may consist of one or more of the following:

- Decision variables
- Objective function coefficients
- Technological coefficients
- Availability of resources.

The components and other terminologies of the linear programming model are explained with the help of a product-mix problem as described here.

2.2.1 Product Mix Problem

This pertains to determining the levels of production activities to be carried out during a pre-decided time-frame so as to gain the maximum profit. Since the requirement of inputs for the different production processes varies, their profitabilities also differ. This may be best understood from the following example:

Example 2.1 A company manufactures two different types of products: P_1 and P_2. Each product requires processing on milling machine and drilling machine. But each type of machines has limited hours available per week. The net profit per unit of the products, resource requirements of the products and availability of resources are summarized in Table 2.1.

Table 2.1 A Sample Data of a Product Mix Problem

Machine type	Processing time (hours)		Machine hours available per week
	Product P_1	Product P_2	
Milling machine	2	5	200
Drilling machine	4	2	240
Profit/unit (Rs.)	250	400	

Develop a linear programming model to determine the optimal production volume of each of the products such that the profit is maximized subject to the availability of machine hours.

Solution Let n be the number of products to be manufactured and m be the number of different machine types used.

The linear programming (LP) model includes a few basic elements which have been described below.

Decision variable. A decision variable is used to represent the level of achievement of a particular course of action. The solution of the linear programming problem will provide the optimal value for each and every decision variable of the model. A generalized definition of the decision variables is presented below.

From the Example 2.1, X_1 and X_2 are the production volumes of the products, P_1 and P_2, respectively.

Objective function coefficient. It is a constant representing the profit per unit or cost per unit of carrying out an activity.

Let, $C_1, C_2, \ldots, C_n$, be the profit per unit of products $P_1, P_2, \ldots, P_n$, respectively. Then from Example 2.1, C_1 be the profit per unit of the product P_1, which is equal to Rs. 250, and C_2 be the profit per unit of the product P_2 which is equal to Rs. 400.

Objective function. It is an expression representing the total profit or cost of carrying out a set of activities at some levels. The objective function will be either a maximization type or minimization type. The benefit-related objective functions will come under maximization type whereas the cost-related objective functions will come under the minimization type. A generalized format of the objective function is presented below.

$$\text{Maximize or Minimize } Z = C_1 X_1 + C_2 X_2 + C_3 X_3 + \cdots + C_n X_n$$

From Example 2.1, we have

$$\text{Maximize } Z = C_1 X_1 + C_2 X_2 = 250 X_1 + 400 X_2$$

Technological coefficient (a_{ij}). The *technological coefficient*, a_{ij} is the amount of resource i required for the activity j, where i varies from 1 to m and j varies from 1 to n. A generalized format of the technological coefficient matrix is:

$$\begin{bmatrix} a_{11} & a_{12} & \cdots & a_{1n} \\ a_{21} & a_{22} & \cdots & a_{2n} \\ \vdots & \vdots & & \vdots \\ a_{m1} & a_{m2} & \cdots & a_{mn} \end{bmatrix}$$

Considering Example 2.1, the technological coefficient matrix of the example problem can be stated as:

$$\begin{bmatrix} 2 & 5 \\ 4 & 2 \end{bmatrix}$$

Resource availability (b_i). The constant b_i is the amount of resource i available during the planning period. A generalized format of the resource availability matrix is:

$$\begin{bmatrix} b_1 \\ b_2 \\ \vdots \\ b_m \end{bmatrix}$$

Now from Table 2.1, we get $b_1 = 200$ hours and $b_2 = 240$ hours.

Set of constraints. A *constraint* is a kind of restriction on the total amount of a particular resource required to carry out the activities at various levels. In a model, there will be many such constraints. So, these constraints will limit the levels of achievement of different decision variables.

In the product mix problem, the profit of the organization can be increased to a great extent if there is no constraint in the system. But unfortunately, there are some practical constraints on the resources like, material, machine hour, money, etc., in terms of their availability during the planning period, which will automatically limit the levels of achievement of different activities.

A generalized format of the constraints can be given as:

$$a_{11}X_1 + a_{12}X_2 + \cdots + a_{1n}X_n \leq, = \text{ or } \geq b_1$$
$$a_{21}X_1 + a_{22}X_2 + \cdots + a_{2n}X_n \leq, = \text{ or } \geq b_2$$
$$\vdots \qquad \vdots \qquad \quad \vdots \qquad \qquad \vdots$$
$$a_{m1}X_1 + a_{m2}X_2 + \cdots + a_{mn}X_n \leq, = \text{ or } \geq b_m$$

Now from the product mix problem, we get

$$2X_1 + 5X_2 \leq 200$$
$$4X_1 + 2X_2 \leq 240$$

Non-negativity constraints. Each and every decision variable in the linear programming model is a non-negative variable. This condition is represented as:

$$X_1, X_2, X_3,..., X_n \geq 0$$

From the product mix problem, we get

$$X_1 \text{ and } X_2 \geq 0$$

Complete model of the product mix problem. The complete linear programming model of the given problem is now shown as under:

$$\text{Maximize } Z = 250X_1 + 400X_2$$

subject to

$$2X_1 + 5X_2 \leq 200$$

$$4X_1 + 2X_2 \leq 240$$

$$X_1 \text{ and } X_2 \geq 0$$

2.2.2 Assumptions in Linear Programming

The following four assumptions are made in the linear programming problems.

Linearity. The amount of resource required for a given activity level is directly proportional to the level of that activity. For example, if the number of hours required on a particular machine (for a given activity level) is 5 hours per unit of that activity, then the total number of hours required on that machine to produce 10 units of that activity is 50 hours.

Divisibility. This means that fractional values of the decision variables are permitted.

Non-negativity. This means that the decision variables are permitted to have only the values which are greater than or equal to zero.

Additivity. This means that the total output for a given combination of activity levels is the algebraic sum of the output of each individual process.

2.2.3 Properties of Linear Programming Solution

Feasible solution. If all the constraints of the given linear programming model are satisfied by the solution of the model, then that solution is known as a *feasible* solution. Several such solutions are possible for a given linear programming model.

Optimal solution. If there is no other superior solution to the solution obtained for a given linear programming model, then the solution obtained is treated as the *optimal* solution.

Alternate optimum solution. For some linear programming model, there may be more than one combination of values of the decision variables yielding the best objective function value. Such combinations of the values of the decision variables are known as *alternate optimum* solutions.

Unbounded solution. For some linear programming model, the objective function value can be increased/decreased infinitely without any limitation. Such solution is known as *unbounded* solution.

Infeasible solution. If there is no combination of the values of the decision variables satisfying all the constraints of the linear programming model, then that model is said to have *infeasible* solution. This means that there is no solution for the given model which can be implemented.

Degenerate solution. In linear programming problems, intersection of two constraints will define a corner point of the feasible region. But if more than two constraints pass through any one of the corner points of the feasible region, excess constraints will not serve any purpose, and therefore they act as redundant constraints. Under such situation, *degeneracy* will occur. This means that some iterations will be carried out in simplex method without any improvement in the objective function.

2.3 DEVELOPMENT OF LP MODELS

Modelling is an art. One can develop this expertise only by seeing more and more models. In this section, the concept of model building is demonstrated using some example problems.

Example 2.2 A company manufactures two types of products, P_1 and P_2. Each product uses lathe and milling machine. The processing time per unit of P_1 on the lathe is 5 hours and on the milling machine is 4 hours. The processing time per unit of P_2 on the lathe is 10 hours and on the milling machine, 4 hours. The maximum number of hours available per week on the lathe and the milling machine are 60 hours and 40 hours, respectively. Also the profit per unit of selling P_1 and P_2 are Rs. 6.00 and Rs. 8.00, respectively. Formulate a linear programming model to determine the production volume of each of the products such that the total profit is maximized.

    Solution The data of the problem are summarized in Table 2.2.

Table 2.2 Details of Products

(in hour)

Machine	Machine hours/unit		Limit on machine hours
	Product P_1	Product P_2	
Lathe	5	10	60
Milling machine	4	4	40
Profit/unit (Rs.)	6	8	

Let X_1 and X_2 be the production volumes of the products P_1 and P_2 respectively. The corresponding linear programming model to determine the production volume of each of the products, such that the total profit is maximized, is presented below.

$$\text{Maximize } Z = 6X_1 + 8X_2$$

subject to

$$5X_1 + 10X_2 \leq 60$$

$$4X_1 + 4X_2 \leq 40$$

$$X_1 \text{ and } X_2 \geq 0$$

Example 2.3 A nutrition scheme for babies is proposed by a committee of doctors. Babies can be given two types of food (I and II) which are available in standard sized packets weighing 50 grams. The cost per packet of these foods are Rs. 2 and Rs. 3, respectively. The vitamin availability in each type of food per packet and the minimum vitamin requirement for each type of vitamin are summarized in Table 2.3. Develop a linear programming model to determine the optimal combination of food types with the minimum cost such that the minimum requirement of vitamin in each type is satisfied.

Table 2.3 Details of Food Types

| Vitamin | Vitamin availability per packet | | Minimum daily required vitamin |
	Food Type I	Food Type II	
1	1	1	6
2	7	1	14
Cost/packet (Rs.)	2	3	

Solution Let, X_1 and X_2 are the number of packets of food Type I and Type II, respectively to be suggested for babies. A linear programming model of this situation is presented below. This model determines the number of packets of each food type with the minimum cost to be suggested for babies such that the minimum daily required vitamins are satisfied.

$$\text{Minimize } Z = 2X_1 + 3X_2$$

subject to

$$X_1 + X_2 \geq 6$$
$$7X_1 + X_2 \geq 14$$
$$X_1 \text{ and } X_2 \geq 0$$

Example 2.4 (Manpower scheduling problem) In a multi-speciality hospital, nurses report to duty at the end of every 4 hour as shown in Table 2.4. Each nurse, after reporting, will work for 8 hours continuously. The minimum number of nurses required during various periods are summarized in Table 2.4. Develop a mathematical model to determine the number of nurses to report at the beginning of each period such that the total number of nurses who have to report to duty in a day is minimized.

Table 2.4 Data Showing the Minimum Number of Nurses to Report for Duty

| Interval number | Time period | | Minimum number of nurses required |
	From	To	
1	12 midnight	4 a.m.	20
2	4 a.m.	8 a.m.	25
3	8 a.m.	12 noon	35
4	12 noon	4 p.m.	32
5	4 p.m.	8 p.m.	22
6	8 p.m.	12 midnight	15

Solution Let, X_j be the number of nurses to report for duty at the beginning of the jth period, where j varies from 1 to 6. The model for this problem is shown below.

$$\text{Minimize } Z = X_1 + X_2 + X_3 + X_4 + X_5 + X_6$$

subject to

$$X_6 + X_1 \geq 20$$
$$X_1 + X_2 \geq 25$$

$$X_2 + X_3 \geq 35$$

$$X_3 + X_4 \geq 32$$

$$X_4 + X_5 \geq 22$$

$$X_5 + X_6 \geq 15$$

$$X_1, X_2, X_3, X_4, X_5, X_6 \geq 0 \text{ and integers}$$

Example 2.5 A textile company can use any or all of the three different processes for weaving its standard white polyester fabric. Each of these production processes has a weaving machine setup cost and per-square-metre processing cost. These costs and the capacities of each of the three production processes are shown as in Table 2.5.

Table 2.5 Costs of Production Processes

Process number	Weaving machine setup cost (Rs.)	Processing cost/sq. m (Rs.)	Maximum daily capacity (sq. m)
1	150	15	2000
2	240	10	3000
3	300	8	3500

The daily demand forecast for its white polyester fabric is 4000 sq. metre. The company's production manager wants to make a decision concerning which combination of production processes is to be utilized to meet the daily demand forecast and at what production level of each selected production process to be operated to minimize total production costs. Develop a linear programming model to assist the production manager.

Solution Let X_j be the production level in square metre for the process j ($j = 1, 2, 3$), where

$$Y_j = 1, \quad \text{if the process } j \text{ is used}$$
$$= 0, \quad \text{otherwise}$$

The model is as follows:

$$\text{Minimize } Z = 15X_1 + 10X_2 + 8X_3 + 150Y_1 + 240Y_2 + 300Y_3$$

subject to

$$X_1 + X_2 + X_3 = 4000$$

$$X_1 - 2000 \, Y_1 \leq 0$$

$$X_2 - 3000 \, Y_2 \leq 0$$

$$X_3 - 3500 \, Y_3 \leq 0$$

$$X_1, X_2, X_3 \geq 0 \quad \text{and} \quad Y_1, Y_2, Y_3 = 0 \text{ or } 1$$

In this model, if the value of Y_j is equal to 1, then the corresponding overhead charges will be included in the objective function and at the same time, the production quantity of the jth process will be limited to its maximum capacity; otherwise, the corresponding overhead cost will not be included in the objective function and at the same time, the production quantity of the jth process will be set to 0.

Example 2.6 A company is planning to determine its product mix out of three different products:

P_1, P_2 and P_3. The monthly sales of the product P_1 is limited to a maximum of 500 units. For every two units of P_2 produced, there will be one unit of by-product which can be sold at the rate of Rs. 20 per unit. The highest monthly demand for this by-product is 200 units. The contributions per unit of the products P_1, P_2 and P_3 are Rs. 50, Rs. 70 and Rs. 60, respectively. The processing requirements of these products are shown in Table 2.6.

Table 2.6 Example 2.6

Process	Hours per unit			Available hours
	P_1	P_2	P_3	
I	3	5	2	1000
II	4	–	3	700
III	4	3	2	1300

Formulate a linear programming model of this problem to find the optimum product mix such that the total contribution is maximized.

Solution   Let, X_1, X_2 and X_3 be the production volumes of the products P_1, P_2 and P_3, respectively. Also let, $X_2/2$ be the production volume of the by-product from the product P_2. Then, a linear programming model of the problem is presented below.

$$\text{Maximize } Z = 50X_1 + 70X_2 + 60X_3 + 20\frac{X_2}{2}$$

subject to

$$3X_1 + 5X_2 + 2X_3 \leq 1000$$
$$4X_1 + 3X_3 \leq 700$$
$$4X_1 + 3X_2 + 2X_3 \leq 1300$$
$$X_1 \leq 500$$
$$\frac{X_2}{2} \leq 200$$
$$X_1, X_2 \text{ and } X_3 \geq 0$$

The final form of the above model is shown below:

$$\text{Maximize } Z = 50X_1 + 80X_2 + 60X_3$$

subject to

$$3X_1 + 5X_2 + 2X_3 \leq 1000$$
$$4X_1 + 3X_3 \leq 700$$
$$4X_1 + 3X_2 + 2X_3 \leq 1300$$
$$X_1 \leq 500$$
$$X_2 \leq 400$$
$$X_1, X_2 \text{ and } X_3 \geq 0$$

***Example 2.7* (Cargo loading problem)** Consider the cargo loading problem, where five items are to be loaded on a vessel. The weight (w_i) and volume (v_i) of each unit of the different items as well as their corresponding returns per unit (r_i) are tabulated in Table 2.7.

Table 2.7 Example 2.7

Item-i	w_i	v_i	r_i
1	5	1	4
2	8	8	7
3	3	6	6
4	2	5	5
5	7	4	4

The maximum cargo weight (W) and volume (V) are given as 112 and 109, respectively. It is required to determine the optimal cargo load in discrete units of each item such that the total return is maximized. Formulate the problem as an integer programming model.

Solution Let, X_i be the number of units of the ith item to be loaded in the cargo, where i varies from 1 to 5. A model to maximize the return is as follows:

$$\text{Maximize } Z = 4X_1 + 7X_2 + 6X_3 + 5X_4 + 4X_5$$

subject to

$$5X_1 + 8X_2 + 3X_3 + 2X_4 + 7X_5 \leq 112$$

$$X_1 + 8X_2 + 6X_3 + 5X_4 + 4X_5 \leq 109$$

$$X_1, X_2, X_3, X_4, \text{ and } X_5 \geq 0 \text{ and integers}$$

Example 2.8 A company manufactures three products, P_1, P_2 and P_3. The machine hour requirements and labour hour requirements to process the three products and the maximum available machine hours per week for each machine type and the maximum available labour hours per week are summarized in Table 2.8. The selling price per unit and product cost per unit are also summarized in the same table. The company wants to limit the production volume per week of the product P_3 to utmost 35 units. Formulate a linear programming model to find the production volume per week of each product such that the total profit is maximized.

Table 2.8 Details of Products for Example 2.8

	Product			Maximum available hours per week
	Machine hours required			
	P_1	P_2	P_3	
Machine 1	4	6	3	500
Machine 2	3	—	2	300
Machine 3	5	7	8	600
Labour	3	2	4	200
Selling price per unit (Rs.)	500	400	550	
Product cost per unit (Rs.)	350	280	390	

Solution   Let, X_j be the production volume per week of the product P_j, for $j = 1, 2$ and 3. The profit per unit of the products P_1, P_2 and P_3 are Rs. 150, Rs. 120 and Rs. 160, respectively. A linear programming model to determine the production volume per week of the products such that the total profit is maximized is presented below.

$$\text{Maximize } Z = 150\, X_1 + 120X_2 + 160X_3$$

subject to

$$4X_1 + 6X_2 + 3X_3 \leq 500$$
$$3X_1 + 2X_3 \leq 300$$
$$5X_1 + 7X_2 + 8X_3 \leq 600$$
$$3X_1 + 2X_2 + 4X_3 \leq 200$$
$$X_3 \leq 35$$
$$X_1, X_2, X_3 \geq 0$$

Example 2.9   In a metro transport corporation, crews (each crew consists of a driver and a conductor) to operate buses report for duty at the beginning of each 2-hour period and work continuously for 8 hours. The different time intervals in a day and the minimum required number of buses in each time interval are shown in Table 2.9. Find the number of crews to report for duty at the beginning of each time interval such that the total number of crews reported in a day is minimized.

(*Note:* The crews which report at 5 p.m., 7 p.m., and 9 p.m. are idle between 11 p.m. and 5 a.m. Hence the crew members are given idle time and stay allowances).

Table 2.9   Details of Time Intervals

Interval number	Time period	Minimum number of buses required
1	05 a.m. to 07 a.m.	a_1
2	07 a.m. to 09 a.m.	a_2
3	09 a.m. to 11 a.m.	a_3
4	11 a.m. to 01 p.m.	a_4
5	01 p.m. to 03 p.m.	a_5
6	03 p.m. to 05 p.m.	a_6
7	05 p.m. to 07 p.m.	a_7
8	07 p.m. to 09 p.m.	a_8
9	09 p.m. to 11 p.m.	a_9

Solution   The total number of time intervals is 9. Each reported crew works continuously for 8 consecutive hours. This means that the crew will work for 4 continuous time intervals. Let, X_i be the number of crews to be reported at the beginning of the time interval i, for $i = 1$ to 9. As per these definitions, the following linear programming model aims to determine the number of crews to be reported for duty at the beginning of each time interval such that the total number of crews to be reported in a day is minimized.

$$\text{Minimize } Z = X_1 + X_2 + X_3 + X_4 + X_5 + X_6 + X_7 + X_8 + X_9$$

subject to

$$X_1 + X_7 + X_8 + X_9 \geq a_1$$
$$X_1 + X_2 + X_8 + X_9 \geq a_2$$

$$X_1 + X_2 + X_3 + X_9 \geq a_3$$
$$X_1 + X_2 + X_3 + X_4 \geq a_4$$
$$X_2 + X_3 + X_4 + X_5 \geq a_5$$
$$X_3 + X_4 + X_5 + X_6 \geq a_6$$
$$X_4 + X_5 + X_6 + X_7 \geq a_7$$
$$X_5 + X_6 + X_7 + X_8 \geq a_8$$
$$X_6 + X_7 + X_8 + X_9 \geq a_9$$
$$X_1, X_2, X_3, X_4, X_5, X_6, X_7, X_8 \text{ and } X_9 \geq 0$$

Example 2.10   A computer company procures cabinets from three different suppliers (A, B and C) located in three different cities. The company has production plants (P, Q and R) in three other cities. The cost of transportation per cabinet for different combinations of supplier and production plant are summarized in Table 2.10. The purchase price per cabinet from different suppliers are also indicated in the same table. The weekly demand at different production plants and the weekly availability of cabinets at different suppliers are also given in the table. Formulate a linear programming model to find the optimal procurement plan for the cabinets such that the total cost is minimized.

Table 2.10   Details of Unit Transportation Cost and Purchase Cost in Rupees

		Production plant				
		P	Q	R	Supply	Price/Cabinet
	A	10	15	8	300	85
Supplier	B	20	21	15	500	90
	C	12	16	13	100	75
	Demand	100	200	300		

Solution   Table 2.10 summarizes the transportation cost/cabinet and purchase price/cabinet. The purchase price per cabinet from each supplier is added to different cells of the row corresponding to that supplier. The modified totals of the cost of transportation per unit and purchase cost per unit, for different combinations of supplier and production plant are summarized in Table 2.11.

Table 2.11   Details of Totals of Transportation Cost per Unit and Purchase Cost per Unit

		Production plant			
		P	Q	R	Supply
	A	95	100	93	300
Supplier	B	110	111	105	500
	C	87	91	88	100
	Demand	100	200	300	

Let, Y_{ij} be the number of cabinets to be procured from the supplier i for the production plant j for $i = A, B, C$ and $j = P, Q, R$.

A linear programming model to find the optimal procurement plan for the cabinets to minimize the total cost of procurement is presented below.

$$\text{Minimize } Z = 95Y_{AP} + 100Y_{AQ} + 93Y_{AR} + 110Y_{BP} + 111Y_{BQ} + 105Y_{BR} + 87Y_{CP} + 91Y_{CQ} + 88Y_{CR}$$

subject to

$$Y_{AP} + Y_{AQ} + Y_{AR} \leq 300$$
$$Y_{BP} + Y_{BQ} + Y_{BR} \leq 500$$
$$Y_{CP} + Y_{CQ} + Y_{CR} \leq 100$$
$$Y_{AP} + Y_{BP} + Y_{CP} \geq 100$$
$$Y_{AQ} + Y_{BQ} + Y_{CQ} \geq 200$$
$$Y_{AR} + Y_{BR} + Y_{CR} \geq 300$$
$$Y_{AP}, Y_{AQ}, Y_{AR}, Y_{BP}, Y_{BQ}, Y_{BR}, Y_{CP}, Y_{CQ}, \text{ and } Y_{CR} \geq 0$$

Example 2.11   A company wants to engage casual labours to assemble its product daily. The company works for only one shift which consists of 8 hours and 6 days a week. The casual labours consist of two categories, viz. skilled and semi-skilled. The daily production per skilled labour is 80 assemblies and that of the semi-skilled labour is 60 assemblies. The rejection rate of the assemblies produced by the skilled labours is 5% and that of the semi-skilled labours is 10%. The loss to the company for rejecting an assembly is Rs. 25. The daily wage per labour of the skilled and semi-skilled labours are Rs. 240 and Rs. 160, respectively. The required weekly production is 1,86,000 assemblies. The company wants to limit the number of semi-skilled labours per day to utmost 400. Develop a linear programming model to determine the optimal mix of the casual labours to be employed so that the total cost (total wage + total cost of rejections) is minimized.

Solution

Daily wage per skilled labour   = Rs. 240
Daily wage per semi-skilled labour = Rs. 160
Weekly required production   = 1,86,000 assemblies
Number of working days per week = 6 days
Therefore, daily required production = 31,000 assemblies

Number of assemblies produced per skilled labour in a day is 80. Rejection rate of assemblies produced by skilled labours is 5% and hence, his number of rejected assemblies in a day is 4. Therefore, the acceptable number of assemblies produced per skilled labour in a day is 76.

Number of assemblies produced per semi-skilled labour in a day is 60. Rejection rate of assemblies produced by a semi-skilled labour is 10% and hence, his number of rejected assemblies in a day is 6. Therefore, the acceptable number of assemblies produced per semi-skilled labour in a day is 54.

The loss per rejected assembly   = Rs. 25
The loss due to rejections per skilled labour in a day   = 4 × Rs. 25 = Rs. 100
The loss due to rejections per semi-skilled labour in a day = 6 × Rs. 25 = Rs. 150

Let, X_1 be the number of skilled labours to be employed per day; X_2 be the number of semi-skilled labours to be employed per day.

A linear programming model to determine the number of labours to be employed per day under each category of casual labours to minimize the sum of the total wages and penalty of rejections in a day is presented below.

$$\text{Minimize } Z = 240X_1 + 160X_2 + (100X_1 + 150X_2)$$
$$= 340X_1 + 310X_2$$

subject to

$$76X_1 + 54X_2 \geq 31000$$
$$X_2 \leq 400$$
$$X_1 \text{ and } X_2 \geq 0$$

2.4 GRAPHICAL METHOD

As stated earlier, if the number of variables in any linear programming problem is only two, one can use *graphical method* to solve it. In this section, the graphical method is demonstrated with three example problems.

Example 2.12 Solve the following LP problem using graphical method.

$$\text{Maximize } Z = 6X_1 + 8X_2$$

subject to

$$5X_1 + 10X_2 \leq 60$$
$$4X_1 + 4X_2 \leq 40$$
$$X_1 \text{ and } X_2 \geq 0$$

Solution In graphical method, the introduction of the non-negative constraints ($X_1 \geq 0$ and $X_2 \geq 0$) will eliminate the second, third and fourth quadrants of the X_1X_2 plane, as shown in Figure 2.1.

Now, we compute the coordinates on the X_1X_2 plane. From the first constraint

$$5X_1 + 10X_2 = 60$$

we get $X_2 = 6$, when $X_1 = 0$; and $X_1 = 12$, when $X_2 = 0$. Now, plot the first constraint as shown in Figure 2.1.

From the second constraint

$$4X_1 + 4X_2 = 40$$

we get $X_2 = 10$, when $X_1 = 0$; and $X_1 = 10$, when $X_2 = 0$. Now, plot the second constraint as shown in Figure 2.1.

The closed polygon *A-B-C-D* is the feasible region. The objective function value at each of the corner points of the closed polygon is computed by substituting its coordinates in the objective function as:

$$Z(A) = 6 \times 0 + 8 \times 0 = 0$$
$$Z(B) = 6 \times 10 + 8 \times 0 = 60$$
$$Z(C) = 6 \times 8 + 8 \times 2 = 48 + 16 = 64$$
$$Z(D) = 6 \times 0 + 8 \times 6 = 48$$

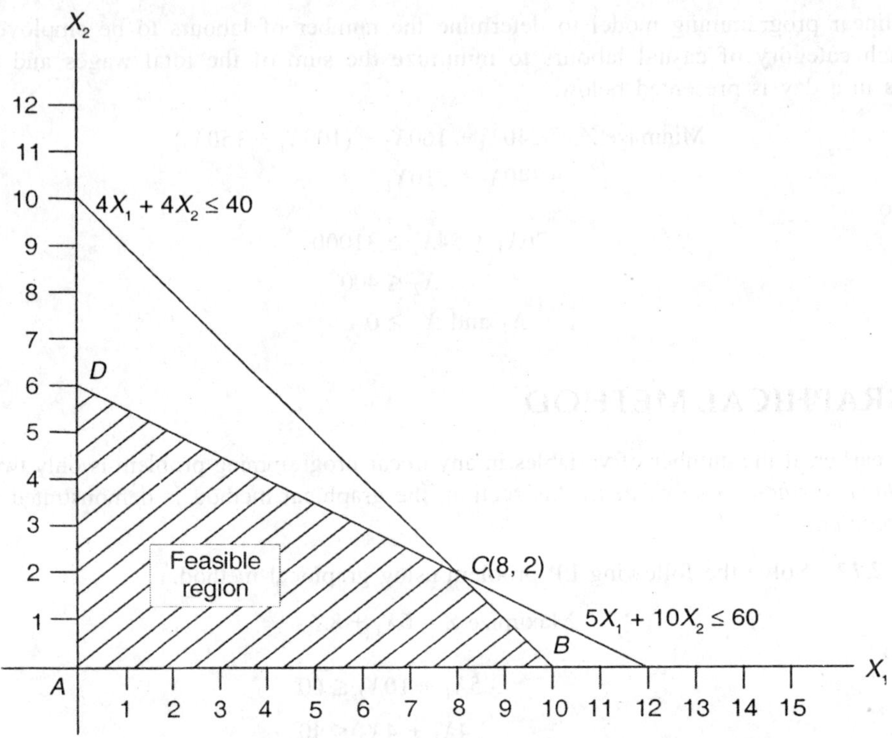

Figure 2.1 Feasible region of Example 2.12.

Since the type of the objective function here is maximization, the solution corresponding to the maximum Z value is to be selected as the optimum solution. The Z value is maximum for the corner point C. Hence, the corresponding solution is presented below.

$$X_1^* = 8, \qquad X_2^* = 2, \qquad Z(\text{optimum}) = 64$$

Example 2.13 Solve the following LP problem using graphical method:

$$\text{Minimize } Z = 2X_1 + 3X_2$$

subject to

$$X_1 + X_2 \geq 6$$

$$7X_1 + X_2 \geq 14$$

$$X_1 \text{ and } X_2 \geq 0$$

Solution The introduction of the non-negative constraints ($X_1 \geq 0$ and $X_2 \geq 0$) will eliminate the second, third and fourth quadrants of the X_1X_2 plane as shown in Figure 2.2.

Now, we compute the coordinates to plot on the X_1X_2 plane relating to different constraints. From the first constraint

$$X_1 + X_2 = 6$$

we get $X_2 = 6$, when $X_1 = 0$; and $X_1 = 6$, when $X_2 = 0$. Now, plot the constraint 1 as shown in Figure 2.2.

Second constraint is given as

$$7X_1 + X_2 = 14$$

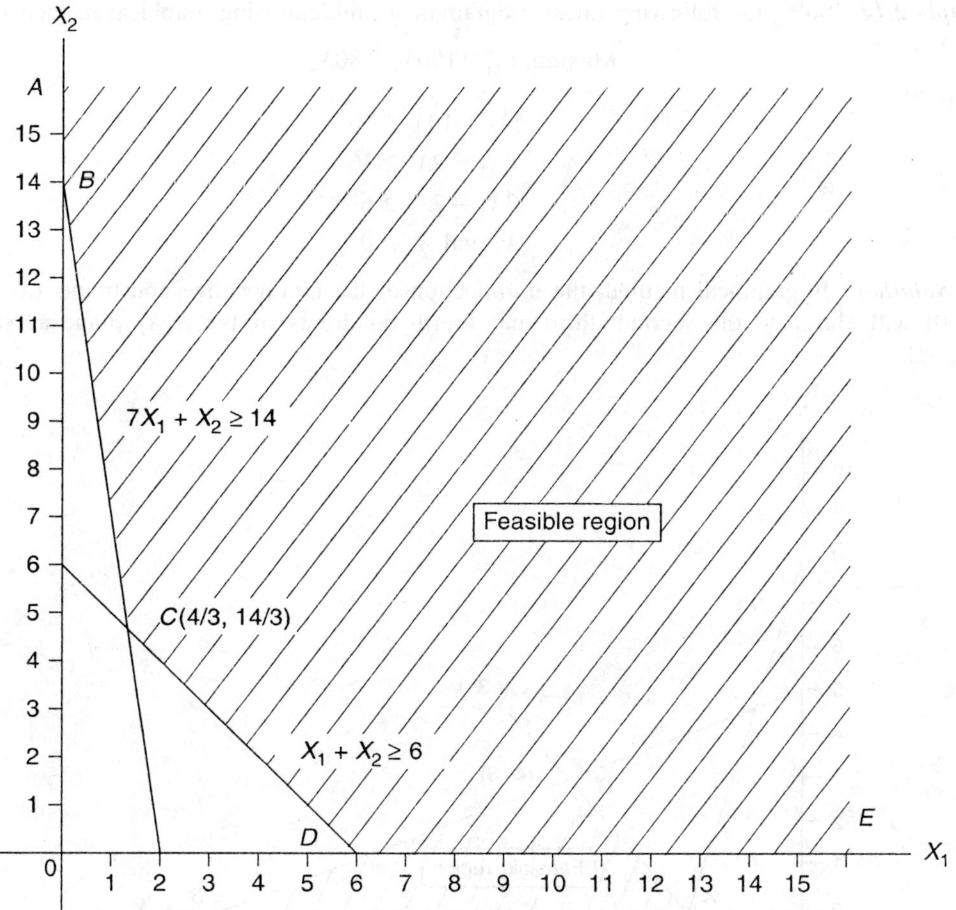

Figure 2.2 Feasible region of Example 2.13.

we get $X_2 = 14$, when $X_1 = 0$; and $X_1 = 2$, when $X_2 = 0$. Now, **plot** the constraint 2 as shown in Figure 2.2.

In Figure 2.2, *A-B-C-D-E* is the feasible region. The optimum solution will be in any one of the corner points, *B*, *C* and *D*. The objective function value at **each** of these corner points is computed as follows by substituting its coordinates in the objective function.

$$Z(B) = 2 \times 0 + 3 \times 14 = 42$$

$$Z(C) = 2 \times \frac{4}{3} + 3 \times \frac{14}{3} = \frac{50}{3} = 16.67$$

$$Z(D) = 2 \times 6 + 3 \times 0 = 12$$

Since the type of the objective function is minimization, the solution corresponding to the minimum Z value is to be selected as the optimum solution. The Z value is minimum for the corner point *D*. Hence, the corresponding optimum solution is:

$$X_1^* = 6, \qquad X_2^* = 0, \qquad Z(\text{optimum}) = 12$$

Example 2.14 Solve the following linear programming problem using graphical method.

$$\text{Maximize } Z = 100X_1 + 80X_2$$

subject to

$$5X_1 + 10X_2 \le 50$$
$$8X_1 + 2X_2 \ge 16$$
$$3X_1 - 2X_2 \ge 6$$
$$X_1 \text{ and } X_2 \ge 0$$

Solution In graphical method, the introduction of the non-negative constraints ($X_1 \ge 0$ and $X_2 \ge 0$) will eliminate the second, third and fourth quadrants of the X_1X_2 plane as shown in Figure 2.3.

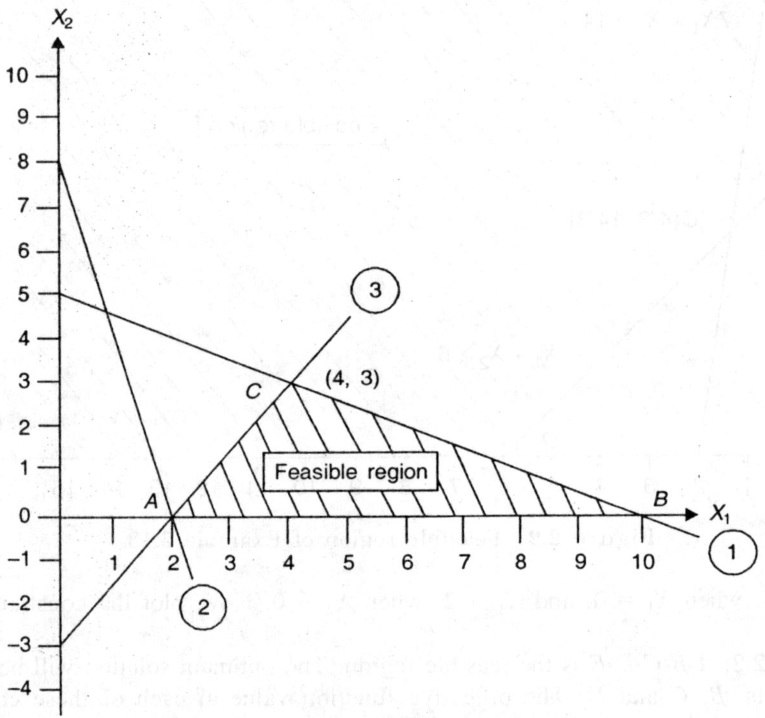

Figure 2.3 Feasible region of Example 2.14.

Now, we compute the coordinates on the X_1X_2 plane as shown below.
From the first constraint

$$5X_1 + 10X_2 = 50$$

we get, $X_2 = 5$ when $X_1 = 0$; and $X_1 = 10$ when $X_2 = 0$.
Now, plot the first constraint as shown in Figure 2.3.
From the second constraint

$$8X_1 + 2X_2 = 16$$

we get, $X_2 = 8$ when $X_1 = 0$; and $X_1 = 2$ when $X_2 = 0$.

Now, plot the second constraint as shown in Figure 2.3.

From the third constraint

$$3X_1 - 2X_2 = 6$$

we get, $X_2 = -3$ when $X_1 = 0$; and $X_1 = 2$ when $X_2 = 0$.

Now, plot the third constraint as shown in Figure 2.3.

The closed polygon *A-B-C* is the feasible region. The objective function value at each of the corner points of the closed polygon is computed as follows by substituting its coordinates in the objective function.

$$Z(A) = 100 \times 2 + 80 \times 0 = 200$$
$$Z(B) = 100 \times 10 + 80 \times 0 = 1000^*$$
$$Z(C) = 100 \times 4 + 80 \times 3 = 640$$

Since, the type of the objective function is maximization, the solution corresponding to the maximum Z value is to be selected as the optimum solution. The Z value is maximum for the corner point *B*. Hence, the corresponding solution is: $X_1^* = 10$, $X_2^* = 0$, Z(optimum) = 1000.

2.5 LINEAR PROGRAMMING METHODS

There are many algorithmic methods to solve linear programming problems. They are discussed in the following sections.

2.5.1 Simplex Method

Simplex method is the basic building block for all other methods. This method is devised based on the concept of solving simultaneous equations. It is demonstrated using a suitable numerical problem.

Example 2.15 Consider the linear programming model of Example 2.2 (as reproduced below) and solve it using the simplex method.

$$\text{Maximize } Z = 6X_1 + 8X_2$$

subject to

$$5X_1 + 10X_2 \leq 60$$
$$4X_1 + 4X_2 \leq 40$$
$$X_1 \text{ and } X_2 \geq 0$$

Solution The standard form of the above LP problem is shown below:

$$\text{Maximize } Z = 6X_1 + 8X_2 + 0S_1 + 0S_2$$

subject to

$$5X_1 + 10X_2 + S_1 = 60$$
$$4X_1 + 4X_2 + S_2 = 40$$
$$X_1, X_2, S_1 \text{ and } S_2 \geq 0$$

where S_1 and S_2 are *slack variables*, which are introduced to balance the constraints.

Canonical form is the form in which each constraint has a basic variable.

Definition of basic variable. A variable is said to be a *basic variable* if it has unit coefficient in one of the constraints and zero coefficient in the remaining constraints. If all the constraints are '$\leq$' type, then the standard form is to be treated as the canonical form. The canonical form is generally used to prepare the initial simplex table. The initial simplex table of the above problem is shown in Table 2.12.

Table 2.12 Initial Simplex Table (Example 2.15)

CB_i	C_j Basic variable	6 X_1	8 X_2	0 S_1	0 S_2	Solution	Ratio
0	S_1	5	10	1	0	60	60/10 = 6**
0	S_2	4	4	0	1	40	40/4 = 10
	Z_j	0	0	0	0	0	
	$C_j - Z_j$	6	8 *	0	0		

*Key column.   **Key row.

Here, C_j is the coefficient of the jth term of the objective function and CB_i is the coefficient of the ith basic variable. The value at the intersection of the key row and the key column is called the *key element*. The value of Z_j is computed using the following formula.

$$Z_j = \sum_{i=1}^{2} (CB_i)(a_{ij})$$

where a_{ij} is the technological coefficient for the ith row and jth column of the table. $C_j - Z_j$ is the relative contribution. In this term, C_j is the objective function coefficient for the jth variable. The value of Z_j against the solution column is the value of the objective function and in this iteration, it is zero.

Optimality condition.   For maximization problem, if all $C_j - Z_j$ are less than or equal to zero, then optimality is reached; otherwise select the variable with the maximum $C_j - Z_j$ value as the entering variable. (For minimization problem, if all $C_j - Z_j$ are greater than or equal to zero, the optimality is reached; otherwise select the variable with the most negative value as the entering variable.)

In Table 2.12, all the values for $C_j - Z_j$ are either equal to or greater than zero. Hence, the solution can be improved further. $C_j - Z_j$ is the maximum for the variable X_2. So, X_2 enters the basis. This is known as *entering variable*, and the corresponding column is called *key column*.

Feasibility condition.   To maintain the feasibility of the solution in each iteration, the following steps need to be followed:

1. In each row, find the ratio between the solution column value and the value in the key column.
2. Then, select the variable from the present set of basic variables with respect to the minimum ratio (break tie randomly). Such variable is the leaving variable and the corresponding row is called the *key row*. The value at the intersection of the key row and key column is called *key element* or *pivot element*.

In Table 2.12, the *leaving variable* is S_1 and the row 1 is the key row. Key element is 10.

The next iteration is shown in Table 2.13. In this table, the basic variable S_1 of the previous table is replaced by X_2. The formula to compute the new values of Table 2.13 is as shown below:

Table 2.13   Iteration 1

CB_i	C_j Basic variable	6 X_1	8 X_2	0 S_1	0 S_2	Solution	Ratio
8	X_2	1/2	1	1/10	0	6	6/(1/2) = 12
0	S_2	$\boxed{2}$	0	–2/5	1	16	16/2 = 8**
	Z_j	4	8	4/5	0	48	
	$C_j - Z_j$	2 *	0	– 4/5	0		

Here

$$\text{New value} = \text{Old value} - \frac{\text{Key column value} \times \text{Key row value}}{\text{Key value}}$$

As a sample calculation, the computation of the new value of row 2 and column X_1 is shown below:

$$\text{New value} = 4 - \frac{4 \times 5}{10} = 4 - \frac{20}{10} = 4 - 2 = 2$$

Computation of the cell values of different tables using this formula is a cumbersome process. So, a different procedure can be used as explained below.

Let the first and second rows in Table 2.12 be L_1 and L_2, respectively; and the first and second rows in Table 2.13 be L_3 and L_4, respectively. The coefficient of the first row of Table 2.13 can now be obtained by using the following formula.

$$L_3 = \frac{L_1}{\text{Pivot element}} = \frac{L_1}{10}$$

This operation makes the value of the cell with respect to the first row and the second column in Table 2.13 as unity. Since the new basic variable is becoming X_2, the cell value with respect to the second row and the second column in Table 2.13 should be made equal to 0.

This can be achieved by multiplying/dividing the value of the first row and the second column in Table 2.13 by a suitable constant and then by adding/subtracting the resultant value to/from the value of the second row and second column in Table 2.12 such that the net value is zero. The necessary formula to achieve this result is shown below.

$$L_4 = L_2 - 4L_3$$

The entries of the second row in Table 2.13 are obtained by using the above formula.

The solution in Table 2.13 is not optimal. The criterion row value for the variable X_1 is the maximum positive value. Hence, the variable X_1 is selected as the entering variable and after computing the ratios, S_2 is selected as the leaving variable. The next iteration is shown in Table 2.14.

Table 2.14 Iteration 2

CB_i	Basic variable	C_j				Solution
		6	8	0	0	
		X_1	X_2	S_1	S_2	
8	X_2	0	1	1/5	−1/4	2
6	X_1	1	0	−1/5	1/2	8
	Z_j	6	8	2/5	1	64
	$C_j - Z_j$	0	0	−2/5	−1	

In Table 2.14, all the values for $C_j - Z_j$ are either 0 or negative. Hence, the optimality is reached. The corresponding optimal solution is as follows:

$$X_1 \text{ (production volume of } P_1) = 8 \text{ units}$$
$$X_2 \text{ (production volume of } P_2) = 2 \text{ units}$$

and the optimal objective function value, Z (total profit) is Rs. 64.

Example 2.16 Solve the following LP problem using simplex method.

$$\text{Maximize } Z = 10X_1 + 15X_2 + 20X_3$$

subject to

$$2X_1 + 4X_2 + 6X_3 \leq 24$$
$$3X_1 + 9X_2 + 6X_3 \leq 30$$
$$X_1, X_2, \text{ and } X_3 \geq 0$$

Solution The standard form of this problem is

$$\text{Maximize } Z = 10X_1 + 15X_2 + 20X_3$$

subject to

$$2X_1 + 4X_2 + 6X_3 + S_1 = 24$$
$$3X_1 + 9X_2 + 6X_3 + S_2 = 30$$
$$X_1, X_2, X_3, S_1, \text{ and } S_2 \geq 0$$

where S_1 and S_2 are slack variables. Here, all the constraints are '$\leq$' type, so the canonical form of the given LP problem is same as the standard form represented below:

$$\text{Maximize } Z = 10X_1 + 15X_2 + 20X_3 + 0S_1 + 0S_2$$

subject to

$$2X_1 + 4X_2 + 6X_3 + S_1 = 24$$
$$3X_1 + 9X_2 + 6X_3 + S_2 = 30$$
$$X_1, X_2, X_3, S_1, \text{ and } S_2 \geq 0$$

The initial simplex table of the above problem is shown in Table 2.15.

In Table 2.15, all the values of $C_j - Z_j$ are not less than or equal to zero. Hence, the initial

Table 2.15 Initial Simplex Table (Example 2.16)

CB_i	C_j	10	15	20	0	0		
	Basic variable	X_1	X_2	X_3	S_1	S_2	Solution	Ratio
0	S_1	2	4	6	1	0	24	4**
0	S_2	3	9	6	0	1	30	5
	Z_j	0	0	0	0	0	0	
	$C_j - Z_j$	10	15	20*	0	0		

solution is not optimum. The variable X_3 is the entering variable because the column with respect to this variable has the highest $C_j - Z_j$ value. The variable S_1 is the leaving variable since the ratio with respect to this row is the least ratio. Hence, the key column is the column corresponding to the variable X_3 and the key row is row 1. The corresponding key element is 6.

The next iteration is shown in Table 2.16. In this table, the basic variable S_1 is replaced by X_3. In Table 2.16, the solution is not optimal. The variable X_1 is selected as the entering variable since the column with respect to this variable has the highest $C_j - Z_j$ value. Row 2 is selected as the key row and the corresponding variable S_2 is treated as the leaving variable, because it has the least ratio. The next iteration is shown in Table 2.17.

Table 2.16 Iteration 1

CB_i	C_j	10	15	20	0	0		
	Basic variable	X_1	X_2	X_3	S_1	S_2	Solution	Ratio
20	X_3	1/3	2/3	1	1/6	0	4	12
0	S_2	1	5	0	−1	1	6	6**
	Z_j	20/3	40/3	20	10/3	0	80	
	$C_j - Z_j$	10/3*	5/3	0	−10/3	0		

Table 2.17 Iteration 2

CB_i	C_j	10	15	20	0	0	
	Basic variable	X_1	X_2	X_3	S_1	S_2	Solution
20	X_3	0	−1	1	1/2	−1/3	2
10	X_1	1	5	0	−1	1	6
	Z_j	10	30	20	0	10/3	100
	$C_j - Z_j$	0	−15	0	0	−10/3	

In Table 2.17, since all the values of $C_j - Z_j$ are less than or equal to zero, the optimality is reached and the corresponding optimal solution is presented as:

$$X_1 = 6, \ X_2 = 0, \ X_3 = 2 \quad \text{and} \quad Z(\text{optimum}) = 100$$

2.5.2 Big *M* Method

If some of the constraints are of '=' or '≥' type, then they will not contain any basic variables. Just to have a basic variable in each of them, a new variable called *artificial variable* will be introduced in each of such constraints with a positive unit coefficient. If the objective function is a maximization type, then the coefficient of the artificial variable in the objective function should be $-M$; otherwise, it should be $+M$, where M is a very large value.

Example 2.17 Consider the following LP model (of Example 2.3), and solve it by using Big *M* method.

$$\text{Minimize } Z = 2X_1 + 3X_2$$

subject to

$$X_1 + X_2 \geq 6$$
$$7X_1 + X_2 \geq 14$$
$$X_1 \text{ and } X_2 \geq 0$$

Solution The standard form of the given problem is presented below.

$$\text{Minimize } Z = 2X_1 + 3X_2$$

subject to

$$X_1 + X_2 - S_1 = 6$$
$$7X_1 + X_2 - S_2 = 14$$
$$X_1, X_2, S_1 \text{ and } S_2 \geq 0$$

where S_1 and S_2 are called *surplus variables* which are introduced to balance the constraints. The canonical form of the above model is presented below after introducing necessary artificial variables.

$$\text{Minimize } Z = 2X_1 + 3X_2 + 0S_1 + 0S_2 + MR_1 + MR_2$$

subject to

$$X_1 + X_2 - S_1 + R_1 = 6$$
$$7X_1 + X_2 - S_2 + R_2 = 14$$
$$X_1, X_2, S_1, S_2, R_1 \text{ and } R_2 \geq 0$$

These artificial variables are included in the model just to solve the model. Therefore in the final solution, these artificial variables should not be available. This is achieved by including these variables in the objective function with a very high positive coefficient M, since the objective function is a minimization type. Even a small value of R_1 or R_2 will increase the value of the objective function infinitely which is against the objective of minimization. So, the solution procedure should necessarily assign zero value to each of the artificial variables in the final solution, except for the problems which are having infeasible solution. The initial table of the model is shown in Table 2.18.

Optimality condition. For minimization problem, if all $C_j - Z_j$ are greater than or equal to zero, optimality is reached; otherwise select the entering variable with the most negative value.

Table 2.18 Initial Table (Example 2.17)

CB_i	C_j	2	3	0	0	M	M		
	Basic variable	X_1	X_2	S_1	S_2	R_1	R_2	Solution	Ratio
M	R_1	1	1	-1	0	1	0	6	6
M	R_2	7	1	0	-1	0	1	14	2**
	Z_j	$8M$	$2M$	$-M$	$-M$	M	M	$20M$	
	$C_j - Z_j$	$2 - 8M$ *	$3 - 2M$	M	M	0	0		

In Table 2.18, all the values for $C_j - Z_j$ are not greater than or equal to zero. Therefore, the solution can be improved further. Since $C_j - Z_j$ has the maximum negative value for the variable X_1, X_1 enters the basis. This is known as *entering variable*, and the corresponding column is called *key column*. If $C_j - Z_j$ is a function of M, then ignore the constant numeric terms in it while making comparison with another $C_j - Z_j$. Feasibility condition is the same as in the maximization problem. The ratio is the minimum for the second row R_2. Hence it is called as the *key row* and the basic variable R_2 leaves the basis. The modified table, after applying the pivot operations, is shown in Table 2.19.

Table 2.19 Iteration 1

CB_i	C_j	2	3	0	0	M	M		
	Basic variable	X_1	X_2	S_1	S_2	R_1	R_2	Solution	Ratio
M	R_1	0	6/7	-1	1/7	1	$-1/7$	4	14/3**
2	X_1	1	1/7	0	$-1/7$	0	1/7	2	14
	Z_j	2	$\frac{6}{7}M + \frac{2}{7}$	$-M$	$\frac{M}{7} - \frac{2}{7}$	M	$-\frac{M}{7} + \frac{2}{7}$	$4M + 4$	
	$C_j - Z_j$	0	$-\frac{6}{7}M + \frac{19}{7}$ *	M	$-\frac{M}{7} + \frac{2}{7}$	0	$\frac{8}{7}M - \frac{2}{7}$		

In Table 2.19, the optimum is not reached. Hence, the column containing X_2 becomes the key column and the row R_1 becomes as the key row. So, R_1 is replaced by X_2 in the next table, i.e. Table 2.20. The modified table after applying the pivot operations is shown in Table 2.20.

In Table 2.20 the optimality is not reached. Hence the column with S_2 becomes the key column and the row containing X_2 becomes the key row. So, X_2 is replaced by S_2 as shown in Table 2.21. The results of the corresponding pivot operations are also shown in Table 2.21.

Since all the $C_j - Z_j$ are equal to 0 and above, the optimality is reached. The corresponding optimal solution is presented as:

$$X_1 = 6,\ X_2 = 0,\ S_1 = 0,\ S_2 = 28,\ \text{and } Z(\text{optimum}) = 12$$

Table 2.20 Iteration 2

CB_i	C_j	2	3	0	0	M	M	Solution	Ratio
	Basic variable	X_1	X_2	S_1	S_2	R_1	R_2		
3	X_2	0	1	$-7/6$	$\boxed{1/6}$	$7/6$	$-1/6$	$14/3$	28^{**}
2	X_1	1	0	$1/6$	$-1/6$	$-1/6$	$1/6$	$4/3$	–
	Z_j	2	3	$-19/6$	$1/6$	$19/6$	$-1/6$	$50/3$	
	$C_j - Z_j$	0	0	$19/6$	$-1/6$	$M - 19/6$	$M + 1/6$		
					*				

Table 2.21 Iteration 3

CB_i	C_j	2	3	0	0	M	M	Solution
	Basic variable	X_1	X_2	S_1	S_2	R_1	R_2	
0	S_2	0	6	-7	1	7	-1	28
2	X_1	1	1	-1	0	1	0	6
	Z_j	2	2	-2	0	2	0	12
	$C_j - Z_j$	0	1	2	0	$M - 2$	M	

Now the results are interpreted as:

$$X_1 \text{ (Number of packets of food Type I)} = 6$$
$$X_2 \text{ (Number of packets of food Type II)} = 0$$

The corresponding daily total cost of food is Rs. 12.00. So, the babies are to be given 6 packets of food type I daily.

Example 2.18 Solve the following linear programming problem using Big M method.

$$\text{Minimize } Z = 10X_1 + 15X_2 + 20X_3$$

subject to

$$2X_1 + 4X_2 + 6X_3 \geq 24$$
$$3X_1 + 9X_2 + 6X_3 \geq 30$$
$$X_1, X_2, X_3 \geq 0$$

Solution The standard form of this problem is as shown below. In this form, S_1, S_2 are called as *surplus* variables which are introduced to balance the constraints.

$$\text{Minimize } Z = 10X_1 + 15X_2 + 20X_3$$

subject to

$$2X_1 + 4X_2 + 6X_3 - S_1 = 24$$
$$3X_1 + 9X_2 + 6X_3 - S_2 = 30$$
$$X_1, X_2, X_3, S_1, S_2 \geq 0$$

The canonical form of the above standard form, which consists of the artificial variables R_1 and R_2, is presented below:

$$\text{Minimize } Z = 10X_1 + 15X_2 + 20X_3 + MR_1 + MR_2$$

subject to

$$2X_1 + 4X_2 + 6X_3 - S_1 + R_1 = 24$$

$$3X_1 + 9X_2 + 6X_3 - S_2 + R_2 = 30$$

$$X_1, X_2, X_3, S_1, S_2, R_1 \text{ and } R_2 \geq 0$$

Here, R_1 and R_2 are introduced to have a basic variable in each of the constraints. Table 2.22 represents the initial table of the canonical form. In Table 2.22, the optimality is not reached. Considering that the entering variable is X_2 and the leaving variable is R_2, the corresponding pivot operations can be presented as shown in Table 2.23. In Table 2.23, the optimality is again not

Table 2.22 Initial Table (Example 2.18)

CB_i	C_j Basic variable	10 X_1	15 X_2	20 X_3	0 S_1	0 S_2	M R_1	M R_2	Solution	Ratio
M	R_1	2	4	6	-1	0	1	0	24	6
M	R_2	3	9	6	0	-1	0	1	30	10/3**
	Z_j	$5M$	$13M$	$12M$	$-M$	$-M$	M	M	$54M$	
	$C_j - Z_j$	$10 - 5M$	$15 - 13M$ *	$20 - 12M$	M	M	0	0		

Table 2.23 Iteration 1

CB_i	C_j Basic variable	10 X_1	15 X_2	20 X_3	0 S_1	0 S_2	M R_1	M R_2	Solution	Ratio
M	R_1	2/3	0	10/3	-1	4/9	1	$-4/9$	32/3	16/5**
15	X_2	1/3	1	2/3	0	$-1/9$	0	1/9	10/3	5
	Z_j	$\dfrac{2}{3}M + 5$	15	$\dfrac{10}{3}M + 10$	$-M$	$\dfrac{4}{9}M - \dfrac{5}{3}$	M	$-\dfrac{4}{9}M + \dfrac{5}{3}$	$\dfrac{32}{3}M + 50$	
	$C_j - Z_j$	$-\dfrac{2}{3}M + 5$	0	$-\dfrac{10}{3}M + 10$ *	M	$-\dfrac{4}{9}M + \dfrac{5}{9}$	0	$\dfrac{13}{9}M - \dfrac{5}{3}$		

reached. With the entering variable as X_3 and the leaving variable as R_1, the corresponding pivot operations are shown in Table 2.24.

In Table 2.24, the optimality is reached and the corresponding optimal result is presented below:

$$X_1 = 0, \ X_2 = \frac{6}{5}, \ X_3 = \frac{16}{5} \quad \text{and} \quad Z(\text{optimum}) = 82$$

Table 2.24 Iteration 2

CB_i	C_j / Basic variable	10	15	20	0	0	M	M	Solution
		X_1	X_2	X_3	S_1	S_2	R_1	R_2	
20	X_3	1/5	0	1	$-3/10$	2/15	3/10	$-2/15$	16/5
15	X_2	1/5	1	0	1/5	$-1/5$	$-1/5$	1/5	6/5
	Z_j	7	15	20	-3	$-1/3$	3	1/3	82
	$C_j - Z_j$	3	0	0	3	1/3	$M-3$	$M - \dfrac{1}{3}$	

Example 2.19 Consider the following linear programming problem and solve it using the simplex method.

$$\text{Maximize } Z = 20X_1 + 10X_2 + 15X_3$$

subject to

$$8X_1 + 6X_2 + 2X_3 \leq 60$$
$$5X_1 + X_2 + 6X_3 \geq 40$$
$$2X_1 + 6X_2 + 3X_3 \leq 30$$
$$X_1,\ X_2 \text{ and } X_3 \geq 0$$

Solution The canonical form of the given problem is shown below:

$$\text{Maximize } Z = 20X_1 + 10X_2 + 15X_3 + 0S_1 + 0S_2 + 0S_3 - MR_1$$

subject to

$$8X_1 + 6X_2 + 2X_3 + S_1 \qquad\qquad = 60$$
$$5X_1 + X_2 + 6X_3 - S_2 + R_1 \quad = 40$$
$$2X_1 + 6X_2 + 3X_3 + S_3 \qquad\quad = 30$$
$$X_1,\ X_2,\ X_3,\ S_1,\ S_2,\ S_3 \text{ and } R_1 \geq 0$$

In the above model, S_1 and S_3 are slack variables; S_2 is a surplus variable and R_1 is an artificial variable.

The different iterations of the simplex method applied to this problem till the optimality is reached are shown in Table 2.25. The optimal results from the last iteration of the Table 2.25 are: $X_1 = 6$, $X_3 = 6$, $S_2 = 26$ and Z(optimum) = 210.

2.5.3 Dual Simplex Method

Dual simplex method is a specialized form of simplex method in which optimality is maintained in all the iterations. Initially, the solution may not be feasible. Successive iterations will remove the infeasibility. If the problem is feasible in an iteration, then the procedure will be stopped, because the solution obtained is feasible and optimal at that stage.

This method is an essential subroutine for integer programming method in which it is repeatedly used to remove the infeasibility due to additional constraints known as *Gomory's cuts*.

In this section, the dual simplex method is demonstrated through a numerical problem.

Table 2.25 Initial Table and Different Iterations of Example 2.19

CB_i	C_j Basic variable	20 X_1	10 X_2	15 X_3	0 S_1	0 S_2	0 S_3	$-M$ R_1	Solution	Ratio
0	S_1	8	6	2	1	0	0	0	60	30
$-M$	R_1	5	1	6	0	-1	0	1	40	20/3
0	S_3	2	6	3	0	0	1	0	30	10
	Z_j	$-5M$	$-M$	$-6M$	0	M	0	$-M$	$-40M$	
	$C_j - Z_j$	$20 + 5M$	$10 + M$	$15 + 6M$	0	$-M$	0	0		
0	S_1	19/3	17/3	0	1	1/3	0	$-1/3$	140/3	140/19
15	X_3	5/6	1/6	1	0	$-1/6$	0	1/6	20/3	8
0	S_3	$-1/2$	11/2	0	0	1/2	1	$-1/2$	10	$-$
	Z_j	25/2	5/2	15	0	$-5/2$	0	5/2	100	
	$C_j - Z_j$	15/2	15/2	0	0	5/2	0	$-M-5/2$		
20	X_1	1	17/19	0	3/19	1/19	0	$-1/19$	140/19	140
15	X_3	0	$-11/19$	1	$-5/38$	$-4/19$	0	4/19	10/19	$-$
0	S_3	0	113/19	0	3/38	10/19	1	$-10/19$	260/19	26
	Z_j	20	175/19	15	45/38	$-40/19$	0	40/19	2950/19	
	$C_j - Z_j$	0	15/19	0	$-45/38$	40/19	0	$-M-40/19$		
20	X_1	1	3/10	0	3/20	0	$-1/10$	0	6	
15	X_3	0	9/5	1	$-1/10$	0	2/5	0	6	
0	S_2	0	113/10	0	3/20	1	19/10	-1	26	
	Z_j	20	33	15	3/2	0	4	0	210	
	$C_j - Z_j$	0	-23	0	$-3/2$	0	-4	$-M$		

Example 2.20 Solve the following linear programming problem using dual simplex method.

$$\text{Minimize } Z = 2X_1 + 4X_2$$

subject to

$$2X_1 + X_2 \geq 4$$
$$X_1 + 2X_2 \geq 3$$
$$2X_1 + 2X_2 \leq 12$$
$$X_1 \text{ and } X_2 \geq 0$$

Solution Convert the constraints of the given linear programming problem into '$\leq$' type, wherever necessary, as shown below:

$$\text{Minimize } Z = 2X_1 + 4X_2$$

subject to

$$-2X_1 - X_2 \leq -4$$
$$-X_1 - 2X_2 \leq -3$$
$$2X_1 + 2X_2 \leq 12$$
$$X_1 \text{ and } X_2 \geq 0$$

The canonical form of the above model is shown below in which S_1, S_2 and S_3 are slack variables.

$$\text{Minimize } Z = 2X_1 + 4X_2$$

subject to

$$-2X_1 - X_2 + S_1 = -4$$
$$-X_1 - 2X_2 + S_2 = -3$$
$$2X_1 + 2X_2 + S_3 = 12$$
$$X_1, X_2, S_1, S_2 \text{ and } S_3 \geq 0$$

The initial table based on the canonical form of the given problem is shown in Table 2.26. For minimization problem, if all $C_j - Z_j$ are greater than or equal to 0, then optimality is reached. In

Table 2.26   Iteration 1

CB_i	C_j	2	4	0	0	0	
	Basic variable	X_1	X_2	S_1	S_2	S_3	Solution
0	S_1	-2	-1	1	0	0	-4**
0	S_2	-1	-2	0	1	0	-3
0	S_3	2	2	0	0	1	12
	Z_j	0	0	0	0	0	0
	$C_j - Z_j$	2*	4	0	0	0	

Table 2.26, $C_j - Z_j$ row clearly shows that the problem is optimal. But some of the values under the solution column are negative. These negative values will retain infeasibility in the solution. The guidelines for removing the infeasibility are presented below.

Feasibility condition.   The leaving variable is the variable which is having the most negative value (break ties arbitrarily). If all the basic variables are non-negative, then the feasible (optimal) solution is reached. Hence, the procedure ends here.

Optimality condition.   The entering variable is selected from among the non-basic variables as follows:

1. Find the ratios of the negative of the coefficients of the criterion row ($C_j - Z_j$) equation to the corresponding left-hand side coefficients of the equation associated with the leaving variable. Ignore the ratios associated with positive or zero denominators.

2. The entering variable is the one with the smallest ratio if the problem is minimization, or the smallest absolute value of the ratios if the problem pertains to maximization (break tie arbitrarily). If all the denominators are zero or positive, then the problem has no feasible solution.

In Table 2.26, the leaving variable is S_1 which has the most negative right-hand side value. The entering variable is determined as shown in Table 2.27.

Table 2.27 Determination of Entering Variable

Variable	X_1	X_2	S_1	S_2	S_3
$-(C_j - Z_j)$	−2	−4	0	0	0
S_1 equation	−2	−1	1	0	0
Ratio	1	4	–	–	–

Since the problem is of the minimization type, entering variable is the one which has the smallest ratio, and the smallest ratio is 1 which is corresponding to the variable X_1. Therefore, the entering variable is X_1. The next iteration is shown in Table 2.28.

Table 2.28 Iteration 2

CB_i	Basic variable	C_j					Solution
		2	4	0	0	0	
		X_1	X_2	S_1	S_2	S_3	
2	X_1	1	1/2	−1/2	0	0	2
0	S_2	0	−3/2	−1/2	1	0	−1**
0	S_3	0	1	1	0	1	8
	Z_j	2	1	−1	0	0	4
	$C_j - Z_j$	0	3	1*	0	0	

In Table 2.28, the solution is optimal but it is not feasible, as the solution value of the row S_2 is negative. So, the leaving variable is S_2 which has negative right-hand side value. The entering variable is determined as in Table 2.29.

Table 2.29 Determination of Entering Variable

Variable	X_1	X_2	S_1	S_2	S_3
$-(C_j - Z_j)$	0	−3	−1	0	0
S_2 equation	0	−3/2	−1/2	1	0
Ratio	–	2	2	–	–

Since the problem is concerned with minimization, the entering variable is the one which has the smallest ratio. The smallest ratio is 2 corresponding to the variable X_2 and S_1. By breaking the tie randomly, S_1 is selected as the entering variable. The next iteration is shown in Table 2.30.

In Table 2.30, the solution values (right-hand side values) are feasible and at the same time, the solution is optimal. The corresponding results are:

$$X_1 = 3, \ X_2 = 0, \ S_1 = 2, \ S_3 = 6 \ \text{ and } \ Z(\text{optimum}) = 6$$

where all other variables are zero.

Table 2.30   Iteration 3

	C_j	2	4	0	0	0	
CB_i	Basic variable	X_1	X_2	S_1	S_2	S_3	Solution
2	X_1	1	2	0	−1	0	3
0	S_1	0	3	1	−2	0	2
0	S_3	0	−2	0	2	1	6
	Z_j	2	4	0	−2	0	6
	$C_j - Z_j$	0	0	0	2	0	

2.5.4   Two-phase Method

When Big M method is computerized, the value of M is assumed to be very large number. As a result, there may be rounding/truncation of coefficients while carrying out different iterations that end up with misleading results. To overcome this difficulty, two-phase simplex method is used for problems involving '$\geq$' or '$=$' type constraints.

Algorithm for two-phase method is discussed as follows:

Step 0:   Obtain the canonical form of the given problem. (*Note:* Whatever may be the type of the objective function in the original problem, only the minimization type of objective function is used in the two-phase method.)

Phase 1

Step 1:   Form a modified problem for phase 1 from the canonical form of the original problem by replacing the objective function with the sum of only the artificial variables along with the same set of constraints of the canonical form of the original problem.

Step 2:   Prepare the initial table for phase 1.

Step 3:   Apply the usual simplex method till the optimality is reached.

Step 4:   Check whether the objective function value is zero in the optimal table of phase 1. If yes, go to phase 2; otherwise, conclude that the original problem has no feasible solution and stop.

Phase 2

Step 5:   Obtain a modified table using the following steps:

    5.1   Drop the columns in the optimum table of phase 1 corresponding to the artificial variables which are currently non-basic.

    5.2   If some artificial variable(s) is (are) present at zero level in the basic solution of the optimal table of phase 1, substitute its objective function coefficient(s) with zero.

    5.3   Substitute the coefficients of the original objective function in the optimal table of the phase 1 for the remaining variables and make necessary changes in CB column.

Step 6:   Carry out further iterations till the optimality is reached and then stop.

Example 2.21   Consider the following linear programming model and solve it using the two-phase method.

$$\text{Minimize } Z = 12X_1 + 18X_2 + 15X_3$$

subject to

$$4X_1 + 8X_2 + 6X_3 \geq 64$$
$$3X_1 + 6X_2 + 12X_3 \geq 96$$
$$X_1, X_2 \text{ and } X_3 \geq 0$$

Solution The canonical form of the given problem is shown below:

$$\text{Minimize } Z = 12X_1 + 18X_2 + 15X_3 + MR_1 + MR_2$$

subject to

$$4X_1 + 8X_2 + 6X_3 - S_1 + R_1 = 64$$
$$3X_1 + 6X_2 + 12X_3 - S_2 + R_2 = 96$$
$$X_1, X_2, X_3, S_1, S_2, R_1, \text{ and } R_2 \geq 0$$

Phase 1

The model for phase 1 with its revised objective function is shown below. The corresponding initial table is presented in Table 2.31. Further iterations towards optimality are shown in Tables 2.32 and 2.33.

$$\text{Minimize } Z = R_1 + R_2$$

subject to

$$4X_1 + 8X_2 + 6X_3 - S_1 + R_1 = 64$$
$$3X_1 + 6X_2 + 12X_3 - S_2 + R_2 = 96$$
$$X_1, X_2, X_3, S_1, S_2, R_1, R_2 \geq 0$$

Table 2.31 Initial Table (Example 2.21)

CB_i	C_j / Basic variable	0 / X_1	0 / X_2	0 / X_3	0 / S_1	0 / S_2	1 / R_1	1 / R_2	Solution
1	R_1	4	8	6	−1	0	1	0	64
1	R_2	3	6	[12]	0	−1	0	1	96**
	Z_j	7	14	18	−1	−1	1	1	160
	$C_j - Z_j$	−7	−14	−18*	1	1	0	0	

Table 2.32 Iteration 1

CB_i	C_j / Basic variable	0 / X_1	0 / X_2	0 / X_3	0 / S_1	0 / S_2	1 / R_1	1 / R_2	Solution
1	R_1	5/2	[5]	0	−1	1/2	1	−1/2	16**
0	X_3	1/4	1/2	1	0	−1/12	0	1/12	8
	Z_j	5/2	5	0	−1	1/2	1	−1/2	16
	$C_j - Z_j$	−5/2	−5*	0	1	−1/2	0	3/2	

Table 2.33   Iteration 2 (Optimal Table of Phase 1)

CB_i	C_j	0	0	0	0	0	1	1	
	Basic variable	X_1	X_2	X_3	S_1	S_2	R_1	R_2	Solution
0	X_2	1/2	1	0	$-1/5$	1/10	1/5	$-1/10$	16/5
0	X_3	0	0	1	1/10	$-2/15$	$-1/10$	2/15	32/5
	Z_j	0	0	0	0	0	0	0	0
	$C_j - Z_j$	0	0	0	0	0	1	1	

The set of basic variables in the optimal table of phase 1 does not contain artificial variables. So, the given problem has a feasible solution.

Phase 2

The only one iteration of phase 2 is shown in Table 2.34. One can verify that Table 2.34, itself gives the optimal solution. The solution in Table 2.34 is optimal and feasible. The optimal results are presented below.

$$X_1 = 0, \; X_2 = \frac{16}{5}, \; X_3 = \frac{32}{5} \quad \text{and} \quad Z(\text{optimum}) = \frac{768}{5} = 153.6$$

Table 2.34   Initial and Optimal Tables of Phase 2

CB_i	C_j	12	18	15	0	0	
	Basic variable	X_1	X_2	X_3	S_1	S_2	Solution
18	X_2	1/2	1	0	$-1/5$	1/10	16/5
15	X_3	0	0	1	1/10	$-2/15$	32/5
	Z_j	9	18	15	$-21/10$	$-1/5$	768/5
	$C_j - Z_j$	3	0	0	21/10	1/5	

2.6   SPECIAL CASES OF LINEAR PROGRAMMING

The special cases of the solution of the linear programming problem can be categorized as shown below.

    (a) Infeasible solution
    (b) Unbounded solution
    (c) Unbounded solution space with finite solution
    (d) Alternate optimum solution
    (e) Degenerate solution.

In this section, the special cases of the linear programming problems are explained with suitable illustrations.

Example 2.22 **(Infeasible solution)** Solve the following LP problem graphically:

$$\text{Maximize } Z = 10X_1 + 3X_2$$

subject to

$$2X_1 + 3X_2 \leq 18$$
$$6X_1 + 5X_2 \geq 60$$
$$X_1 \text{ and } X_2 \geq 0$$

Solution The introduction of the non-negative constraints ($X_1 \geq 0$ and $X_2 \geq 0$) of the given problem will eliminate the second, third and fourth quadrants of the X_1X_2 plane, as shown in Figure 2.4. Compute the coordinates to plot equations relating to different constraints on the X_1X_2 plane, as shown below.

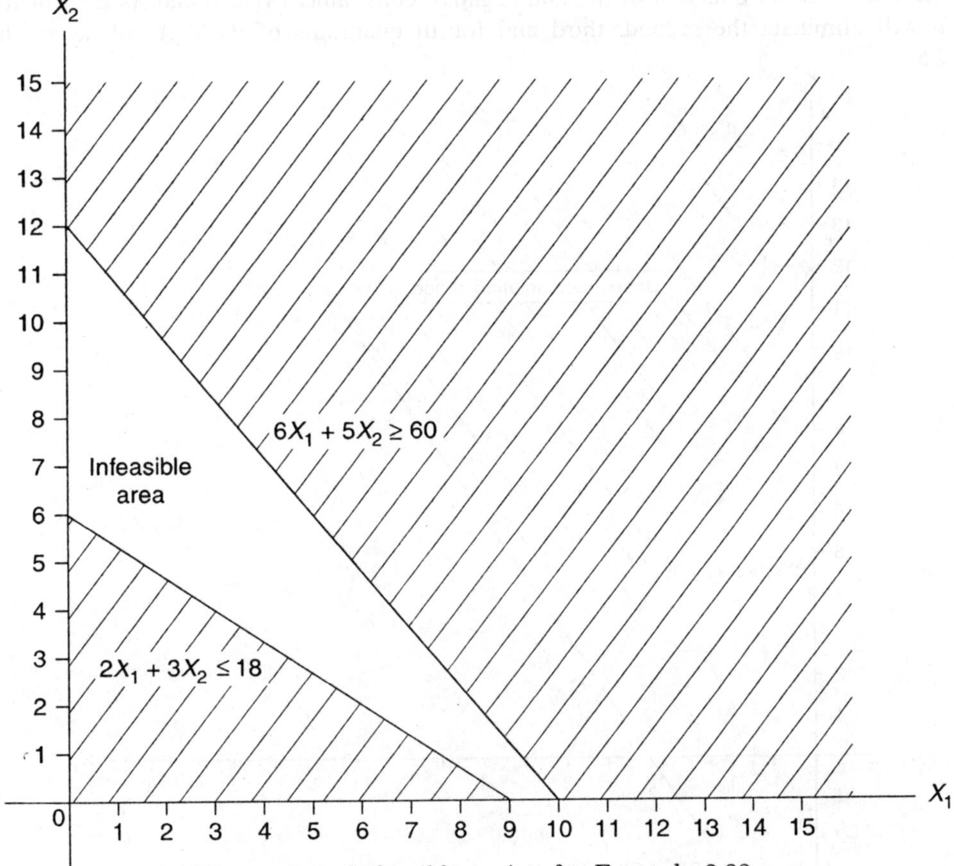

Figure 2.4 Infeasible region for Example 2.22.

From the first constraint in equation form

$$2X_1 + 3X_2 = 18$$

we get $X_2 = 6$, when $X_1 = 0$; and $X_1 = 9$, when $X_2 = 0$. Also, from the second constraint in equation form

$$6X_1 + 5X_2 = 60$$

we have $X_2 = 12$, when $X_1 = 0$; and $X_1 = 10$, when $X_2 = 0$.

Now, plot the constraints 1 and 2 as shown in Figure 2.4. The feasible side of each constraint is shaded. From Figure 2.4, it is clear that there is no common intersecting area of the shaded regions. This means that the problem has infeasible solution.

Example 2.23 **(Unbounded solution)**   Solve the following LP problem using graphical method.

$$\text{Maximize } Z = 12X_1 + 25X_2$$

subject to

$$12X_1 + 3X_2 \geq 36$$
$$15X_1 - 5X_2 \leq 30$$
$$X_1 \text{ and } X_2 \geq 0$$

Solution   The introduction of the non-negative constraints ($X_1 \geq 0$ and $X_2 \geq 0$) of the given problem will eliminate the second, third and fourth quadrants of the X_1X_2 plane as shown in Figure 2.5.

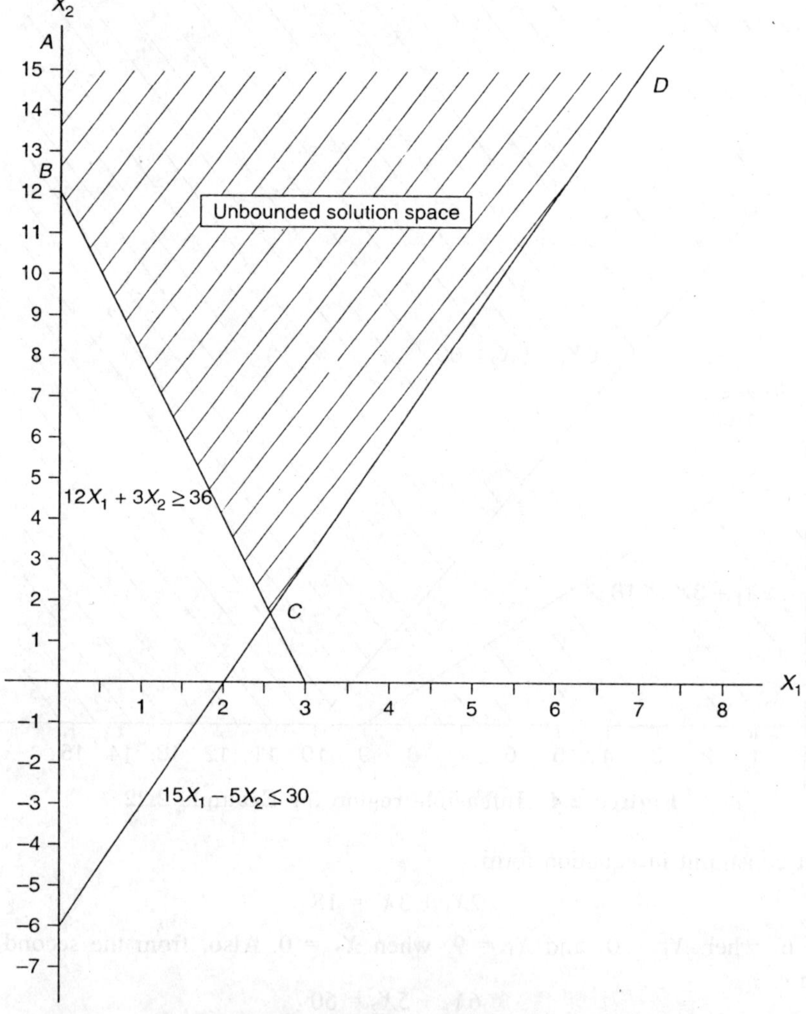

Figure 2.5   Unbounded solution space for Example 2.23.

Compute the coordinates to plot the equations relating to different constraints on the X_1X_2 plane as shown below. From the first constraint in equation form

$$12X_1 + 3X_2 = 36$$

we get $X_2 = 12$, when $X_1 = 0$; and $X_1 = 3$, when $X_2 = 0$. Also, from the second constraint in equation form

$$15X_1 - 5X_2 = 30$$

we have $X_2 = -6$, when $X_1 = 0$; and $X_1 = 2$, when $X_2 = 0$.

Now, plot constraints 1 and 2 as shown in Figure 2.5. In this figure, the solution space is denoted by A, B, C and D. One of the sides of this shaded region is not closed. This indicates the unbounded nature of the solution of the given problem. The objective function can be increased to infinity.

Example 2.24 **(Unbounded solution space with finite solution)**   Solve the following LP problem graphically.

$$\text{Maximize } Z = 5X_1 - 2X_2$$

subject to

$$X_1 \le 2$$

$$-X_1 + 2X_2 \ge 4$$

$$X_1 \text{ and } X_2 \ge 0$$

Solution   The introduction of the non-negative constraints ($X_1 \ge 0$ and $X_2 \ge 0$) will eliminate the second, third and fourth quadrants of the X_1X_2 plane as shown in Figure 2.6.

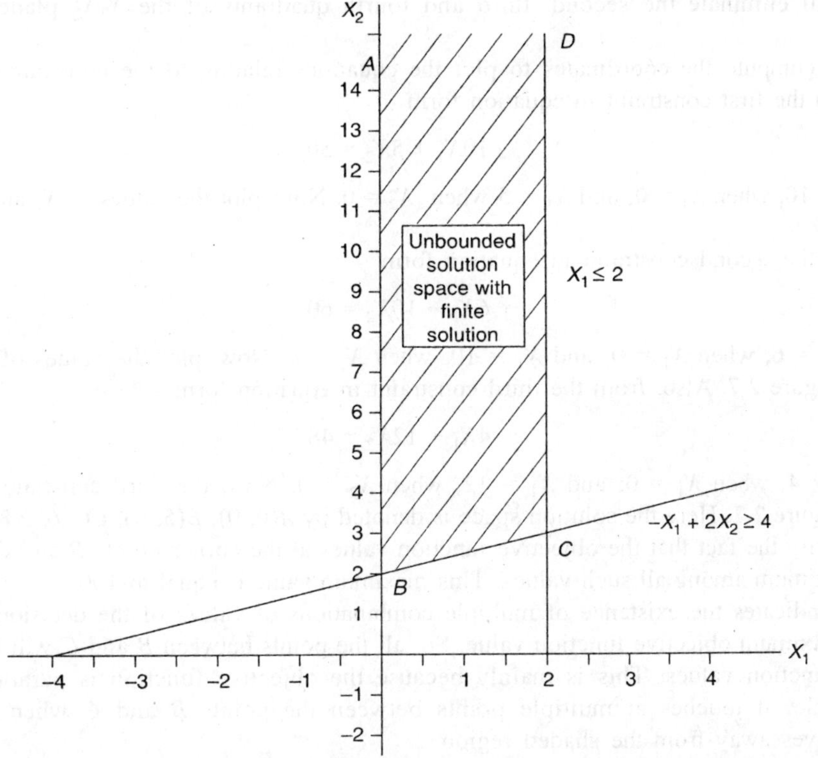

Figure 2.6   Unbounded solution space with finite solution of Example 2.24.

Now, from the first constraint, we get $X_1 = 2$, and from the second constraint in equation form

$$-X_1 + 2X_2 = 4$$

we get $X_2 = 2$, when $X_1 = 0$; and $X_1 = -4$, when $X_2 = 0$. We plot the constraints 1 and 2 as shown in Figure 2.6. In Figure 2.6, the solution space $ABCD$ is unbounded because the value of the variable X_2 is unlimited. Since the coefficient of the variable X_2 in the objective function is negative, any increase in X_2 will decrease the objective function value. But the value of the variable X_1 is limited. So, the solution space has a feasible and optimal solution at C. The corresponding results are:

$$X_1 = 2, X_2 = 3 \quad \text{and} \quad Z(\text{optimum}) = 4$$

Example 2.25 **(Alternate optima/multiple optimum solution)** Solve the following LP problem using graphical method:

$$\text{Maximize } Z = 20X_1 + 10X_2$$

subject to

$$10X_1 + 5X_2 \leq 50$$

$$6X_1 + 10X_2 \leq 60$$

$$4X_1 + 12X_2 \leq 48$$

$$X_1 \text{ and } X_2 \geq 0$$

Solution The introduction of the non-negative constraints ($X_1 \geq 0$ and $X_2 \geq 0$) of the given problem will eliminate the second, third and fourth quadrants of the X_1X_2 plane as shown in Figure 2.7.

Now, compute the coordinates to plot the equations relating to the constraints on the X_1X_2 plane. From the first constraint in equation form

$$10X_1 + 5X_2 = 50$$

we get $X_2 = 10$, when $X_1 = 0$; and $X_1 = 5$ when $X_2 = 0$. Now, plot the values of X_1 and X_2 as shown in Figure 2.7.

From the second constraint in equation form

$$6X_1 + 10X_2 = 60$$

we have $X_2 = 6$, when $X_1 = 0$; and $X_1 = 10$, when $X_2 = 0$. Now, plot the values of X_1 and X_2 as shown in Figure 2.7. Also, from the third constraint in equation form

$$4X_1 + 12X_2 = 48$$

we get $X_2 = 4$, when $X_1 = 0$; and $X_1 = 12$, when $X_2 = 0$. Now, the third constraint is plotted as shown in Figure 2.7. Here the solution space is denoted by $A(0, 0)$, $B(5, 0)$, $C(3.6, 2.8)$ and $D(0, 4)$. One can verify the fact that the objective function values at the corner points B and C are the same and the maximum among all such values. This maximum value is equal to 100.

This indicates the existence of multiple combinations of values of the decision variables for the same maximum objective function value. So, all the points between B and C will have the same objective function values. This is mainly because the objective function is parallel to the first constraint. So, it touches at multiple points between the points B and C when the objective function moves away from the shaded region.

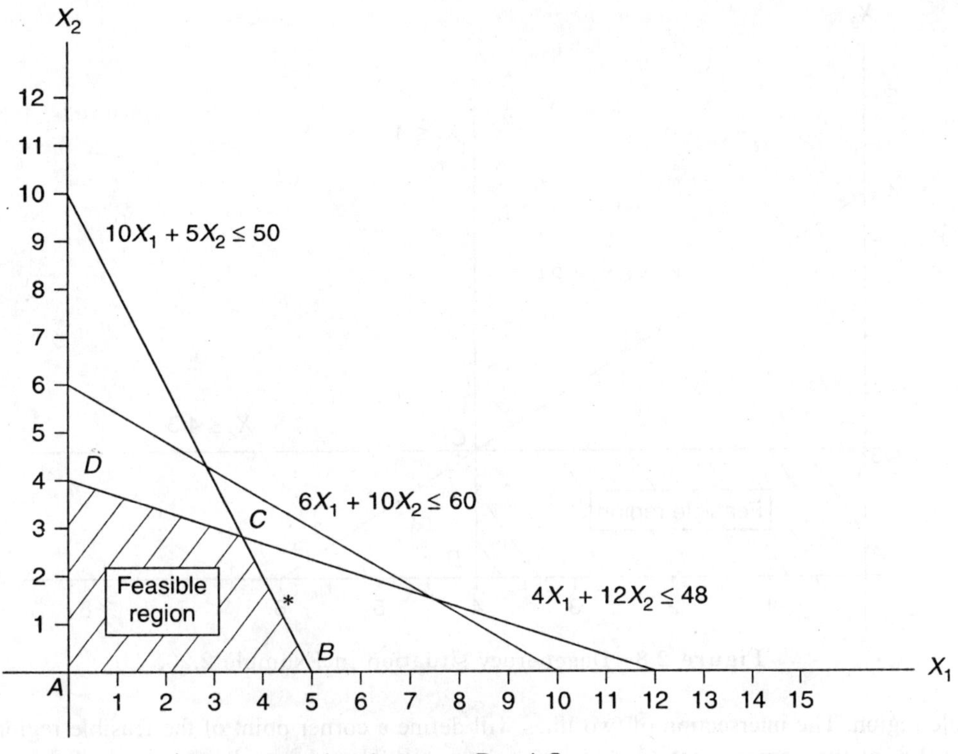

*Alternate optima between *B* and *C*

Figure 2.7 Alternate optima of Example 2.25.

***Example 2.26* (Degeneracy)**   Solve the following LP problem graphically.

$$\text{Maximize } Z = 100X_1 + 50X_2$$

subject to

$$4X_1 + 6X_2 \leq 24$$

$$X_1 \leq 4$$

$$X_2 \leq \frac{4}{3}$$

$$X_1,\ X_2 \geq 0$$

Solution   The introduction of the non-negative constraints ($X_1 \geq 0$ and $X_2 \geq 0$) will eliminate the second, third and fourth quadrants of the X_1X_2 plane, as shown in Figure 2.8.

From the first constraint in equation form

$$4X_1 + 6X_2 = 24$$

we have $X_2 = 4$, when $X_1 = 0$; and $X_1 = 6$, when $X_2 = 0$. Now, plot the constraint 1. The second constraint in equation form $X_1 = 4$ helps plot the constraint 2. Also the third constraint in equation form $X_2 = 4/3$, plots the constraint 3 as shown in Figure 2.8. The closed polygon *ABCD* is the

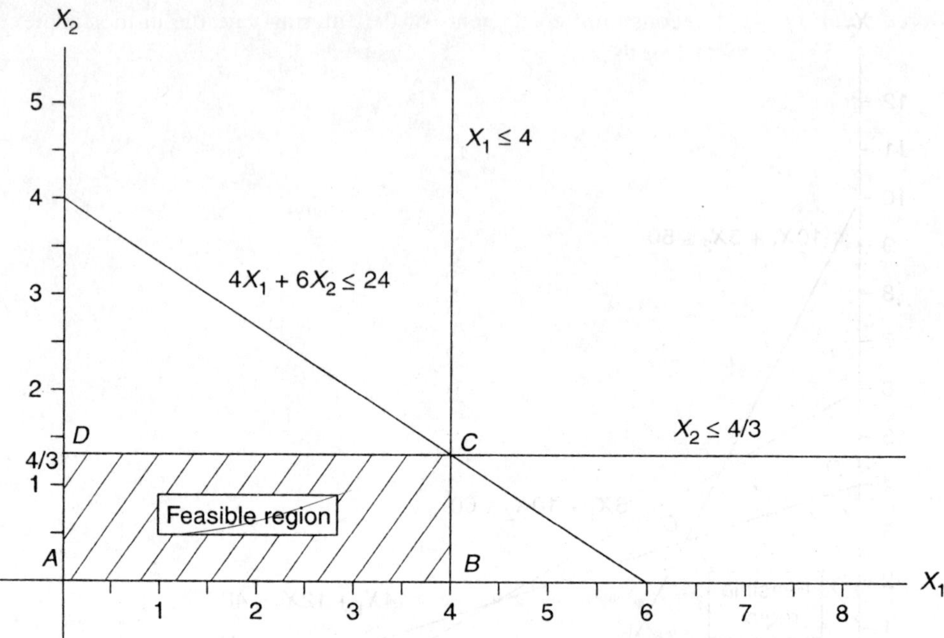

Figure 2.8 Degeneracy situation in Example 2.26.

feasible region. The intersection of two lines will define a corner point of the feasible region. But in Figure 2.8, at the corner point C, three lines intersect. This shows the presence of degeneracy in the problem.

The objective function value at each of the corner points of the closed polygon is computed by substituting its coordinates in the objective function. The coordinates of the corner point C are $(4, 4/3)$. The values of Z for the different points are:

$$Z(A) = 0$$
$$Z(B) = 400$$
$$Z(C) = \frac{1400}{3} = 466.67$$
$$Z(D) = 66.67$$

The Z value is maximum for the corner point C. Hence, the corresponding solution is presented as follows:

$$X_1^* = 4, \qquad X_2^* = \frac{4}{3} = 1.33, \qquad Z(\text{optimum}) = 466.67$$

2.6.1 Identification of Special Cases from Simplex Table

In this section, the guidelines for identifying the special cases from simplex tables are discussed.

Infeasible solution. In the optimal simplex table, if the set of basic variables contains atleast one artificial variable, then the given problem is said to have infeasible solution space and hence it has infeasible solution.

Unbounded solution. If the constraint coefficients of the entering variable in the simplex table are either less than or equal to 0, then the solution space is unbounded and has no finite optimum solution.

Alternate optimum solution. In the optimum simplex table, if $C_j - Z_j$ value for atleast one non-basic variable is equal to 0, then the problem is said to have alternate optimum solutions.

Solution for degeneracy. In any iteration, if there is a tie for the minimum ratio to maintain the feasibility, the objective function of the next iteration will be same as that of the current iteration. In any of the corner points of the feasible regions, if more than two lines pass through it, then this phenomenon will occur. This will force us to carry out some iterations without any improvement in the objective function value.

2.7 DUALITY

A generalized format of the linear programming problem is represented here.

$$\text{Maximize or minimize } Z = C_1X_1 + C_2X_2 + \cdots + C_nX_n$$

subject to

$$a_{11}X_1 + a_{12}X_2 + \cdots + a_{1n}X_n \leq, = \text{ or } \geq b_1$$
$$a_{21}X_1 + a_{22}X_2 + \cdots + a_{2n}X_n \leq, = \text{ or } \geq b_2$$
$$\vdots \qquad \vdots \qquad\qquad \vdots \qquad\qquad \vdots$$
$$a_{i1}X_1 + a_{i2}X_2 + \cdots + a_{in}X_n \leq, = \text{ or } \geq b_i$$
$$\vdots \qquad \vdots \qquad\qquad \vdots \qquad\qquad \vdots$$
$$a_{m1}X_1 + a_{m2}X_2 + \cdots + a_{mn}X_n \leq, = \text{ or } \geq b_m$$

where, $X_1, X_2, X_3, \ldots, X_n \geq 0$.

Let this problem be called as a *primal linear programming* problem. If the constraints in the primal problem are too many, then the time taken to solve the problem is expected to be higher. Under such situation, the primal linear programming problem can be converted into its dual linear programming problem which requires relatively lesser time to solve. Then the solution of the primal problem can be obtained from the optimal table of its dual problem by following certain rules.

2.7.1 Formulation of Dual Problem

The primal problem is again reproduced below:

$$\text{Maximize or Minimize } Z = C_1X_1 + C_2X_2 + \cdots + C_nX_n$$

subject to

$$a_{11}X_1 + a_{12}X_2 + \cdots + a_{1n}X_n \lesseqgtr b_1 \quad\longleftarrow\quad Y_1$$
$$a_{21}X_1 + a_{22}X_2 + \cdots + a_{2n}X_n \lesseqgtr b_2 \quad\longleftarrow\quad Y_2$$
$$\vdots \quad\quad \vdots \quad\quad\quad \vdots \quad\quad \vdots \qquad\qquad \vdots$$
$$a_{i1}X_1 + a_{i2}X_2 + \cdots + a_{in}X_n \lesseqgtr b_i \quad\longleftarrow\quad Y_i$$
$$\vdots \quad\quad \vdots \quad\quad\quad \vdots \quad\quad \vdots \qquad\qquad \vdots$$
$$a_{m1}X_1 + a_{m2}X_2 + \cdots + a_{mn}X_n \lesseqgtr b_m \quad\longleftarrow\quad Y_m$$

where, $X_1, X_2, X_3, \ldots, X_n \geq 0$.

In the above model, the variable Y_i is called as the *dual variable* associated with the constraint i.

Objective function. The number of variables in the dual problem is equal to the number of constraints in the primal problem. The objective function of the dual problem is constructed by adding the multiples of the right-hand side constants of the constraints of the primal problem with the respective dual variables.

Constraints. The number of constraints in the dual problem is equal to the number of variables in the primal problem. Each dual constraint corresponds to each primal variable. The left-hand side of the dual constraint corresponding to the jth primal variable is the sum of the multiples of the left-hand side constraint coefficients of the jth primal variable with the corresponding dual variables. The right-hand side constant of the dual constraint corresponding to the jth primal variable is the objective function coefficient of the jth primal variable.

Some more guidelines for forming the dual problem are presented in Table 2.35.

Table 2.35 Guidelines for Dual Formation

Type of problem	Objective function	Constraints type	Nature of variables
Primal	Maximize	$\leq$	Restricted in sign
Dual	Minimize	$\geq$	Restricted in sign
Primal	Minimize	$\geq$	Restricted in sign
Dual	Maximize	$\leq$	Restricted in sign
Primal	Maximize	$=$	Restricted in sign
Dual	Minimize	$\geq$	Unrestricted in sign
Primal	Minimize	$=$	Restricted in sign
Dual	Maximize	$\leq$	Unrestricted in sign
Primal	Maximize	$\leq$	Unrestricted in sign
Dual	Minimize	$=$	Restricted in sign
Primal	Minimize	$\geq$	Unrestricted in sign
Dual	Maximize	$=$	Restricted in sign

Example 2.27 Form the dual of the following primal problem.

$$\text{Maximize } Z = 4X_1 + 10X_2 + 25X_3$$

subject to

$$2X_1 + 4X_2 + 8X_3 \leq 25$$

$$4X_1 + 9X_2 + 8X_3 \leq 30$$

$$6X_1 + 8X_2 + 2X_3 \leq 40$$

$$X_1, X_2 \text{ and } X_3 \geq 0$$

Solution The given problem is termed as a primal problem which is as shown below. Let Y_i be the dual variable associated with the ith constraint of the primal problem as shown on next page.

$$\text{Maximize } Z = \boxed{4}\ X_1 + \boxed{10}\ X_2 + \boxed{25}\ X_3$$

subject to

$$\boxed{2}\ X_1 + \boxed{4}\ X_2 + \boxed{8}\ X_3 \le 25 \longleftarrow Y_1$$
$$\boxed{4}\ X_1 + \boxed{9}\ X_2 + \boxed{8}\ X_3 \le 30 \longleftarrow Y_2$$
$$\boxed{6}\ X_1 + \boxed{8}\ X_2 + \boxed{2}\ X_3 \le 40 \longleftarrow Y_3$$

$$X_1, X_2, X_3 \ge 0$$

The corresponding dual problem may be presented as:

$$\text{Minimize } Y = 25Y_1 + 30Y_2 + 40Y_3$$

subject to

$$2Y_1 + 4Y_2 + 6Y_3 \ge 4$$

$$4Y_1 + 9Y_2 + 8Y_3 \ge 10$$

$$8Y_1 + 8Y_2 + 2Y_3 \ge 25$$

$$Y_1,\ Y_2 \text{ and } Y_3 \ge 0$$

Example 2.28 Form the dual of the following primal problem.

$$\text{Minimize } Z = 20X_1 + 40X_2$$

subject to

$$2X_1 + 20X_2 \ge 40$$

$$20X_1 + 3X_2 \ge 20$$

$$4X_1 + 15X_2 \ge 30$$

$$X_1 \text{ and } X_2 \ge 0$$

Solution Let, Y_i be the dual variable associated with the ith constraint of the given primal problem. The dual of the given above primal problem is:

$$\text{Maximize } Y = 40Y_1 + 20Y_2 + 30Y_3$$

subject to

$$2Y_1 + 20Y_2 + 4Y_3 \le 20$$

$$20Y_1 + 3Y_2 + 15Y_3 \le 40$$

$$Y_1,\ Y_2 \text{ and } Y_3 \ge 0$$

Example 2.29 Form the dual of the following primal problem.

$$\text{Maximize } Z = 4X_1 + 10X_2 + 25X_3$$

subject to

$$2X_1 + 4X_2 + 8X_3 = 25$$

$$4X_1 + 9X_2 + 8X_3 = 30$$

$$6X_1 + 8X_2 + 2X_3 = 40$$

$$X_1, X_2 \text{ and } X_3 \ge 0$$

Solution Let Y_i be the dual variable associated with the ith constraint of the given primal problem. The dual of the above primal problem is:

$$\text{Minimize } Y = 25Y_1 + 30Y_2 + 40Y_3$$

subject to

$$2Y_1 + 4Y_2 + 6Y_3 \geq 4$$
$$4Y_1 + 9Y_2 + 8Y_3 \geq 10$$
$$8Y_1 + 8Y_2 + 2Y_3 \geq 25$$

Y_1, Y_2 and Y_3 are unrestricted in sign.

Example 2.30 Form the dual of the following primal problem.

$$\text{Minimize } Z = 20X_1 + 40X_2$$

subject to

$$2X_1 + 20X_2 = 40$$
$$20X_1 + 3X_2 = 20$$
$$4X_1 + 15X_2 = 30$$
$$X_1 \text{ and } X_2 \geq 0$$

Solution Let Y_i be the dual variable associated with the ith constraint of the given primal problem. The dual of the primal problem is shown below.

$$\text{Maximize } Y = 40Y_1 + 20Y_2 + 30Y_3$$

subject to

$$2Y_1 + 20Y_2 + 4Y_3 \leq 20$$
$$20Y_1 + 3Y_2 + 15Y_3 \leq 40$$

where Y_1, Y_2 and Y_3 are unrestricted in sign.

Example 2.31 Form the dual of the following primal problem.

$$\text{Minimize } Z = 5X_1 + 8X_2$$

subject to

$$4X_1 + 9X_2 \geq 100$$
$$2X_1 + X_2 \leq 20$$
$$2X_1 + 5X_2 \geq 120$$
$$X_1 \text{ and } X_2 \geq 0$$

Solution Let, Y_i be the dual variable associated with the ith constraint of the primal problem as shown below.

$$\text{Minimize } Z = 5X_1 + 8X_2$$

subject to

$$4X_1 + 9X_2 \geq 100 \quad \longleftarrow Y_1$$
$$-2X_1 - X_2 \geq -20 \quad \longleftarrow Y_2^*$$
$$2X_1 + 5X_2 \geq 120 \quad \longleftarrow Y_3$$
$$X_1 \text{ and } X_2 \geq 0$$

Note that the original constraint 2 of the primal problem is multiplied throughout by -1 and hence the relation of this constraint is changed into $\geq$ type. The dual of the above primal problem is shown below:

$$\text{Maximize } Y = 100Y_1 - 20Y_2 + 120Y_3$$

subject to

$$4Y_1 - 2Y_2 + 2Y_3 \leq 5$$
$$9Y_1 - Y_2 + 5Y_3 \leq 8$$
$$Y_1,\ Y_2 \text{ and } Y_3 \geq 0$$

Example 2.32 Form the dual of the following primal problem.

$$\text{Minimize } Z = 2X_1 + 6X_2$$

subject to

$$9X_1 + 3X_2 \geq 20$$
$$2X_1 + 7X_2 = 40$$
$$X_1 \text{ and } X_2 \geq 0$$

Solution A modified form of the given problem is presented below. In this model, constraint 2 of the original problem is represented by two constraints constraint 2 and constraint 3.

$$\text{Minimize } Z = 2X_1 + 6X_2$$

subject to

$$9X_1 + 3X_2 \geq 20$$
$$2X_1 + 7X_2 \geq 40$$
$$2X_1 + 7X_2 \leq 40$$
$$X_1 \text{ and } X_2 \geq 0$$

Now, to have a model with minimization objective function with all $\geq$ type constraints, the third constraint is modified as $\geq$ type. Then

$$\text{Minimize } Z = 2X_1 + 6X_2$$

subject to

$$9X_1 + 3X_2 \geq 20 \quad \longleftarrow \quad Y_1$$
$$2X_1 + 7X_2 \geq 40 \quad \longleftarrow \quad Y_2'$$
$$-2X_1 - 7X_2 \geq -40 \quad \longleftarrow \quad Y_2''$$
$$X_1,\ X_2 \geq 0$$

Also, let Y_1, Y_2' and Y_2'' be the dual variables associated with the first, second and third constraints of the primal problem. The dual of the above primal problem is given as:

$$\text{Maximize } Y = 20Y_1 + 40Y_2' - 40Y_2''$$

subject to

$$9Y_1 + 2Y_2' - 2Y_2'' \leq 2$$
$$3Y_1 + 7Y_2' - 7Y_2'' \leq 6$$
$$Y_1,\ Y_2' \text{ and } Y_2'' \geq 0$$

By, substituting $Y_2 = Y_2' - Y_2''$, we get

$$\text{Maximize } Y = 20Y_1 + 40Y_2$$

subject to

$$9Y_1 + 2Y_2 \leq 2$$
$$3Y_1 + 7Y_2 \leq 6$$

where $Y_1 \geq 0$ and Y_2 is unrestricted in sign.

Example 2.33 Form the dual of the following primal problem.

$$\text{Maximize } Z = 2X_1 + 3X_2$$

subject to

$$5X_1 + 2X_2 \leq 40$$
$$6X_1 + 12X_2 \leq 80$$

X_1 and X_2 are unrestricted in sign.

Solution Let $X_1 = X_1' - X_1''$ and $X_2 = X_2' - X_2''$. The given model can now be presented as follows:

$$\text{Maximize } Z = 2(X_1' - X_1'') + 3(X_2' - X_2'')$$

subject to

$$5(X_1' - X_1'') + 2(X_2' - X_2'') \leq 40 \longleftarrow Y_1$$
$$6(X_1' - X_1'') + 12(X_2' - X_2'') \leq 80 \longleftarrow Y_2$$
$$X_1', X_1'', X_2' \text{ and } X_2'' \geq 0$$

Let Y_i be the dual variable associated with the ith constraint of the primal problem as shown above. The dual of the above primal problem is:

$$\text{Minimize } Y = 40Y_1 + 80Y_2$$

subject to

$$5Y_1 + 6Y_2 \geq 2$$
$$-5Y_1 - 6Y_2 \geq -2$$
$$2Y_1 + 12Y_2 \geq 3$$
$$-2Y_1 - 12Y_2 \geq -3$$
$$Y_1 \text{ and } Y_2 \geq 0$$

The rearranged version of the above model is:

$$\text{Minimize } Y = 40Y_1 + 80Y_2$$

subject to

$$5Y_1 + 6Y_2 \geq 2 \tag{2.1}$$
$$5Y_1 + 6Y_2 \leq 2 \tag{2.2}$$
$$2Y_1 + 12Y_2 \geq 3 \tag{2.3}$$
$$2Y_1 + 12Y_2 \leq 3 \tag{2.4}$$
$$Y_1 \text{ and } Y_2 \geq 0$$

In the foregoing model, constraints (2.1) and (2.2) are combined together to yield a single constraint with = type constraint. Similarly, constraints (2.3) and (2.4) of the above model are also combined to form a single constraint with = type constraint as shown below:

$$\text{Minimize } Y = 40Y_1 + 80Y_2$$

subject to

$$5Y_1 + 6Y_2 = 2$$
$$2Y_1 + 12Y_2 = 3$$
$$Y_1 \text{ and } Y_2 \geq 0$$

2.7.2 Application of Duality

In this section, the application of duality concept for problem solving is demonstrated using two examples. The necessary guidelines are explained within the example problems which are presented in this section.

Example 2.34 Solve the following linear programming problem using the result of its dual problem.

$$\text{Minimize } Z_1 = 24X_1 + 30X_2$$

subject to

$$2X_1 + 3X_2 \geq 10$$
$$4X_1 + 9X_2 \geq 15$$
$$6X_1 + 6X_2 \geq 20$$
$$X_1 \text{ and } X_2 \geq 0$$

Solution   Let, Y_1, Y_2 and Y_3 be the dual variables with respect to the constraints 1, 2 and 3 respectively. Then the corresponding dual problem is:

$$\text{Maximize } Z = 10Y_1 + 15Y_2 + 20Y_3$$

subject to

$$2Y_1 + 4Y_2 + 6Y_3 \leq 24$$
$$3Y_1 + 9Y_2 + 6Y_3 \leq 30$$
$$Y_1, Y_2 \text{ and } Y_3 \geq 0$$

The canonical form of the above dual problem is shown below, where S_1 and S_2 are the slack variables in constraint 1 and constraint 2, respectively:

$$\text{Maximize } Z = 10Y_1 + 15Y_2 + 20Y_3 + 0S_1 + 0S_2$$

subject to

$$2Y_1 + 4Y_2 + 6Y_3 + S_1 = 24$$
$$3Y_1 + 9Y_2 + 6Y_3 + S_2 = 30$$
$$Y_1, Y_2, Y_3, S_1 \text{ and } S_2 \geq 0$$

The different iterations of the above problem are shown in Table 2.36 through Table 2.38.

Since all the values of $C_j - Z_j$ in Table 2.38 are negative or zero, the optimality is reached.

Table 2.36 Iteration 1 (Initial Table)

CB_i	Basic variable	C_j					Solution	Ratio
		10	15	20	0	0		
		Y_1	Y_2	Y_3	S_1	S_2		
0	S_1	2	4	6	1	0	24	4**
0	S_2	3	9	6	0	1	30	5
	Z_j	0	0	0	0	0	0	
	$C_j - Z_j$	10	15	20*	0	0		

Table 2.37 Iteration 2

CB_i	Basic variable	C_j					Solution	Ratio
		10	15	20	0	0		
		Y_1	Y_2	Y_3	S_1	S_2		
20	Y_3	1/3	2/3	1	1/6	0	4	12
0	S_2	1	5	0	−1	1	6	6**
	Z_j	20/3	40/3	20	10/3	0	80	
	$C_j - Z_j$	10/3*	5/3	0	−10/3	0		

Table 2.38 Iteration 3 (Final Table)

CB_i	Basic variable	C_j					Solution
		10	15	20	0	0	
		Y_1	Y_2	Y_3	S_1	S_2	
20	Y_3	0	−1	1	1/2	−1/3	2
10	Y_1	1	5	0	−1	1	6
	Z_j	10	30	20	0	10/3	100
	$C_j - Z_j$	0	−15	0	0	−10/3	

Determination of the solution of the primal problem: The solution of the primal problem is inferred as shown in Table 2.39.

Table 2.39 Determination of Solution of Primal Problem

Basic variables in the initial table of dual problem	S_1	S_2
$-(C_j - Z_j)$ of the final table of dual problem	0	10/3
Corresponding primal variables	X_1	X_2

From Tables 2.38 and 2.39, the results of the primal problem can be obtained as:

$$X_1 = 0, \qquad X_2 = \frac{10}{3}, \qquad Z_1(\text{optimum}) = 100$$

Example 2.35 Consider the following linear programming problem and solve it using its dual solution.

$$\text{Minimize } Z = 40X_1 + 30X_2 + 25X_3$$

subject to

$$4X_1 + 2X_2 + 5X_3 \geq 30$$
$$3X_1 + 6X_2 + X_3 \geq 20$$
$$X_1 + 3X_2 + 6X_3 \geq 36$$
$$X_1, X_2 \text{ and } X_3 \geq 0$$

Solution Let Y_1, Y_2 and Y_3 be the dual variables with respect to the constraints 1, 2 and 3, respectively of the primal problem as shown below.

Primal Problem:

$$\text{Minimize } Z = 40X_1 + 30X_2 + 25X_3$$

subject to

$$4X_1 + 2X_2 + 5X_3 \geq 30 \quad \longleftarrow \quad Y_1$$
$$3X_1 + 6X_2 + X_3 \geq 20 \quad \longleftarrow \quad Y_2$$
$$X_1 + 3X_2 + 6X_3 \geq 36 \quad \longleftarrow \quad Y_3$$
$$X_1, X_2 \text{ and } X_3 \geq 0$$

Dual Problem:

The dual of the above primal problem is presented below.

$$\text{Maximize } Z = 30Y_1 + 20Y_2 + 36Y_3$$

subject to

$$4Y_1 + 3Y_2 + Y_3 \leq 40$$
$$2Y_1 + 6Y_2 + 3Y_3 \leq 30$$
$$5Y_1 + Y_2 + 6Y_3 \leq 25$$
$$Y_1, Y_2 \text{ and } Y_3 \geq 0$$

The canonical form of this dual problem is as follows:

$$\text{Maximize } Z = 30Y_1 + 20Y_2 + 36Y_3 + 0S_1 + 0S_2 + 0S_3$$

subject to

$$4Y_1 + 3Y_2 + Y_3 + S_1 \leq 40$$
$$2Y_1 + 6Y_2 + 3Y_3 + S_2 \leq 30$$
$$5Y_1 + Y_2 + 6Y_3 + S_3 \leq 25$$
$$Y_1, Y_2, Y_3, S_1, S_2 \text{ and } S_3 \geq 0$$

The initial table and different iterations of simplex method applied to the dual problem are shown in Table 2.40. From the final iteration of the dual problem shown in Table 2.40, the solution of the primal problem is obtained as shown in Table 2.41. From the Table 2.41, the optimal values of the decision variables and the objective function of the primal problem are: $X_1 = 0$, $X_2 = 5/2$, $X_3 = 5$ and $Z(\text{optimum}) = 200$.

Table 2.40 Initial Table and Different Iterations of Example 2.35

CB_i	C_j Basic variable	30 Y_1	20 Y_2	36 Y_3	0 S_1	0 S_2	0 S_3	Solution	Ratio
0	S_1	4	3	1	1	0	0	40	40
0	S_2	2	6	3	0	1	0	30	10
0	S_3	5	1	6	0	0	1	25	25/6
	Z_j	0	0	0	0	0	0	0	
	$C_j - Z_j$	30	20	36	0	0	0		
0	S_1	19/6	17/6	0	1	0	−1/6	215/6	215/17
0	S_2	−1/2	11/2	0	0	1	−1/2	35/2	35/11
36	Y_3	5/6	1/6	1	0	0	1/6	25/6	25
	Z_j	30	6	36	0	0	6	150	
	$C_j - Z_j$	0	14	0	0	0	−6		
0	S_1	113/33	0	0	1	−17/33	1/11	885/33	885/113
20	Y_2	−1/11	1	0	0	2/11	−1/11	35/11	−
36	Y_3	28/33	0	1	0	−1/33	2/11	40/11	30/7
	Z_j	316/11	20	36	0	28/11	52/11	2140/11	
	$C_j - Z_j$	14/11	0	0	0	−28/11	−52/11		
0	S_1	0	0	−113/28	1	−11/28	−9/14	85/7	
20	Y_2	0	1	3/28	0	55/308	−1/14	275/77	
30	Y_1	1	0	33/28	0	−1/28	3/14	30/7	
	Z_j	30	20	525/14	0	5/2	5	200	
	$C_j - Z_j$	0	0	−3/2	0	−5/2	−5		

Table 2.41 Determination of Solution of Primal

Basic variable in the initial table of the dual problem	S_1	S_2	S_3
− $(C_j - Z_j)$ of the final table of the dual problem	0	5/2	5
Corresponding primal variable	X_1	X_2	X_3

2.8 SENSITIVITY ANALYSIS

In many situations, the parameters and characteristics of a linear programming model may change over a period of time. Also, the analyst may be interested to know the effect of changing the parameters and characteristics of the model on the optimality. This kind of sensitivity analysis can be carried out in the following ways:

1. Making changes in the right-hand side constants of the constraints
2. Making changes in the objective function coefficients
3. Adding a new constraint
4. Adding a new variable.

These are discussed in the following sections.

2.8.1   Changes in the Right-hand Side Constants of Constraints

The right-hand side constant (resource availability) of one or more constraints of a linear programming model may change over a period of time. So, the analyst may be interested in knowing the revised optimum solution based on the optimum table of the original problem after incorporating the new changes in the right-hand side constants. The changes bring in the following results.

(a) Same set of basic variables with modified right-hand side constants in the optimal table.
(b) Different set of basic variables in the optimal table.

The following example is considered to demonstrate the above two cases.

Example 2.36   Consider Example 2.15 as reproduced below:

$$\text{Maximize } Z = 6X_1 + 8X_2$$

subject to

$$5X_1 + 10X_2 \leq 60$$
$$4X_1 + 4X_2 \leq 40$$
$$X_1 \text{ and } X_2 \geq 0$$

The optimum solution of this problem is shown in Table 2.42. Solve the problem,

Table 2.42   Optimal Table of Example 2.36

CB_i	C_j Basic variable	6 X_1	8 X_2	0 S_1	0 S_2	Solution
8	X_2	0	1	1/5	−1/4	2
6	X_1	1	0	−1/5	1/2	8
	Z_j	6	8	2/5	1	64
	$C_j - Z_j$	0	0	−2/5	−1	

(a) if the right-hand side constants of constraint 1 and constraint 2 are changed from 60 and 40 to 40 and 20, respectively.
(b) if the right-hand side constants of the constraints are changed from 60 and 40 to 20 and 40 respectively.

Solution (a) The revised right-hand side constants after incorporating the changes in the constraints are obtained by using the following formula.

$$\begin{bmatrix} \text{Basic variables in} \\ \text{the optimal table} \end{bmatrix} = \begin{bmatrix} \text{Technological coefficient} \\ \text{columns in the optimal} \\ \text{table with respect to the basic} \\ \text{variables in initial table} \end{bmatrix} \begin{bmatrix} \text{New R.H.S.} \\ \text{constants} \end{bmatrix}$$

Applying the formula, we have

$$\begin{bmatrix} X_2 \\ X_1 \end{bmatrix} = \begin{bmatrix} \dfrac{1}{5} & -\dfrac{1}{4} \\ -\dfrac{1}{5} & \dfrac{1}{2} \end{bmatrix} \begin{bmatrix} 40 \\ 20 \end{bmatrix} = \begin{bmatrix} 3 \\ 2 \end{bmatrix}$$

Solving them, we get X_1 is 2 and that of X_2 is 3. Since, these values are non-negative, the revised solution is feasible and optimal. The corresponding optimal objective function value is 36.

(b) The revised solution of the basic variables in Table 2.42 after incorporating the changes in the right-hand side values of the constraints are obtained as shown below.

$$\begin{bmatrix} X_2 \\ X_1 \end{bmatrix} = \begin{bmatrix} \dfrac{1}{5} & -\dfrac{1}{4} \\ -\dfrac{1}{5} & \dfrac{1}{2} \end{bmatrix} \begin{bmatrix} 20 \\ 40 \end{bmatrix} = \begin{bmatrix} -6 \\ 16 \end{bmatrix}$$

which gives the values of X_1 and X_2 as 16 and –6, respectively. Since, the value of the basic variable X_2 is negative, the solution is infeasible. This infeasibility can be removed using the dual simplex method. Based on Table 2.42, the table for the dual simplex method is shown in Table 2.43.

Table 2.43   Table for Dual Simplex Method

CB_i	C_j	6	8	0	0	
	Basic variable	X_1	X_2	S_1	S_2	Solution
8	X_2	0	1	1/5	–1/4	–6
6	X_1	1	0	–1/5	1/2	16
	Z_j	6	8	2/5	1	48
	$C_j - Z_j$	0	0	–2/5	–1	

Application of the dual simplex method to Table 2.43 yields the following optimum result in the next iteration itself:

$$X_1 = 4, \qquad X_2 = 0, \qquad S_1 = 0, \qquad S_2 = 24, \qquad Z(\text{optimum}) = 24$$

2.8.2   Changes in the Objective Function Coefficients

In reality, the profit coefficients or the cost coefficients of the objective function undergo changes over a period of time. Under such a situation, one can obtain the revised optimum solution from the optimal table of the original problem by following certain steps. Also, one will be interested to know the range of the coefficient of a variable in the objective function over which the optimality is unaffected. These are illustrated using the following example.

Example 2.37   Consider Example 2.16 which is reproduced below:

$$\text{Maximize } Z = 10X_1 + 15X_2 + 20X_3$$

subject to

$$2X_1 + 4X_2 + 6X_3 \leq 24$$
$$3X_1 + 9X_2 + 6X_3 \leq 30$$
$$X_1, X_2 \text{ and } X_3 \geq 0$$

The optimum table of the above problem is given as in Table 2.44.

Table 2.44   Optimal Table for Example 2.37

CB_i	C_j Basic variable	10 X_1	15 X_2	20 X_3	0 S_1	0 S_2	Solution
20	X_3	0	−1	1	1/2	−1/3	2
10	X_1	1	5	0	−1	1	6
	Z_j	10	30	20	0	10/3	100
	$C_j - Z_j$	0	−15	0	0	−10/3	

(a) Find the range of the objective function coefficient C_1 of the variable X_1 such that the optimality is unaffected.
(b) Find the range of the objective function coefficient C_2 of the variable X_2 such that the optimality is unaffected.
(c) Check whether the optimality is affected, if the profit coefficients are changed from (10, 15, 20) to (7, 14, 15). If so, find the revised optimum solution.

Solution   (a) *Determination of the range of C_1 of the basic variable X_1.*   After making some changes in the objective function coefficients, if the optimality is not affected, then the present set of basic variables will continue, and in that case the $C_j - Z_j$ values of the basic variables will be equal to 0; but the $C_j - Z_j$ values of the non-basic variables will change. Hence, care should be taken in establishing the range for each of C_j values such that the corresponding $C_j - Z_j$ value of that non-basic variable is limited to at most 0.

Since, the variable X_1 with respect to the coefficient C_1 is a basic variable in the optimal table of the original problem, the $C_j - Z_j$ value will change for the non-basic variables: X_2, S_1 and S_2. The values can be computed in terms of C_1. Then, these expressions can be restricted to at most 0 to maintain optimality. By solving the above inequalities for C_1, its range can be determined.

The expressions of $C_2 - Z_2$, $C_4 - Z_4$ and $C_5 - Z_5$ for the non-basic variables, X_2, S_1 and S_2, respectively are:

$$C_2 - Z_2 = 15 - [20 \ C_1] \begin{bmatrix} -1 \\ 5 \end{bmatrix} = 35 - 5C_1$$

$$C_4 - Z_4 = 0 - [20 \ C_1] \begin{bmatrix} 1 \\ 2 \\ -1 \end{bmatrix} = -10 + C_1$$

$$C_5 - Z_5 = 0 - [20 \ C_1] \begin{bmatrix} -\dfrac{1}{3} \\ 1 \end{bmatrix} = \dfrac{20}{3} - C_1$$

The above relative contributions are restricted to at most 0 to maintain the optimality as shown below.

$$35 - 5C_1 \leq 0 \quad \text{or} \quad C_1 \geq 7$$
$$-10 + C_1 \leq 0 \quad \text{or} \quad C_1 \leq 10$$
$$\dfrac{20}{3} - C_1 \leq 0 \quad \text{or} \quad C_1 \geq \dfrac{20}{3}$$

Here the value of C_1 ranges from 7 to 10 (i.e. $7 \leq C_1 \leq 10$). In this interval of C_1, the optimality is unaffected.

(b) *Determination of the range of C_2 of the non-basic variable X_2.* Since, C_2 corresponds to one of the non-basic variables X_2, the range of C_2 can be obtained by just restricting the $C_2 - Z_2$ to at most 0. Therefore,

$$C_2 - Z_2 = C_2 - [20 \ 10] \begin{bmatrix} -1 \\ 5 \end{bmatrix} = C_2 - 30$$

This relative contribution is restricted to atmost 0 to maintain the optimality as shown below.

$$C_2 - 30 \leq 0 \quad \text{or} \quad C_2 \leq 30$$

Hence, it is clear that the optimality will remain the same as long as the value of C_2 is less than or equal to 30.

(c) *Checking the optimality.* The new values of the objective function coefficients, C_1, C_2 and C_3 of the variables, X_1, X_2 and X_3 are 7, 14 and 15, respectively. The corresponding modified relative contributions of all the variables are computed as follows:

$$C_1 - Z_1 = 7 - [15 \ 7] \begin{bmatrix} 0 \\ 1 \end{bmatrix} = 0$$

$$C_2 - Z_2 = 14 - [15 \ 7] \begin{bmatrix} -1 \\ 5 \end{bmatrix} = -6$$

$$C_3 - Z_3 = 15 - [15\ 7] \begin{bmatrix} 1 \\ 0 \end{bmatrix} = 0$$

$$C_4 - Z_4 = 0 - [15\ 7] \begin{bmatrix} \frac{1}{2} \\ -1 \end{bmatrix} = -\frac{1}{2}$$

$$C_5 - Z_5 = 0 - [15\ 7] \begin{bmatrix} -\frac{1}{3} \\ 1 \end{bmatrix} = -2$$

Since all the $C_j - Z_j$ values are less than or equal to 0, the optimality is unaffected. The solution in the optimal table of the original problem, which is shown in Table 2.44, is the optimal solution for the problem with the modified objective function coefficients of this section.

2.8.3 Adding a New Constraint

Sometimes, a new constraint may be added to an existing linear programming model as per changing realities. Under such situation, each of the basic variables in the new constraint is substituted with the corresponding expression based on the current optimal table. This will yield a modified version of the new constraint in terms of only the current non-basic variables.

If the new constraint is satisfied by the values of the current basic variables, the constraint is said to be a redundant one. So, the optimality of the original problem will not be affected even after including the new constraint into the existing model.

If the new constraint is not satisfied by the values of the current basic variables, the optimality of the original problem will be affected. So, the modified version of the new constraint is to be augmented to the optimal table of the original problem and iterated till the optimality is reached.

Example 2.38 Consider Example 2.15 to demonstrate the impact of adding a new constraint into an existing model:

$$\text{Maximize } Z = 6X_1 + 8X_2$$

subject to

$$5X_1 + 10X_2 \leq 60$$

$$4X_1 + 4X_2 \leq 40$$

$$X_1 \text{ and } X_2 \geq 0$$

The optimal table of the aforementioned example is reproduced as Table 2.45.

    (a) Check whether the addition of the constraint $7X_1 + 2X_2 \leq 65$ affects the optimality. If it does, find the new optimum solution.

    (b) Check whether the addition of the constraint $6X_1 + 3X_2 \leq 48$ affects the optimality. If it does, find the new optimum solution.

Solution   (a) The new constraint is:

$$7X_1 + 2X_2 \leq 65$$

Table 2.45 Optimal Table for Example 2.38

CB_i	C_j	6	8	0	0	
	Basic variable	X_1	X_2	S_1	S_2	Solution
8	X_2	0	1	1/5	−1/4	2
6	X_1	1	0	−1/5	1/2	8
	Z_j	6	8	2/5	1	64
	$C_j - Z_j$	0	0	−2/5	−1	

This is satisfied by the values of the current basic variables ($X_1 = 8$ and $X_2 = 2$). Hence, the new constraint is said to be a redundant constraint and the optimality will not be affected even after including the new constraint into the existing model.

    (b) The new constraint is:

$$6X_1 + 3X_2 \le 48$$

This is not satisfied by the values of the current basic variables ($X_1 = 8$ and $X_2 = 2$). So, the modified form of the new constraint in terms of only non-basic variables is obtained.

    The standard form of the new constraint after including a slack variable S_3 is as follows:

$$6X_1 + 3X_2 + S_3 = 48$$

From Table 2.45, the expressions with respect to X_1 and X_2 rows can be written as:

$$X_2 + \frac{1}{5}S_1 - \frac{1}{4}S_2 = 2$$

$$X_1 - \frac{1}{5}S_1 + \frac{1}{2}S_2 = 8$$

Substitution of the expressions for X_1 and X_2 in the standard form of the new constraint, yields the following.

$$\frac{3}{5}S_1 - \frac{9}{4}S_2 + S_3 = -6$$

    Now, the above constraint is included in the optimal Table 2.45 and the result is shown in Table 2.46.

    In Table 2.46, S_3 row contains a negative right-hand side constant. Hence, the solution is infeasible. This infeasibility can be removed by using dual simplex method. Application of the dual simplex method to Table 2.46 yields the following results:

$$X_1 = \frac{20}{3}, X_2 = \frac{8}{3}, S_1 = 0, S_2 = \frac{8}{3}, S_3 = 0, Z(\text{optimum}) = \frac{184}{3} = 61.33$$

Table 2.46 Augmented Version of Table 2.45

CB_i	C_j Basic variable	6 X_1	8 X_2	0 S_1	0 S_2	0 S_3	Solution
8	X_2	0	1	1/5	−1/4	0	2
6	X_1	1	0	−1/5	1/2	0	8
0	S_3	0	0	3/5	−9/4	1	−6
	Z_j	6	8	2/5	1	0	64
	$C_j - Z_j$	0	0	−2/5	−1	0	

2.8.4 Adding a New Variable

In a problem like the product mix problem, over a period of time, a new product may be added to the existing product mix. Under such a situation, one will be interested in finding the revised optimal solution from the optimal table of the original problem.

In this analysis, the following items are to be determined after incorporating the data of the new variable (new product).

The $C_j - Z_j$ value

$$C_j - Z_j = C_j - [CB]_{1 \times m} \begin{bmatrix} \text{Technological coefficients of} \\ \text{optimal table with respect to the basic} \\ \text{variables of the initial table} \end{bmatrix}_{m \times m} \times \begin{bmatrix} \text{Constraint coefficients} \\ \text{of new variable} \end{bmatrix}_{m \times 1}$$

where, m is the number of constraints in the problem. If the $C_j - Z_j$ value of the new variable indicates the optimality as per the nature of optimization (maximization or minimization), the optimality of the problem after including the new variable is not affected. Otherwise, the constraint coefficients (technological coefficients) of the new variable are to be computed.

The constraint coefficients (technological coefficients) of the column corresponding to the new variable

$$\begin{bmatrix} \text{Revised constraint} \\ \text{coefficients of the} \\ \text{new variable} \end{bmatrix}_{m \times 1} = \begin{bmatrix} \text{Technological coefficients of} \\ \text{optimal table with respect to the} \\ \text{basic variables of the initial table} \end{bmatrix}_{m \times m} \times \begin{bmatrix} \text{Constraints coefficients} \\ \text{of new variable} \end{bmatrix}_{m \times 1}$$

where, m is the number of constraints in the problem. These coefficients are incorporated in the current optimal table and the necessary number of iterations is to be carried out from the current table till the optimality is reached.

Example 2.39 Consider Example 2.15 which is reproduced as

$$\text{Maximize } Z = 6X_1 + 8X_2$$

subject to

$$5X_1 + 10X_2 \leq 60$$

$$4X_1 + 4X_2 \leq 40$$

$$X_1 \text{ and } X_2 \geq 0$$

The optimal table of Example 2.15 is reproduced in Table 2.47.

Table 2.47 Optimal Table of Example 2.15

CB_i	C_j Basic variable	6 X_1	8 X_2	0 S_1	0 S_2	Solution
8	X_2	0	1	1/5	−1/4	2
6	X_1	1	0	−1/5	1/2	8
	Z_j	6	8	2/5	1	64
	$C_j - Z_j$	0	0	−2/5	−1	

A new product P_3 is included in the existing product mix. The profit per unit of the new product is Rs. 20. The processing requirements of the new product on lathe and milling machines are 6 hours per unit and 5 hours per unit, respectively.

(a) Check whether the inclusion of the product P_3 changes the optimality.
(b) If it changes the optimality, find the revised optimal solution.

Solution The LP problem after incorporating the data of the new product P_3 is shown below.

$$\text{Maximize } Z = 6X_1 + 8X_2 + 20X_3$$

subject to

$$5X_1 + 10X_2 + 6X_3 \leq 60$$

$$4X_1 + 4X_2 + 5X_3 \leq 40$$

$$X_1, X_2, X_3 \geq 0$$

(a) Determination of $C_3 - Z_3$. The relative contribution of the new product, P_3 is computed using the following formula.

$$C_j - Z_j = C_j - [CB] \begin{bmatrix} \text{Technological coefficients of} \\ \text{optimal table with respect to the} \\ \text{basic variables of the initial table} \end{bmatrix} \times \begin{bmatrix} \text{Constraint coefficients} \\ \text{of new variable} \end{bmatrix}$$

or

$$C_3 - Z_3 = 20 - [8 \ 6] \begin{bmatrix} \dfrac{1}{5} & -\dfrac{1}{4} \\ -\dfrac{1}{5} & \dfrac{1}{2} \end{bmatrix} \times \begin{bmatrix} 6 \\ 5 \end{bmatrix} = \dfrac{63}{5}$$

Since, the $C_3 - Z_3$ value is greater than zero, the solution of the modified problem is not optimal. This means that the inclusion of the new product (new variable) in the original problem changes the optimality.

(b) Optimization of the modified problem. The constraint coefficients of the new variable X_3 are determined using the following formula:

$$\begin{bmatrix} \text{Revised constraint} \\ \text{coefficients of the} \\ \text{new variable} \end{bmatrix} = \begin{bmatrix} \text{Technological coefficients of} \\ \text{optimal table with respect to the} \\ \text{basic variables of the initial table} \end{bmatrix} \times \begin{bmatrix} \text{Constraint coefficients} \\ \text{of new variable} \end{bmatrix}$$

$$= \begin{bmatrix} \dfrac{1}{5} & -\dfrac{1}{4} \\ -\dfrac{1}{5} & \dfrac{1}{2} \end{bmatrix} \times \begin{bmatrix} 6 \\ 5 \end{bmatrix}$$

$$= \begin{bmatrix} -\dfrac{1}{20} \\ \dfrac{13}{10} \end{bmatrix}$$

Table 2.47 of the original problem is modified as per the above details and shown in Table 2.48.

Table 2.48 Modified Table for Table 2.47

CB_i	C_j	6	8	20	0	0	
	Basic variable	X_1	X_2	X_3	S_1	S_2	Solution
8	X_2	0	1	−1/20	1/5	−1/4	2
6	X_1	1	0	13/10	−1/5	1/2	8
	Z_j	6	8	37/5	2/5	1	64
	$C_j - Z_j$	0	0	63/5	−2/5	−1	

The following optimal result is obtained after carrying out two more iterations from Table 2.48.

$$X_1 = 0, \quad X_2 = 0, \quad X_3 = 8, \quad S_1 = 12, \quad S_2 = 0, \quad Z(\text{optimum}) = 160$$

QUESTIONS

1. Explain the terminologies of linear programming model.
2. List and explain the assumptions of linear programming problems.
3. Define the following:
   (a) Alternate optimum solution
   (b) Unbounded solution

    (c) Infeasible solution

    (d) Degenerate solution

    (e) Slack variable

    (f) Surplus variable

    (g) Artificial variable

    (h) Basic variable

    (i) Criterion value.

4. A small manufacturer employs 5 skilled men and 10 semi-skilled men for making a product in two qualities: a deluxe model and an ordinary model. The production of a deluxe model requires 2-hour work by a skilled man and 1-hour work by a semi-skilled man. The ordinary model requires 2-hour work by a skilled man and 3-hour work by a semi-skilled man. According to worker union's rules, no man can work more than 8 hours per day. The profit of the deluxe model is Rs. 1000 per unit and that of the ordinary model is Rs. 800 per unit. Formulate a linear programming model for this manufacturing situation to determine the production volume of each model such that the total profit is maximized.

5. A firm manufactures three products A, B and C. Their profits per unit are Rs. 300, Rs. 200 and Rs. 400, respectively. The firm has two machines and the required processing time in minutes on each machine for each product is given in the following table:

		Product		
		A	B	C
Machine	1	4	3	5
	2	2	2	4

Machines 1 and 2 have 2000 and 2500 machine-minute, respectively. The upper limits for the production volumes of the product A, B and C are 100 units, 200 units and 50 units, respectively. But, the firm must produce a minimum of 50 units of the product A. Develop a LP model for this manufacturing situation to determine the production volume of each product such that the total profit is maximized.

6. The manager of an oil refinery has to decide on the optimal mix of two possible blending processes. The inputs and the outputs per production run of the blending process are as follows:

Process	Input		Output	
	Crude A	Crude B	Gasoline G_1	Gasoline G_2
1	5	3	5	8
2	4	5	4	4

The maximum amounts of availability of crude A and B are 200 units and 150 units, respectively. Market requirements show that at least 100 units of gasoline G_1 and 80 units of gasoline G_2 must be produced. The profits per production run from process 1 and process 2 are Rs. 3,00,000 and Rs. 4,00,000, respectively. Formulate this problem as a LP model to determine the number of production runs of each process such that the total profit is maximized.

7. Solve the following LP problem graphically:

$$\text{Maximize } Z = 20X_1 + 80X_2$$

subject to

$$4X_1 + 6X_2 \leq 90$$
$$8X_1 + 6X_2 \leq 100$$
$$5X_1 + 4X_2 \leq 80$$
$$X_1 \text{ and } X_2 \geq 0$$

8. Solve the following LP problem graphically:

$$\text{Minimize } Z = 20X_1 + 10X_2$$

subject to

$$X_1 + 2X_2 \leq 40$$
$$3X_1 + X_2 \geq 30$$
$$4X_1 + 3X_2 \geq 60$$
$$X_1 \text{ and } X_2 \geq 0$$

9. Solve the following LP problem graphically:

$$\text{Maximize } Z = 60X_1 + 90X_2$$

subject to

$$X_1 + 2X_2 \leq 40$$
$$2X_1 + 3X_2 \leq 90$$
$$X_1 - X_2 \geq 10$$
$$X_1 \text{ and } X_2 \geq 0$$

10. Solve the following LP problem graphically:

$$\text{Minimize } Z = 45X_1 + 55X_2$$

subject to

$$X_1 + 2X_2 \leq 30$$
$$2X_1 + 3X_2 \leq 80$$
$$X_1 - X_2 \geq 8$$
$$X_1 \text{ and } X_2 \geq 0$$

11. Solve the following LP problem graphically:

$$\text{Maximize } Z = 3X_1 + 2X_2$$

subject to

$$-2X_1 + 3X_2 \leq 9$$
$$X_1 - 5X_2 \geq -20$$
$$X_1 \text{ and } X_2 \geq 0$$

12. Solve the following LP problem using simplex method:

$$\text{Maximize } Z = 3X_1 + 2X_2 + 5X_3$$

subject to

$$X_1 + X_2 + X_3 \leq 9$$
$$2X_1 + 3X_2 + 5X_3 \leq 30$$
$$2X_1 - X_2 - X_3 \leq 8$$
$$X_1, X_2 \text{ and } X_3 \geq 0$$

13. Solve the following LP problem using simplex method:

$$\text{Maximize } Z = 5X_1 + 3X_2 + 7X_3$$

subject to

$$X_1 + X_2 + 2X_3 \leq 22$$
$$3X_1 + 2X_2 + X_3 \leq 26$$
$$X_1 + X_2 + X_3 \leq 18$$
$$X_1, X_2 \text{ and } X_3 \geq 0$$

14. Solve the following LP problem:

$$\text{Maximize } Z = 6X_1 + 4X_2$$

subject to

$$2X_1 + 3X_2 \leq 30$$
$$3X_1 + 2X_2 \leq 24$$
$$X_1 + X_2 \geq 3$$
$$X_1 \text{ and } X_2 \geq 0$$

Is the solution unique? If not, give other alternate optimum solutions.

15. Solve the following LP problem using dual simplex method:

$$\text{Minimize } Z = X_1 + X_2$$

subject to

$$2X_1 + X_2 \geq 2$$
$$-X_1 - X_2 \geq 1$$
$$X_1 \text{ and } X_2 \geq 0$$

16. Solve the following LP problem using dual simplex method:

$$\text{Minimize } Z = X_1 + 2X_2 + 3X_3$$

subject to

$$2X_1 - X_2 + X_3 \geq 4$$
$$X_1 + X_2 + 2X_3 \leq 8$$
$$X_2 - X_3 \geq 2$$
$$X_1, X_2 \text{ and } X_3 \geq 0$$

17. Solve the following LP problem using two-phase method:

$$\text{Minimize } Z = 10X_1 + 6X_2 + 2X_3$$

subject to

$$-X_1 + X_2 + X_3 \geq 1$$
$$3X_1 + X_2 - X_3 \geq 2$$
$$X_1, X_2 \text{ and } X_3 \geq 0$$

18. Write the dual of the following LP problem:

$$\text{Minimize } Z = 3X_1 - 2X_2 + 4X_3$$

subject to

$$3X_1 + 5X_2 + 4X_3 \geq 7$$
$$6X_1 + X_2 + 3X_3 \geq 4$$
$$7X_1 - 2X_2 - X_3 \leq 10$$
$$X_1 - 2X_2 + 5X_3 \geq 3$$
$$4X_1 + 7X_2 - 2X_3 \geq 2$$
$$X_1, X_2 \text{ and } X_3 \geq 0$$

19. Write the dual of the following LP problem:

$$\text{Maximize } Z = 5X_1 + 6X_2$$

subject to

$$4X_1 + 7X_2 = 20$$
$$5X_1 + 2X_2 = 10$$
$$6X_1 + 8X_2 = 25$$
$$X_1 \text{ and } X_2 \geq 0$$

20. Use duality to solve the following LP problem:

$$\text{Maximize } Z = 2X_1 + X_2$$

subject to

$$X_1 + 2X_2 \leq 10$$
$$X_1 + X_2 \leq 6$$
$$X_1 - X_2 \leq 2$$
$$X_1 - 2X_2 \leq 1$$
$$X_1 \text{ and } X_2 \geq 0$$

21. Consider the following LP problem:

$$\text{Maximize } Z = 3X_1 + 2X_2 - 5X_3$$

subject to

$$X_1 + X_2 \leq 2$$
$$2X_1 + X_2 + 6X_3 \leq 6$$
$$X_1 - X_2 + 3X_3 = 0$$
$$X_1, X_2 \text{ and } X_3 \geq 0$$

Solve the above LP problem. If the right-hand side of the primal is changed from (2, 6, 0) to (2, 10, 5), find the new optimal solution.

22. Solve the following LP problem:

$$\text{Maximize } Z = X_1 + 5X_2 + 3X_3$$

subject to

$$X_1 + 2X_2 + X_3 = 3$$
$$2X_1 - X_2 = 4$$
$$X_1,\ X_2 \text{ and } X_3 \geq 0$$

If the objective function is changed to (2, 5, 2), find the new optimum solution.

23. Solve the following LP problem:

$$\text{Maximize } Z = 3X_1 + 5X_2$$

subject to

$$X_1 + X_2 \leq 4$$
$$3X_1 + 2X_2 \leq 18$$
$$X_1 \text{ and } X_2 \geq 0$$

If a new variable is included in the above LP problem with a profit of 7, and 2 and 4 as the coefficients of the first and the second constraints, respectively, find the solution to the new problem.

24. Solve the following LP problem using simplex methods:

$$\text{Maximize } Z = 20X_1 + 80X_2$$

subject to

$$4X_1 + 6X_2 \leq 90$$
$$8X_1 + 6X_2 \leq 100$$
$$X_1 \text{ and } X_2 \geq 0$$

If the following new constraint is added to this model, find the solution to the new problem.

$$5X_1 + 4X_2 \leq 80$$

TRANSPORTATION PROBLEM

3

3.1 INTRODUCTION

Transportation problem is a special kind of linear programming problem in which goods are transported from a set of sources to a set of destinations subject to the supply and demand of the source and destination, respectively, such that the total cost of transportation is minimized. Some examples of transportation problem are summarized in Table 3.1.

Table 3.1 Examples of Transportation Problem

Source	Destination	Commodity	Objective
Plants	Markets	Finished goods	Minimizing total cost of shipping
Plants	Finished goods warehouses	Finished goods	Minimizing total cost of shipping
Finished goods warehouses	Markets	Finished goods	Minimizing total cost of shipping
Suppliers	Plants	Raw materials	Minimizing total cost of shipping
Suppliers	Raw material warehouses	Raw materials	Minimizing total cost of shipping
Raw material warehouses	Plants	Raw materials	Minimizing total cost of shipping

Let

    m be the number of sources

    n be the number of destinations

    a_i be the supply at the source i

    b_j be the demand at the destination j

    c_{ij} be the cost of transportation per unit from source i to destination j

    X_{ij} be the number of units to be transported from the source i to the destination j.

A schematic representation of transportation problem is shown in Figure 3.1.

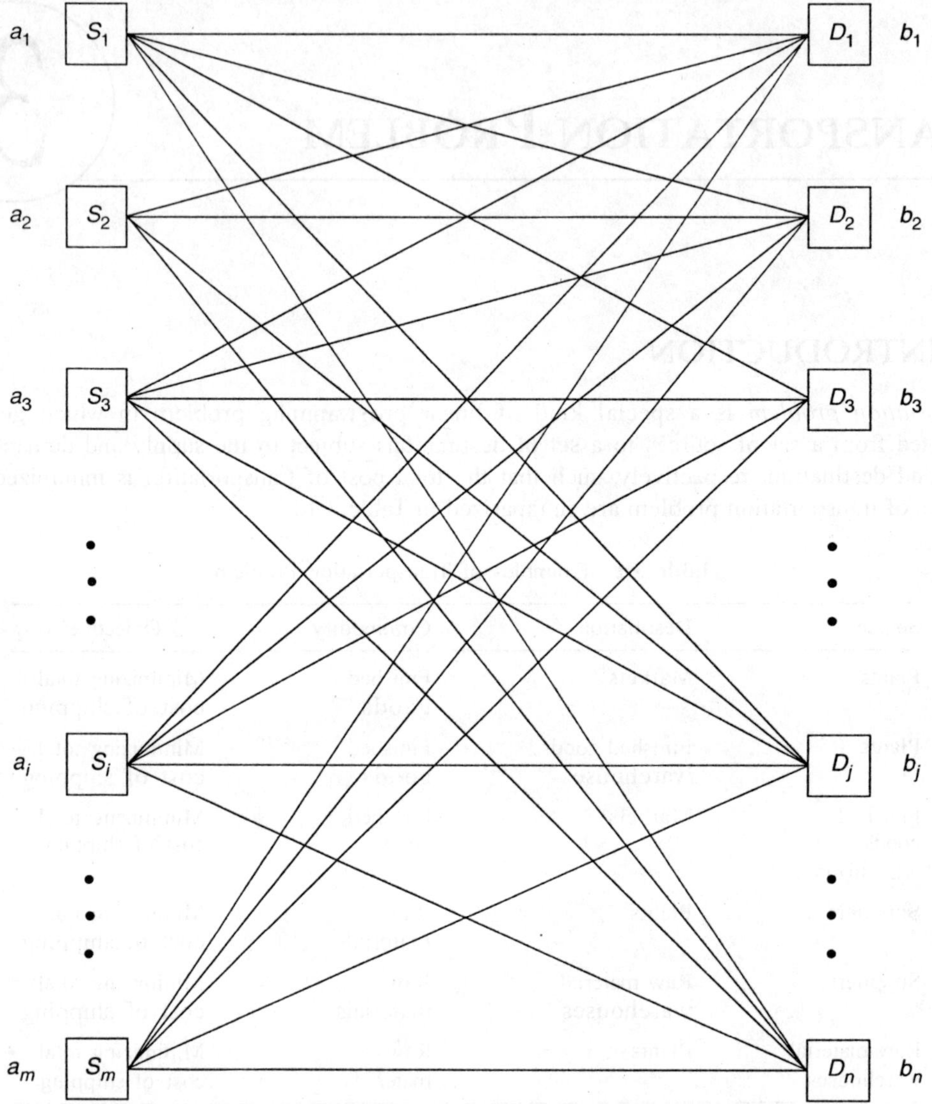

Figure 3.1 Schematic diagram of simple transportation problem.

A generalized format of the transportation problem is presented in Table 3.2.

3.2 MATHEMATICAL MODEL FOR TRANSPORTATION PROBLEM

In this section, a linear programming model for the transportation problem is presented.

$$\text{Minimize } Z = \sum_{i=1}^{m} \sum_{j=1}^{n} c_{ij} X_{ij}$$

Table 3.2 Generalized Format of the Transportation Problem

Destination (j)

	1	2	$\cdots$	j	$\cdots$	n	Supply
1	c_{11}	c_{12}	$\cdots$	c_{1j}	$\cdots$	c_{1n}	a_1
2	c_{21}	c_{22}	$\cdots$	c_{2j}	$\cdots$	c_{2n}	a_2
$\vdots$	$\vdots$	$\vdots$		$\vdots$		$\vdots$	$\vdots$
i	c_{i1}	c_{i2}	$\cdots$	c_{ij}	$\cdots$	c_{in}	a_i
$\vdots$	$\vdots$	$\vdots$		$\vdots$		$\vdots$	$\vdots$
m	c_{m1}	c_{m2}	$\cdots$	c_{mj}	$\cdots$	c_{mn}	a_m
Demand	b_1	b_2	$\cdots$	b_j	$\cdots$	b_n	

Source (i)

subject to

$$\sum_{j=1}^{n} X_{ij} \le a_i, \qquad i = 1, 2, 3, \ldots, m$$

and

$$\sum_{i=1}^{m} X_{ij} \ge b_j, \qquad j = 1, 2, 3, \ldots, n$$

where

$$X_{ij} \ge 0, \qquad i = 1, 2, 3, \ldots, m \quad \text{and} \quad j = 1, 2, 3, \ldots, n$$

The objective function minimizes the total cost of transportation (Z) between various sources and destinations. The constraint i in the first set of constraints ensures that the total units transported from the source i is less than or equal to its supply. The constraint j in the second set of constraints ensures that the total units transported to the destination j is greater than or equal to its demand.

Example 3.1 Consider the following transportation problem (Table 3.3) involving 3 sources and 3 destinations. Develop a linear programming (LP) model for this problem and solve it.

Table 3.3 Example 3.1

		Destination			
		1	2	3	Supply
	1	20	10	15	200
Source	2	10	12	9	300
	3	25	30	18	500
	Demand	200	400	400	1000

Solution Let X_{ij} be the number of units to be transported from the source i to the destination j, where $i = 1, 2, 3$ and $j = 1, 2, 3$. An LP model of this problem is:

Minimize $Z = 20X_{11} + 10X_{12} + 15X_{13} + 10X_{21} + 12X_{22} + 9X_{23} + 25X_{31} + 30X_{32} + 18X_{33}$

subject to

$$X_{11} + X_{12} + X_{13} \leq 200$$
$$X_{21} + X_{22} + X_{23} \leq 300$$
$$X_{31} + X_{32} + X_{33} \leq 500$$
$$X_{11} + X_{21} + X_{31} \geq 200$$
$$X_{12} + X_{22} + X_{32} \geq 400$$
$$X_{13} + X_{23} + X_{33} \geq 400$$
$$X_{ij} \geq 0, \quad i = 1, 2, 3 \quad \text{and} \quad j = 1, 2, 3$$

Application of simplex method to the above LP model yields the optimal shipping plan as presented in Table 3.4.

Table 3.4 Optimal Shipping Plan (Example 3.1)

Source	Destination	Quantity shipped
1	2	200
2	1	100
2	2	200
3	1	100
3	3	400

Total minimum cost = Rs. 15,100.*

3.3 TYPES OF TRANSPORTATION PROBLEM

The transportation problem can be classified into balanced transportation problem and unbalanced transportation problem.

3.3.1 Balanced Transportation Problem

If the sum of the supplies of all the sources is equal to the sum of the demands of all the destinations, then the problem is termed as *balanced* transportation problem. This may be represented by the relation:

$$\sum_{i=1}^{m} a_i = \sum_{j=1}^{n} b_j$$

Example 3.1 represents a balanced transportation problem.

3.3.2 Unbalanced Transportation Problem

If the sum of the supplies of all the sources is not equal to the sum of the demands of all the destinations, then the problem is termed as *unbalanced* transportation problem. That means, for any unbalanced transportation problem, we have

$$\sum_{i=1}^{m} a_i \neq \sum_{j=1}^{n} b_j$$

---

*Steps for calculating the total cost are not shown.

Example 3.2 Convert the transportation problem shown in Table 3.5 into a balanced transportation problem.

Table 3.5 Example 3.2

	Destination 1	2	3	Supply
Source 1	30	50	15	300
Source 2	35	70	20	200
Source 3	20	45	60	500
Demand	300	200	400	900/1000

Solution For the given problem,

$$\sum_{i=1}^{3} a_i = 1000 \quad \text{and} \quad \sum_{j=1}^{3} b_j = 900$$

Here

$$\sum_{i=1}^{3} a_i \neq \sum_{j=1}^{3} b_j$$

Hence it is an unbalanced transportation problem. Under this situation, an additional source or destination is to be included in the table as per the guidelines discussed now.

If $\sum_{i=1}^{m} a_i > \sum_{j=1}^{n} b_j$, then include a dummy destination to absorb the excess supply. The demand of the dummy destination is equal to $\sum_{i=1}^{m} a_i - \sum_{j=1}^{n} b_j$. The cost coefficients in the dummy destination are assumed as zeros. If $\sum_{j=1}^{n} b_j > \sum_{i=1}^{m} a_i$, then include a dummy source to supply the excess demand. The supply of the dummy source is equal to

$$\sum_{j=1}^{n} b_j - \sum_{i=1}^{m} a_i$$

The cost coefficients in the dummy source are assumed as zeros.

Table 3.5 is modified by including a dummy destination with a demand of 100 units. This is shown in Table 3.6, which is now a balanced transportation table.

Table 3.6 Balanced Transportation Problem (Example 3.2)

	Destination 1	2	3	4	Supply
Source 1	30	50	15	0	300
Source 2	35	70	20	0	200
Source 3	20	45	60	0	500
Demand	300	200	400	100	1000

Example 3.3 Convert the transportation problem shown as in Table 3.7 into a balanced transportation problem.

Table 3.7   Example 3.3

Destination

	1	2	3	4	Supply
1	5	12	6	10	300
Source 2	7	8	10	3	400
3	9	4	9	2	300
Demand	200	300	450	250	1200/1000

Solution   We have

$$\sum_{i=1}^{3} a_i = 1000 \quad \text{and} \quad \sum_{j=1}^{4} b_j = 1200$$

Since

$$\sum_{j=1}^{4} b_j > \sum_{i=1}^{3} a_i$$

it is an unbalanced transportation problem. This is converted into a balanced transportation problem by including a dummy source as shown in Table 3.8.

Table 3.8   Balanced Transportation Problem of Example 3.3

Destination

	1	2	3	4	Supply
1	5	12	6	10	300
Source 2	7	8	10	3	400
3	9	4	9	2	300
4	0	0	0	0	200
Demand	200	300	450	250	1200

3.4   METHODS TO SOLVE TRANSPORTATION PROBLEM

The solution procedure for the transportation problem consists of two phases:

1. Finding the initial basic feasible solution is the first phase.

2. Second phase involves optimization of the initial basic feasible solution which is obtained in Phase 1.

These are discussed in the following sections.

3.4.1 Finding the Initial Basic Solution

There are three types of techniques available to find the initial basic feasible solution. The solution using these techniques may not be optimal:

1. Northwest corner cell method
2. Least cost cell method
3. Vogel's approximation method (VAM)/penalty method.

The three techniques mentioned above are in the increasing order of their solution accuracy. The cost of the initial basic feasible solution through VAM will be the least among all the three techniques. Algorithm for each of the three techniques for finding the initial basic feasible solution is presented now.

Algorithm for northwest corner cell method

Step 1: Find the minimum of the supply and demand values with respect to the current northwest corner cell of the cost matrix.

Step 2: Allocate this minimum value to the current northwest corner cell and subtract this minimum from the supply and demand values with respect to the current northwest corner cell.

Step 3: Check whether exactly one of the row/column corresponding to the northwest corner cell has zero supply/demand, respectively. If so, go to step 4 otherwise, go to step 5.

Step 4: Delete that row/column with respect to the current northwest corner cell which has the zero supply/demand and go to step 6.

Step 5: Delete both the row and the column with respect to the current northwest corner cell.

Step 6: Check whether exactly one row or column is left out. If yes, go to step 7 otherwise go to step 1.

Step 7: Match the supply/demand of that row/column with the remaining demands/supplies of the undeleted columns/rows.

Step 8. Go to phase 2.

Algorithm for least cost cell method

Step 1: Find the minimum of the (undeleted) values in the cost matrix (i.e. find the matrix minimum).

Step 2: Find the minimum of the supply and demand values (X) with respect to the cell corresponding to the matrix minimum.

Step 3: Allocate X units to the cell with the matrix minimum. Also, subtract X units from the supply and the demand values corresponding to the cell with the matrix minimum.

Step 4: Check whether exactly one of the row/column corresponding to the cell with the matrix minimum has zero supply/zero demand, respectively. If yes, go to step 5 otherwise, go to step 6.

Step 5: Delete that row/column with respect to the cell with the matrix minimum which has the zero supply/zero demand and go to step 7.

Step 6: Delete both the row and the column with respect to the cell with the matrix minimum.

Step 7: Check whether exactly one row or column is left out. If yes, go to step 8 otherwise, go to step 1.

Step 8: Match the supply/demand of that row/column with the remaining demands/supplies of the undeleted columns/rows.

Step 9: Go to phase 2.

Algorithm for Vogel's approximation method

Step 1: Find row penalties, i.e. the difference between the first minimum and the second minimum in each row. If the two minimum values are equal, then the row penalty is zero.

Step 2: Find column penalties, i.e. the difference between the first minimum and the second minimum in each column. If the two minimum values are equal, then the column penalty is zero.

Step 3: Find the maximum amongst the row penalties and column penalties and identify whether it occurs in a row or in a column (break tie randomly). If the maximum penalty is in a row, go to step 4 otherwise, go to step 7.

Step 4: Identify the cell for allocation which has the least cost in that row.

Step 5: Find the minimum of the supply and demand values with respect to the selected cell.

Step 6: Allocate this minimum value to that cell and subtract this minimum from the supply and demand values with respect to the selected cell and go to step 10.

Step 7: Identify the cell for allocation which has the least cost in that column.

Step 8: Find the minimum of the supply and demand values with respect to the selected cell.

Step 9: Allocate this minimum value to the selected cell and subtract this minimum from the supply and demand values with respect to the selected cell.

Step 10: Check whether exactly one of the row/column corresponding to the selected cell has zero supply/zero demand, respectively. If yes, go to step 11; otherwise go to step 12.

Step 11: Delete the row/column which has the zero supply/zero demand and revise the corresponding row/column penalties. Then, go to step 13.

Step 12: Delete both the row and the column with respect to the selected cell. Then, revise the row and the column penalties.

Step 13: Check whether exactly one row/column is left out. If yes, go to step 14, otherwise, go to step 3.

Step 14: Match the supply/demand of the left-out row/column with the remaining demands/supplies of the undeleted columns/rows.

Step 15: Go to phase 2.

3.4.2 Optimizing the Basic Feasible Solution Applying U–V Method

Step 1: Row 1, row 2,..., row m of the cost matrix are assigned with variables U_1, U_2,..., U_m, respectively and the column 1, column 2,..., column n are assigned with variables V_1, V_2,..., V_n, respectively.

Step 2: Check whether the number of basic cells in the set of initial basic feasible solution is equal to $m + n - 1$. If yes, go to step 4, otherwise, go to step 3.

Step 3: Convert the necessary number of non-basic cells into basic cells to satisfy the condition stated in step 2 (while doing this, sufficient care should be taken such that there is no closed loop formation with the inclusion of the new basic cell(s)). The concept of the closed loop is explained in step 8.

Step 4: Compute the values for $U_1, U_2,..., U_m$, and $V_1, V_2,..., V_n$ by applying the following formula to all the basic cells only.

$$U_i + V_j = c_{ij} \qquad (\text{assume } U_1 = 0)$$

Step 5: Compute penalties P_{ij} for the non-basic cells by using the formula:

$$P_{ij} = U_i + V_j - c_{ij}$$

Step 6: Check whether all P_{ij} values are less than or equal to zero. If yes, go to step 12, otherwise, go to step 7.

Step 7: Identify the non-basic cell which has the maximum positive penalty, and term that cell as the new basic cell.

Step 8: Starting from the new cell, draw a closed loop consisting of only horizontal and vertical lines passing through some basic cells. (*Note:* Change of direction of the loop should be with 90 degrees only at some basic cell.)

Step 9: Starting from the new basic cell, alternatively assign positive (+) and negative (–) signs at the corners of the closed loop.

Step 10: Find the minimum of the allocations made amongst the negatively signed cells.

Step 11: Obtain the table for the next iteration by doing the following steps and then go to step 2.

(i) Add the minimum allocation obtained in the previous step to all the positively signed cells and subtract minimum allocation from all the negatively signed cells and then treat the net allocations as the allocations in the corresponding cells of the next iteration.

(ii) Copy the allocations which are on the closed loop but not at the corner points of the closed loop, as well as the allocations which are not on the loop as such without any modifications to the corresponding cells of the next iteration.

Step 12. The optimality is reached. Treat the present allocations to the set of basic cells as the optimum allocations.

Step 13. Stop.

Example 3.4 Consider the following transportation problem (Table 3.9) involving three sources and four destinations. The cell entries represent the cost of transportation per unit.

Table 3.9 Example 3.4

		1	2	3	4	Supply
	1	3	1	7	4	300
Source	2	2	6	5	9	400
	3	8	3	3	2	500
Demand		250	350	400	200	1200

Obtain the initial basic feasible solution using the following methods:

(a) Northwest corner cell method
(b) Least cost cell method
(c) Vogel's approximation method (VAM)/penalty method.

Solution   In Table 3.9, the sum of the supplies is equal to the sum of the demands. Hence, the given transportation problem is a balanced one. The process of obtaining the initial basic feasible solution for the given problem using each of the three techniques is demonstrated in detail in the following text.

By northwest corner cell method.   In Table 3.10, the supply and the demand values corresponding to the northwest corner cell (1, 1) are 300 and 250, respectively. The minimum of these values is 250. Hence, allocate 250 units to the cell (1, 1) and subtract the same from the supply and demand values of the cell (1, 1). Now the supply to Destination 1 is fully met. Hence, this column is deleted and the resultant data is shown in Table 3.11.

Table 3.10   Data of the Given Problem with Initial Allocation

	Destination 1	2	3	4	Supply
Source 1	250 3	1	7	4	~~300~~ 50
Source 2	2	6	5	9	400
Source 3	8	3	3	2	500
Demand	~~250~~ 0	350	400	200	

In Table 3.11, the supply and the demand values corresponding to the northwest corner cell (1, 2) are 50 and 350, respectively. The minimum of these values is 50. Hence, we should allocate 50 units to the cell (1, 2) and subtract the same from the supply and demand values of the cell (1, 2).

Table 3.11   Result after Deleting Column 1

	Destination 2	3	4	Supply
Source 1	50 1	7	4	~~50~~ 0
Source 2	6	5	9	400
Source 3	3	3	2	500
Demand	~~350~~ 300	400	200	

In this process, the supply of the source 1 is fully exhausted. Hence, this row is deleted and the resultant data is shown in Table 3.12.

Table 3.12 Result after Deleting Row 1

| | Destination | | | |
Source	2	3	4	Supply
2	6 **[300]**	5	9	~~400~~ 100
3	3	3	2	500
Demand	~~300~~ 0	400	200	

In Table 3.12, the supply and the demand values corresponding to the northwest corner cell (2, 2) are 400 and 300, respectively. The minimum of these values is 300. Hence, allocate 300 units to the cell (2, 2) and subtract the same from the supply and demand values of the cell (2, 2).

In this process, the demand of the destination 2 is fully satisfied. Hence, after deleting this column, the resultant data is shown in Table 3.13.

Table 3.13 Result after Deleting Column 2

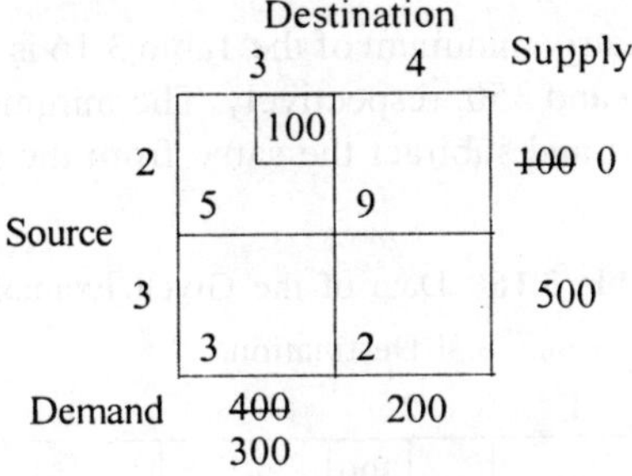

| | Destination | | |
Source	3	4	Supply
2	5 **[100]**	9	~~100~~ 0
3	3	2	500
Demand	~~400~~ 300	200	

In Table 3.13, the supply and the demand values corresponding to the northwest corner cell (2, 3) are 100 and 400, respectively. The minimum of these values is 100. Hence, we should allocate 100 units to the cell (2, 3) and subtract the same from the supply and demand values of the cell (2, 3).

In this process, the supply of the source 2 is fully exhausted. Hence, this row is deleted and the resultant data is shown in Table 3.14.

Table 3.14 Result after Deleting Row 2

| | Destination | | |
Source	3	4	Supply
3	3 **[300]**	2 **[200]**	~~500~~ 0
Demand	~~300~~ 0	~~200~~ 0	

In Table 3.14, only one source is left out. Hence, the demands of the destinations 3 and 4 need to be matched with the supply of the source 3. The initial basic feasible solution for the given problem using the northwest corner cell method is shown in Table 3.15.

Table 3.15   Initial Basic Feasible Solution Using Northwest Corner Cell Method

	Destination 1	2	3	4	Supply
1	250 / 3	50 / 1	/ 7	/ 4	300
Source 2	/ 2	300 / 6	100 / 5	/ 9	400
3	/ 8	/ 3	300 / 3	200 / 2	500
Demand	250	350	400	200	

The total cost of the solution is Rs. 4400. The total cost, is calculated by adding the products of the transportation cost per unit in each and every basic cell and the corresponding number of units allocated to it. A *basic cell* is one which has a positive allocation. Thus,

$$\text{Total cost} = 3 \times 250 + 1 \times 50 + 6 \times 300 + 5 \times 100 + 3 \times 300 + 2 \times 200 = \text{Rs. } 4400$$

By least cost cell method.   The matrix minimum of the Table 3.16 is 1 at cell (1, 2). The corresponding supply and demand values are 300 and 350, respectively. The minimum of these values is 300. Hence, allocate 300 units to the cell (1, 2) and subtract the same from the supply and demand values of the cell (1, 2).

Table 3.16   Data of the Given Example

	Destination 1	2	3	4	Supply
1	/ 3	300 / 1	/ 7	/ 4	~~300~~ 0
Source 2	/ 2	/ 6	/ 5	/ 9	400
3	/ 8	/ 3	/ 3	/ 2	500
Demand	250	~~350~~ 50	400	200	

In this process, the supply of the source 1 is fully exhausted. Hence, deleting this row, we get the resultant data as shown in Table 3.17.

In Table 3.17, the matrix minimum is 2 which occurs at cell (2, 1) and (3, 4). The cell (2, 1) is selected randomly for allocation. The supply and the demand values corresponding to the cell (2, 1) are 400 and 250, respectively. The minimum of these values is 250. Hence, allocate 250 units to the cell (2, 1) and subtract the same from the supply and demand values of the cell (2, 1).

Table 3.17 Result after Deleting Row 1

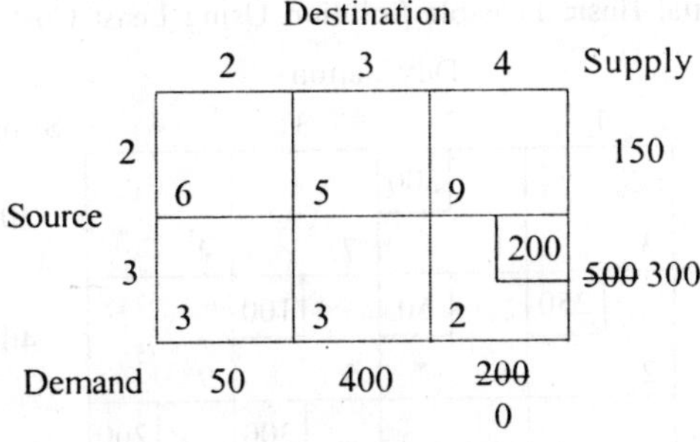

In this process, the demand at the destination 1 is fully satisfied. Therefore, we delete this column and the resultant data is shown in Table 3.18.

Table 3.18 Result after Deleting Column 1

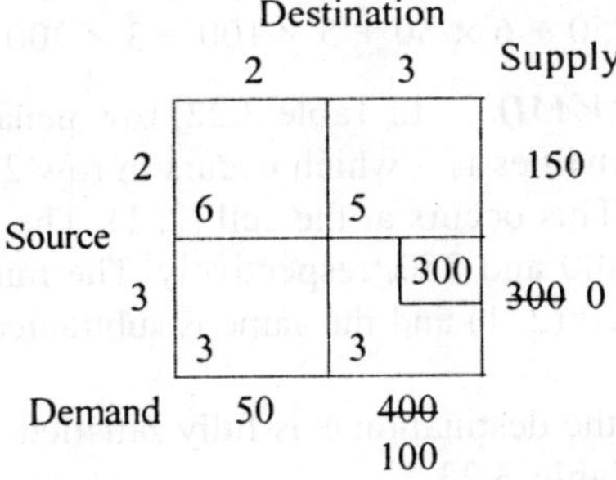

In Table 3.18, the matrix minimum is 2 at the cell (3, 4). The supply and the demand values corresponding to the cell (3, 4) are 500 and 200, respectively. The minimum of these values is 200. Hence we need to allocate 200 units to the cell (3, 4) and subtract the same from the supply and demand values of the cell (3, 4). In this process, the demand of the destination 4 is fully satisfied. After, deleting this column, the resultant data is given as shown in Table 3.19.

Table 3.19 Result after Deleting Column 4

In Table 3.19, the matrix minimum is 3 which occurs at cells, (3, 2) and (3, 3). The tie is broken randomly and the cell (3, 3) is selected for allocation. The supply and the demand values corresponding to the cell (3, 3) are 300 and 400, respectively. The minimum of these values is 300. Hence, allocate 300 units to the cell (3, 3) and subtract the same from the supply and demand values of the cell (3, 3). In this process, the supply of the source 3 is fully exhausted. Hence, this row is deleted and the resultant data is shown in Table 3.20.

Table 3.20   Result after Deleting Row 3

Destination

	2	3	Supply
Source 2	50 / 6	100 / 5	~~150~~ 0
Demand	~~50~~ 0	~~100~~ 0	

In Table 3.20, only one source is left out. Therefore, we should match the demands of destinations 2 and 3 with the supply of source 2. The initial basic feasible solution for Example 3.4 using the least cost cell method is shown in Table 3.21.

Table 3.21   Initial Basic Feasible Solution Using Least Cost Cell Method

Destination

	1	2	3	4	Supply
1	3	300 / 1	7	4	300
Source 2	250 / 2	50 / 6	100 / 5	9	400
3	8	3	300 / 3	200 / 2	500
Demand	250	350	400	200	

The total cost of the solution obtained using the least cost cell method is Rs. 2900. The total cost is calculated by adding the products of the cost of transportation per unit in each cell and the corresponding number of units allocated to it as shown below (a basic cell always has a positive allocation). Therefore,

$$\text{Total cost} = 1 \times 300 + 2 \times 250 + 6 \times 50 + 5 \times 100 + 3 \times 300 + 2 \times 200 = \text{Rs. } 2900$$

By Vogel's Approximation Method (VAM).   In Table 3.22, row penalties and column penalties are computed. The maximum of these penalties is 3 which occurs in row 2. Hence, the cell with the least cost in the row 2 is to be identified. This occurs at the cell (2, 1). The supply and the demand values corresponding to the cell (2, 1) are 400 and 250, respectively. The minimum of these values is 250. Thus 250 units are allocated to the cell (2, 1) and the same is subtracted from the supply and demand values of the cell (2, 1).

In this process, the demand at the destination 1 is fully satisfied. Hence, this column is deleted and the resultant data is shown in Table 3.23.

Since a column has been deleted, it is important to revise the row penalties as shown in Table 3.23. The maximum of these penalties is 3 which occurs in row 1. Therefore, the cell with the least cost is identified in row 1. This occurs at the cell (1, 2). The supply and the demand values corresponding to the cell (1, 2) are 300 and 350, respectively. The minimum of these values is 300.

Table 3.22 Data of the Given Problem

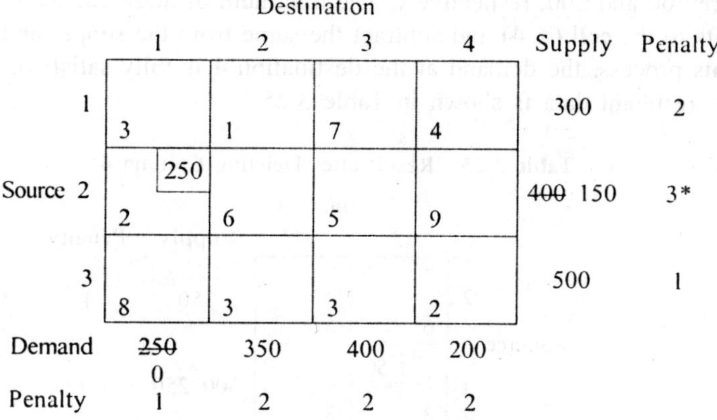

	Destination 1	2	3	4	Supply	Penalty
1	3	1	7	4	300	2
Source 2	2 [250]	6	5	9	~~400~~ 150	3*
3	8	3	3	2	500	1
Demand	~~250~~ 0	350	400	200		
Penalty	1	2	2	2		

Table 3.23 Result after Deleting Column 1

	Destination 2	3	4	Supply	Penalty
1	1 [300]	7	4	~~300~~ 0	3*
Source 2	6	5	9	150	1
3	3	3	2	500	1
Demand	~~350~~ 50	400	200		
Penalty	2	2	2		

Hence, allocate 300 units to the cell (1, 2) and subtract the same from the supply and demand values of the cell (1, 2). In this process, the supply at the source 1 is fully exhausted. Hence, delete this row and the resultant data is shown in Table 3.24.

Since a row has been deleted, the next step is to revise the column penalties as shown in Table 3.24. The maximum of these penalties is 7 which occurs in column 4. Next, the cell with the least

Table 3.24 Result after Deleting Row 1

	Destination 2	3	4	Supply	Penalty
Source 2	6	5	9	150	1
3	3	3	2 [200]	~~500~~ 300	1
Demand	50	400	~~200~~ 0		
Penalty	3	2	7*		

cost in column 4 is identified, which is (3, 4). The supply and the demand values corresponding to the cell (3, 4) are 500 and 200, respectively. The minimum of these values is 200. Thus, we need to allocate 200 units to the cell (3, 4) and subtract the same from the supply and demand values of the cell (3, 4). In this process, the demand at the destination 4 is fully satisfied. Hence, this column is deleted and the resultant data is shown in Table 3.25.

Table 3.25 Result after Deleting Column 4

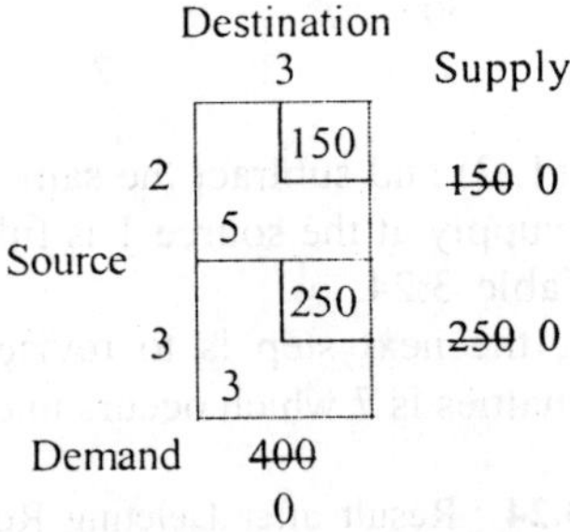

Since a column has been deleted, revise the row penalties as shown in Table 3.25. The maximum of these penalties is 3 which occurs in column 2. Hence, identify the cell with the least cost in the column 2. This occurs at the cell (3, 2). The supply and the demand values corresponding to the cell (3, 2) are 300 and 50, respectively. The minimum of these values is 50. Hence, allocate 50 units to the cell (3, 2) and subtract the same from the supply and demand values of the cell (3, 2). In this process, the demand at destination 2 is fully satisfied. Hence this column is deleted and the resultant data is given in Table 3.26.

Table 3.26 Result after Deleting Column 2

Since only one column is left out, the supplies of the sources 2 and 3 are matched with the demand of the destination 3 as shown in Table 3.26.

The set of basic feasible solution by applying the VAM to the given problem is shown in Table 3.27. The total cost of transportation for the solution is Rs. 2850.

As shown in Table 3.27, the total cost of the solution is obtained by adding the products of the cost of transportation per unit given in each and every basic cell and the corresponding number of units allocated to it.

Total cost = $1 \times 300 + 2 \times 250 + 5 \times 150 + 3 \times 50 + 3 \times 250 + 2 \times 200 = $ Rs. 2850

Table 3.27 Initial Basic Feasible Solution Using VAM

Destination

	1	2	3	4	Supply
1	3	300 1	7	4	300
Source 2	250 2	6	150 5	9	400
3	8	50 3	250 3	200 2	500
Demand	250	350	400	200	

Example 3.5 Consider Example 3.4 involving three sources and four destinations as reproduced below (Table 3.28). The cell entries represent the cost of transportation per unit.

Table 3.28 Example 3.5

Destination

	1	2	3	4	Supply
1	3	1	7	4	300
Source 2	2	6	5	9	400
3	8	3	3	2	500
Demand	250	350	400	200	

Obtain the initial basic feasible solution using the northwest corner cell method and then optimize the solution using *U-V* method.

Solution Application of the northwest corner cell method to this problem yields the result as shown in Table 3.29 (same as in Table 3.15). The total cost of the solution obtained, using the northwest corner cell method is Rs. 4400.

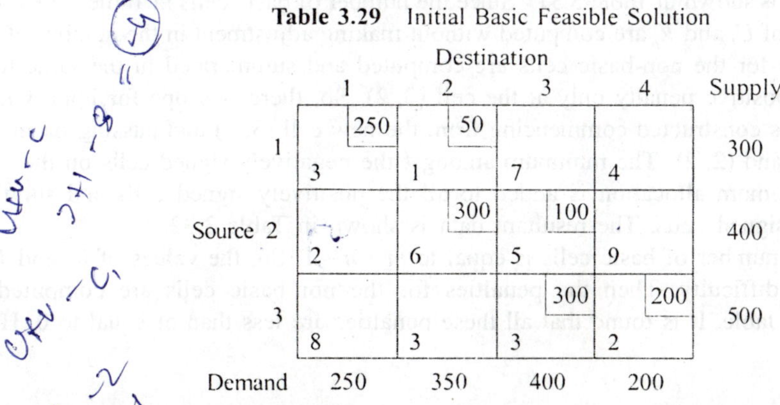

Table 3.29 Initial Basic Feasible Solution

Destination

	1	2	3	4	Supply
1	250 3	50 1	7	4	300
Source 2	2	300 6	100 5	9	400
3	8	3	300 3	200 2	500
Demand	250	350	400	200	

88 • *Operations Research*

Now, the initial basic feasible solution obtained, using northwest corner cell method, is optimized using *U-V* method. First, verify whether the number of basic cells is equal to $m + n - 1$. For this problem, the number of basic cells is 6 which is equal to $m + n - 1$ (m is the number of sources and n, the number of destinations). In Table 3.30, the values for U_i and V_j are computed by applying the formula, $U_i + V_j = c_{ij}$, to all the basic cells.

Table 3.30 Iteration 1

Destination

Source	$V_1 = 3$ 1	$V_2 = 1$ 2	$V_3 = 0$ 3	$V_4 = -1$ 4	Supply
$U_1 = 0$ 1	− 250 3	+ 50 1	7 −ve	4 −ve	300
$U_2 = 5$ 2	+ X 6 2	− 300 6	100 5	9 −ve	400
$U_3 = 3$ 3	8 −ve	3 1	300 3	200 2	500
Demand	250	350	400	200	

Then, the penalty for each of the non-basic cells is computed using the following formula and summarized within the squares at the bottom-left corners of the respective cells as shown in Table 3.30.

$$P_{ij} = U_i + V_j - c_{ij}$$

In Table 3.30, if all P_{ij} are less than or equal to 0 (for minimization type), then the optimality is reached. Otherwise the cell which has the most positive penalty, is to be selected. Here, the cell (2, 1) has the most positive penalty. So, there is a scope to improve the solution. This non-basic cell is to be converted into a basic cell without affecting the supply and demand restrictions. Hence, we should construct a closed loop starting from this new basic cell and passing through basic cells, (2, 2), (1, 2) and (1, 1) as shown in this table. Then alternatively, the '+' sign and the '−' sign are assigned in the basic cells on the closed loop commencing from the new basic cell. The minimum of the existing allocations amongst the negatively signed cells on the loop is identified, which is equal to 250 units. This minimum allocation is now added to all the positively signed cells but is subtracted from all the negatively signed cells on the closed loop.

The resulting table is shown in Table 3.31. Since the number of basic cells in Table 3.31 is equal to $m + n - 1$, the values of U_i and V_j are computed without making adjustment in the number of basic cells. Then the penalties for the non-basic cells are computed and summarized in the same table.

Table 3.31 has a positive penalty only at the cell (3, 2). So, there is scope for improving the solution. A closed loop is constructed commencing from the new cell (3, 2) and passing through the basic cells (3, 3), (2, 3) and (2, 2). The minimum amongst the negatively signed cells on the closed loop is 50. So, this minimum allocation is added to all the positively signed cells and subtracted from all the negatively signed cells. The resultant data is shown in Table 3.32.

In Table 3.32, the number of basic cells is equal to $m + n - 1$. So, the values of U_i and V_j are computed without any difficulty. Then the penalties for the non-basic cells are computed and summarized in the same table. It is found that all these penalties are less than or equal to 0. Hence,

Table 3.31 Iteration 2

	$V_1=-3$	$V_2=1$	$V_3=0$	$V_4=-1$	Supply
	1	2	3	4	
$U_1=0$ 1	3 (–ve)	1 [300]	7 (–ve)	4 (–ve)	300
$U_2=5$ 2	2 [250]	6 [50] (–)	5 [100] (+)	9 (–ve)	400
$U_3=3$ 3	8 (–ve)	1 (+) X	3 [300] (–)	2 [200]	500
Demand	250	350	400	200	

Table 3.32 Iteration 3

	$V_1=-2$	$V_2=1$	$V_3=1$	$V_4=0$	Supply
	1	2	3	4	
$U_1=0$ 1	3 (–ve)	1 [300]	7 (–ve)	4 (–ve)	300
$U_2=4$ 2	2 [250]	6 (–ve)	5 [150]	9 (–ve)	400
$U_3=2$ 3	8 (–ve)	3 [50]	3 [250]	2 [200]	500
Demand	250	350	400	200	

the optimum is reached. The allocations in Table 3.32 form the optimal solution. The corresponding total cost of transportation is Rs. 2850.

The total cost of the optimal solution using U-V method shown in Table 3.32 is calculated by adding the products of the cost of transportation per unit in each basic cell and the corresponding number of units allocated to it. Therefore,

$$\text{Total cost} = 1 \times 300 + 2 \times 250 + 5 \times 150 + 3 \times 50 + 3 \times 250 + 2 \times 200 = \text{Rs. } 2850$$

Example 3.6 Consider the problem of allocating raw materials from four different warehouses to five different plants. The availability of the raw material in the four warehouses are 25 ton, 30 ton, 20 ton and 30 ton. The demand of the raw material in the five plants are 20 ton, 20 ton, 30 ton, 10 ton and 25 ton. It is not possible to ship the raw material from warehouse 4 to plant 4 because of steep road. From the unit costs of transportation (in hundreds) given in Table 3.33, find the optimal shipping plan for the raw material (use least cost cell method to obtain the initial basic feasible solution).

Solution The transportation table, for determining of the initial basic feasible solution using least cost cell method, is given as in Table 3.33.

In the table, the sum of the supply values is equal to the sum of the demand values. So, the given problem is a balanced transportation problem. Application of the least cost cell method to this problem yields the following results which are shown in Tables 3.34 through 3.39.

Table 3.33 Example 3.6

Plant

Warehouse	1	2	3	4	5	Supply
1	10	2	3	15	9	25
2	5	10	15	2	4	30
3	15	5	14	7	15	20
4	20	15	13	—	8	30
Demand	20	20	30	10	25	105

Note: Cost figures are given in Rupees hundred.

The matrix minimum of Table 3.34 is 2 at cells (1, 2) and (2, 4). The cell (1, 2) is randomly selected. The corresponding supply and demand values are 25 and 20, respectively. The minimum of these values is 20. Hence, allocate 20 units to the cell (1, 2) and subtract the same from the supply and

Table 3.34 Initial Table

Plant

Warehouse	1	2	3	4	5	Supply
1	10	20 2	3	15	9	~~25~~ 5
2	5	10	15	2	4	30
3	15	5	14	7	15	20
4	20	15	13	∞	8	30
Demand	20	~~20~~ 0	30	10	25	

demand values of the cell (1, 2) as shown in Table 3.34. In this process, the demand of the plant 2 is fully met. Hence, delete the column 2 and the resultant data is shown in Table 3.35.

In Table 3.35, the matrix minimum is 2 which occurs at the cell (2, 4). The supply and the demand values corresponding to the cell (2, 4) are 30 and 10, respectively. The minimum of these values is 10. Hence, allocate 10 units to the cell (2, 4) and subtract the same from the supply and demand values of the cell (2, 4). In this process, the demand at the plant 4 is fully satisfied, and so this column is deleted. The resultant data is shown in Table 3.36.

Table 3.35 Result after Deleting Column 2

Plant

Warehouse	1	3	4	5	Supply
1	10	3	15	9	5
2	5	15	2 [10]	4	~~30~~ 20
3	15	14	7	15	20
4	20	13	∞	8	30
Demand	20	30	~~10~~ 0	25	

Table 3.36 Result after Deleting Column 4

Plant

Warehouse	1	3	5	Supply
1	10	3 [5]	9	~~5~~ 0
2	5	15	4	20
3	15	14	15	20
4	20	13	8	30
Demand	20	~~30~~ 25	25	

In Table 3.36, the matrix minimum is 3 at the cell (1, 3). The supply and the demand values corresponding to the cell (1, 3) are 5 and 30, respectively. The minimum of these values is 5. Hence, allocate 5 units to the cell (1, 3) and subtract the same from the supply and demand values of the cell (1, 3). In this process, the supply of the warehouse 1 is fully exhausted. Hence, this row is deleted and the resultant data is shown in Table 3.37. Here, the matrix minimum is 4 which occurs at cells, (2, 5). The supply and the demand values corresponding to the cell (2, 5) are 20 and 25, respectively. The minimum of these values is 20. Hence, allocate 20 units to the cell (2, 5) and subtract the same from the supply and demand values of the cell (2, 5).

In this process, the supply of the source 2 is fully exhausted. Hence, after deleting this row the resultant data is obtained as shown in Table 3.38.

In Table 3.38, the matrix minimum is 8 which occurs at cell (4, 5). The supply and the demand

Table 3.37   Result after Deleting Row 1

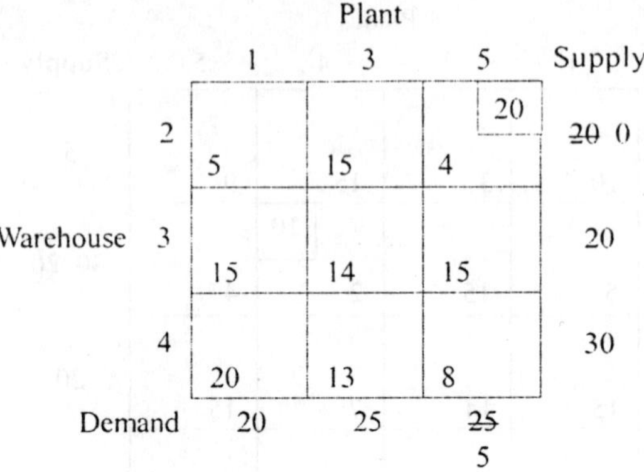

Table 3.38   Result after Deleting Row 2

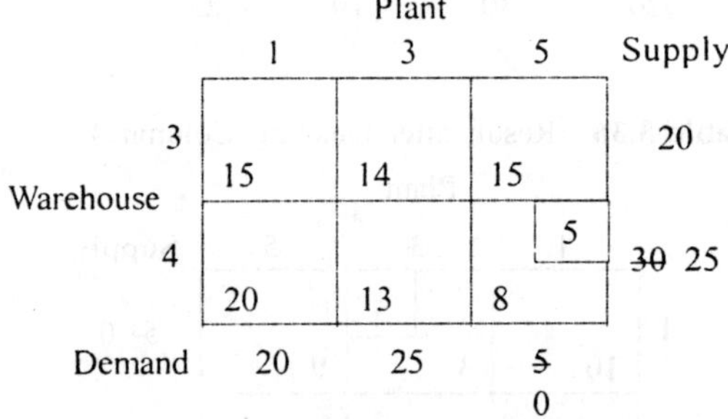

values corresponding to the cell (4, 5) are 30 and 5, respectively. The minimum of these values is 5. Thus we need to allocate 5 units to the cell (4, 5) and subtract the same from the supply and demand values of the cell (4, 5). In this process, the demand of the plant 5 is fully met. Hence, delete this column and the resultant data is shown in Table 3.39.

Table 3.39   Result after Deleting Column 5

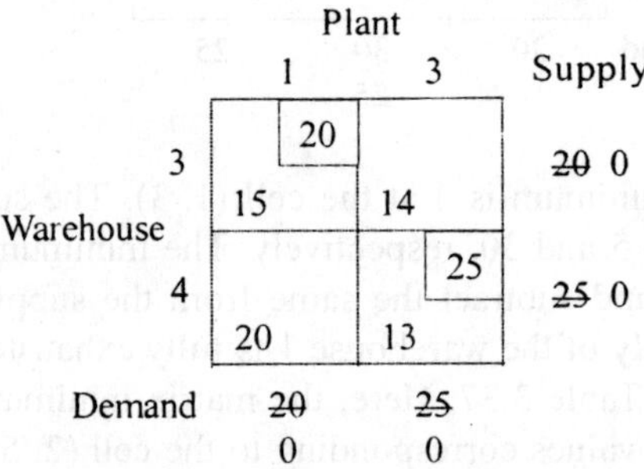

In Table 3.39, the matrix minimum is 13 which occurs at the cell (4, 3). The supply and the demand values corresponding to the cell (4, 3) are 25 and 25, respectively. The minimum of these values is 25. Hence, 25 units would be allocated to cell (4, 3) and the same would be subtracted from the supply and demand values of the cell (4, 3). Now, the supply and the demand values corresponding to the cell (3, 1) are 20 and 20 respectively. This value is allocated to the cell (3,1).

In this process, the supplies of warehouses 3 and 4 are fully exhausted and the demands of plants 1 and 3 are also fully met. Hence, we should delete rows 3 and 4 and columns 1 and 3. The resultant data is shown in Table 3.40.

Table 3.40 Initial Basic Feasible Solution

Plant

Warehouse \ Plant	1	2	3	4	5	Supply
1	10	[20] 2	[5] 3	15	9	25
2	5	10	15	[10] 2	[20] 4	30
3	[20] 15	5	14	7	15	20
4	20	15	[25] 13	∞	[5] 8	30
Demand	20	20	30	10	25	

The total cost, using least cost cell method as shown in Table 3.40, is calculated by adding the product of the transportation cost per unit in each basic cell and the corresponding number of units allocated to it. Therefore, we have

$$\text{Total cost} = 2 \times 20 + 3 \times 5 + 2 \times 10 + 4 \times 20 + 15 \times 20 + 13 \times 25 + 8 \times 5 = \text{Rs. } 820$$

Application of U–V method   In this phase, the initial basic feasible solution obtained using the least cost cell method is optimized using *U-V* method. First, verify whether the number of basic cells in Table 3.40 is equal to $m + n - 1$. In this problem, the number of basic cells is 7 which is not equal to $m + n - 1(=8)$; m being the number of sources (rows), which is equal to 4 and n, the number of destinations (columns), which is equal to 5.

In Table 3.40, each and every basic cell is indicated by certain allocation quantity written within a square at the top right corner of the respective cell. The seven basic cells, which are presented in Table 3.40, are given here with the corresponding row and column combinations: (1, 2), (1, 3), (2, 4), (2, 5), (3, 1), (4, 3) and (4, 5).

Now, convert the cell (3, 2) into a basic cell with a small allocation, ε to satisfy the above condition as shown in Table 3.41 such that its conversion does not form a closed loop.

In Table 3.41, the values for U_i and V_j are computed by applying the following formula to all the basic cells.

$$U_i + V_j = c_{ij}$$

The penalty for each of the non-basic cells is computed using the following formula and summarized as shown in Table 3.41.

$$P_{ij} = U_i + V_j - c_{ij}$$

In Table 3.41, the cell (2, 1) has the most positive penalty. This non-basic cell is to be converted into a basic cell without affecting the supply and demand restrictions. Hence, construct a closed loop

Table 3.41 Iteration 1

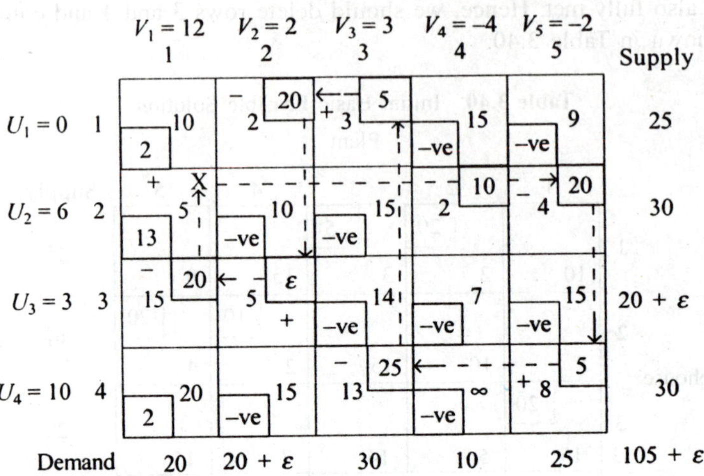

	$V_1 = 12$ 1	$V_2 = 2$ 2	$V_3 = 3$ 3	$V_4 = -4$ 4	$V_5 = -2$ 5	Supply
$U_1 = 0$ 1	10 / 2	2 / 20 (−)	3 / 5 (+)	15 / −ve	9 / −ve	25
$U_2 = 6$ 2	5 / 13 (+) X	10 / −ve	15 / −ve	2 / 10	4 / 20	30
$U_3 = 3$ 3	15 / 20 (−)	5 / ε (+)	14 / −ve	7 / −ve	15 / −ve	20 + ε
$U_4 = 10$ 4	2 / 20	15 / −ve	13 / 25 (−)	∞ / −ve	8 / 5 (+)	30
Demand	20	20 + ε	30	10	25	105 + ε

starting from this new basic cell and passing through basic cells (2, 5), (4, 5), (4, 3), (1,3), (1, 2), (3, 2) and (3, 1) as shown in this Table. Then, alternatively assign '+' sign and '−' signs in the basic cells on the closed loop commencing from the new basic cell. Then, identify the minimum of the existing allocations amongst the negatively signed cells on the loop. It is equal to 20. Add this minimum allocation to all the positively signed cells and subtract it from all the negatively signed cells on the closed loop. The resulting data is shown in Table 3.42.

Table 3.42 Iteration 2

		1	2	Plant 3	4	5	Supply
	1	10	2	3 / 25	15	9	25
	2	5 / 20	10	15	2 / 10	4	30
Warehouse	3	15	5 / 20+ε	14	7	15	20 + ε
	4	20	15	13 / 5	∞	8 / 25	30
	Demand	20	20 + ε	30	10	25	105 + ε

The row and column combinations of the basic cells are: (1, 3), (2, 1), (2, 4), (3, 2), (4, 3) and (4, 5). Since, the number of basic cells in Table 3.42 is 6 which is not equal to $m + n - 1$ (=8), the non-basic cells (1, 2) and (3, 4) are converted into basic cells with a small allocation, ε as shown in Table 3.43, such that their conversion does not form a closed loop. Then, the values of U_i and V_j are computed as shown in Table 3.43. The penalties for the non-basic cells are computed and summarized in the same table. It is found that all these penalties are less than or equal to 0. Hence,

Table 3.43 Iteration 3

	$V_1 = 7$	$V_2 = 2$	$V_3 = 3$	$V_4 = 4$	$V_5 = -2$	
	1	2	3	4	5	Supply
$U_1 = 0$ 1	10 (−ve)	2 [ε]	3 [25]	15 (−ve)	9 (−ve)	$25 + \varepsilon$
$U_2 = -2$ 2	5 [20]	10 (−ve)	15 (−ve)	2 [10]	4 (−ve)	30
$U_3 = 3$ 3	15 (−ve)	5 [20+ε]	14 (−ve)	7 [ε]	15 (−ve)	$20 + 2\varepsilon$
$U_4 = 10$ 4	20 (−ve)	15 (−ve)	13 [5]	∞ (−ve)	8 [25]	30
Demand	20	$20 + 2\varepsilon$	30	$10 + \varepsilon$	25	$105 + 3\varepsilon$

the optimum is reached. The allocations in the Table 3.43 form the optimal solution. Based on this table the optimal solution is represented in Table 3.44 with only necessary details after ignoring ε. The corresponding total cost of transportation is Rs. 560 hundreds or Rs. 56,000 (since the cost of transportation per unit has been taken in hundreds of Rs.). The total cost of the optimal solution using *U-V* method:

$$\text{Total cost} = 3 \times 25 + 5 \times 20 + 2 \times 10 + 5 \times 20 + 13 \times 5 + 8 \times 25$$
$$= \text{Rs. 560 hundreds}$$
$$= \text{Rs. 56,000}$$

Table 3.44 Final Solution

		Plant				
	1	2	3	4	5	Supply
Warehouse 1	10	2	3 [25]	15	9	25
2	5 [20]	10	15	2 [10]	4	30
3	15	5 [20]	14	7	15	20
4	20	15	13 [5]	∞	8 [25]	30
Demand	20	20	30	10	25	105

Example 3.7 Consider the transportation problem shown in Table 3.45.

Table 3.45 Example 3.7

Market

		1	2	3	4	5	Supply
	1	10	2	16	14	10	300
Plant	2	6	18	12	13	16	500
	3	8	4	14	12	10	825
	4	14	22	20	8	18	375
Demand		350	400	250	150	400	

Find the initial basic feasible solution using each of the following methods and compare their total costs.

(a) Northwest corner method
(b) Least cost cell method
(c) Vogel's approximation method

Solution The given problem is unbalanced because the total demand (1550) is less than the total supply (2000). Hence, this problem is converted into a balanced transportation problem as shown in Table 3.46.

Table 3.46 Balanced Version of Example 3.7

Market

		1	2	3	4	5	6	Supply
	1	10	2	16	14	10	0	300
Plant	2	6	18	12	13	16	0	500
	3	8	4	14	12	10	0	825
	4	14	22	20	8	18	0	375
Demand		350	400	250	150	400	450	

(a) Initial Basic Feasible Solution Using Northwest Corner Cell Method

The different iterations to get the initial basic feasible solution using the Northwest corner cell method are shown from Table 3.47 to Table 3.54.

Table 3.47 Initial Table

Market

		1	2	3	4	5	6	Supply
		[300]						0
	1	10	2	16	14	10	0	300
Plant	2	6	18	12	13	16	0	500
	3	8	4	14	12	10	0	825
	4	14	22	20	8	18	0	375
Demand		350 50	400	250	150	400	450	

Table 3.48 Results after Deleting Row 1

Market

		1	2	3	4	5	6	Supply
	2	[50]						450 500
		6	18	12	13	16	0	
Plant	3	8	4	14	12	10	0	825
	4	14	22	20	8	18	0	375
Demand		50 0	400	250	150	400	450	

Table 3.49 Results after Deleting Column 1

Market

		2	3	4	5	6	Supply
	2	[400]					50 450
		18	12	13	16	0	
Plant	3	4	14	12	10	0	825
	4	22	20	8	18	0	375
Demand		400 0	250	150	400	450	

Table 3.50 Results after Deleting Column 2

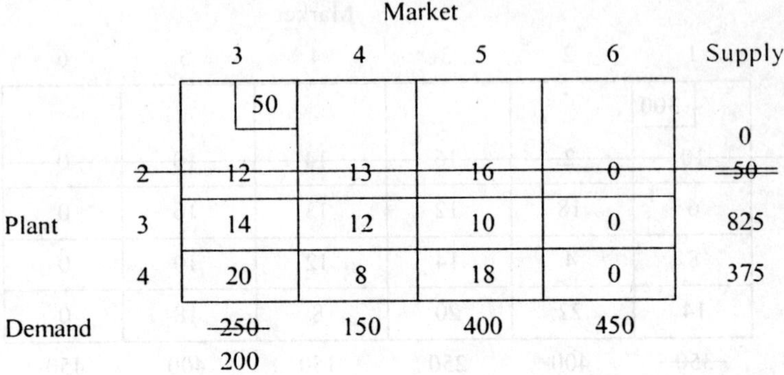

Table 3.51 Results after Deleting Row 2

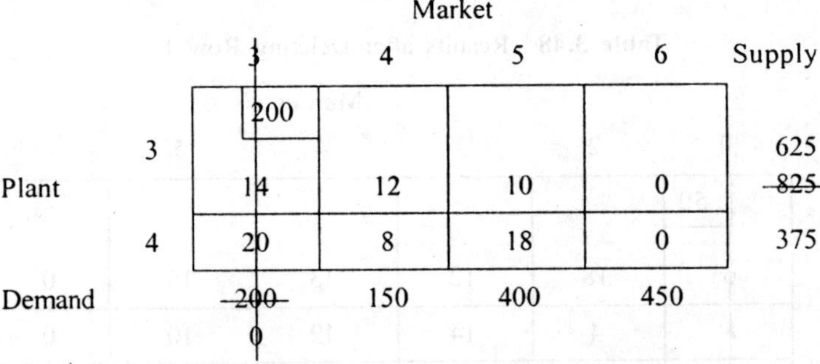

Table 3.52 Results after Deleting Column 3

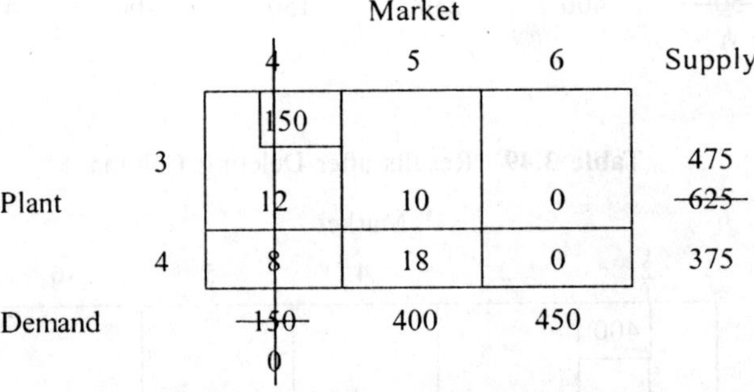

Table 3.53 Results after Deleting Column 4

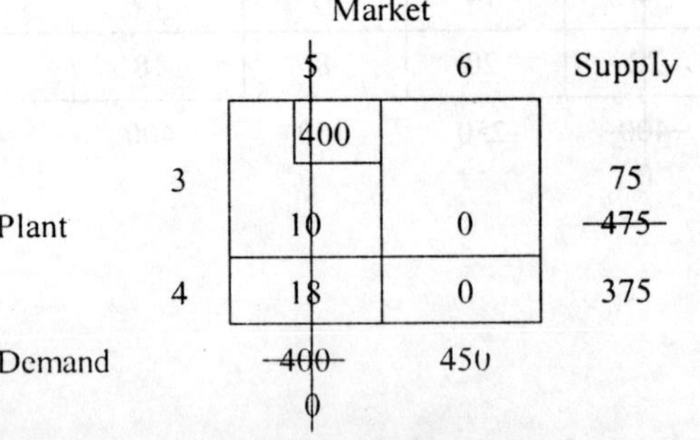

Table 3.54 Results after Deleting Column 5

<table>
<tr><td></td><td></td><td>Market</td><td></td><td></td></tr>
<tr><td></td><td></td><td>6</td><td>Supply</td><td></td></tr>
<tr><td rowspan="2">Plant</td><td>3</td><td>75
0</td><td>7̶5̶</td><td>0</td></tr>
<tr><td>4</td><td>375
0</td><td>3̶7̶5̶</td><td>0</td></tr>
<tr><td>Demand</td><td></td><td>4̶5̶0̶
0</td><td></td><td></td></tr>
</table>

The initial basic feasible solution for the given problem using the northwest corner cell method is shown in Table 3.55. The corresponding total cost is Rs. 19,700.

Table 3.55 Initial Basic Feasible Solution Using Northwest Corner Cell Method

				Market				Supply
		1	2	3	4	5	6	
	1	300 10	2	16	14	10	0	300
Plant	2	50 6	400 18	50 12	13	16	0	500
	3	8	4	200 14	150 12	400 10	75 0	825
	4	14	22	20	8	18	375 0	375
Demand		350	400	250	150	400	450	

(b) Initial Basic Feasible Solution Using Least Cost Cell Method

The different iterations of the least cost cell method applied to the Example 3.7 are shown from Table 3.56 to Table 3.62.

Table 3.56 Initial Table

				Market				Supply
		1	2	3	4	5	6	
	1̶	1̶0̶	2̶	1̶6̶	1̶4̶	1̶0̶	300 0̶	0 3̶0̶0̶
Plant	2	6	18	12	13	16	0	500
	3	8	4	14	12	10	0	825
	4	14	22	20	8	18	0	375
Demand		350	400	250	150	400	4̶5̶0̶ 150	

Table 3.57 Results after Deleting Row 1

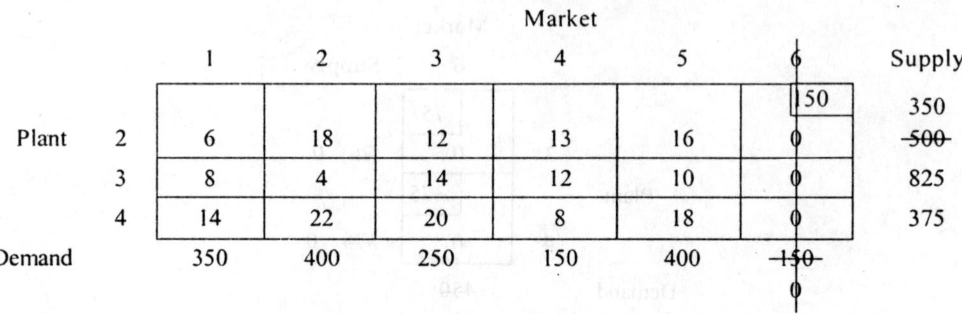

	Market						Supply
	1	2	3	4	5	6	
Plant 2	6	18	12	13	16	150 / 0	350 / ~~500~~
3	8	4	14	12	10	0	825
4	14	22	20	8	18	0	375
Demand	350	400	250	150	400	~~150~~ 0	

Table 3.58 Results after Deleting Column 6

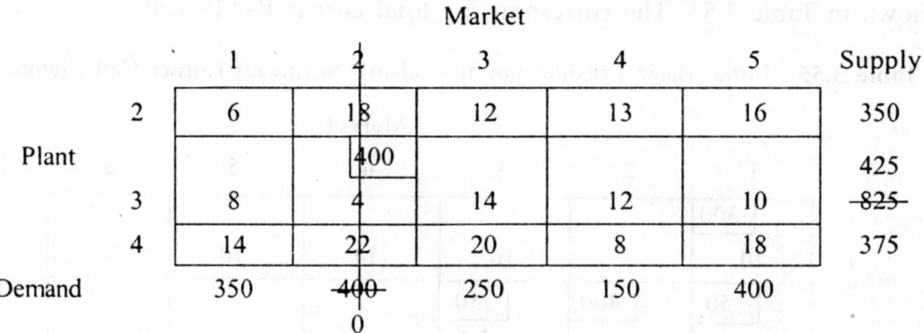

	Market					Supply
	1	2	3	4	5	
Plant 2	6	18	12	13	16	350
		400				425
3	8	4	14	12	10	~~825~~
4	14	22	20	8	18	375
Demand	350	~~400~~ 0	250	150	400	

Table 3.59 Results after Deleting Column 2

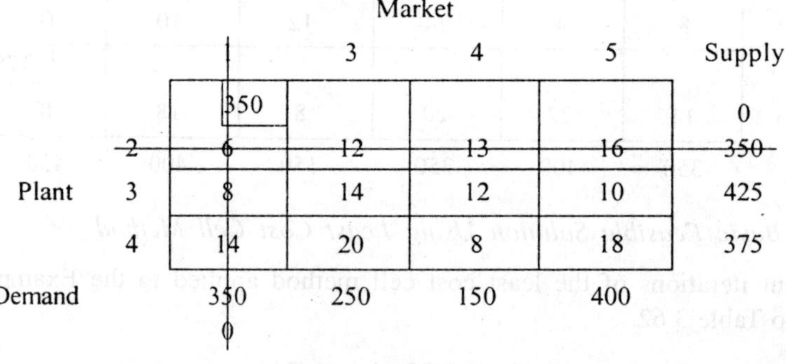

	Market				Supply
	1	3	4	5	
	350				0
2	6	12	13	16	~~350~~
Plant 3	8	14	12	10	425
4	14	20	8	18	375
Demand	350 / 0	250	150	400	

Table 3.60 Results after Deleting Row 2 Column 1

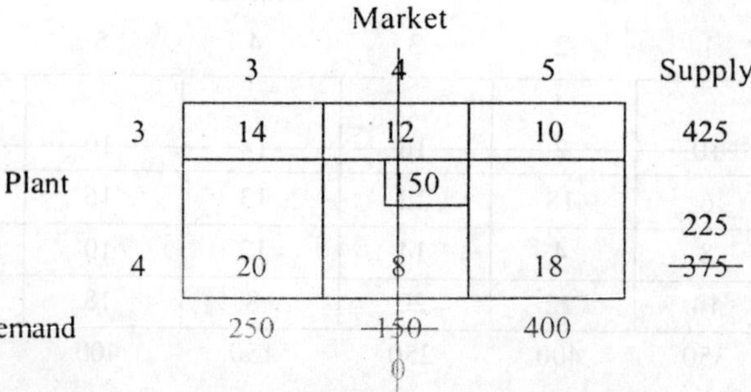

	Market			Supply
	3	4	5	
3	14	12	10	425
Plant 4	20	8 / 150	18	225 / ~~375~~
Demand	250	~~150~~ 0	400	

Table 3.61 Results after Deleting Column 4

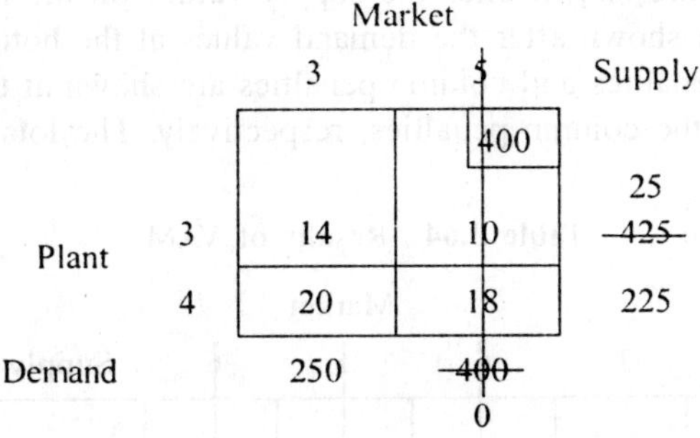

Table 3.62 Results after Deleting Column 5

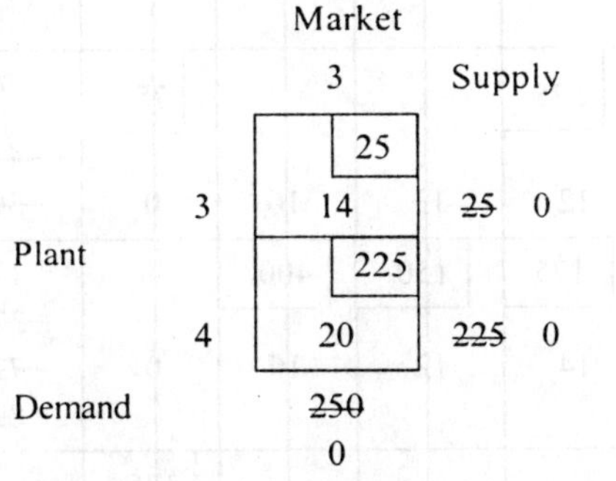

The initial basic feasible solution for the given problem using the least cost cell method is shown in Table 3.63. The corresponding total cost is Rs. 13,750.

Table 3.63 Initial Basic Feasible Solution Using Least Cost Cell Method

		Market					
	1	**2**	**3**	**4**	**5**	**6**	**Supply**
1	10	2	16	14	10	300 0	300
Plant **2**	350 6	18	12	13	16	150 0	500
3	8	400 4	25 14	12	400 10	0	825
4	14	22	225 20	150 8	18	0	375
Demand	350	400	250	150	400	450	

(c) Initial Basic Feasible Solution Using Vogel's Approximation Method

The results of applying Vogel's approximation method (VAM) to the Example 3.7 are shown in Table 3.64. In the Table 3.64, the sequence by which the deletions of rows/columns are shown by the numbers from 1 to 7 with encircling either at the end of the rows or at the bottom of the columns. The row penalties are shown after the supply values on the right-hand side of the table and the column penalties are shown after the demand values at the bottom of the table. The order of computation of the row penalties and column penalties are shown at the top of the row penalties and at the left-hand side of the column penalties, respectively. The total cost using this method is Rs. 12,250.

Table 3.64 Results of VAM

Market

Plant	1	2	3	4	5	6	Supply	1	4	6	7	8
1	10	300 2	16	14	10	0	300	2	8			(3)
2	350 6	18	75 12	13	16	75 0	75 425 500	6	6	6	1	1
3	8	100 4	175 14	150 12	400 10	0	175 325 725 825	4	4	2	2	2
4	14	22	20	8	18	375 0	375 0	8				(1)
Demand	350 0	400 100 0	250	150 0	400 0	450 75 0						
2	2	2	2	4	0	0						
3	2	2	2	1	0	0						
5	2	14	2	1	6							
	(5)	(4)		(7)	(6)	(2)						

The summary of the total costs using the three methods is shown in Table 3.65. From this table, it is clear that VAM gives the best initial basic feasible solution for the given problem, because its total cost is the minimum when compared to that of the other two methods.

Table 3.65 Comparison of Total Costs of the Three Methods

Method	Total cost
Northwest Corner Cell Method	19,700
Least Cost Cell Method	13,750
Vogel's Approximation Method	12,250

Example 3.8 Consider the Example 3.7 as reproduced in Table 3.66 and find the optimal distribution by using Vogel's approximation method in the first stage to get the initial basic feasible solution.

Table 3.66 Example 3.8

Plant	Market 1	2	3	4	5	Supply
1	10	2	16	14	10	300
2	6	18	12	13	16	500
3	8	4	14	12	10	825
4	14	22	20	8	18	375
Demand	350	400	250	150	400	

Solution The initial basic feasible solution using the Vogel's approximation method is shown in Table 3.67 which is reproduced from Table 3.64.

(*Note:* Dummy column 6 is introduced to balanced the problem).

Table 3.67 Initial Basic Feasible Solution of Example 3.8 Using VAM

Plant	Market 1	2	3	4	5	6	Supply
1	10	[300] 2	16	14	10	0	300
2	[350] 6	18	[75] 12	13	16	[75] 0	500
3	8	[100] 4	[175] 14	[150] 12	[400] 10	0	825
4	14	22	20	8	18	[375] 0	375
Demand	350	400	250	150	400	450	

Application of U-V method to optimize the solution

Iteration 1: The computed values of U_i and V_j, as well as the values of the penalties are shown in Table 3.68. The highest penalty is 2 (positive) which occurs in two cells. So, there is a scope of improving the solution. By randomly breaking the tie, the cell (3, 6) is considered for conversion into a basic cell. Starting from this cell, a closed loop is constructed.

Table 3.68 Iteration 1

Market

	$V_1 = 6$ (1)	$V_2 = 2$ (2)	$V_3 = 12$ (3)	$V_4 = 10$ (4)	$V_5 = 8$ (5)	$V_6 = 0$ (6)	Supply
$U_1 = 0$ 1	10 −ve	**300** 2	16 −ve	14 −ve	10 −ve	0 0	300
Plant $U_2 = 0$ 2	**350** 6	18 −ve	**75** + 12 ↓	13 −ve	16 −ve	**75** − 0 ↑	500
$U_3 = 2$ 3	8 0	**100** 4	− **175** 14	**150** 12	**400** 10	**X** + 0 2	825
$U_4 = 0$ 4	14 −ve	22 −ve	20 −ve	8 2	18 −ve	**375** 0	375
Demand	350	400	250	150	400	450	

Iteration 2: The revised allocations based on the calculations in Table 3.68 are presented in Table 3.69. The solution in Table 3.69 is also not optimal because the highest of the penalties of the non-basic cells is positive (4) for the cell (4, 4). Starting from this cell, a closed loop is constructed.

Table 3.69 Iteration 2

Market

	$V_1 = 6$ (1)	$V_2 = 2$ (2)	$V_3 = 12$ (3)	$V_4 = 10$ (4)	$V_5 = 8$ (5)	$V_6 = -2$ (6)	Supply
$U_1 = 0$ 1	10 −ve	**300** 2	16 −ve	14 −ve	10 −ve	0 −ve	300
Plant $U_2 = 0$ 2	**350** 6	18 −ve	**150** 12	13 −ve	16 −ve	0 −ve	500
$U_3 = 2$ 3	8 0	**100** 4	**100** 14	**150** − 12 ↓	**400** ← 10	**75** + 0 ↑ 2	825
$U_4 = 2$ 4	14 −ve	22 −ve	20 −ve	**X** + 8 4	18 −ve	**375** − 0	375
Demand	350	400	250	150	400	450	

Iteration 3: The revised solution based on the calculations in Table 3.69 is shown in Table 3.70. In this table, all the penalties of the non-basic cells are zero/negative. Hence, the optimality is reached. The corresponding total cost is Rs. 11,500 and final distribution plan is given in Table 3.71.

Table 3.70 Iteration 3

Market

		1 ($V_1 = 6$)	2 ($V_2 = 2$)	3 ($V_3 = 12$)	4 ($V_4 = 6$)	5 ($V_5 = 8$)	6 ($V_6 = -2$)	Supply
	1 ($U_1 = 0$)	10 –ve	2 300	16 –ve	14 –ve	10 –ve	0 –ve	300
Plant	2 ($U_2 = 0$)	6 350	18 –ve	12 150	13 –ve	16 –ve	0 –ve	500
	3 ($U_3 = 2$)	8 0	4 100	14 100	12 –ve	10 400	0 225	825
	4 ($U_4 = 2$)	14 –ve	22 –ve	20 –ve	8 150	18 –ve	0 225	375
Demand		350	400	250	150	400	450	

Table 3.71 Final Solution of Example 3.8

Market

		1	2	3	4	5	6	Supply
	1	10	2 300	16	14	10	0	300
Plant	2	6 350	18	12 150	13	16	0	500
	3	8	4 100	14 100	12	10 400	0 225	825
	4	14	22	20	8 150	18	0 225	375
Demand		350	400	250	150	400	450	

3.5   TRANSSHIPMENT MODEL

In generalized transshipment model, items are supplied from different sources to different destinations. It is sometimes economical if the shipment passes through some transient nodes in between the sources and destinations. Unlike in transportation problem, where shipments are sent directly from a particular source to a particular destination, in transshipment problem, the objective is to minimize the total cost of shipments, and thus the shipment passes through one or more intermediate nodes before it reaches its desired destination.

There are mainly two types of the transshipment problem discussed in the following sections.

3.5.1   Transshipment Problem with Sources and Destinations Acting as Transient Nodes

A schematic diagram of a simple form of transshipment problem in which the sources and destinations act as transient nodes is shown in Figure 3.2.

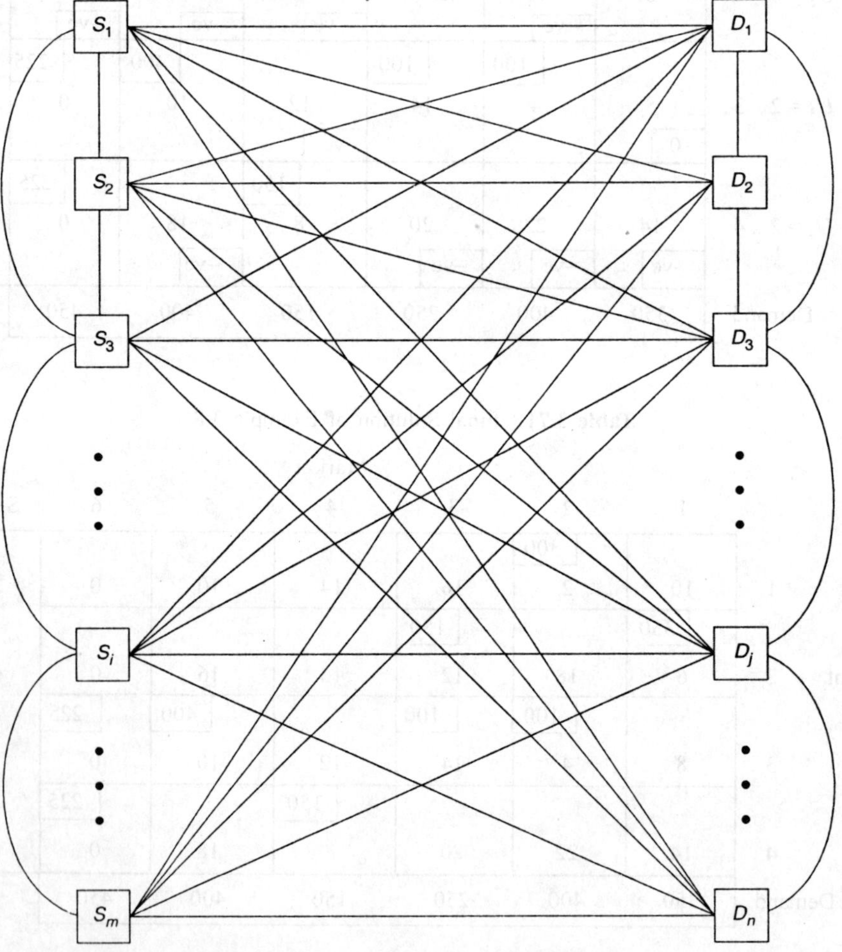

Figure 3.2   Schematic diagram of simple transshipment model.

In the figure, consider the shipment of items from source 1 to destination 2. The shipment from the source 1 can pass through the source 2 and the destination 1 before it reaches the specified destination 2. Since, in this case the shipment passes through some transient nodes, the arrangement is termed as *transshipment model.* The objective of the transshipment problem is to find the optimal shipping pattern such that the total cost of transportation is minimized.

A different view of the Figure 3.2 is shown in Figure 3.3 in which the number of starting nodes as well as the number of ending nodes is the sum of the number of sources and the number of destinations of the original problem.

Let B be the buffer which must be maintained at each of the transient sources and transient destinations. At the minimum the buffer, B can be equal to the sum of the supplies or the sum of the demands, assuming that it is a balanced problem. So, a constant B is added to all the starting nodes and all the ending nodes as shown in Figure 3.3. Thus we have

$$B = \sum_{i=1}^{m} a_i = \sum_{j=1}^{n} b_j$$

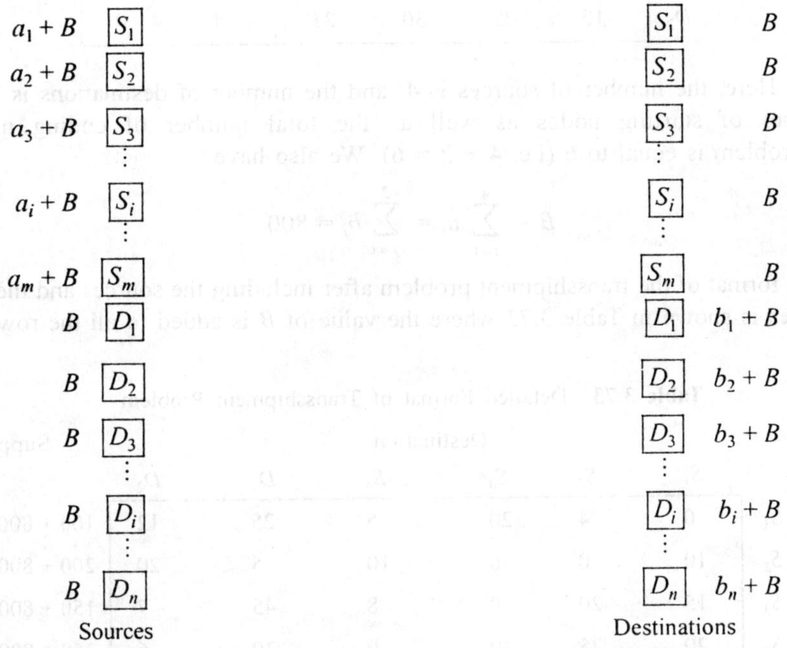

Figure 3.3 — Sources (left): $a_1 + B$ at S_1; $a_2 + B$ at S_2; $a_3 + B$ at S_3; $a_i + B$ at S_i; $a_m + B$ at S_m; B at D_1; B at D_2; B at D_3; B at D_i; B at D_n.

Destinations (right): B at S_1; B at S_2; B at S_3; B at S_i; B at S_m; $b_1 + B$ at D_1; $b_2 + B$ at D_2; $b_3 + B$ at D_3; $b_i + B$ at D_i; $b_n + B$ at D_n.

Figure 3.3 Modified view of simple transshipment problem.

The destinations D_1, D_2, D_3,..., D_i,..., D_n are included as additional starting nodes in Figure 3.3 mainly to act as transient nodes. So, they are not having any original supply. The supply of each of these transient nodes should be at least equal to B. Hence, each of these transient nodes is assigned with B units as the supply value. Similarly, the sources S_1, S_2, S_3,..., S_i,..., S_m are included as additional ending nodes in Figure 3.3, mainly to act as transient nodes. These nodes are not having any original demand. But, each of these transient nodes is assigned with B units as the demand value. So, just to have a balance, B is added to each a_i of the starting nodes and to each b_i of the ending nodes in Figure 3.3. Then the problem in Figure 3.3 is similar to any conventional transportation problem for which one can use *U-V* method to get the optimum shipping plan.

Example 3.9 Consider the following transshipment problem involving 4 sources and 2 destinations. The supply values of the sources S_1, S_2, S_3 and S_4 are 100 units, 200 units, 150 units and 350 units, respectively. The demand values of destinations D_1 and D_2 are 350 units and 450 units, respectively. The transportation cost per unit between different sources and destinations are summarized as in Table 3.72. Solve the transshipment problem.

Table 3.72 C_{ij} Values for Example 3.9

		S_1	S_2	S_3	S_4	D_1	D_2
	S_1	0	4	20	5	25	12
	S_2	10	0	6	10	5	20
Source	S_3	15	20	0	8	45	7
	S_4	20	25	10	0	30	6
	D_1	20	18	60	15	0	10
	D_2	10	25	30	23	4	0

Solution Here, the number of sources is 4, and the number of destinations is 2. Therefore, the total number of starting nodes as well as the total number of ending nodes of the transshipment problem is equal to 6 (i.e. 4 + 2 = 6). We also have

$$B = \sum_{i=1}^{4} a_i = \sum_{j=1}^{2} b_j = 800$$

A detailed format of the transshipment problem after including the sources and the destinations as transient nodes is shown in Table 3.73 where the value of B is added to all the rows and all the columns.

Table 3.73 Detailed Format of Transshipment Problem

		S_1	S_2	S_3	S_4	D_1	D_2	Supply
	S_1	0	4	20	5	25	12	100 + 800 = 900
	S_2	10	0	6	10	5	20	200 + 800 = 1000
Source	S_3	15	20	0	8	45	7	150 + 800 = 950
	S_4	20	25	10	0	30	6	350 + 800 = 1150
	D_1	20	18	60	15	0	10	800
	D_2	10	25	30	23	4	0	800
Demand		800	800	800	800	450 + 800 = 1250	350 + 800 = 1150	

The solution to the problem in Table 3.73 is shown in Table 3.74 and the corresponding total cost of transportation is Rs. 5,250. The allocations in the main diagonal cells are to be ignored. The shipping pattern is diagrammatically presented in Figure 3.4 which shows the shipments related to the off-diagonal cells alone.

Table 3.74 Solution

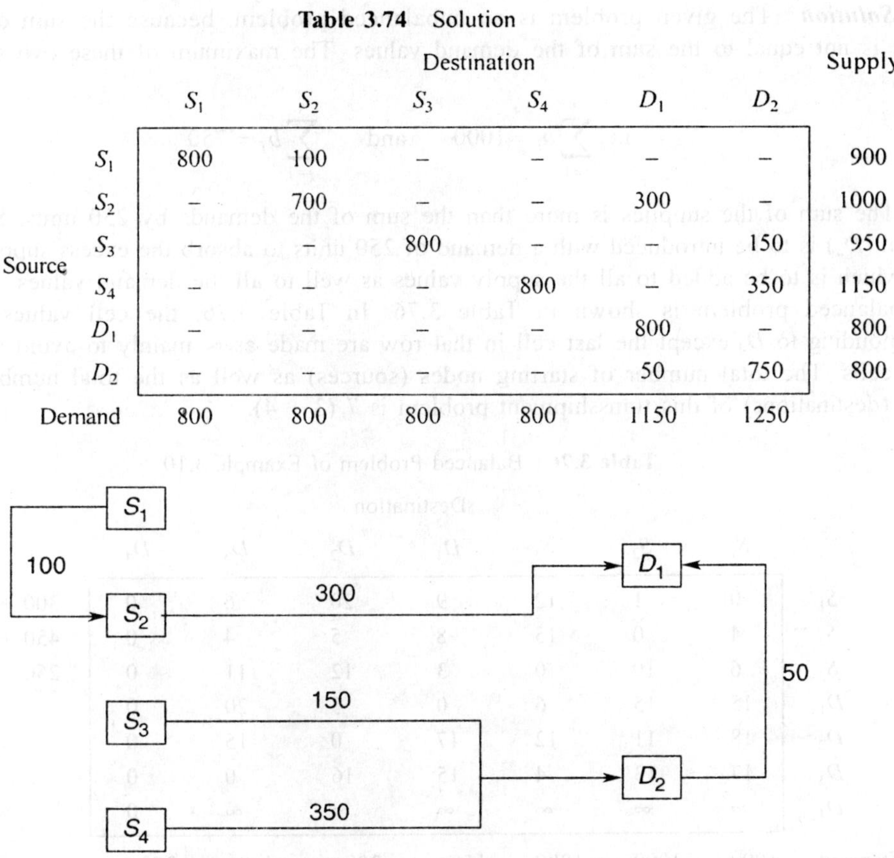

	Destination						Supply
	S_1	S_2	S_3	S_4	D_1	D_2	
S_1	800	100	–	–	–	–	900
S_2	–	700	–	–	300	–	1000
S_3	–	–	800	–	–	150	950
S_4	–	–	–	800	–	350	1150
D_1	–	–	–	–	800	–	800
D_2	–	–	–	–	50	750	800
Demand	800	800	800	800	1150	1250	

Figure 3.4 Optimal shipping pattern.

Example 3.10   The supply values of the sources S_1, S_2 and S_3 are 300 units, 450 units and 250 units, respectively. The demands of the destinations D_1, D_2 and D_3 are 150 units, 200 units and 400 units, respectively. The cost of transportation (in rupees) per unit between different source and destination combinations are shown in Table 3.75. Find the optimal shipping plan for this transshipment problem.

Table 3.75 Data for Example 3.10

	Destination					
	S_1	S_2	S_3	D_1	D_2	D_3
S_1	0	1	12	9	20	6
S_2	4	0	15	8	5	4
S_3	6	10	0	3	12	11
D_1	15	15	6	0	3	20
D_2	18	11	12	17	0	15
D_3	17	13	4	15	16	0

Source (label for the left column of Table 3.75)

Solution The given problem is an unbalanced problem, because the sum of the supply values is not equal to the sum of the demand values. The maximum of these two sums is 1000 units.

$$\text{i.e. } \sum_{i=1}^{3} a_i = 1000 \quad \text{and} \quad \sum_{j=1}^{3} b_j = 750$$

The sum of the supplies is more than the sum of the demands by 250 units. So, a dummy column (D_4) is to be introduced with a demand of 250 units to absorb the excess supply. The value of B which is to be added to all the supply values as well to all the demand values is 1000 units. The balanced problem is shown in Table 3.76. In Table 3.76, the cell values in the row corresponding to D_4 except the last cell in that row are made as ∞ mainly to avoid allocations to those cells. The total number of starting nodes (sources) as well as the total number of ending nodes (destinations) of this transshipment problem is 7 (3 + 4).

Table 3.76 Balanced Problem of Example 3.10

Destination

Source	S_1	S_2	S_3	D_1	D_2	D_3	D_4	Supply
S_1	0	1	12	9	20	6	0	300 + 1000 = 1300
S_2	4	0	15	8	5	4	0	450 + 1000 = 1450
S_3	6	10	0	3	12	11	0	250 + 1000 = 1250
D_1	15	15	6	0	3	20	0	1000
D_2	18	11	12	17	0	15	0	1000
D_3	17	13	4	15	16	0	0	1000
D_4	∞	∞	∞	∞	∞	∞	0	1000
Demand	1000	1000	1000	150+1000 =1150	200+1000 =1200	400+1000 =1400	250+1000 =1250	

The solution of the problem shown in Table 3.76 is presented in Table 3.77. The optimal shipment plan as per the solution given in Table 3.77 is shown in Figure 3.5. The corresponding total cost of shipment is Rs. 3,200.

Table 3.77 Balanced Problem of Example 3.10

Destination

Source	S_1	S_2	S_3	D_1	D_2	D_3	D_4	Supply
S_1	1000	150	–	–	–	–	150	1300
S_2	–	850	–	–	200	400	–	1450
S_3	–	–	1000	150	–	–	100	1250
D_1	–	–	–	1000	–	–	–	1000
D_2	–	–	–	–	1000	–	–	1000
D_3	–	–	–	–	–	1000	–	1000
D_4	–	–	–	–	–	–	1000	1000
Demand	1000	1000	1000	1150	1200	1400	1250	

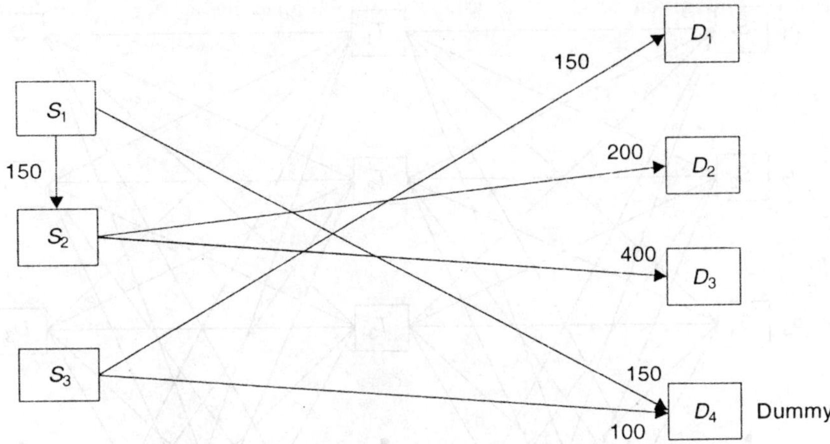

Figure 3.5 Optimal shipping pattern.

3.5.2 Transportation Problem with some Transient Nodes between Sources and Destinations

Consider the case of shipping items from different plants to different market places through some intermediate finished goods warehouses. This is an example of the transshipment problem involving the conventional transportation problem with some transient nodes as shown in Figure 3.6.

Let m, be the number of sources; n, the number of destinations; o, the number of transient nodes in between sources and destinations; S_i, the source i, where $i = 1, 2, 3,..., m$; D_j, the destination j, where $j = 1, 2, 3,..., n$; T_k, the transient node k, $k = 1, 2, 3,..., o$; a_i, the supply of the source i; b_j, the demand of the destination j; and c_{ij}, the cost of transportation per unit between different possible routes in the problem. Also let

$$B = \sum_{i=1}^{m} a_i = \sum_{j=1}^{n} b_j$$

Therefore, B can be termed as the capacity of each transient node.

Guidelines to solve the problem

Add the transient nodes as additional sources as well as additional destinations and form a regular transportation table with necessary cost details. The supply of each of the transient nodes as well as the demand of each of the transient nodes is assumed as B. Assume, infinity a very large value, for the c_{ij} values between different sources and destinations. Then solve the problem using regular transportation method.

Example 3.11 A multi-plant organization has three plants (P_1, P_2 and P_3) and three market places (M_1, M_2 and M_3). The items from the plants are transported to the market places through two intermediate finished goods warehouses. The details on cost of transportation per unit for different combinations between the plants and warehouses, between warehouses and markets, between warehouses, supply values of the plants and demand values of the markets are summarized in Table 3.78. The c_{ij} values between the plants and markets are assumed as infinity. (While solving this

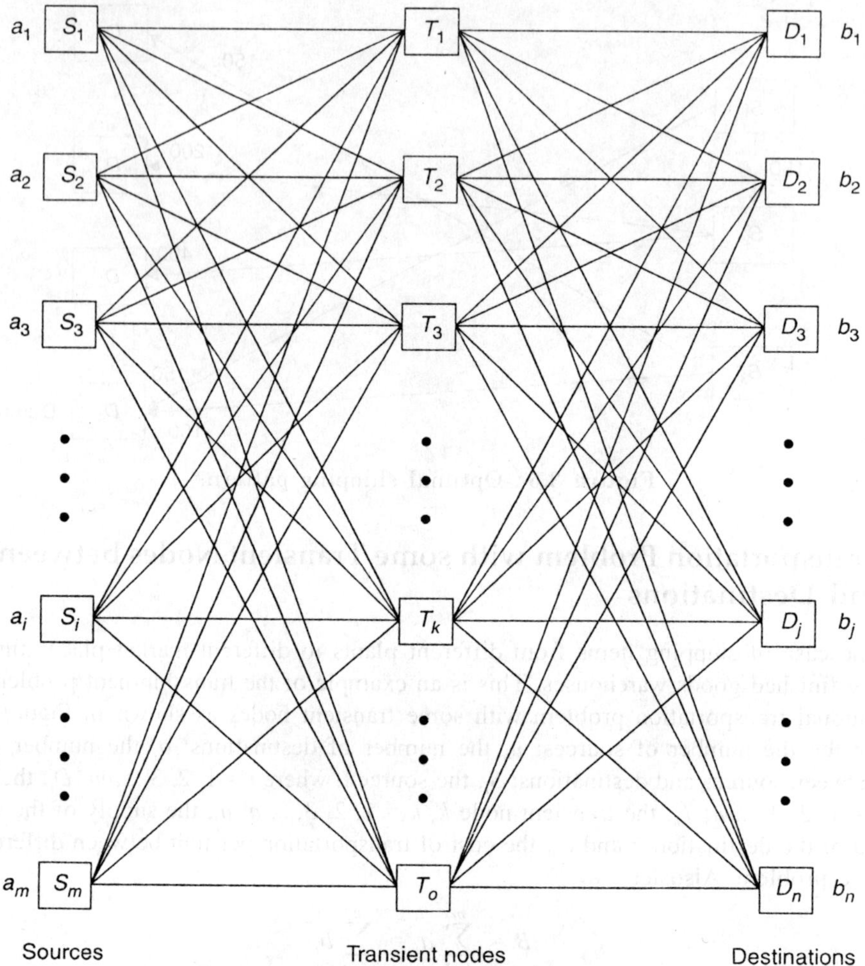

Figure 3.6 Conventional transportation problem with some transient nodes between sources and destinations.

problem using computer, instead of infinity, a very large value may be assumed.) Find the optimal shipping plan such that total cost of transportation is minimized.

Table 3.78 Example 3.11

		Terminal nodes					Supply
		M_1	M_2	M_3	W_1	W_2	
	P_1	∞	∞	∞	15	30	200
	P_2	∞	∞	∞	28	10	300
Starting nodes	P_3	∞	∞	∞	30	15	400
	W_1	10	40	30	0	20	—
	W_2	25	15	35	25	0	—
Demand		100	400	400	—	—	

Solution Here, number of plants = number of markets = 3 and the number of warehouse (or transient nodes) is 2. Therefore, the total number of starting nodes as well as the total number of ending nodes of the transshipment problem is equal to 5 (i.e. 3 + 2 = 5). We also have

$$B = \sum_{i=1}^{3} a_i = \sum_{j=1}^{3} b_j = 900$$

A revised format of the transshipment problem is shown as in Table 3.79. In the table, the value of B is assumed as the sum of the supply values of all the plants or the sum of the demand values of all the markets.

Table 3.79 Revised Table 3.78

		Terminal nodes				Supply
	M_1	M_2	M_3	W_1	W_2	
P_1	∞	∞	∞	15	30	200
P_2	∞	∞	∞	28	10	300
Starting nodes P_3	∞	∞	∞	30	15	400
W_1	10	40	30	0	20	900
W_2	25	15	35	25	0	900
Demand	100	400	400	900	900	

The solution of the problem shown in the Table 3.79 is shown in Table 3.80. A diagrammatic view of the optimal shipping pattern is shown in Figure 3.7. In Table 3.80, shipments in the main diagonal cells are to be ignored. The shipments in the off-diagonal cells in the Table 3.80 are alone shown in the Figure 3.7. The corresponding total cost of the optimal solution is Rs. 32,500.

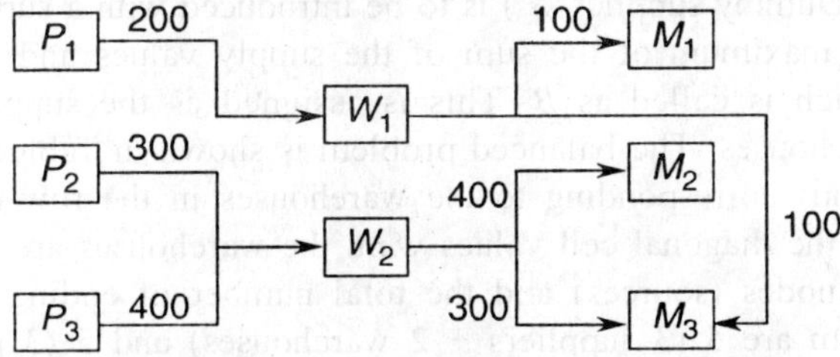

Figure 3.7 Diagrammatic view of optimal shipping pattern.

Table 3.80 Optimal Shipping Plan

		Terminal nodes				Supply
	M_1	M_2	M_3	W_1	W_2	
P_1	–	–	–	200	–	200
P_2	–	–	–	–	300	300
Starting nodes P_3	–	–	–	–	400	400
W_1	100	–	100	700	–	900
W_2	–	400	300	–	200	900
Demand	100	400	400	900	900	

Example 3.12 A multi-plant organization sources its raw materials from two suppliers located in different cities (S_1 and S_2) to meet the requirements at three different plants located in different cities (P_1, P_2 and P_3). The organization maintains two different raw material warehouses (W_1 and W_2) to receive and stock raw materials from the two suppliers and then ship them to the three plants. The supply values of the suppliers S_1 and S_2 are 1000 units and 1500 units, respectively. The demands of the plants P_1, P_2 and P_3 are 800 units, 1200 units and 1000 units, respectively. The cost of transportation (in rupees) per unit between different source and destination combinations are shown in Table 3.81. Find the optimal shipping plan for this transshipment problem.

Table 3.81 Data for Example 3.12

Destination

		P_1	P_2	P_3	W_1	W_2	Supply
	S_1	∞	∞	∞	17	16	1000
Source	S_2	∞	∞	∞	11	13	1500
	W_1	15	8	20	∞	13	
	W_2	18	10	9	14	∞	
Demand		800	1200	1000			

Solution The given problem is an unbalanced problem, because the sum of the demands of the plants (3000 units) is more than the sum of the supplies from the suppliers (2500 units).

$$\text{i.e. } \sum_{i=1}^{3} a_i = 2500 \quad \text{and} \quad \sum_{j=1}^{3} b_j = 3000$$

So, a dummy row (Dummy supplier, S_3) is to be introduced with a supply of 500 units to meet the excess demand. The maximum of the sum of the supply values and the sum of the demand values is 3000 units which is called as B. This is assigned as the supply value as well as the demand value of the warehouses. The balanced problem is shown in Table 3.82. In Table 3.82, the cell values for the columns corresponding to the warehouses in the row corresponding to S_3 are assigned as ∞. Similarly, the diagonal cell values w.r.t. the warehouses are also assigned as ∞. The total number of starting nodes (sources) and the total number of ending nodes (destinations) of this transshipment problem are 5 (3 suppliers + 2 warehouses) and 5 (3 plants + 2 warehouses), respectively.

Table 3.82 Balanced Problem of Example 3.12

Destination

		P_1	P_2	P_3	W_1	W_2	Supply
	S_1	∞	∞	∞	17	16	1000
	S_2	∞	∞	∞	11	13	1500
Source	S_3	0	0	0	∞	∞	500
	W_1	15	8	20	∞	13	3000
	W_2	18	10	9	14	∞	3000
Demand		800	1200	1000	3000	3000	

The solution of the problem shown in Table 3.82 is presented in Table 3.83. The optimal shipment pattern as per the solution given in Table 3.83 is shown in Figure 3.8. The corresponding total cost of shipments is Rs. 63,100. In Table 3.83, the shipment from W_1 to W_2 is 2000 units and the shipment from W_2 to W_1 is 1500 units. So, the net shipment from W_1 to W_2 is 500 units which is shown in Figure 3.8.

Table 3.83 Optimal Solution of Example 3.12

		P_1	P_2	P_3	W_1	W_2	Supply
	S_1	–	–	–	–	1000	1000
	S_2	–	–	–	1500	–	1500
Source	S_3	500	–	–	–	–	500
	W_1	300	700	–	–	2000	3000
	W_2	–	500	1000	1500	–	3000
Demand		800	1200	1000	3000	3000	

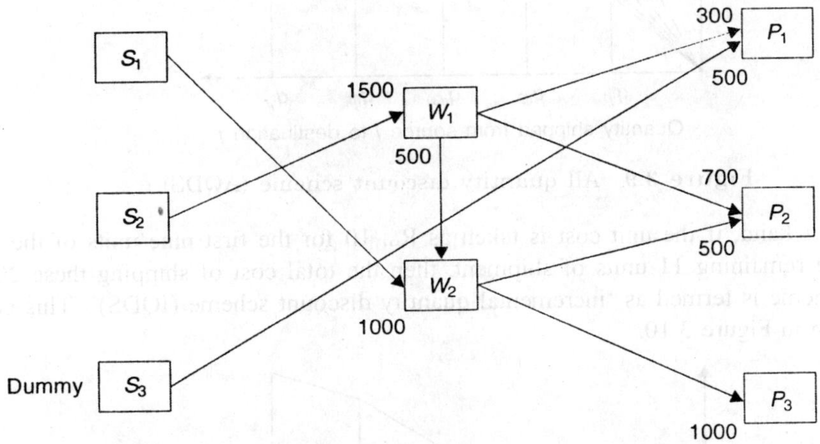

The top column header "Destination" spans P_1, P_2, P_3, W_1, W_2.

Figure 3.8 Optimal shipping pattern.

3.6 MODELLING THE TRANSPORTATION PROBLEM WITH QUANTITY DISCOUNTS

In the classical transportation problem, the unit cost of shipment from a given source to a given destination is assumed to be a constant irrespective of the quantity shipped. However, in reality, the normal practice in most of the transport agencies is to offer a discount on the unit cost for the increased volume of shipment.

Further, the quantity discount can be offered in two ways: *all* quantity discounts and *incremental* quantity discounts.

Let, C_{ij} be the unit cost of shipment and X_{ij} be the quantity shipped from the ith source to the jth destination. The cost structure (price breaks) under quantity discount scheme is as follows:

$$C_{ij} = \text{Rs. } 10, \qquad \text{if } 0 \leq X_{ij} < \text{Rs. } 10$$
$$= \text{Rs. } 8, \qquad \text{if } 10 \leq X_{ij} < \text{Rs. } 25$$
$$= \text{Rs. } 5, \qquad \text{if } 25 \leq X_{ij} < \infty$$

Consider a shipment of 20 units from the source i to the destination j under the above cost structure. If the unit cost is taken as Rs. 8 for all the 20 units, then the total cost of shipping these 20 units is Rs. 160. This scheme is termed as 'all quantity discounts scheme (AQDS)'. This concept is graphically shown in Figure 3.9.

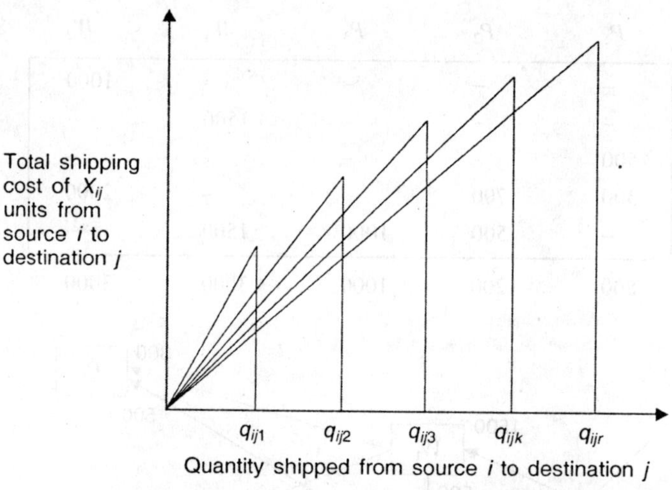

Figure 3.9 All quantity discount scheme (AQDS).

On the other hand, if the unit cost is taken as Rs. 10 for the first nine units of the shipment and Rs. 8 for the remaining 11 units of shipment, then the total cost of shipping these 20 units is Rs. 178. This scheme is termed as 'incremental quantity discount scheme (IQDS)'. This concept is graphically shown in Figure 3.10.

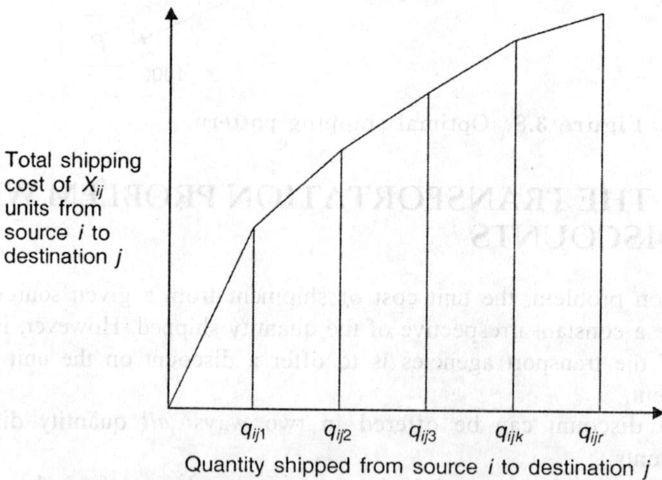

Figure 3.10 Incremental quantity discount scheme (IQDS).

A generalized representation of the intervals of quantity discounts of the transportation problem in a tabular form is shown in Table 3.84.

Table 3.84 Generalized Format

Destination

Source	1	2	j	n	
1	$0 \le X_{111} < q_{111}{:}p_{111}$ $q_{111} \le X_{112} < q_{112}{:}p_{112}$ $\vdots$ $q_{11(r-1)} \le X_{11r} < \infty{:}p_{11r}$	$0 \le X_{121} < q_{121}{:}p_{121}$ $q_{121} \le X_{122} < p_{122}{:}p_{122}$ $\vdots$ $q_{12(r-1)} \le X_{12r} < \infty{:}p_{12r}$	$\cdots$	$0 \le X_{1n1} < q_{1n1}{:}p_{1n1}$ $q_{1n1} \le X_{1n2} < q_{1n2}{:}p_{1n2}$ $\vdots$ $q_{1n(r-1)} \le X_{1nr} < \infty{:}p_{1nr}$	a_1
2	$0 \le X_{211} < q_{211}{:}p_{211}$ $q_{211} \le X_{212} < q_{212}{:}p_{212}$ $\vdots$ $q_{21(r-1)} \le X_{21r} < \infty{:}p_{21r}$	$0 \le X_{221} < q_{221}{:}p_{221}$ $q_{221} \le X_{222} < q_{222}{:}p_{222}$ $\vdots$ $q_{22(r-1)} \le X_{22r} < \infty{:}p_{22r}$	$\cdots$	$0 \le X_{2n1} < q_{2n1}{:}p_{2n1}$ $q_{2n1} \le X_{2n2} < q_{2n2}{:}p_{2n2}$ $\vdots$ $q_{2n(r-1)} \le X_{2nr} < \infty{:}p_{2nr}$	a_2
i	$\vdots$	$\vdots$	$\cdots$	$\vdots$	a_i
m	$0 \le X_{m11} < q_{m11}{:}p_{m11}$ $q_{m11} \le X_{m12} < q_{m12}{:}p_{m12}$ $\vdots$ $q_{m1(r-1)} \le X_{m1r} < \infty{:}p_{m1r}$	$0 \le X_{m21} < q_{m21}{:}p_{m21}$ $q_{m21} \le X_{m22} < q_{m22}{:}p_{m22}$ $\vdots$ $q_{m2(r-1)} \le X_{m2r} < \infty{:}p_{m2r}$	$\cdots$	$0 \le X_{mn1} < q_{mn1}{:}p_{mn1}$ $q_{mn1} \le X_{mn2} < q_{mn2}{:}p_{mn2}$ $\vdots$ $q_{mn(r-1)} \le X_{mnr} < \infty{:}p_{mnr}$	a_m
	b_1	b_2	b_j	b_n	

Here, a_i be the capacity of source i, $i = 1, 2, 3\ldots, m$; b_j, the demand of the destination j, $j = 1, 2, 3\ldots, n$; r, the total number of price breaks in any given combination of source and destination; c_{ijk}, the cost/unit of shipping from the source i to destination j under the kth price break, $k = 1, 2, 3,\ldots, r$; q_{ijr}, the upper bound in the last price break in any given cell can be either finite or infinite; X_{ijk}, the number of units to be shipped from the source i to the destination j under the kth price break; and, p_{ijk} is the price per unit of transportation for the kth price break from the source i to the destination j.

3.6.1 Model for AQDS

As per these guidelines, a model to minimize the total cost of shipment under AQDS, developed by Panneerselvam et al. (1991), is presented below:

Let

$$Y_{ijk} = 1, \qquad \text{if } X_{ijk} > 0$$
$$= 0, \qquad \text{otherwise}$$

$$\text{Minimize } Z = \sum_{i=1}^{m}\sum_{j=1}^{n}\sum_{k=1}^{r} C_{ijk} X_{ijk}$$

subject to

$$\sum_{j=1}^{n}\sum_{k=1}^{r} X_{ijk} = a_i, \qquad i = 1, 2,..., m \tag{1}$$

$$\sum_{i=1}^{m}\sum_{k=1}^{r} X_{ijk} = b_j, \qquad j = 1, 2,..., n \tag{2}$$

$$\sum_{k=1}^{r} Y_{ijk} \leq 1, \qquad i = 1, 2,..., m, \quad j = 1, 2, 3,..., n \tag{3}$$

$$X_{ijk} \leq (q_{ijk} - 1)Y_{ijk}, \; i = 1, 2,..., m, j = 1, 2,..., n \text{ and } k = 1, 2,..., r \tag{4}$$

$$X_{ijk} \geq q_{ij(k-1)}Y_{ijk}, \; i = 1, 2,..., m, j = 1, 2,..., n \text{ and } k = 2, 3,..., r \tag{5}$$

where $X_{ijk} \geq 0$ and $Y_{ijk} = 0$ or $1, i = 1, 2,..., m; j = 1, 2,..., n; k = 1, 2,..., r$.

In this model, the objective is to minimize the total cost of shipping under AQDS.

The constraint set 1 and constraint set 2 are same as the supply and demand constraint sets in the classical transportation model.

The constraint set 3 makes sure that the allocation is made under one and only price break within any given combination of the source i and the destination j.

The constraint set 4 limits the allocation of units under any price break to the respective upper bound within any given combination of the source i and the destination j.

The constraint set 5 assures that the units allocated under any price break is more than or equal to the respective lower bound within any given combination of the source i and the destination j.

3.6.2 Model for IQDS

The model for IQDS is presented in this section.

$$\text{Minimize } Z = \sum_{i=1}^{m}\sum_{j=1}^{n}\sum_{k=1}^{r} C_{ijk} X_{ijk}$$

subject to

$$\sum_{j=1}^{n}\sum_{k=1}^{r} X_{ijk} = a_i, \qquad i = 1, 2,..., m \tag{1}$$

$$\sum_{i=1}^{m}\sum_{k=1}^{r} X_{ijk} = b_j, \qquad j = 1, 2,..., n \tag{2}$$

$$X_{ij1} \leq (q_{ij1} - 1)Y_{ij1}, \qquad i = 1, 2,..., m, \qquad j = 1, 2,..., n \tag{3a}$$

$$X_{ijk} \leq [q_{ijk} - q_{ij(k-1)}]Y_{ijk}, \qquad i = 1, 2..., m, \qquad j = 1, 2,..., n, k = 2, 3,..., r \tag{3b}$$

$$X_{ij1} \geq (q_{ij1} - 1)Y_{ij2}, \qquad i = 1, 2,... m, \qquad j = 1, 2,..., n \tag{4a}$$

$$X_{ijk} \geq [q_{ijk} - q_{ij(k-1)}]Y_{ij(k+1)}, \qquad i = 1, 2,..., m, \qquad j = 1, 2,..., n, k = 2, 3,..., (r-1) \tag{4b}$$

$$X_{ijk} \geq 0$$

$$Y_{ijk} = 0 \text{ or } 1 \qquad i = 1, 2,..., m, \qquad j = 1, 2,..., n, k = 1, 2,..., r \tag{5}$$

In this model, the objective is to minimize the total cost of shipping under IQDS.

The constraint set 1 and constraint set 2 are same as the supply and demand constraints in the classical transportation model.

The constraint sets (3a) and (3b) limit the allocation of units in any cell to its upper bound on the incremental quantity within any given combination of the source i and destination j.

The constraint sets (4a) and (4b) assure the following:

For a given source i and destination j, if the allocation is made with respect to the $(k + 1)$th price break $(k \geq 2)$, then the allocation with respect to the kth price break must be equal to the respective upper bound on the incremental quantity, i.e. $[q_{ijk} - q_{ij(k-1)}]$.

Example 3.13 Consider a problem with two sources and two destinations. Assume that the number of price breaks is equal to three for each combination of the sources and destinations. The corresponding data is shown in Table 3.85.

Table 3.85 Data

Destination

		1	2	Supply
		$0 \leq X_{111} < 10{:}5$	$0 \leq X_{121} < 5{:}8$	
	1	$10 \leq X_{112} < 20{:}4$	$5 \leq X_{122} < 10{:}7$	45
		$20 \leq X_{113} < \infty{:}3$	$10 \leq X_{123} < \infty{:}6$	
Source				
		$0 \leq X_{211} < 20{:}8$	$0 \leq X_{221} < 5{:}10$	
	2	$20 \leq X_{212} < 65{:}6$	$5 \leq X_{222} < 10{:}9$	45
		$65 \leq X_{213} < \infty{:}5$	$10 \leq X_{223} < \infty{:}8$	
Demand		60	30	90/90

Solution (a) Solution for AQDS:

The model for AQDS is presented below and the corresponding optimal result is shown in Table 3.86. The total cost of the optimal shipping plan is Rs. 475.

Model formulation for AQDS:

$$\text{Minimize } Z = 5X_{111} + 4X_{112} + 3X_{113} + 8X_{121} + 7X_{122} + 6X_{123} + 8X_{211}$$
$$+ 6X_{212} + 5X_{213} + 10X_{221} + 9X_{222} + 8X_{223}$$

subject to

$$X_{111} + X_{112} + X_{113} + X_{121} + X_{122} + X_{123} = 45$$
$$X_{211} + X_{212} + X_{213} + X_{221} + X_{222} + X_{223} = 45$$
$$X_{111} + X_{112} + X_{113} + X_{211} + X_{212} + X_{213} = 60$$
$$X_{121} + X_{122} + X_{123} + X_{221} + X_{222} + X_{223} = 30$$
$$Y_{111} + Y_{112} + Y_{113} \leq 1$$
$$Y_{121} + Y_{122} + Y_{123} \leq 1$$
$$Y_{211} + Y_{212} + Y_{223} \leq 1$$

$$Y_{221} + Y_{222} + Y_{223} \leq 1$$
$$-9Y_{111} + X_{111} \leq 0$$
$$-19Y_{112} + X_{112} \leq 0$$
$$-1000Y_{113} + X_{113} \leq 0$$
$$-4Y_{121} + X_{121} \leq 0$$
$$-9Y_{122} + X_{122} \leq 0$$
$$-1000Y_{123} + X_{123} \leq 0$$
$$-19Y_{211} + X_{211} \leq 0$$
$$-64Y_{212} + X_{212} \leq 0$$
$$-1000Y_{213} + X_{213} \leq 0$$
$$-4Y_{221} + X_{221} \leq 0$$
$$-9Y_{222} + X_{222} \leq 0$$
$$-1000Y_{223} + X_{223} \leq 0$$
$$-10Y_{112} + X_{112} \geq 0$$
$$-20Y_{113} + X_{113} \geq 0$$
$$-5Y_{122} + X_{122} \geq 0$$
$$-10Y_{123} + X_{123} \geq 0$$
$$-20Y_{212} + X_{212} \geq 0$$
$$-65Y_{213} + Y_{213} \geq 0$$
$$-5Y_{222} + X_{222} \geq 0$$
$$-10Y_{223} + X_{223} \geq 0$$

where $X_{ijk} \geq 0$ and $Y_{ijk} = 0$ or   1 for $i = 1, 2$, $j = 1, 2$ and $k = 1, 2, 3$.

Table 3.86   Optimal Shipping Strategy under AQDS

		Destination		
		1	2	Supply
Source	1	40	5	45
	2	20	25	45
Demand		60	30	90

(b) Solution for IQDS:

The model for IQDS is presented below and its optimal solution is presented in Table 3.87. The total cost of the optimal shipping plan is Rs. 536.

Model formulation for IQDS:

$$\text{Minimize } Z = 5X_{111} + 4X_{112} + 3X_{113} + 8X_{121} + 7X_{122} + 6X_{123} + 8X_{211}$$
$$+ 6X_{212} + 5X_{213} + 10X_{221} + 9X_{222} + 8X_{223}$$

subject to

$$X_{111} + X_{112} + X_{113} + X_{121} + X_{122} + X_{123} = 45$$
$$X_{211} + X_{212} + X_{213} + X_{221} + X_{222} + X_{223} = 45$$
$$X_{111} + X_{112} + X_{113} + X_{211} + X_{212} + X_{213} = 60$$
$$X_{121} + X_{122} + X_{123} + X_{221} + X_{222} + X_{223} = 30$$
$$-9Y_{111} + X_{111} \leq 0$$
$$-10Y_{112} + X_{112} \leq 0$$
$$-1000Y_{113} + X_{113} \leq 0$$
$$-4Y_{121} + X_{121} \leq 0$$
$$-5Y_{122} + X_{122} \leq 0$$
$$-1000Y_{123} + X_{123} \leq 0$$
$$-19Y_{211} + X_{211} \leq 0$$
$$-45Y_{212} + X_{212} \leq 0$$
$$-1000Y_{213} + X_{213} \leq 0$$
$$-4Y_{221} + X_{221} \leq 0$$
$$-5Y_{222} + X_{222} \leq 0$$
$$-1000Y_{223} + X_{223} \leq 0$$
$$-9Y_{112} + X_{111} \geq 0$$
$$-10Y_{113} + X_{112} \geq 0$$
$$-4Y_{122} + X_{121} \geq 0$$
$$-5Y_{123} + X_{122} \geq 0$$
$$-19Y_{212} + X_{211} \geq 0$$
$$-45Y_{213} + X_{212} \geq 0$$
$$-4Y_{222} + X_{221} \geq 0$$
$$-5Y_{223} + X_{222} \geq 0$$

where $X_{ijk} \geq 0$ and $Y_{ijk} = 0$ or   1 for $i = 1, 2, j = 1, 2$ and $k = 1, 2, 3$.

Table 3.87 Optimal Shipping Strategy under IQDS

<table>
<tr><td></td><td></td><td colspan="2" align="center">Destination</td><td></td></tr>
<tr><td></td><td></td><td align="center">1</td><td align="center">2</td><td>Supply</td></tr>
<tr><td rowspan="2">Source</td><td>1</td><td>45</td><td>–</td><td>45</td></tr>
<tr><td>2</td><td>15</td><td>30</td><td>45</td></tr>
<tr><td></td><td>Demand</td><td>60</td><td>30</td><td>90</td></tr>
</table>

QUESTIONS

1. Give different practical applications of transportation problem.

2. What are types of transportation problem? Explain them with suitable examples.

3. Write a linear programming model of the transportation problem.

4. Write the procedure for each of the following:

   (a) Northwest corner cell method
   (b) Least cost cell method
   (c) Vogel's approximation method
   (d) *U-V* method.

5. Determine an initial basic feasible solution to the following transportation problem using northwest corner cell method.

<table>
<tr><td rowspan="2"></td><td rowspan="2"></td><td colspan="5" align="center">To</td><td rowspan="2">Availability</td></tr>
<tr><td>1</td><td>2</td><td>3</td><td>4</td><td>5</td></tr>
<tr><td rowspan="4">From</td><td>1</td><td>3</td><td>4</td><td>6</td><td>8</td><td>9</td><td>20</td></tr>
<tr><td>2</td><td>2</td><td>10</td><td>1</td><td>5</td><td>8</td><td>30</td></tr>
<tr><td>3</td><td>7</td><td>11</td><td>20</td><td>40</td><td>3</td><td>15</td></tr>
<tr><td>4</td><td>2</td><td>1</td><td>9</td><td>14</td><td>16</td><td>13</td></tr>
<tr><td>Demand</td><td></td><td>40</td><td>6</td><td>8</td><td>18</td><td>6</td><td></td></tr>
</table>

6. Find the initial basic feasible solution to the following transportation problem by

   (a) Northwest corner cell method and
   (b) Least cost cell method.

<table>
<tr><td rowspan="2"></td><td rowspan="2"></td><td colspan="3" align="center">To</td><td rowspan="2">Supply</td></tr>
<tr><td>1</td><td>2</td><td>3</td></tr>
<tr><td rowspan="4">From</td><td>1</td><td>2</td><td>7</td><td>4</td><td>5</td></tr>
<tr><td>2</td><td>3</td><td>3</td><td>1</td><td>8</td></tr>
<tr><td>3</td><td>5</td><td>4</td><td>7</td><td>7</td></tr>
<tr><td>4</td><td>1</td><td>6</td><td>2</td><td>14</td></tr>
<tr><td>Demand</td><td></td><td>2</td><td>9</td><td>18</td><td></td></tr>
</table>

   State which of the methods is better.

7. Find the initial basic feasible solution of the following transportation problem by Vogel's approximation method:

<table>
<tr><td rowspan="2"></td><td rowspan="2"></td><td colspan="4" align="center">Warehouses</td><td rowspan="2">Capacity</td></tr>
<tr><td>W_1</td><td>W_2</td><td>W_3</td><td>W_4</td></tr>
<tr><td rowspan="3">Factory</td><td>F_1</td><td>10</td><td>30</td><td>50</td><td>10</td><td>7</td></tr>
<tr><td>F_2</td><td>70</td><td>30</td><td>40</td><td>60</td><td>9</td></tr>
<tr><td>F_3</td><td>40</td><td>8</td><td>70</td><td>20</td><td>18</td></tr>
<tr><td>Requirement</td><td></td><td>5</td><td>8</td><td>7</td><td>14</td><td>34</td></tr>
</table>

8. Determine an initial basic feasible solution to the following transportation problem using:

(a) Northwest corner cell method and

(b) Vogel's approximation method.

Destination

Origin	A_1	B_1	C_1	D_1	E_1	Supply
A	2	11	10	3	7	4
B	1	4	7	2	1	8
C	3	9	4	8	12	9
Demand	3	3	4	5	6	

9. A manufacturing company has three factories F_1, F_2 and F_3 with monthly manufacturing capacities of 7000, 4000 and 10,000 units of a product. The product is to be supplied to seven stores. The manufacturing costs in these factories are slightly different but the important factor is the shipping cost from each factory to a particular store. The following table represents the factory capacities, store requirements and unit cost (in rupees) of shipping from each factory to each store. Here, slack is the difference between the total factory capacity and the total requirement.

Factory	S_1	S_2	S_3	S_4	S_5	S_6	S_7	Factory capacity
F_1	5	6	4	3	7	5	4	7000
F_2	9	4	3	4	3	2	1	4000
F_3	8	4	2	5	4	8	3	10,000
Store demand	1500	2000	4500	4000	2500	3500	3000	

Find the optimal transportation plan so as to minimize the transportation cost.

10. A company has received a contract to supply gravel for three new construction projects located in towns A, B and C. Construction engineers have estimated the required amounts of gravel which will be needed at these construction projects as shown below:

Project location	Weekly requirement (truck loads)
A	72
B	102
C	41

The company has three gravel plants X, Y and Z located in three different towns. The gravel required by the construction projects can be supplied by these three plants. The amount of gravel which can be supplied by each plant is as follows:

Plant	Amount available/week (truck loads)
X	76
Y	62
Z	77

The company has computed the delivery cost from each plant to each project site. These costs (in rupees) are shown in the following table:

Cost per truck load

Plant	A	B	C
X	4	8	8
Y	16	24	16
Z	8	16	35

(a) Schedule the shipment from each plant to each project in such a manner so as to minimize the total transportation cost within the constraints imposed by plant capacities and project requirements.

(b) Find the minimum cost.

(c) Is the solution unique? If it is not, find alternative schedule with the same minimum cost.

11. A company has factories at four different places (1, 2, 3 and 4) which supply items to warehouses *A, B, C, D* and *E*. Monthly factory capacities are 200, 175, 150 and 325, respectively. Monthly warehouse requirements are 110, 90, 120, 230 and 160, respectively. Unit shipping costs (in rupees) are given in the following table:

To

	A	B	C	D	E
1	13	–	31	8	20
2	14	9	17	26	10
3	25	11	12	17	15
4	10	21	13	–	17

Shipments from 1 to *B* and from 4 to *D* are not possible. Determine the optimum distribution plan to minimize the shipping cost.

12. A company has plants at *A, B* and *C* which have capacities to produce 300 kg, 200 kg and 500 kg respectively of a particular chemical per day. The production costs (per kg) in these plants are Rs. 70, Rs. 60 and Rs. 66, respectively. Four bulk consumers have placed orders for the product on the following basis:

Consumer	kg required per day	Price offered (Rs./kg)
I	400	100
II	250	100
III	350	102
IV	150	103

Shipping costs (in rupees per kg) from plants to consumers are given in the table below:

		To		
	I	II	III	IV
A	3	5	4	6
From B	8	11	9	12
C	4	6	2	8

Find the optimal schedule for the above situation.

13. Distinguish between transportation problem and transshipment problem. What are the types of transshipment problem? Explain them with suitable sketches.

14. Consider the following transshipment problem involving 4 sources and 2 destinations. The supply values of the sources S_1, S_2, S_3 and S_4 are 200 units, 250 units, 200 units and 450 units, respectively. The demand values of the destinations D_1 and D_2 are 550 units and 550 units, respectively. The transportation cost per unit between different sources and destinations are summarized in the following table. Solve the transshipment problem.

		Destination					
		S_1	S_2	S_3	S_4	D_1	D_2
	S_1	0	6	24	7	24	10
	S_2	10	0	6	12	5	20
	S_3	15	20	0	8	45	7
Source	S_4	18	25	10	0	30	6
	D_1	15	20	60	15	0	10
	D_2	10	25	25	23	4	0

15. Consider the following transshipment problem with two sources and three destinations. The unit cost of transportation between different possible nodes is given in the following table. Find the optimal shipping plan such that the total cost is minimized.

		Destination					
		S_1	S_2	D_1	D_2	D_3	Supply
	S_1	0	3	12	4	12	800
	S_2	5	0	3	6	10	700
Source	D_1	8	10	0	4	20	–
	D_2	20	12	5	0	15	–
	D_3	8 ·	10	30	8	0	–
Demand		–	–	500	400	600	

16. A multi-plant organization has three plants (*A, B* and *C*) and three market places (*X, Y* and *Z*). The items from the plants are transported to the market places through two intermediate finished goods warehouses, W_1 and W_2. The details on cost of transportation per unit for different combinations between the plants and warehouses, between warehouses and markets, between warehouses, supply values of the plants and demand values of the markets are summarized in the following table. The c_{ij} values between the plants and markets are assumed as infinity. (While solving this problem using computer, instead of infinity, a very large value may be assumed.) Find the optimal shipping plan such that total cost of transportation is minimized.

		Terminal nodes				Supply
	X	Y	Z	W_1	W_2	
A	∞	∞	∞	25	40	400
B	∞	∞	∞	38	20	500
Starting nodes C	∞	∞	∞	40	25	600
W_1	20	45	25	0	25	–
W_2	30	20	40	40	0	–
Demand	300	700	500	–	–	

17. Discuss the modelling of AQDS problem of transportation problem.

18. Discuss the modelling of IQDS problem of transportation problem.

ASSIGNMENT PROBLEM

4

4.1 INTRODUCTION

Assignment problem is a special kind of transportation problem in which each source should have the capacity to fulfil the demand of any of the destinations. In other words, any operator should be able to perform any job regardless of his skills, although the cost will be more if the job does not match with the worker's skill. An example of assigning operators to jobs in a shop floor situation is shown as in Table 4.1.

Table 4.1 Generalized Format of Assignment Problem

		Operator				
		1	2	...	j	...	m
Job	1	t_{11}	t_{12}	...	t_{1j}	...	t_{1m}
	2	t_{21}	t_{22}	...	t_{2j}	...	t_{2m}
	i	t_{i1}	t_{i2}	...	t_{ij}	...	t_{im}
	m	t_{m1}	t_{m2}	...	t_{mj}	...	t_{mm}

In Table 4.1, m be the number of jobs as well as the number of operators, and t_{ij} be the processing time of the job i if it is assigned to the operator j. Here, the objective is to assign the jobs to the operators such that the total processing time is minimized. Table 4.2 summarizes different examples (applications) of the assignment problem.

Table 4.2 Examples of Assignment Problem

Row entity	Column entity	Cell entry
Jobs	Operators	Processing time
Operators	Machines	Processing time
Teachers	Subjects	Students pass percentage
Drivers of company vehicles	Routes	Travel time
Physicians	Treatments	Number of cases handled

4.2 ZERO-ONE PROGRAMMING MODEL FOR ASSIGNMENT PROBLEM

A zero-one programming model for the assignment problem is presented below:

Minimize $Z = C_{11}X_{11} + C_{12}X_{12} + \cdots + C_{1m}X_{1m} + C_{21}X_{21} + C_{22}X_{22} + \cdots + C_{2m}X_{2m} + \cdots + C_{i1}X_{i1}$
$\qquad + C_{i2}X_{i2} + \cdots + C_{im}X_{im} + \cdots + C_{m1}X_{m1} + C_{m2}X_{m2} + \cdots + C_{mm}X_{mm}$

subject to

$$X_{11} + X_{12} + \cdots + X_{1j} + \cdots + X_{1m} = 1$$
$$X_{21} + X_{22} + \cdots + X_{2j} + \cdots + X_{2m} = 1$$
$$\vdots \qquad \vdots \qquad \qquad \vdots \qquad \qquad \vdots$$
$$X_{i1} + X_{i2} + \cdots + X_{ij} + \cdots + X_{im} = 1$$
$$\vdots \qquad \vdots \qquad \qquad \vdots \qquad \qquad \vdots$$
$$X_{m1} + X_{m2} + \cdots + X_{mj} + \cdots + X_{mm} = 1$$
$$X_{11} + X_{21} + \cdots + X_{i1} + \cdots + X_{m1} = 1$$
$$X_{12} + X_{22} + \cdots + X_{i2} + \cdots + X_{m2} = 1$$
$$\vdots \qquad \vdots \qquad \qquad \vdots \qquad \qquad \vdots$$
$$X_{1j} + X_{2j} + \cdots + X_{ij} + \cdots + X_{mj} = 1$$
$$\vdots \qquad \vdots \qquad \qquad \vdots \qquad \qquad \vdots$$
$$X_{1m} + X_{2m} + \cdots + X_{im} + \cdots + X_{mm} = 1$$
$$X_{ij} = 0 \text{ or } 1, \text{ for } i = 1, 2,\ldots, m \text{ and}$$
$$j = 1, 2,\ldots, m$$

The above model is presented in a short form as:

$$\text{Minimize } Z = \sum_{i=1}^{m} \sum_{j=1}^{m} C_{ij}X_{ij}$$

subject to

$$\sum_{j=1}^{m} X_{ij} = 1, \, i = 1, 2, 3,\ldots, m$$

and

$$\sum_{i=1}^{m} X_{ij} = 1, \, j = 1, 2, 3,\ldots, m$$

where $X_{ij} = 0$ or 1 for $i = 1, 2, 3,\ldots, m$ and $j = 1, 2, 3,\ldots, m$, m being the number of rows (jobs) as well as the number of columns (operators) and C_{ij} the time/cost (processing time, travel time, etc.) of assigning the row i to the column j. Thus,

$$X_{ij} = 1, \text{ if the row } i \text{ is assigned to the column } j$$
$$= 0, \text{ otherwise.}$$

In this model, the objective function minimizes the total cost of assigning the rows to the columns. The first set of constraints ensures that each row (job) is assigned to only one column (operator). The second set of constraints ensures that each column (operator) is assigned to only one row (job).

Example 4.1 Consider the assignment problem as shown in Table 4.3. In this problem, 5 different jobs are to be assigned to 5 different operators such that the total processing time is minimized. The matrix entries represent processing times in hours.

Table 4.3   Example 4.1

Operator

		1	2	3	4	5
	1	10	12	15	12	8
	2	7	16	14	14	11
Job	3	13	14	7	9	9
	4	12	10	11	13	10
	5	8	13	15	11	15

Develop a zero–one programming model.

Solution   Let

$$X_{ij} = 1, \quad \text{if the job } i \text{ is assigned to the operator } j$$
$$= 0, \quad \text{otherwise.}$$

A zero–one programming model for the assignment problem to minimize the total processing time is presented below:

$$\text{Minimize } Z = 10X_{11} + 12X_{12} + 15X_{13} + 12X_{14} + 8X_{15}$$
$$+ 7X_{21} + 16X_{22} + 14X_{23} + 14X_{24} + 11X_{25}$$
$$+ 13X_{31} + 14X_{32} + 7X_{33} + 9X_{34} + 9X_{35}$$
$$+ 12X_{41} + 10X_{42} + 11X_{43} + 13X_{44} + 10X_{45}$$
$$+ 8X_{51} + 13X_{52} + 15X_{53} + 11X_{54} + 15X_{55}$$

subject to

$$X_{11} + X_{12} + X_{13} + X_{14} + X_{15} = 1$$
$$X_{21} + X_{22} + X_{23} + X_{24} + X_{25} = 1$$
$$X_{31} + X_{32} + X_{33} + X_{34} + X_{35} = 1$$
$$X_{41} + X_{42} + X_{43} + X_{44} + X_{45} = 1$$
$$X_{51} + X_{52} + X_{53} + X_{54} + X_{55} = 1$$
$$X_{11} + X_{21} + X_{31} + X_{41} + X_{51} = 1$$
$$X_{12} + X_{22} + X_{32} + X_{42} + X_{52} = 1$$
$$X_{13} + X_{23} + X_{33} + X_{43} + X_{53} = 1$$
$$X_{14} + X_{24} + X_{34} + X_{44} + X_{54} = 1$$
$$X_{15} + X_{25} + X_{35} + X_{45} + X_{55} = 1$$

$$X_{ij} = 0 \text{ or } 1, \quad i = 1, 2, 3, 4, 5 \quad \text{and} \quad j = 1, 2, 3, 4, 5$$

The optimal solution of the above model is presented in Table 4.4.

Table 4.4 Optimal Solution for Example 4.1

Operator

		1	2	3	4	5
	1	10	12	15	12	$X_{15}=1$ 8
	2	$X_{21}=1$ 7	16	14	14	11
Job	3	13	14	$X_{33}=1$ 7	9	9
	4	12	$X_{42}=1$ 10	11	13	10
	5	8	13	15	$X_{54}=1$ 11	15

The corresponding total processing time is 43 hours.

4.3 TYPES OF ASSIGNMENT PROBLEM

The assignment problem is classified into balanced assignment problem and unbalanced assignment problem. If the number of rows (jobs) is equal to the number of columns (operators), then the problem is termed as a balanced assignment problem; otherwise, an unbalanced assignment problem.

If the problem is unbalanced, like an unbalanced transportation problem, then necessary number of dummy row(s)/column(s) are added such that the cost matrix is a square matrix. The values for the entries in the dummy row(s)/column(s) are assumed to be zero. Under such a condition, while implementing the solution, the dummy row(s) or job(s) (or the dummy column(s) or operator(s)) will not have assignment(s).

4.4 HUNGERIAN METHOD

An assignment problem can be easily solved by applying Hungerian method which consists of two phases. In the first phase, row reductions and column reductions are carried out. In the second phase, the solution is optimized on iterative basis.

Phase 1: Row and column reductions

Step 0: Consider the given cost matrix.

Step 1: Obtain the next matrix by subtracting the minimum value of each row from the entries of that row.

Step 2: Obtain the next matrix by subtracting the minimum value of each column from the entries of that column. Now, treat this matrix as the input for phase 2.

Phase 2: Optimation of the problem

Step 3: Draw a minimum number of lines to cover all the zeros of the matrix. The procedure for drawing minimum number of lines involves the following steps:

3.1 Row scanning

1. Starting from the first row, ask the following question. Is there exactly one zero in that row? If yes, mark a square around that zero entry and draw a vertical line passing through that zero; otherwise skip that row.
2. After scanning the last row, check whether all the zeros are covered with lines. If yes, go to step 4; otherwise, do column scanning (i.e. go to step 3.2).

3.2 Column scanning

1. Starting from the first column, ask the following question: Is there exactly one zero in that column? If yes, mark a square around that zero entry and draw a horizontal line passing through that zero; otherwise skip that column.
2. After scanning the last column, check whether all the zeros are covered with lines. If yes, go to step 4; Otherwise, do row scanning (i.e. go to step 3.1).

Step 4: Check whether the number of squares marked is equal to the number of rows of the matrix. If yes, go to step 7; otherwise, go to step 5.

Step 5: Identify the minimum value of the undeleted cell values. Obtain the next matrix by following the steps mentioned below.

5.1 Copy the entries on the lines but not on the intersection points of the present matrix as such without any modification to the corresponding positions of the next matrix.

5.2 Copy the entries at the intersection points of the present matrix after adding the minimum undeleted cell value to the corresponding positions of the next matrix.

5.3 Subtract the minimum undeleted cell value from all the undeleted cell values and then copy them to the corresponding positions of the next matrix.

Step 6: Go to step 3.

Step 7: Treat the solution as marked by the squares as the optimal solution.

Note: While performing step 3, sometimes it will repeat endlessly when the number of zeros in the applicable rows as well as columns is more than one. Under such a situation, one should mark squares on diagonally opposite cells having zeros. This means multiple optimal solutions exist.

Example 4.2 Solve Example 4.1 using Hungerian method. The matrix entries represent the processing times in hours.

Solution In this phase, row reductions and column reductions are carried out as shown in Tables 4.5 through 4.7.

Table 4.5 Matrix Showing the Minimum Value of Each Row

	Operator					Row minimum
	1	2	3	4	5	
1	10	12	15	12	8	8
2	7	16	14	14	11	7
Job 3	13	14	7	9	9	7
4	12	10	11	13	10	10
5	8	13	15	11	15	8

The row reductions are carried out as shown in Table 4.6.

Table 4.6 Matrix after Row Reductions

	Operator				
	1	2	3	4	5
1	2	4	7	4	0
2	0	9	7	7	4
Job 3	6	7	0	2	2
4	2	0	1	3	0
5	0	5	7	3	7
Column minimum	0	0	0	2	0

The column reductions are carried out as shown in Table 4.7

Table 4.7 Matrix after Column Reductions

	Operator				
	1	2	3	4	5
1	2	4	7	2	0
2	0	9	7	5	4
Job 3	6	7	0	0	2
4	2	0	1	1	0
5	0	5	7	1	7

In the next phase, the optimum solution is obtained in an interactive manner.

Iteration 1: Table 4.8 shows the minimum required number of lines which are drawn to cover all the zeros.

Table 4.8 Matrix with Minimum Number of Lines (Iteration 1)

Operator

	1	2	3	4	5
1	2	4	7	2	[0]
2	[0]	9	7	5	4
Job 3	6	7	[0]	0	2
4	2	[0]	1	1	0
5	0	5	7	1	7

In Table 4.8, the number of squares marked is 4 which is not equal to the number of rows (5). Hence, the solution is not feasible and optimal.

Iteration 2: The minimum among the unselected entries of Table 4.8 is 1. The entries of Table 4.9 are obtained from Table 4.8 by applying step 5. Table 4.9 also shows the minimum required number of lines which are drawn to cover all the zeros. In Table 4.9, the number of squares marked is 5 which is equal to the number of rows of the matrix. Hence, the solution is optimal and feasible. The corresponding solution is summarized in Table 4.10.

Table 4.9 Matrix with Minimum Number of Lines (Iteration 2)

Operator

	1	2	3	4	5
1	2	4	6	1	[0]
2	[0]	9	6	4	4
Job 3	7	8	[0]	0	3
4	2	[0]	0	0	0
5	0	5	6	[0]	7

Table 4.10 Optimal Solution

Job	Operator	Time
1	5	8
2	1	7
3	3	7
4	2	10
5	4	11

Total processing time = 43 hours

Example 4.3 A college is having a degree programme for which the effective semester time available is very less and the programme requires field work. Hence, a few hours can be saved from the total number of class hours, and can be utilized for the field work. Based on past experience, the college has estimated the number of hours required to teach each subject by each faculty. The course in its present semester has 5 subjects and the college has considered 6 existing faculty members to teach these courses. The objective is to assign the best 5 teachers out of these 6 faculty members to teach 5 different subjects so that the total number of class hours required is minimized. The data of this problem is summarized in Table 4.11. Solve this assignment problem optimally.

Table 4.11   Data for Example 4.3

| | | Subject | | | |
	1	2	3	4	5
1	30	39	31	38	40
2	43	37	32	35	38
3	34	41	33	41	34
4	39	36	43	32	36
5	32	49	35	40	37
6	36	42	35	44	42

(Faculty indexes the rows 1–6.)

Solution   In the given problem, the number of rows is not equal to the number of columns. Hence, it is an unbalanced assignment problem. So, this problem should be converted into a balanced assignment problem by introducing a dummy column with all zero cell entries as shown in Table 4.12.

Table 4.12   Modified Data for Example 4.3

| | | Subject | | | | | Row minimum |
	1	2	3	4	5	6	
1	30	39	31	38	40	0	0
2	43	37	32	35	38	0	0
3	34	41	33	41	34	0	0
4	39	36	43	32	36	0	0
5	32	49	35	40	37	0	0
6	36	42	35	44	42	0	0

(Faculty indexes the rows 1–6.)

Phase 1

Row reduction is carried out as shown in Table 4.13.

Table 4.13 Matrix after Row Reductions

Subject

	1	2	3	4	5	6
1	30	39	31	38	40	0
2	43	37	32	35	38	0
3	34	41	33	41	34	0
4	39	36	43	32	36	0
5	32	49	35	40	37	0
6	36	42	35	44	42	0
Column minimum	30	36	31	32	34	0

(Faculty labels rows 1–6)

The column reduction is carried out as shown in Table 4.14. The matrix in Table 4.14 is the input for the phase 2.

Table 4.14 Matrix after Column Reductions

Subject

	1	2	3	4	5	6
1	0	3	0	6	6	0
2	13	1	1	3	4	0
3	4	5	2	9	0	0
4	9	0	12	0	2	0
5	2	13	4	8	3	0
6	6	6	4	12	8	0

(Faculty labels rows 1–6)

Phase 2

Table 4.15 shows the minimum required number of lines which are drawn to cover all the zeros. The number of squares marked in Table 4.15 is 4 which is not equal to the number of rows. Hence, go to next iteration.

Table 4.15 Matrix with Minimum Number of Lines (Iteration 1)

Subject

	1	2	3	4	5	6
1	[0]	3	0	6	6	0
2	13	1	1	3	4	[0]
3	4	5	2	9	[0]	0
4	9	[0]	12	0	2	0
5	2	13	4	8	3	0
6	6	6	4	12	8	0

(Faculty labels rows 1–6)

The minimum among the undeleted entries in Table 4.15 is 1. The entries in Table 4.16 are obtained from Table 4.15 by applying step 5. Table 4.16 also shows the minimum required number of lines which are drawn to cover all the zeros.

In Table 4.16, the total number of cells marked with squares is 5 which is not equal to the number of rows of the matrix. Hence, go to the next iteration.

Table 4.16 Matrix with Minimum Number of Lines (Iteration 2)

Faculty \ Subject	1	2	3	4	5	6
1	[0]	3	0	6	7	1
2	12	0	[0]	2	4	0
3	3	4	1	8	[0]	0
4	9	0	12	[0]	3	1
5	1	12	3	7	3	[0]
6	5	5	3	11	8	0

The minimum among the unselected entries in Table 4.16 is 1. The entries in the Table 4.17 are obtained from Table 4.16 by applying step 5. Table 4.17 also shows the minimum required number of lines which are drawn to cover all the zeros.

Table 4.17 Matrix with Minimum Number of Lines (Iteration 3)

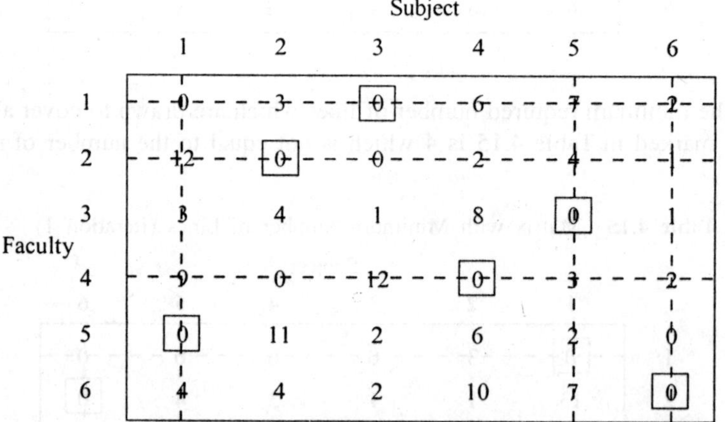

Faculty \ Subject	1	2	3	4	5	6
1	0	3	[0]	6	7	2
2	12	[0]	0	2	4	1
3	3	4	1	8	[0]	1
4	9	0	12	[0]	3	2
5	[0]	11	2	6	2	0
6	4	4	2	10	7	[0]

In Table 4.17, the total number of cells marked with squares is 6, which is equal to the number of rows of the square matrix. So, the solution of this iteration is feasible and optimal and the corresponding results are summarized in Table 4.18.

Table 4.18 Final Solution of Example 4.3

Faculty	Subject	Time
1	3	31
2	2	37
3	5	34
4	4	32
5	1	32
6	6 (Dummy)*	0

Total time is 166 hours, where faculty 6 is not assigned any subject.

Example 4.4 Consider the problem of assigning four sales persons to four different sales regions as shown in Table 4.19 such that the total sales is maximized.

Table 4.19 Data for Example 4.4

	Sales region			
	1	2	3	4
Salesman 1	10	22	12	14
Salesman 2	16	18	22	10
Salesman 3	24	20	12	18
Salesman 4	16	14	24	20

The cell entries represent annual sales figures in lakhs of rupees. Find the optimal allocation of the sales persons to different regions.

Solution In the conventional assignment problem, the objective is to minimize the total cost of assignment. But the problem of assigning sales persons to different sales regions has the aim of maximizing the total yearly sales. So, convert this maximization problem into the usual minimization problem by multiplying all the cell entries by -1. Then, apply the usual procedure of the assignment problem with the minimizing objective function.

Phase 1

The modified data in Example 4.4, after multiplying the cell entries of Table 4.19 with -1 is shown in Table 4.20.

Table 4.20 Modified Data for Example 4.4

	Sales region				Row minimum
	1	2	3	4	
Salesman 1	-10	-22	-12	-14	-22
Salesman 2	-16	-18	-22	-10	-22
Salesman 3	-24	-20	-12	-18	-24
Salesman 4	-16	-14	-24	-20	-24

The row reductions are carried out as shown in Table 4.21.

Table 4.21   Matrix after Row Reductions

	Sales region			
Salesman	1	2	3	4
1	12	0	10	8
2	6	4	0	12
3	0	4	12	6
4	8	10	0	4
Column minimum	0	0	0	4

The column reductions are carried out as shown in Table 4.22. The matrix in Table 4.22 is the input for Phase 2.

Table 4.22   Matrix after Column Reductions

	Sales region			
Salesman	1	2	3	4
1	12	0	10	4
2	6	4	0	8
3	0	4	12	2
4	8	10	0	0

Phase 2

Table 4.23 shows the minimum required number of lines which are drawn to cover all the zeros.

Table 4.23   Matrix with Minimum Number of Lines (Iteration 1)

	Sales region			
Salesman	1	2	3	4
1	12	[0]	10	4
2	6	4	[0]	8
3	[0]	4	12	2
4	8	10	0	[0]

In Table 4.23, the number of cells marked with squares is 4, which is equal to the number of rows of the matrix. Hence, the solution is feasible and optimal. The corresponding results are summarized as in Table 4.24.

Table 4.24   Final Solution of Example 4.4

Salesman	Sales region	Sales
1	2	22
2	3	22
3	1	24
4	4	20

Now total sales = Rs. 88 lakhs.

Example 4.5   The flight timings between two cities X and Y, are as given in Tables 4.25 and 4.26, respectively. The minimum layover time of any crew in either of the cities is 2 hours. Determine the base city for each crew so that the sum of the layover times of all the crew members in non-base cities is minimized.

Table 4.25   Flight Timings from City X to City Y

Flight number	Departure time (from City X)	Arrival time (to City Y)
101	5 a.m.	6.15 a.m.
102	9 a.m.	10.15 a.m.
103	1 p.m.	2.15 p.m.
104	6 p.m.	7.15 p.m.

Table 4.26   Flight Timings from City Y to City X

Flight number	Departure time (from City Y)	Arrival time (to City X)
201	6.30 a.m.	7.30 a.m.
202	9.00 a.m.	10.00 a.m.
203	3.30 p.m.	4.30 p.m.
204	10.00 p.m.	11.00 p.m.

Solution   A careful examination of the timings reveals that 15 minutes is the minimum integral time units for the purpose of calculation. As per this assumption, the layover times of the crew for different paring of flights from 101 series to 201 series as well as from 201 series to 101 series are shown in Tables 4.27 and 4.28, spectively.

Table 4.27   Layover Times of Crews in City Y

| | | Flight from City Y | | | |

		201	202	203	204
Flights from City X	101	97	11*	37*	63
	102	81*	91*	21*	47
	103	65	75	101	31*
	104	45	55	81*	11*

Table 4.28   Layover Times of Crews in City X

Flights from City Y

		201	202	203	204
	101	86*	76	50	24*
Flights from	102	102	92	66	40*
City X	103	22*	12*	82*	56
	104	42*	32*	102	76

Now, a new matrix showing the minimum layover times is generated as in Table 4.29 based on the data from Tables 4.27 and 4.28. Each cell entry in Table 4.29 is the minimum of the corresponding cell entries in Tables 4.27 and 4.28. Table 4.29 shows the inputs for the assignment method to determine pairing of different flights.

Table 4.29   Minimum Layover Times of the Crews

City Y

		201	202	203	204
	101	86	11	37	24
	102	81	91	21	40
City X	103	22	12	82	31
	104	42	32	81	11

Phase 1

The minimum of each row is shown in Table 4.30. The row reductions are carried out as shown in Table 4.31. The column reductions are carried out as shown in Table 4.32. The matrix in this table serves as the input for Phase 2.

Table 4.30   Minimum Layover Times of the Crews

		City Y				Row
		201	202	203	204	minimum
	101	86	11	37	24	11
	102	81	91	21	40	21
City X	103	22	12	82	31	12
	104	42	32	81	11	11

Table 4.31   Matrix after Row Reductions

City Y

		201	202	203	204
	101	75	0	26	13
City X	102	60	70	0	19
	103	10	0	70	19
	104	31	21	70	0
Column minimum		10	0	0	0

Table 4.32 Matrix after Column Reductions

City Y

	201	202	203	204
101	65	0	26	13
102	50	70	0	19
103	0	0	70	19
104	21	21	70	0

City X

Phase 2

Table 4.33 shows the minimum required number of lines which are drawn to cover all the zeros in Table 4.32.

Table 4.33 Matrix with Minimum Number of Lines (Iteration 1)

City Y

	201	202	203	204
101	65	0	26	13
102	50	70	0	19
103	0	0	70	19
104	21	21	70	0

City X

Since the Table 4.33 has four squares, the solution is feasible and optimal. The corresponding results are shown in Table 4.34. The optimal total layover time of all the crews is 16 hours 15 minutes or sixty five 15 minutes intervals.

Table 4.34 Optimal Results of Example 4.5

Paired flights			Layover time of crew members in the other city	
Flight number	Flight number	Base city for crew members	No. of 15 min. interval	Hours
101	202	X	11	2 hrs 45 min.
102	203	X	21	5 hrs 15 min.
201	103	Y	22	5 hrs 30 min.
104	204	X	11	2 hrs 45 min.

Example 4.6 Alpha Construction Company has five crews. The skills of the crews differ from one another because of the difference in the composition of the crews. The company has five different projects on hand. The times (in days) taken by different crews to complete different projects are summarized in Table 4.35. Find the best assignment of the crews to different projects such that the total time taken to complete all the projects is minimized.

Table 4.35 Project Execution Times in Days

Project

	A	B	C	D	E
1	20	30	25	15	35
2	25	10	40	12	28
Crew 3	15	18	22	32	24
4	29	8	34	10	40
5	35	23	17	26	45

Solution

Phase 1 The row minimums are shown in Table 4.36. The matrix after row reductions is shown in Table 4.37. In the same table, the column minimums are shown at its bottom. The matrix after column reductions is presented in Table 4.38.

Phase 2 The iterations of the second phase to draw minimum number of lines to cover all the zeros are presented from Table 4.39 to Table 4.41.

Table 4.36 Project Execution Times in Days

Project

	A	B	C	D	E	Row minimum
1	20	30	25	15	35	15
2	25	10	40	12	28	10
Crew 3	15	18	22	32	24	15
4	29	8	34	10	40	8
5	35	23	17	26	45	17

Table 4.37 Matrix after Row Reductions

Project

	A	B	C	D	E
1	5	15	10	0	20
2	15	0	30	2	18
Crew 3	0	3	7	17	9
4	21	0	26	2	32
5	18	6	0	9	28
Column minimum	0	0	0	0	9

Table 4.38 Matrix after Column Reductions

Project

Crew	A	B	C	D	E
1	5	15	10	0	11
2	15	0	30	2	9
3	0	3	7	17	0
4	21	0	26	2	23
5	18	6	0	9	19

Table 4.39 Iteration 1

Project

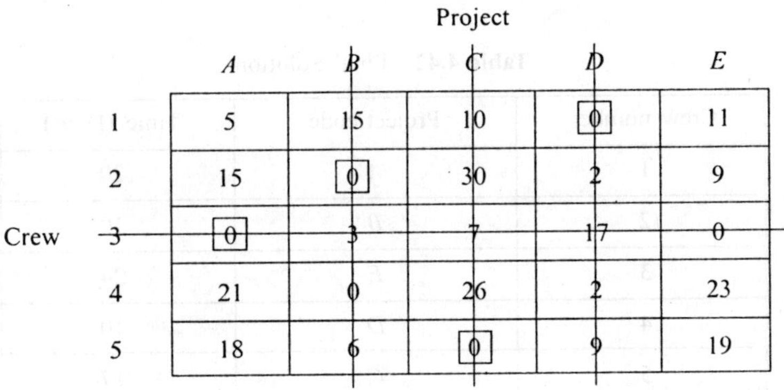

Crew	A	B	C	D	E
1	5	15	10	[0]	11
2	15	[0]	30	2	9
3	[0]	3	7	17	0
4	21	0	26	2	23
5	18	6	[0]	9	19

Table 4.40 Iteration 2

Project

Crew	A	B	C	D	E
1	0	15	10	[0]	6
2	10	[0]	30	2	4
3	0	8	12	22	[0]
4	16	0	26	2	18
5	13	6	[0]	9	14

Table 4.41 Iteration 3

Project

Crew	A	B	C	D	E
1	[0]	17	12	0	6
2	8	[0]	30	(0)	2
3	0	10	14	22	[0]
4	14	(0)	26	[0]	16
5	11	6	[0]	7	12

From Table 4.41, one can notice that the given problem has two feasible and optimal alternate solutions as shown in Table 4.42 and Table 4.43. The total time of completing all the projects is 81 days.

Table 4.42 Final Solution

Crew number	Project code	Time (Days)
1	A	20
2	B	10
3	E	24
4	D	10
5	C	17
Total time		81

Table 4.43 Final Solution (Alternate)

Crew number	Project code	Time (Days)
1	A	20
2	D	12
3	E	24
4	B	8
5	C	17
Total time		81

Example 4.7 A company has six machines which can process six different jobs. The processing time (minutes) of different jobs by different machines is presented in Table 4.44. Find the optimal assignment of the jobs to the machines such that the total processing time is minimized.

Table 4.44   Processing Time of Job (in minutes)

Machine

		A	B	C	D	E	F
	1	10	15	12	18	14	13
	2	17	14	22	16	19	20
Job	3	12	15	13	8	12	9
	4	11	16	15	22	21	18
	5	13	10	17	19	15	10
	6	15	8	14	25	16	18

Solution

Phase 1   The row minimums are shown in Table 4.45. The matrix after row reductions is shown in Table 4.46. In the same table, the column minimums are shown at its bottom. The matrix after column reductions is presented in Table 4.47.

Phase 2   The iterations of the phase 2 to draw minimum number of lines to cover all the zeros are presented in Tables 4.48 and 4.49.

The final solution is given in Table 4.50 and the total time required to process all the six jobs is 68 minutes.

Table 4.45   Processing Time of Job (in minutes)

Machine

		A	B	C	D	E	F	Row minimum
	1	10	15	12	18	14	13	10
	2	17	14	22	16	19	20	14
Job	3	12	15	13	8	12	9	8
	4	11	16	15	22	21	18	11
	5	13	10	17	19	15	10	10
	6	15	8	14	25	16	18	8

Table 4.46 Matrix after Row Reductions

Machine

		A	B	C	D	E	F
	1	0	5	2	8	4	3
	2	3	0	8	2	5	6
	3	4	7	5	0	4	1
Job	4	0	5	4	11	10	7
	5	3	0	7	9	5	0
	6	7	0	6	17	8	10
Column minimum		0	0	2	0	4	0

Table 4.47 Matrix after Column Reductions

Machine

		A	B	C	D	E	F
	1	0	5	0	8	0	3
	2	3	0	6	2	1	6
Job	3	4	7	3	0	0	1
	4	0	5	2	11	6	7
	5	3	0	5	9	1	0
	6	7	0	4	17	4	10

Table 4.48 Iteration 1

Machine

		A	B	C	D	E	F
	1	0	5	[0]	8	0	3
	2	3	[0]	6	2	1	6
Job	3	4	7	3	[0]	0	1
	4	[0]	5	2	11	6	7
	5	3	0	5	9	1	[0]
	6	7	0	4	17	4	10

Table 4.49 Iteration 2

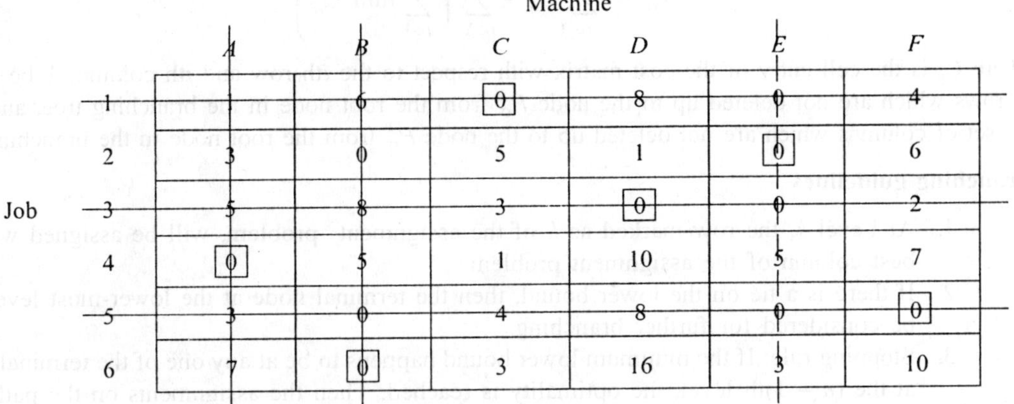

In Table 4.49, the number of squares is 6. Hence, the solution is feasible and optimal. Based on the assignments in the Table 4.49, the final solution is shown in Table 4.50 and the corresponding total time to complete all the six jobs is 68 minutes.

Table 4.50 Final Solution

Job	Machine	Time (in minutes)
1	C	12
2	E	19
3	D	8
4	A	11
5	F	10
6	B	8
Total time		68

4.5 BRANCH-AND-BOUND TECHNIQUE FOR ASSIGNMENT PROBLEM

The assignment problem can also be solved using a branch-and-bound algorithm. It is a curtailed enumeration technique. The terminologies of the branch and bound technique applied to the assignment problem are presented below.

Let k be the level number in the branching tree (for root node, it is 0), σ be an assignment made in the current node of a branching tree. P_σ^k be an assignment at level k of the branching tree, A be the set of assigned cells (partial assignment) up to the node P_σ^k from the root node (set of i and j values with respect to the assigned cells up to the node P_σ^k from the root node), and V_σ be the lower bound of the partial assignment, A up to P_σ^k, such that,

$$V_\sigma = \sum_{i,j \in A} C_{ij} + \sum_{i \in X} \left(\sum_{j \in Y} \min C_{ij} \right)$$

where C_{ij} is the cell entry of the cost matrix with respect to the ith row and jth column, X be the set of rows which are not deleted up to the node P_σ^k from the root node in the branching tree, and Y be the set of columns which are not deleted up to the node P_σ^k from the root node in the branching tree.

Branching guidelines

1. At Level k, the row marked as k of the assignment problem, will be assigned with the best column of the assignment problem.
2. If there is a tie on the lower bound, then the terminal node at the lower-most level is to be considered for further branching.
3. Stopping rule: If the minimum lower bound happens to be at any one of the terminal nodes at the $(n - 1)$th level, the optimality is reached. Then the assignments on the path from the root node to that node along with the missing pair of row–column combination will form the optimum solution.

Example 4.8 Solve the assignment problem (Table 4.51) using the branch-and-bound algorithm. The cell entries represent the processing time in hours (C_{ij}) of the job i if it is assigned to the operator j.

Table 4.51 Data for Example 4.8

Operator j

	1	2	3	4
1	23	20	21	24
Job i 2	19	21	20	20
3	20	18	24	22
4	22	18	21	23

Solution Initially, no job is assigned to any operator. So, the assignment (σ) at the root node (level 0) of the branching tree is a null set and the corresponding lower bound V_σ is also 0, as shown in Figure 4.1.

$$\boxed{P_\phi^0} \quad \begin{array}{l} \sigma = \phi \\ V_\phi = 0 \end{array}$$

Figure 4.1 Branching tree at the root node.

Further branching. The four different sub-problems under the root node are shown as in Figure 4.2. The lower bound for each of the sub-problems is shown on its right-hand side.

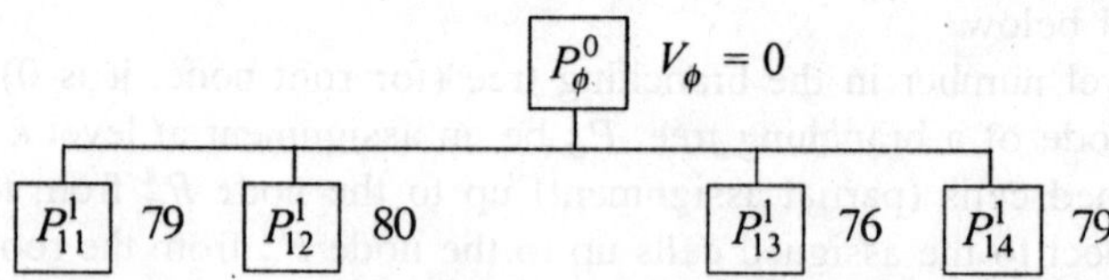

Figure 4.2 Tree with lower bounds after branching from P_ϕ^0.

Sample calculations to compute the lower bound for the first and the third sub-problems are shown below.

Lower bound for P_{11}^1

$$V_\sigma = \sum_{i \text{ and } j \in A} C_{ij} + \sum_{i \in X} \left(\sum_{j \in Y} \min\ C_{ij} \right)$$

where

$$\sigma = \{(11)\}, \qquad A = \{(11)\}, \qquad X = \{2, 3, 4\}, \qquad Y = \{2, 3, 4\}$$

Then

$$V_{(11)} = C_{11} + \sum_{i \in (2,3,4)} \left(\sum_{j \in (2,3,4)} \min\ C_{ij} \right) = 23 + (20 + 18 + 18) = 79$$

Lower bound for P_{13}^1

$$\sigma = \{(13)\}, \qquad A = \{(13)\}, \qquad X = \{2, 3, 4\}, \qquad Y = \{1, 2, 4\},$$

Then

$$V_{(13)} = C_{13} + \sum_{i \in (2,3,4)} \left(\sum_{j \in (1,2,4)} \min\ C_{ij} \right) = 21 + (19 + 18 + 18) = 76$$

Further branching.   Further branching is done from the terminal node which has the least lower bound. At this stage, the nodes P_{11}^1, P_{12}^1, P_{13}^1 and P_{14}^1 are the terminal nodes. Among these nodes, the node P_{13}^1 has the least-lower bound. Hence, further branching from this node is shown as in Figure 4.3. The lower bound of each of the newly created nodes is shown by the side of it. As an example, the calculation pertaining to the lower bound of the node P_{22}^2 is presented below.

$$\sigma = \{(22)\}, \qquad A = \{(13), (22)\}, \qquad X = \{3, 4\}, \qquad Y = \{1, 4\}$$

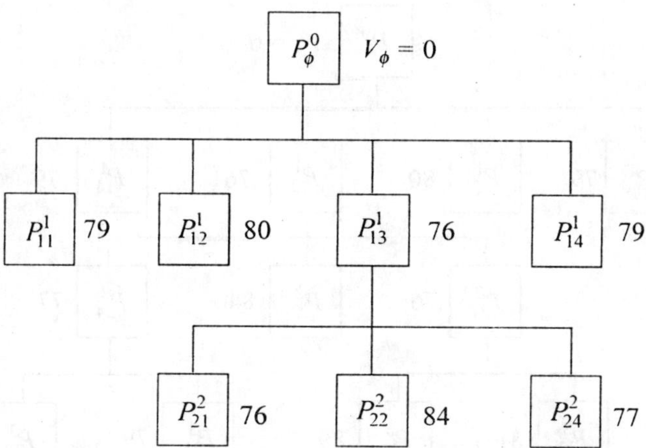

Figure 4.3   Branching tree after branching from P_{13}^1.

Then

$$V_{(22)} = C_{13} + C_{22} + \sum_{i \in (3,4)} \left(\sum_{j \in (1,4)} \min C_{ij} \right) = 21 + 21 + (20 + 22) = 84$$

Further branching. At this stage, the nodes P_{11}^1, P_{12}^1, P_{21}^2, P_{22}^2, P_{24}^2 and P_{14}^1 are the terminal nodes. Among these nodes, the node P_{21}^2 has the least lower bound. Hence further branching from this node is shown as in Figure 4.4. The lower bound of each of the newly created nodes is shown by the side of it.

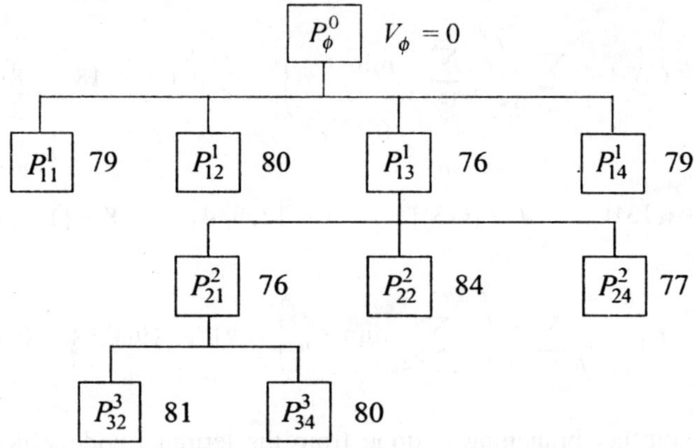

Figure 4.4   Tree with lower bounds after branching from P_{21}^2.

Further branching. At this stage, the nodes P_{11}^1, P_{12}^1, P_{32}^3, P_{34}^3, P_{22}^2, P_{24}^2 and P_{14}^1 are the terminal nodes. Among these nodes, the node P_{24}^2 has the least lower bound. Hence, further branching from this node is shown as in Figure 4.5. The lower bound of each of the newly created nodes is shown by the side of it.

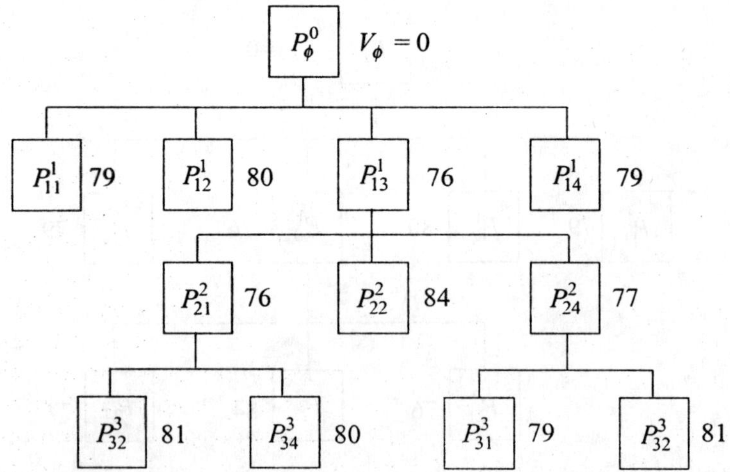

Figure 4.5   Tree with lower bounds after branching from P_{24}^2.

Further branching.   At this stage, the nodes P_{11}^1, P_{12}^1, P_{32}^3, P_{34}^3, P_{22}^2, P_{31}^3, P_{32}^3 and P_{14}^1 are the terminal nodes. Among these nodes, there are three nodes with the least lower bound of 79. So, the node P_{31}^3 which is at the bottom-most level is considered for further branching. Since this node lies at $(n-1)$th level ($k = 3$) of the branching tree, where n is the size of the assignment problem, optimality is reached. The corresponding solution is traced from the root node to the node P_{31}^3 along with the missing pair of job and operator combination, (4, 2) as shown in Table 4.52.

Table 4.52   Optimal Solution of Example 4.8

Job	Operator	Time (in hours)
1	3	21
2	4	20
3	1	20
4	2	18

Hence, total time = 79 hours.

QUESTIONS

1. Discuss the similarity between transportation problem and assignment problem.

2. Discuss practical applications of assignment problem.

3. Develop a zero-one programming model for assignment problem.

4. Discuss the steps of Hungerian method.

5. Consider the assignment problem as shown below. In this problem, five different jobs are to be assigned to five different operators such that the total processing time is minimized. The matrix entries represent processing times in hours.

Operator

		1	2	3	4	5
	1	5	6	8	6	4
	2	4	8	7	7	5
Job	3	7	7	4	5	4
	4	6	5	6	7	5
	5	4	7	8	6	8

Develop a zero-one programming model for the above problem.

6. Solve the following assignment problem using Hungerian method. The matrix entries are processing times in hours.

		Operator			
	1	2	3	4	5
1	20	22	35	22	18
2	4	26	24	24	7
Job 3	23	14	17	19	19
4	17	15	16	18	15
5	16	19	21	19	25

7. A college is having an undergraduate programme for which the effective semester time available is very less and the degree course requires field work. Hence, the savings in the total number of class hours handled can be utilized for such field work. Based on past experience, the college has established the number of hours required by each faculty to teach each subject. The course in its present semester has 4 subjects and the college has considered 6 existing faculty members to teach these courses. The objective is to assign the best 4 teachers, out of these 6 faculty to teach 4 different subjects such that the total number of class hours required is minimized. The data for this problem is summarized below. Solve and optimize the assignment problem.

	Subject			
	1	2	3	4
1	25	44	33	35
2	33	40	40	43
3	40	35	33	30
Faculty 4	44	45	28	35
5	45	35	38	40
6	40	49	40	46

8. Consider the **problem of** assigning four sales persons to four different sales regions as shown below such that **the total sales** is maximized.

	Sales region			
	1	2	3	4
1	5	11	8	9
2	5	7	9	7
Salesman 3	7	8	9	9
4	6	8	11	12

The cell entries represent annual sales figures in crores of rupees. Find the optimal allocation of the sales persons to different regions.

9. The flight timings between two cities, X and Y are as given in the following two tables. The minimum layover time of any crew in either of the cities is 3 hours. Determine the base city for each crew so that the sum of the layover times of all the crews in non-base cities is minimized.

Timings of Flights from City X to City Y

Flight number	Departure time (from City X)	Arrival time (to City Y)
101	6 a.m.	8.00 a.m.
102	10 a.m.	12.00 noon
103	3 p.m.	5.00 p.m.
104	8 p.m.	10.00 p.m.

Timings of Flights from City Y to City X

Flight number	Departure time (from City Y)	Arrival time (to City X)
201	5.30 a.m.	7.00 a.m.
202	9.00 a.m.	10.30 a.m.
203	4.00 p.m.	5.30 p.m.
204	10.00 p.m.	11.30 p.m.

10. Solve the assignment problem which is shown below using the branch-and-bound algorithm. The cell entries represent the processing time in hours (C_{ij}) of the job i if it is assigned to the operator j.

Operator j

Job i	1	2	3	4
1	13	5	8	10
2	9	15	18	10
3	12	14	10	10
4	10	14	9	12

NETWORK TECHNIQUES   5

5.1 INTRODUCTION

A *network* consists of a set of nodes (vertices) and a set of arcs (edges). Each node represents a location (city) and each arc represents the connection (road link) between two different locations (cities). The number on each arc represents the distance between the two locations (cities) (refer to Figure 5.1).

There are three types of network techniques:

1. Shortest-path model
2. Minimum spanning tree model
3. Maximal flow model.

The 'shortest-path model' can be further classified into two: shortest path between only one pair of nodes and shortest path between any pair of nodes in a network.

The shortest path between only one pair of nodes can be determined using a systematic method or Dijkstra's algorithm. The shortest path between any pair of nodes in a given network can be determined using Floyd's algorithm. The minimum spanning tree problem can be solved using either PRIM algorithm or Kruskal's algorithm.

5.2 SHORTEST-PATH MODEL

In transport organizations, one of the objectives is to find the shortest path to a particular node from any of the other nodes in a road network for shipping cargo. Determination of the shortest path using specialized procedures is known as *shortest-path model*. The following methods are used to find the shortest path in a distance network:

1. Systematic method
2. Dijkstra's algorithm
3. Floyd's algorithm

These are discussed in the next sections.

5.2.1 Systematic Method

The algorithm of systematic method for determining the shortest path between two given nodes is presented as follows:

Step 1: Represent the details of the distance network in the form of a table (say first table). For each of the nodes in the network, a column is provided in this table. The arcs that are emanating from a given node are arranged as per the increasing order of their distances and presented in the corresponding column of the first table from top to bottom.

Step 2: Select node 1 and set the cumulative distance covered up to node 1 to 0 as shown at the top of the respective column of the first table.

Step 3: Delete all the arcs in the first table that are pointing towards node 1.

Step 4: Include the recently selected node in List A of another table (say, second table).

Step 5: Find the nearest node for each of the nodes in List A and write it in the third column of the second table. (If all the arcs are deleted in the first table corresponding to any node in the List A, then remove that node from the List A of the second table.)

Step 6: For each node in List A of the second table, calculate the cumulative distance up to its nearest node as shown in the last column of the second table. Then, select the nearest node which has the least cumulative distance (X) and mark a square around it in the third column of the second table. Write the cumulative distance covered up to the selected node as X at the top of the respective column of the first table.

Step 7: Check whether the recently selected node is same as the required destination. If so, go to step 9; otherwise go to step 8.

Step 8: Delete all the arcs in the first table, that are pointing towards the recently selected node, and then go to step 4.

Step 9: Treat the lastly selected node as the last node in the shortest path.

Step 10: Find the node in List A corresponding to the recently selected/prefixed* node and prefix that node in the partially formed shortest path.

Step 11: Check whether the prefixed node is the required source node. If no, go to step 12; otherwise to step 13.

Step 12: Move to the iteration in which the recently prefixed node is selected (found with square) in the third column of the second table. Then, go to step 10.

Step 13: Treat the path which is constructed, based on the above guidelines, as the shortest path and the shortest distance for this shortest path is equal to the least cumulative distance in the last iteration of the second table.

Example 5.1   Consider the road network as in Figure 5.1, where distances between different pairs of adjacent cities are summarized. Find the shortest path from City 1 to City 10.

Solution

Step 1: The distance network which is shown in Figure 5.1 is represented in matrix form as shown in Table 5.1.

---

*The term 'prefixed' will have its meaning while step 10 is repeated.

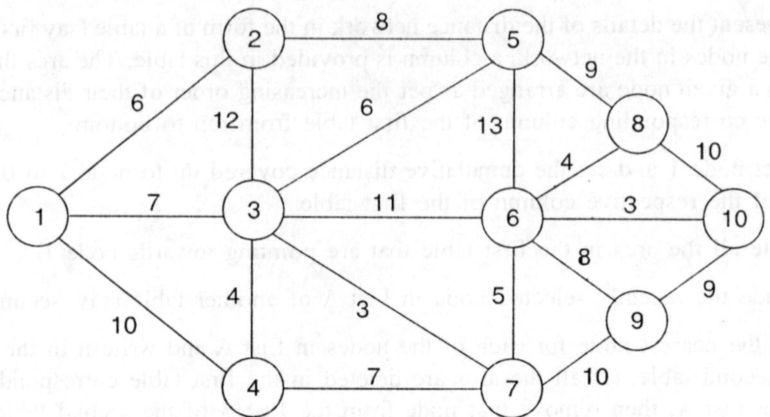

Figure 5.1 Distance network.

Table 5.1 Details of Distances (Iteration 1)

1		2		3		4		5		6		7		8		9		10	
1–2	6	2–1	6	3–7	3	4–3	4	5–3	6	6–10	3	7–3	3	8–6	4	9–6	8	10–6	3
1–3	7	2–5	8	3–4	4	4–7	7	5–2	8	6–8	4	7–6	5	8–5	9	9–10	9	10–9	9
1–4	10	2–3	12	3–5	6	4–1	10	5–8	9	6–7	5	7–4	7	8–10	10	9–7	10	10–8	10
				3–1	7			5–6	13	6–9	8	7–9	10						
				3–6	11					6–3	11								
				3–2	12					6–5	13								

Step 2: City 1 is selected as shown in Table 5.2, and the cumulative distance travelled up to City 1 is set to 0 as shown in the last column of Table 5.2 as well as at the top of the column 1 of Table 5.3 of the next step.

Table 5.2 Details of Selection of Node in Iteration 1

Iteration	List A (nodes included)	Nearest nodes	Distance calculation
1	–	1	0

Step 3: Delete all the arcs in Table 5.1 that are pointing towards City 1 (2–1, 3–1 and 4–1) as shown in Table 5.3 (deletion of arc is indicated using a '*' by the side of that arc).

Table 5.3 Updated Distance (Iteration 2)

0

1		2		3		4		5		6		7		8		9		10	
1–2	6	2–1	6*	3–7	3	4–3	4	5–3	6	6–10	3	7–3	3	8–6	4	9–6	8	10–6	3
1–3	7	2–5	8	3–4	4	4–7	7	5–2	8	6–8	4	7–6	5	8–5	9	9–10	9	10–9	9
1–4	10	2–3	12	3–5	6	4–1	10*	5–8	9	6–7	5	7–4	7	8–10	10	9–7	10	10–8	10
				3–1	7*			5–6	13	6–9	8	7–9	10						
				3–6	11					6–3	11								
				3–2	12					6–5	13								

Steps 4, 5 and 6: The calculations and updations with respect to these steps are shown in Table 5.4.

Table 5.4 Details of Selection of Node in Iteration 2

Iteration	List A (nodes included)	Nearest nodes	Distance calculation
1	–	1	0
2	1	2	0 + 6 = 6*

Step 7: The recently selected City 2 is not the required City 10. So, go to step 8.

Step 8: Delete all the arcs that are pointing towards City 2 (1–2, 3–2 and 5–2) as shown in Table 5.5, and then go to step 4.

Table 5.5 Updated Distance (Iteration 3)

	0	6							
1	**2**	**3**	**4**	**5**	**6**	**7**	**8**	**9**	**10**
1–2 6*	2–1 6*	3–7 3	4–3 4	5–3 6	6–10 3	7–3 3	8–6 4	9–6 8	10–6 3
1–3 7	2–5 8	3–4 4	4–7 7	5–2 8*	6–8 4	7–6 5	8–5 9	9–10 9	10–9 9
1–4 10	2–3 12	3–5 6	4–1 10*	5–8 9	6–7 5	7–4 7	8–10 10	9–7 10	10–8 10
		3–1 7*		5–6 13	6–9 8	7–9 10			
		3–6 11			6–3 11				
		3–2 12*			6–5 13				

Steps 4, 5 and 6: The calculations and updations with respect to these steps are shown in Table 5.6.

Table 5.6 Details of Selection of Node in Iteration 3

Iteration	List A (nodes included)	Nearest nodes	Distance calculation
1	–	1	0
2	1	2	0 + 6 = 6*
3	1, 2	3 , 5	0 + 7 = 7*
			6 + 8 = 14

Step 7: The recently selected city, i.e. City 3 is not the required City 10. So, go to step 8.

Step 8: Delete all the arcs that are pointing towards City 3 (1–3, 2–3, 5–3, 6–3, 7–3 and 4–3) as shown in Table 5.7 and then go to step 4.

Table 5.7 Updated Distance (Iteration 4)

0	6	7							
1	2	3	4	5	6	7	8	9	10
1–2 6*	2–1 6*	3–7 3	4–3 4*	5–3 6*	6–10 3	7–3 3*	8–6 4	9–6 8	10–6 3
1–3 7*	2–5 8	3–4 4	4–7 7	5–2 8*	6–8 4	7–6 5	8–5 9	9–10 9	10–9 9
1–4 10	2–3 12*	3–5 6	4–1 10*	5–8 9	6–7 5	7–4 7	8–10 10	9–7 10	10–8 10
		3–1 7*		5–6 13	6–9 8	7–9 10			
		3–6 11			6–3 11*				
		3–2 12*			6–5 13				

Steps 4, 5 and 6: The calculations and updations with respect to these steps are shown in Table 5.8.

Table 5.8 Details of Selection of Node in Iteration 4

Iteration	List A (nodes included)	Nearest nodes	Distance calculation
1	–	1	0
2	1	2	0 + 6 = 6*
3	1, 2	3 , 5	0 + 7 = 7* 6 + 8 = 14
4	1, 2, 3	4, 5, 7	0 + 10 = 10* 6 + 8 = 14 7 + 3 = 10*

In Table 5.8, City 7 is selected by breaking tie between City 4 and City 7, randomly.

Step 7: The recently selected City 7 is not the required City 10. So, go to step 8.

Step 8: Delete all the arcs that are pointing towards City 7 (4–7, 3–7, 6–7 and 9–7) as shown in Table 5.9 and then go to step 4.

Table 5.9 Updated Distance (Iteration 5)

0	6	7				10			
1	2	3	4	5	6	7	8	9	10
1–2 6*	2–1 6*	3–7 3*	4–3 4*	5–3 6*	6–10 3	7–3 3*	8–6 4	9–6 8	10–6 3
1–3 7*	2–5 8	3–4 4	4–7 7*	5–2 8*	6–8 4	7–6 5	8–5 9	9–10 9	10–9 9
1–4 10	2–3 12*	3–5 6	4–1 10*	5–8 9	6–7 5*	7–4 7	8–10 10	9–7 10*	10–8 10
		3–1 7*		5–6 13	6–9 8	7–9 10			
		3–6 11			6–3 11*				
		3–2 12*			6–5 13				

Steps 4, 5 and 6: The calculations and updations with respect to these steps are shown in Table 5.10.

Table 5.10 Details of Selection of Node in Iteration 5

Iteration	List A (nodes included)	Nearest nodes	Distance calculation
1	–	1	0
2	1	2	$0 + 6 = 6*$
3	1, 2	3 , 5	$0 + 7 = 7*$ $6 + 8 = 14$
4	1, 2, 3	4, 5, 7	$0 + 10 = 10*$ $6 + 8 = 14$ $7 + 3 = 10*$
5	1, 2, 3, 7	4 , 5, 4, 6	$0 + 10 = 10*$ $6 + 8 = 14$ $7 + 4 = 11$ $10 + 5 = 15$

Step 7: City 4, that is selected recently, is not the required City 10. So, go to step 8.

Step 8: Delete all the arcs that are pointing towards City 4 (1–4, 3–4 and 7–4) as shown in Table 5.11 and then go to step 4.

Table 5.11 Updated Distance (Iteration 6)

0	6	7	10			10			
1	2	3	4	5	6	7	8	9	10
1–2 6*	2–1 6*	3–7 3*	4–3 4*	5–3 6*	6–10 3	7–3 3*	8–6 4	9–6 8	10–6 3
1–3 7*	2–5 8	3–4 4*	4–7 7*	5–2 8*	6–8 4	7–6 5	8–5 9	9–10 9	10–9 9
1–4 10*	2–3 12*	3–5 6	4–1 10*	5–8 9	6–7 5*	7–4 7*	8–10 10	9–7 10*	10–8 10
		3–1 7*		5–6 13	6–9 8	7–9 10			
		3–6 11			6–3 11*				
		3–2 12*			6–5 13				

Steps 4, 5 and 6: The calculations and updations with respect to these steps are shown in Table 5.12.

In Table 5.11, all the arcs with respect to node 1 and node 4 are deleted. Hence, in Table 5.12 under List A, these nodes are deleted by marking 'X' against each of them. The cumulative distance up to the nearest neighbour of each of the remaining nodes, i.e. 2, 3 and 7 are shown in the last row of Table 5.12.

Table 5.12 Details of Selection of Node in Iteration 6

Iteration	List A (nodes included)	Nearest nodes	Distance calculation
1	–	1	0
2	1	2	$0 + 6 = 6*$
3	1, 2	3 , 5	$0 + 7 = 7*$
			$6 + 8 = 14$
4	1, 2, 3	4, 5, 7	$0 + 10 = 10*$
			$6 + 8 = 14$
			$7 + 3 = 10*$
5	1, 2, 3, 7	4 , 5, 4, 6	$0 + 10 = 10*$
			$6 + 8 = 14$
			$7 + 4 = 11$
			$10 + 5 = 15$
6	1, 2, 3, 7, 4	5, 5 , 6	$6 + 8 = 14$
			$7 + 6 = 13*$
			$10 + 5 = 15$

Step 7: The recently selected City 5 is not the required destination City 10. So, go to step 8.

Step 8: Delete all the arcs that are pointing towards City 5 (2–5, 3–5, 6–5 and 8–5) as shown in Table 5.13 and then go to step 4.

Table 5.13 Updated Distance (Iteration 7)

0		6		7		10		13				10	
1		2		3		4		5		6		7		8		9		10

1		2		3		4		5		6		7		8		9		10	
1–2	6*	2–1	6*	3–7	3*	4–3	4*	5–3	6*	6–10	3	7–3	3*	8–6	4	9–6	8	10–6	3
1–3	7*	2–5	8*	3–4	4*	4–7	7*	5–2	8*	6–8	4	7–6	5	8–5	9*	9–10	9	10–9	9
1–4	10*	2–3	12*	3–5	6*	4–1	10*	5–8	9	6–7	5*	7–4	7*	8–10	10	9–7	10*	10–8	10
				3–1	7*			5–6	13	6–9	8	7–9	10						
				3–6	11					6–3	11*								
				3–2	12*					6–5	13*								

Steps 4, 5 and 6: The calculations and updations with respect to these steps are shown in Table 5.14.

Step 7: The recently selected City 6 is not the required City 10. So, go to step 8.

Step 8: Delete all the arcs that are pointing towards City 6 (3–6, 5–6, 8–6, 10–6, 9–6 and 7–6) as shown in Table 5.15, and then go to step 4.

Table 5.14 Details of Selection of Node in Iteration 7

Iteration	List A (nodes included)	Nearest nodes	Distance calculation
1	–	1	0
2	1	2	$0 + 6 = 6^*$
3	1, 2	3 , 5	$0 + 7 = 7^*$ $6 + 8 = 14$
4	1, 2, 3	4, 5, 7	$0 + 10 = 10^*$ $6 + 8 = 14$ $7 + 3 = 10^*$
5	1, 2, 3, 7	4 , 5, 4, 6	$0 + 10 = 10^*$ $6 + 8 = 14$ $7 + 4 = 11$ $10 + 5 = 15$
	X X		$6 + 8 = 14$
6	1, 2, 3, 7, 4	5, 5 , 6	$7 + 6 = 13^*$ $10 + 5 = 15$
	X		
7	2, 3, 7, 5	6, 6 , 8	$7 + 11 = 18$ $10 + 5 = 15^*$ $13 + 9 = 22$

Table 5.15 Updated Distance (Iteration 8)

0	6	7	10	13	15	10			
1	2	3	4	5	6	7	8	9	10
1–2 6*	2–1 6*	3–7 3*	4–3 4*	5–3 6*	6–10 3	7–3 3*	8–6 4*	9–6 8*	10–6 3*
1–3 7*	2–5 8*	3–4 4*	4–7 7*	5–2 8*	6–8 4	7–6 5*	8–5 9*	9–10 9	10–9 9
1–4 10*	2–3 12*	3–5 6*	4–1 10*	5–8 9	6–7 5*	7–4 7*	8–10 10	9–7 10*	10–8 10
		3–1 7*		5–6 13*	6–9 8	7–9 10			
		3–6 11*			6–3 11*				
		3–2 12*			6–5 13*				

Steps 4, 5 and 6: The calculations and updations with respect to these steps are shown in Table 5.16.

Table 5.16 Details of Selection of Node in Iteration 8

Iteration	List A (nodes included)	Nearest nodes	Distance calculation
1	–	1	0
2	1	2	$0 + 6 = 6*$
3	1, 2	3 , 5	$0 + 7 = 7*$
			$6 + 8 = 14$
4	1, 2, 3	4, 5, 7	$0 + 10 = 10*$
			$6 + 8 = 14$
			$7 + 3 = 10*$
5	1, 2, 3, 7	4 , 5, 4, 6	$0 + 10 = 10*$
			$6 + 8 = 14$
			$7 + 4 = 11$
			$10 + 5 = 15$
6	1, 2, 3, 7, 4	5, 5 , 6	$6 + 8 = 14$
			$7 + 6 = 13*$
			$10 + 5 = 15$
7	2, 3, 7, 5	6, 6 , 8	$7 + 11 = 18$
			$10 + 5 = 15*$
			$13 + 9 = 22$
8	3, 7, 5, 6	9, 8, 10	$10 + 10 = 20$
			$13 + 9 = 22$
			$15 + 3 = 18*$

[*Note:* The calculations of each of the two types of tables are shown in different tables mainly for better understanding. But each of them can be shown in a single table while solving a problem.]

Step 7: The recently selected City 10 is the required destination city. So, go to step 9.

Step 9: The partial path: 10.

Step 10: The city in List A, corresponding to City 10, is 6. So, prefix City 6 in the partial path. The partial path is 6–10.

Step 11: The recently prefixed city is not the source city. So, go to step 12.

Step 12: Move to Iteration 7 where City 6 is found with square and go to step 10.

Step 10: The city in List A, corresponding to the City 6, is 7. So, prefix City 7 in the partial path. The partial path is 7–6–10.

Step 11: The recently prefixed city is not the source city. So, go to step 12.

Step 12: Move to Iteration 4 where City 7 is found with square and go to step 10.

Step 10: The city in the List A corresponding to the City 7, is 3. So, prefix City 3 in the partial path. Now, the partial path is 3–7–6–10.

Step 11: The recently prefixed city is not the source city. So, go to step 12.

Step 12: Move to Iteration 3 where City 3 is found with square and go to step 10.

Step 10: The city in the List A, corresponding to City 3, is 1. So, prefix City 1 in the partial path. Now, the partial path is 1–3–7–6–10.

Step 11: The recently prefixed city is the source city. So, go to step 13.

Step 13: The shortest path, presented in Figure 5.2 with double lines, covers a distance of 18 units. The root followed from City 1 is 1–3–7–6–10.

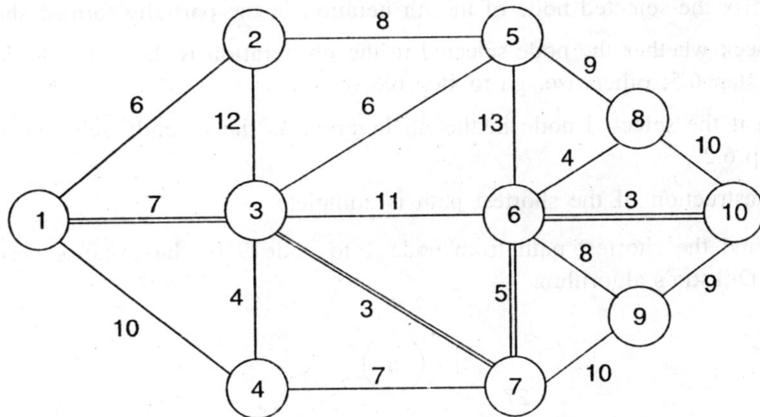

Figure 5.2   Sketch of shortest path indicated using double line.

5.2.2   Dijkstra's Algorithm

Dijkstra's algorithm is another procedure to find the shortest path between any two nodes in a distance network. The arcs in the network may be directed or undirected. The steps of the algorithm are summarized below.

Step 1: Form a distance matrix from the start node to all other nodes. (If there is no direct arc from the start node to another node, the corresponding distance is assumed as ∞.) Let the currently selected node (start node) be K.

Step 2: Find the smallest of the distances from the node K to all other unselected nodes in the distance matrix. Let it be X and the corresponding node be L. Check whether L is a neighbour of the start node. If yes, do step 2.1; otherwise, do step 2.2.

    2.1   Thicken (select) the arc connecting the start node and L. Then, go to step 3.

    2.2   Find the iteration among the preceding iterations in which the distance to L from the start node is different from X. Thicken (select) the arc connecting the selected node of that iteration and the selected node L of the present iteration. Then, go to step 3.

Step 3: Check whether L is the required destination. If not, go to step 4; otherwise, go to step 6.

Step 4: Form the distance matrix by following the guidelines presented below. For each of the unselected neighbours of L, find its new distance by adding X to its distance from L, and update its old distance from the start node in the distance matrix if the new distance is smaller than its old distance. Transfer the data of the last node of the distance matrix to the column corresponding to the recently selected node, L. (This reduces the matrix column size by 1 in every iteration.)

Step 5: set $K = L$ and go to step 2.

Step 6: Trace the shortest path.

    6.1   Fix the selected node of the last iteration as the last node of the shortest path.

    6.2   Traverse back through iterations and identify the iteration (jth iteration) at which the distance to the recently selected node is different.

    6.3   Prefix the selected node of the jth iteration in the partially formed shortest path.

    6.4   Check whether the node selected in the jth iteration is the source node. If not so, go to step 6.5; otherwise, go to step 6.6.

    6.5   Treat the selected node in the jth iteration as the recently selected node and go to step 6.2.

    6.6   Construction of the shortest path is complete.

Example 5.2 Find the shortest path from node 1 to node 9 of the distance network shown in Figure 5.3 using Dijkstra's algorithm.

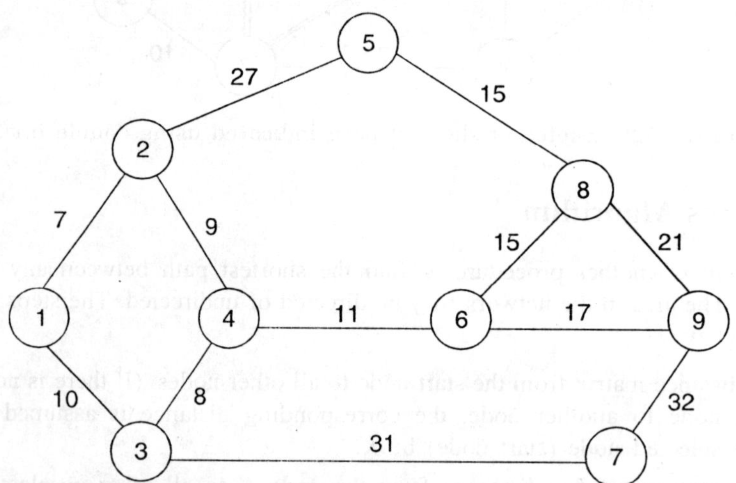

Figure 5.3 Distance network for Example 5.2.

Solution

Iteration 1

Step 1: The distance matrix summarizing the distances from the start node 1 to all other nodes is given in Figure 5.4. In this figure one can see the actual distance from the start node 1 to all its neighbours. The distance for all other nodes from node 1 is assumed as ∞.

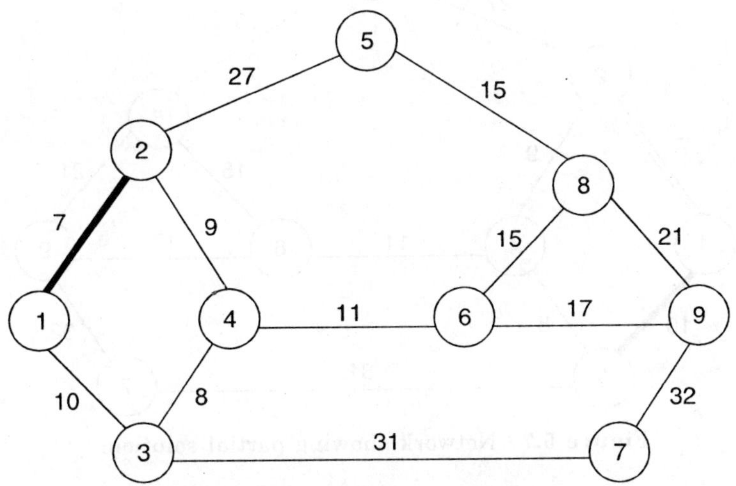

Figure 5.4 Distance matrix of Iteration 1.

Step 2: In Figure 5.4, the smallest distance (X) is 7 and the corresponding node (L) is 2. Since the node 2 is a neighbour of the start node, thicken the arc 1–2 as shown in Figure 5.5.

Figure 5.5 Network showing partial solution.

Step 3: Since node 2 is not the required destination, go to step 4.

Iteration 2

Step 4: For node 2 (L), nodes 4 and 5 are the unselected neighbours. Then $X + d_{24} = 7 + 9 = 16$, where d_{24} is the actual distance from node 2 to node 4. Since this distance is less than ∞ (old distance up to node 4 in Figure 5.4), update the distance to node 4 as 16. Then

$$X + d_{25} = 7 + 27 = 34$$

Again, since this distance is less than ∞ (old distance up to node 5 in Figure 5.4), update the distance to node 5 as 34. Transfer the data of node 9 to the position of node 2. These are summarized in Figure 5.6.

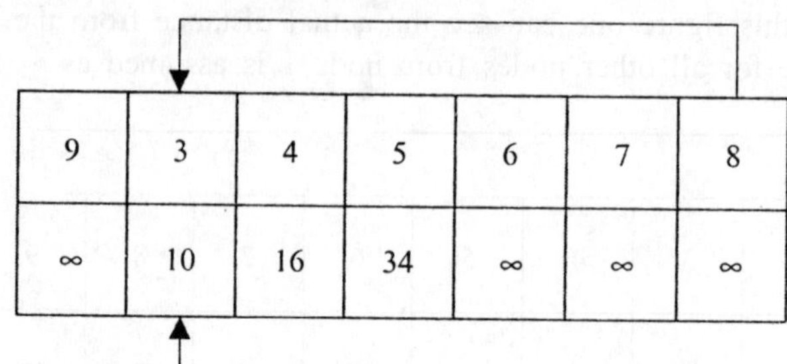

Figure 5.6 Distance matrix of Iteration 2.

Step 5: Set $K = 2$ and go to step 2.

Step 2: In Figure 5.6, the smallest distance (X) is 10, and the corresponding node (L) is 3. Since node 3 is a neighbour of the start node, thicken the arc 1–3 as shown in Figure 5.7.

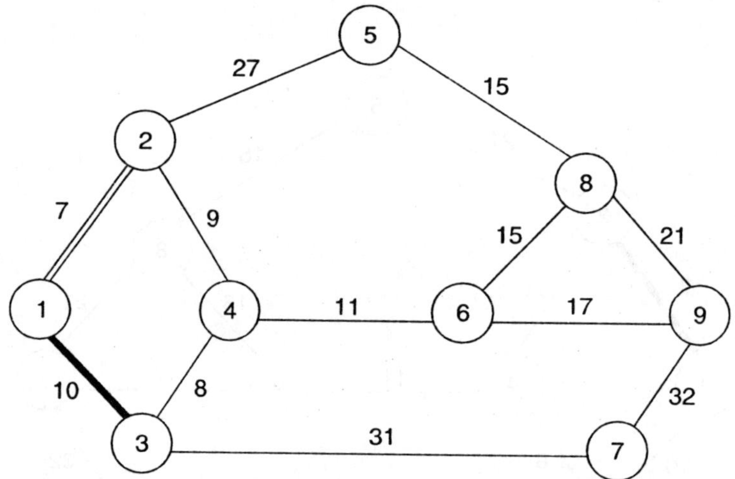

Figure 5.7 Network showing partial solution.

Step 3: Since, the node 3 is not the required destination, go to step 4.

Iteration 3

Step 4: The results of this step are summarized as in Figure 5.8.

Figure 5.8 Distance matrix of Iteration 3.

Step 5: Set $K = 3$ and go to step 2.

Step 2: In Figure 5.8, the smallest distance (X) is 16 and the corresponding node (L) is 4. Since, L is not a neighbour of the start node, perform step 2.2.

2.2 The iteration among the preceding iterations in which the distance to L from the starting node is different from X is the Iteration 1. The node selected in that iteration is 2. So, thicken the arc 2–4 as shown in Figure 5.9.

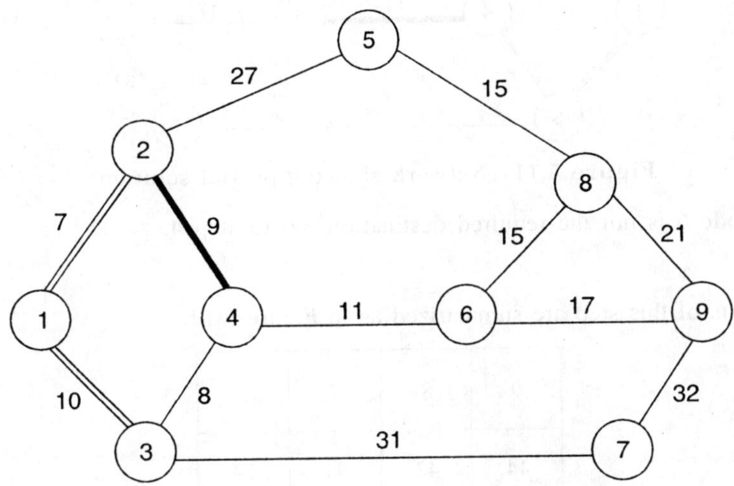

Figure 5.9   Network showing partial solution.

Step 3: Since node 4 is not the required destination, go to step 4.

Iteration 4

Step 4:   The results of this step are summarized as in Figure 5.10.

9	8	7	5	6
∞	∞	41	34	27

Figure 5.10   Distance matrix of Iteration 4.

Step 5: Set $K = 4$ and go to step 2.

Step 2: In Figure 5.10, the smallest distance (X) is 27 and the corresponding node (L) is 6. Since, L is not a neighbour of the start node, perform the step 2.2.

2.2 The iteration among the preceding iterations in which the distance to L from the starting node is different from X is the Iteration 3. The node selected in that iteration is 4. So, thicken the arc 4–6 as shown in Figure 5.11.

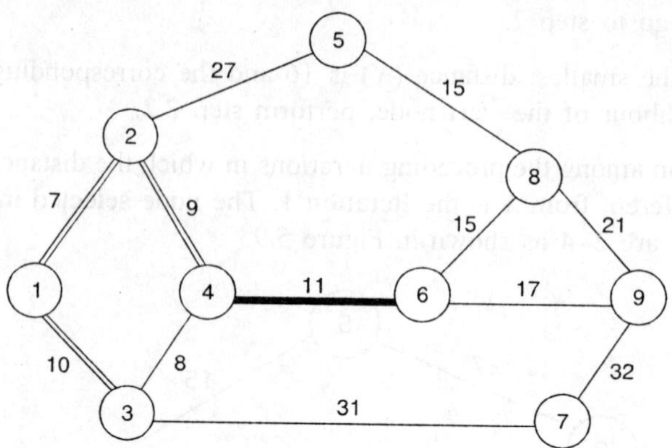

Figure 5.11 Network showing partial solution.

Step 3: Since, node 6 is not the required destination, go to step 4.

Iteration 5

Step 4: The results of this step are summarized as in Figure 5.12.

9	8	7	5
44	42	41	34

Figure 5.12 Distance matrix of Iteration 5.

Step 5: Set $K = 6$ and go to step 2.

Step 2: In Figure 5.12, the smallest distance (X) is 34 and the corresponding node (L) is 5. Since L is not a neighbour of the start node, perform step 2.2.

    2.2 The iteration among the preceding iterations in which the distance to L from the starting node is different from X is the Iteration 1. The node selected in that iteration is 2. So, thicken the arc 2–5 as shown in Figure 5.13.

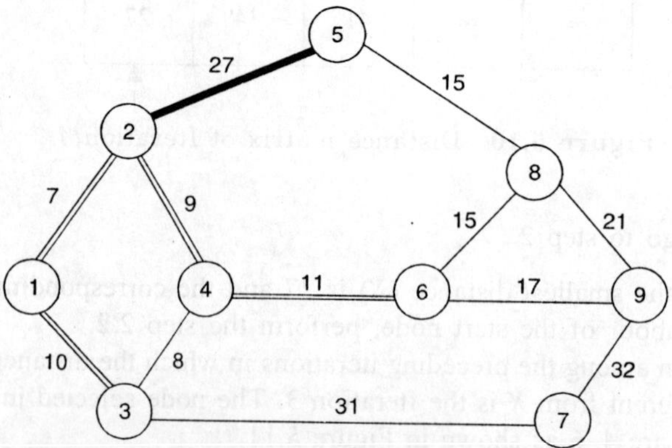

Figure 5.13 Network showing partial solution.

Step 3: Since node 5 is not the required destination, go to step 4.

Iteration 6

Step 4: The results of this step are summarized as in Figure 5.14.

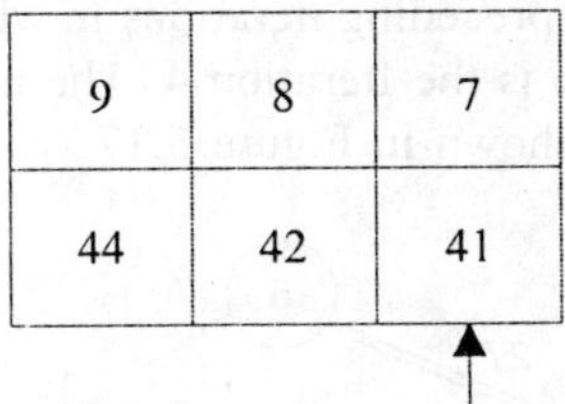

Figure 5.14 Distance matrix of Iteration 6.

Step 5: Set $K = 5$ and go to step 2.

Step 2: In Figure 5.14, the smallest distance (X) is 41 and the corresponding node (L) is 7. Since L is not a neighbour of the start node, perform step 2.2.

    2.2 The iteration among the preceding iterations in which the distance to L from the starting node is different from X is the Iteration 2. The node selected in that iteration is 3. So, thicken the arc 3–7 as shown in Figure 5.15.

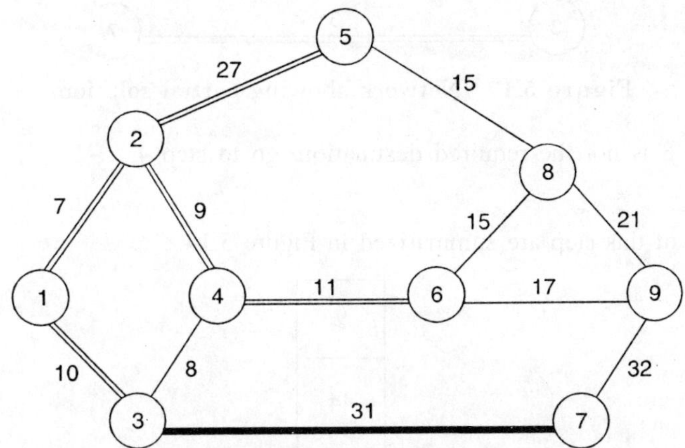

Figure 5.15 Network showing partial solution.

Step 3: Since node 7 is not the required destination, go to step 4.

Iteration 7

Step 4: The results of this step are summarized as in Figure 5.16.

Figure 5.16 Distance matrix of Iteration 7.

Step 5: Set $K = 7$ and go to step 2.

Step 2: In the Figure 5.16, the smallest distance (X) is 42 and the corresponding node (L) is 8. Since, L is not a neighbour of the start node, perform step 2.2.

    2.2 The iteration among the preceding iterations in which the distance to L from the starting node is different from X is the Iteration 4. The node selected in that iteration is 6. So, thicken the arc 6–8 as shown in Figure 5.17.

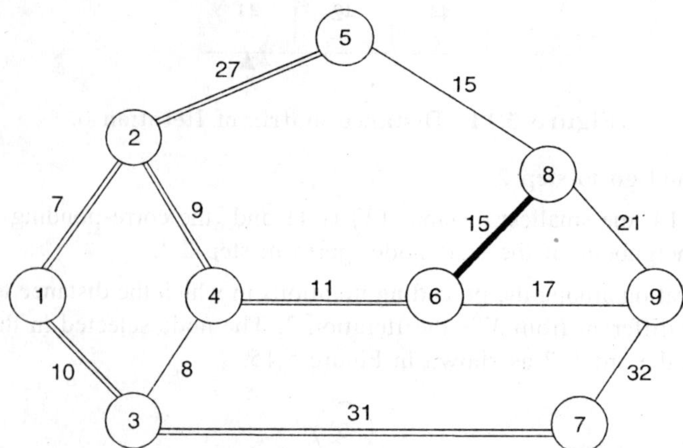

Figure 5.17   Network showing partial solution.

Step 3: Since node 8 is not the required destination, go to step 4.

Iteration 8

Step 4: The results of this step are summarized in Figure 5.18.

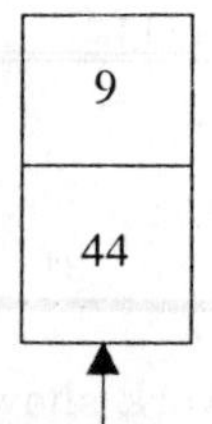

Figure 5.18   Distance matrix of Iteration 8.

Step 5: Set $K = 8$ and go to step 2.

Step 2: In the Figure 5.18, the smallest distance (X) is 44 and the corresponding node (L) is 9. Since, L is not a neighbour of the start node, perform step 2.2.

    2.2 The iteration among the preceding iterations in which the distance to L from the starting node is different from X is Iteration 4. The node selected in that iteration is 6. So, thicken the arc 6–9 as shown in Figure 5.19.

Step 3. Since node 9 is the required destination, go to step 6.

Step 6. Based on the guidelines of this step, the shortest path is 1–2–4–6–9 and the corresponding distance is 44.

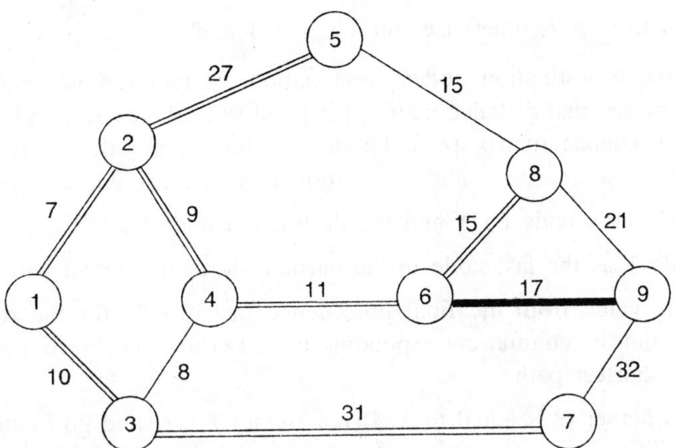

Figure 5.19   Network showing final solution.

5.2.3   Floyd's Algorithm

Floyd's algorithm is used to find the shortest path and the corresponding distance from any source node to any destination node in a given distance network.

This algorithm takes the initial distance matrix $[D^0]$ and initial precedence matrix $[P^0]$ as input. Then, it performs n iterations (n is the number of nodes in the distance matrix) and generates the final distance matrix $[D'']$ and the final precedence matrix $[P'']$. One can find the shortest distance between any two nodes from the final distance matrix $[D'']$ and can trace the corresponding path from the final precedence matrix $[P'']$.

Steps of Floyd's algorithm

Step 1:   Set $k = 0$.

Step 2:   Form the initial distance matrix $[D^0]$ and the initial precedence matrix $[P^0]$ from the distance network.

The distance matrix as well as the precedence matrix are with n rows and n columns. In the distance matrix, between any two nodes, if there is no direct arc, then the corresponding distance is assumed as ∞. The diagonal values of the initial distance matrix are assumed as 0. In the initial precedence matrix, the diagonal values are assumed as '–' and the other entries in ith row are assumed as i, for $i = 1, 2, 3,..., n$.

Step 3:   Set $k = k + 1$.

Step 4:   Obtain the values of the distance matrix, $[D^k]$ for all the cells where i is not equal to j using the following formula.

$$D_{ij}^k = \min \left[D_{ij}^{k-1}, \left(D_{ik}^{k-1} + D_{kj}^{k-1} \right) \right]$$

Step 5:   Obtain the values of the precedence matrix, $[P^k]$, for all the cells where i is not equal to j using the following formula.

$$P_{ij}^k = P_{kj}^{k-1}, \qquad \text{if } D_{ij}^k \text{ is not equal to } D_{ij}^{k-1}$$
$$= P_{ij}^{k-1}, \qquad \text{otherwise.}$$

Step 6: If $k = n$, go to step 7; otherwise, set $k = k + 1$ and go to step 4.

Step 7: For each source-destination nodes combination, as required in reality, find the shortest distance from the final distance matrix, $[D'']$ and trace the corresponding shortest path from the final precedence matrix $[P'']$. Guidelines for tracing the shortest path for a given combination of source node and destination node are presented below.

    7.1 Let the source node be X and the destination node be Y.

    7.2 Fix node Y as the last node in the partially formed shortest path.

    7.3 Find the value from the final precedence matrix $[P'']$ for the row corresponding to node X and the column corresponding to Y. Let it be Q. Prefix node Q in the partially formed shortest path.

    7.4 Check whether Q is equal to X. If not so, set $Y = Q$ and go to step 7.3; otherwise go to step 7.5.

    7.5 The path constructed is the required shortest path from the source node X to the destination node Y.

Example 5.3 Consider the distance network as shown in Figure 5.20.

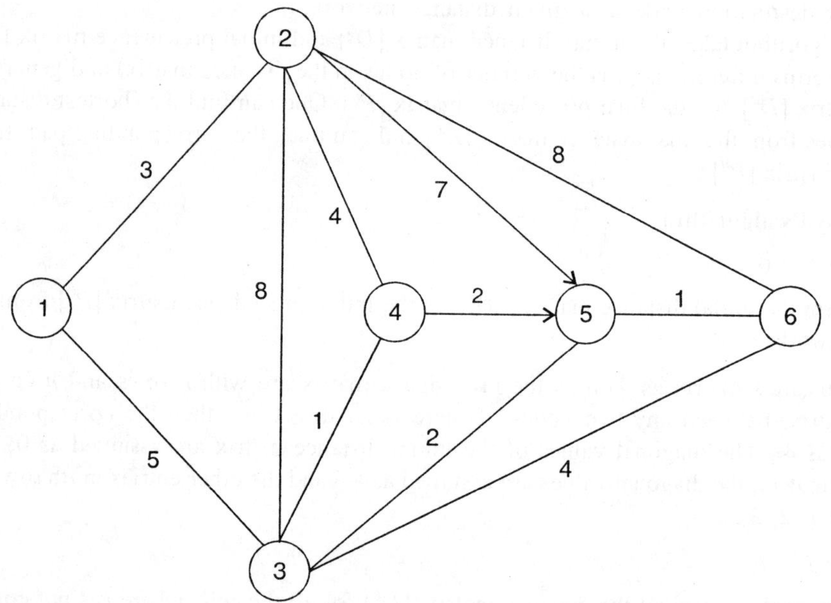

Figure 5.20 Distance network.

(a) Apply Floyd's algorithm to it and generate the final distance matrix and precedence matrix.

(b) Find the shortest path and the corresponding distance from the source node to the destination node as indicated in each of the cases: 1–6, 5–1 and 5–2.

Solution Applying Floyd's algorithm, we get Tables 5.17–5.23.

Iteration 0

Table 5.17(a) Distance Matrix $[D^0]$

	1	2	3	4	5	6
1	0	3	5	∞	∞	∞
2	3	0	8	4	7	8
3	5	8	0	1	2	4
4	∞	4	1	0	2	∞
5	∞	∞	2	∞	0	1
6	∞	8	4	∞	1	0

Table 5.17(b) Precedence Matrix $[P^0]$

	1	2	3	4	5	6
1	–	1	1	1	1	1
2	2	–	2	2	2	2
3	3	3	–	3	3	3
4	4	4	4	–	4	4
5	5	5	5	5	–	5
6	6	6	6	6	6	–

Iteration 1

Table 5.18(a) Distance Matrix $[D^1]$

	1	2	3	4	5	6
1	0	3	5	∞	∞	∞
2	3	0	8	4	7	8
3	5	8	0	1	2	4
4	∞	4	1	0	2	∞
5	∞	∞	2	∞	0	1
6	∞	8	4	∞	1	0

Table 5.18(b) Precedence Matrix $[P^1]$

	1	2	3	4	5	6
1	–	1	1	1	1	1
2	2	–	2	2	2	2
3	3	3	–	3	3	3
4	4	4	4	–	4	4
5	5	5	5	5	–	5
6	6	6	6	6	6	–

Iteration 2

Table 5.19(a) Distance Matrix $[D^2]$

	1	2	3	4	5	6
1	0	3	5	7	10	11
2	3	0	8	4	7	8
3	5	8	0	1	2	4
4	7	4	1	0	2	12
5	∞	∞	2	∞	0	1
6	11	8	4	12	1	0

Table 5.19(b) Precedence Matrix $[P^2]$

	1	2	3	4	5	6
1	–	1	1	2	2	2
2	2	–	2	2	2	2
3	3	3	–	3	3	3
4	2	4	4	–	4	2
5	5	5	5	5	–	5
6	2	6	6	2	6	–

Iteration 3

Table 5.20(a) Distance Matrix $[D^3]$

	1	2	3	4	5	6
1	0	3	5	6	7	9
2	3	0	8	4	7	8
3	5	8	0	1	2	4
4	6	4	1	0	2	5
5	7	10	2	3	0	1
6	9	8	4	5	1	0

Table 5.20(b) Precedence Matrix $[P^3]$

	1	2	3	4	5	6
1	–	1	1	3	3	3
2	2	–	2	2	2	2
3	3	3	–	3	3	3
4	3	4	4	–	4	3
5	3	3	5	3	–	5
6	3	6	6	3	6	–

Iteration 4

Table 5.21(a) Distance Matrix $[D^4]$

	1	2	3	4	5	6
1	0	3	5	6	7	9
2	3	0	5	4	6	8
3	5	5	0	1	2	4
4	6	4	1	0	2	5
5	7	7	2	3	0	1
6	9	8	4	5	1	0

Table 5.21(b) Precedence Matrix $[P^4]$

	1	2	3	4	5	6
1	–	1	1	3	3	3
2	2	–	4	2	4	2
3	3	4	–	3	3	3
4	3	4	4	–	4	3
5	3	4	5	3	–	5
6	3	6	6	3	6	–

Iteration 5

Table 5.22(a) Distance Matrix $[D^5]$

	1	2	3	4	5	6
1	0	3	5	6	7	8
2	3	0	5	4	6	7
3	5	5	0	1	2	3
4	6	4	1	0	2	3
5	7	7	2	3	0	1
6	8	8	3	4	1	0

Table 5.22(b) Precedence Matrix $[P^5]$

	1	2	3	4	5	6
1	–	1	1	3	3	5
2	2	–	4	2	4	5
3	3	4	–	3	3	5
4	3	4	4	–	4	5
5	3	4	5	3	–	5
6	3	6	5	3	6	–

Iteration 6

Table 5.23(a) Distance Matrix $[D^6]$

	1	2	3	4	5	6
1	0	3	5	6	7	8
2	3	0	5	4	6	7
3	5	5	0	1	2	3
4	6	4	1	0	2	3
5	7	7	2	3	0	1
6	8	8	3	4	1	0

Table 5.23(b) Precedence Matrix $[P^6]$

	1	2	3	4	5	6
1	–	1	1	3	3	5
2	2	–	4	2	4	5
3	3	4	–	3	3	5
4	3	4	4	–	4	5
5	3	4	5	3	–	5
6	3	6	5	3	6	–

(b) The shortest path from node 1 to node 6 is 1–3–5–6; the corresponding distance = 8. The shortest path from node 5 to node 1 is 5–3–1; the corresponding distance = 7. The shortest path from node 5 to node 2 is 5–3–4–2; the corresponding distance = 7.

5.3 MINIMUM SPANNING TREE PROBLEM

Let us consider the example of laying telephone cable in a locality as shown in Figure 5.21 which summarizes the distance network of the locality. The number on each arc represents the distance between the nodes connected by that arc.

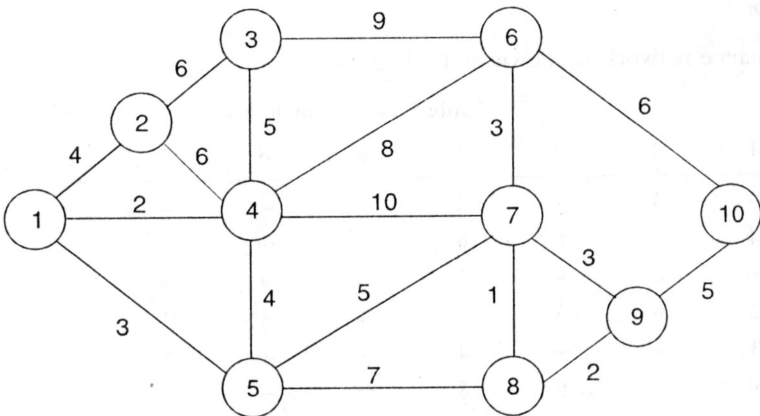

Figure 5.21 Distance network.

The objective of the minimum spanning tree problem is to connect the nodes of the network by a set of arcs such that the total length of all the arcs is minimized. Therefore, in the case of laying the telephone cable, the objective is to connect all the nodes by a set of arcs such that the total length of telephone cable to be laid is minimized.

In this section, the two algorithms for the minimum spanning tree problem are presented: *PRIM algorithm* and *Kruskal's algorithm*. These are discussed in the next sections.

5.3.1 PRIM Algorithm

PRIM algorithm is used to find solution for the minimum spanning tree problem. The steps of this algorithm are presented below.

Step 1: Represent the distance network (Figure 5.21) in a matrix form (Table 5.24).

Step 2: Let Q is a null set, which is the set of selected row numbers of the matrix.

Step 3: Select row 1 and include it in Q. Then, delete column 1 of the matrix.

Step 4: Find the minimum of the undeleted values among the rows in Q and select it by marking a square around it. (In case of tie, break it randomly.)

Step 5: Identify the corresponding column number (K) and then include row K in Q.

Step 6: Delete column K of the matrix.

Step 7: Check whether all the columns are deleted. If yes, go to step 8; otherwise go to step 4.

Step 8: Show the arcs in the spanning tree corresponding to the cells of the matrix marked with squares by thick lines.

Step 9: Find the total of all the values (marked with squares). This is the minimized total length of the arcs to connect all the nodes of the network as per the minimum spanning tree concept.

Example 5.4 Consider the distance network shown in Figure 5.21. Find the minimum spanning tree of this network using the PRIM algorithm.

Solution

Step 1: The distance network is shown in Table 5.24.

Table 5.24   Example 5.4

	1	2	3	4	5	6	7	8	9	10
1	–	4	∞	2	3	∞	∞	∞	∞	∞
2	4	–	6	6	∞	∞	∞	∞	∞	∞
3	∞	6	–	5	∞	9	∞	∞	∞	∞
4	2	6	5	–	4	8	10	∞	∞	∞
5	3	∞	∞	4	–	∞	5	7	∞	∞
6	∞	∞	9	8	∞	–	3	∞	∞	6
7	∞	∞	∞	10	5	3	–	1	3	∞
8	∞	∞	∞	∞	7	∞	1	–	2	∞
9	∞	∞	∞	∞	∞	∞	3	2	–	5
10	∞	∞	∞	∞	∞	6	∞	∞	5	–

Step 2: Let Q is a null set.

Step 3: $Q = \{1\}$. The matrix after deleting column 1 is shown as in Table 5.25, where each of the rows in Q is indicated by '*'.

Table 5.25   Distance Matrix after Deleting Column 1

	2	3	4	5	6	7	8	9	10
*1	4	∞	[2]	3	∞	∞	∞	∞	∞
2	–	6	6	∞	∞	∞	∞	∞	∞
3	6	–	5	∞	9	∞	∞	∞	∞
4	6	5	–	4	8	10	∞	∞	∞
5	∞	∞	4	–	∞	5	7	∞	∞
6	∞	9	8	∞	–	3	∞	∞	6
7	∞	∞	10	5	3	–	1	3	∞
8	∞	∞	∞	7	∞	1	–	2	∞
9	∞	∞	∞	∞	∞	3	2	–	5
10	∞	∞	∞	∞	6	∞	∞	5	–

Step 4: Minimum of the undeleted cell values of the rows in Q is 2. So, it is marked with a square as shown in Table 5.25.

Step 5: $K = 4$ and $Q = \{1, 4\}$.

Step 6: The matrix after deleting column 4 is shown in Table 5.26.

Table 5.26 Distance Matrix after Deleting Column 4

	2	3	5	6	7	8	9	10
*1	4	∞	3	∞	∞	∞	∞	∞
2	–	6	∞	∞	∞	∞	∞	∞
3	6	–	∞	9	∞	∞	∞	∞
*4	6	5	4	8	10	∞	∞	∞
5	∞	∞	–	∞	5	7	∞	∞
6	∞	9	∞	–	3	∞	∞	6
7	∞	∞	5	3	–	1	3	∞
8	∞	∞	7	∞	1	–	2	∞
9	∞	∞	∞	∞	3	2	–	5
10	∞	∞	∞	6	∞	∞	5	–

Step 7: All the columns are not deleted. So, go to step 4.

Step 4: Minimum of the undeleted cell values of the rows in Q is 3. So, it is marked with a square as shown in Table 5.26.

Step 5: $K = 5$ and $Q = \{1, 4, 5\}$.

Step 6: The matrix after deleting column 5 is shown as in Table 5.27.

Table 5.27 Distance Matrix after Deleting Column 5

	2	3	6	7	8	9	10
*1	4	∞	∞	∞	∞	∞	∞
2	–	6	∞	∞	∞	∞	∞
3	6	–	9	∞	∞	∞	∞
*4	6	5	8	10	∞	∞	∞
*5	∞	∞	∞	5	7	∞	∞
6	∞	9	–	3	∞	∞	6
7	∞	∞	3	–	1	3	∞
8	∞	∞	∞	1	–	2	∞
9	∞	∞	∞	3	2	–	5
10	∞	∞	6	∞	∞	5	–

Step 7: All the columns are not deleted. So, go to step 4.

Step 4: Minimum of the undeleted cell values of the rows in Q is 4. So, it is marked with a square as shown in Table 5.27.

Step 5: $K = 2$ and $Q = \{1, 4, 5, 2\}$.

Step 6: The matrix after deleting column 2 is shown as in Table 5.28.

Table 5.28 Distance Matrix after Deleting Column 2

	3	6	7	8	9	10
*1	∞	∞	∞	∞	∞	∞
*2	6	∞	∞	∞	∞	∞
3	–	9	∞	∞	∞	∞
*4	5	8	10	∞	∞	∞
*5	∞	∞	[5]	7	∞	∞
6	9	–	3	∞	∞	6
7	∞	3	–	1	3	∞
8	∞	∞	1	–	2	∞
9	∞	∞	3	2	–	5
10	∞	6	∞	∞	5	–

Step 7: All the columns are not deleted. So, go to step 4.

Step 4: Minimum of the undeleted cell values of the rows in Q is 5. (This minimum is for column 3 and row 4 as well as column 7 and row 5. One of these two locations is selected randomly.) So, it is marked with a square at the cell with respect to the row 5 and the column 7 as shown in Table 5.28.

Step 5: $K = 7$ (random selection) and $Q = \{1, 4, 5, 2, 7\}$.

Step 6: The matrix after deleting column 7 is shown as in Table 5.29.

Table 5.29 Distance Matrix after Deleting Column 7

	3	6	8	9	10
*1	∞	∞	∞	∞	∞
*2	6	∞	∞	∞	∞
3	–	9	∞	∞	∞
*4	5	8	∞	∞	∞
*5	∞	∞	7	∞	∞
6	9	–	∞	∞	6
*7	∞	3	[1]	3	∞
8	∞	∞	–	2	∞
9	∞	∞	2	–	5
10	∞	6	∞	5	–

Step 7: All the columns are not deleted. So, go to step 4.

Step 4: Minimum of the undeleted cell values of the rows in Q is 1. So, it is marked with a square as shown in Table 5.29.

Step 5: $K = 8$ and $Q = \{1, 4, 5, 2, 7, 8\}$.

Step 6: The matrix after deleting column 8 is shown as in Table 5.30.

Table 5.30   Distance Matrix after Deleting Column 8

	3	6	9	10
*1	∞	∞	∞	∞
*2	6	∞	∞	∞
3	–	9	∞	∞
*4	5	8	∞	∞
*5	∞	∞	∞	∞
6	9	–	∞	6
*7	∞	3	3	∞
*8	∞	∞	[2]	∞
9	∞	∞	–	5
10	∞	6	5	–

Step 7: All the columns are not deleted. So, go to step 4.

Step 4: Minimum of the undeleted values of the rows in Q is 2. So, it is marked with a square as shown in Table 5.30.

Step 5: $K = 9$ and $Q = \{1, 4, 5, 2, 7, 8, 9\}$.

Step 6: The matrix after deleting column 9 is shown as in Table 5.31.

Table 5.31   Distance Matrix after Deleting Column 9

	3	6	10
*1	∞	∞	∞
*2	6	∞	∞
3	–	9	∞
*4	5	8	∞
*5	∞	∞	∞
6	9	–	6
*7	∞	[3]	∞
*8	∞	∞	∞
*9	∞	∞	5
10	∞	6	–

Step 7: All the columns are not deleted. So, go to step 4.

Step 4: Minimum of the undeleted cell values of the rows in Q is 3. So, it is marked with a square as shown in Table 5.31.

Step 5: $K = 6$ and $Q = \{1, 4, 5, 2, 7, 8, 9, 6\}$.

Step 6: The matrix after deleting column 6 is shown as in Table 5.32.

Table 5.32   Distance Matrix after Deleting Column 6

	3	10
*1	∞	∞
*2	6	∞
3	–	∞
*4	5	∞
*5	∞	∞
*6	9	6
*7	∞	∞
*8	∞	∞
*9	∞	[5]
10	∞	–

Step 7: All the columns are not deleted. So, go to step 4.

Step 4: Minimum of the undeleted cell values of the rows in Q is 5, for column 3 and row 4 as well as column 10 and row 9. So, it is marked with a square as shown in Table 5.32.

Step 5: $K = 10$ (randomly selected) $Q = \{1, 4, 5, 2, 7, 8, 9, 6, 10\}$.

Step 6: The matrix after deleting column 10 is shown as in Table 5.33.

Table 5.33   Distance Matrix after Deleting Column 10

	3
*1	∞
*2	6
3	–
*4	[5]
*5	∞
*6	9
*7	∞
*8	∞
*9	∞
*10	∞

Step 7: All the columns are not deleted. So, go to step 4.

Step 4: Minimum of the undeleted cell values of the rows in Q is 5. So, it is marked with a square as shown in Table 5.33.

Step 5: $K = 3$ and $Q = \{1, 4, 5, 2, 7, 8, 9, 6, 10, 3\}$.

Step 6: After deleting column 3, the matrix is completely deleted.

Step 7: All the columns are deleted. So, go to step 8.

Step 8: The minimum spanning tree is shown in Figure 5.22 and the corresponding minimum total length of the arcs is 30.

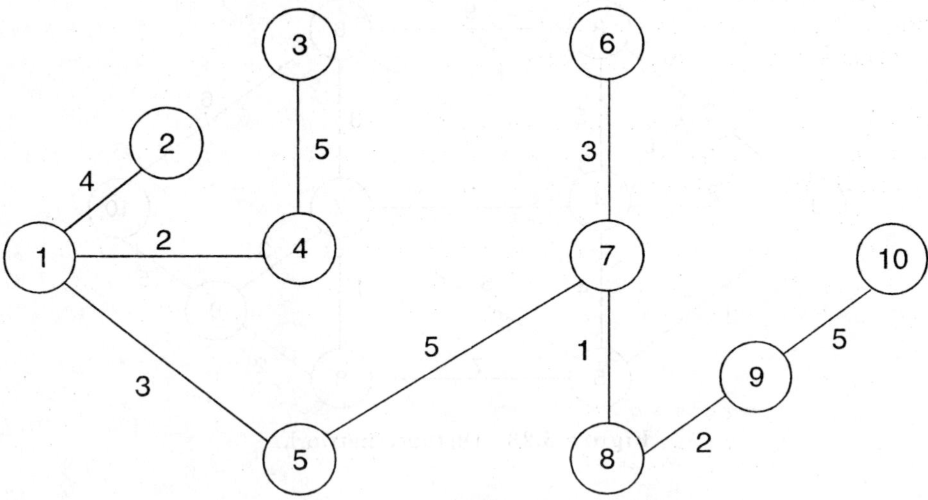

Figure 5.22 Minimum spanning tree.

5.3.2 Kruskal's Algorithm

Consider the arcs of the distance network of the given problem. If the network is undirected, then it is sufficient to form a table summarizing the distance of each arc (i, j) in the network only for $i < j$.

Step 1: Arrange the arcs along with their distances as per ascending order of their distances. Let this arrangement be called as set M.

Step 2: Create a graph with n nodes without arcs.

Step 3: Select the first undeleted arc from the set M.

Step 4: Check whether the inclusion of the selected arc in the previous step into the graph forms a cycle. If yes, go to step 6; otherwise, go to step 5.

Step 5: Include the selected arc in the graph.

Step 6: Delete that arc from the set M.

Step 7: Check whether all the nodes in the graph are connected. If no, go to step 3; otherwise go to step 8.

Step 8: Stop.

Example 5.5   Consider the network given as in Figure 5.23. Find the minimum spanning tree using Kruskal's algorithm.

 Solution   Since, the network is undirected, the distance of each arc (i, j), for $i < j$ is presented as in Table 5.34.

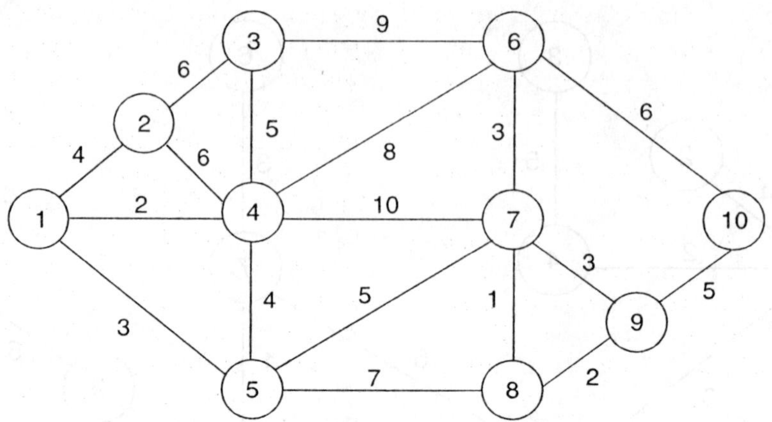

Figure 5.23 Distance network.

Table 5.34 Distances of the Arcs

Arc	Distance	Arc	Distance
1–2	4	4–7	10
1–4	2	5–7	5
1–5	3	5–8	7
2–3	6	6–7	3
2–4	6	6–10	6
3–4	5	7–8	1
3–6	9	7–9	3
4–5	4	8–9	2
4–6	8	9–10	5

Step 1: Sort the distances of the arcs (refer to Table 5.35).

Table 5.35 Sorted Distances of Arcs

Arc	Distance	Arc	Distance
7–8	1	5–7	5
1–4	2	9–10	5
8–9	2	2–3	6
1–5	3	2–4	6
6–7	3	6–10	6
7–9	3	5–8	7
1–2	4	4–6	8
4–5	4	3–6	9
3–4	5	4–7	10

Step 2 onwards: Figures 5.24 through 5.32 show step 2 and onwards. The arcs from 7–8 up to 9–10 except the arcs 7–9 and 4–5 in Table 5.35 are included in the spanning tree which is shown as in Figure 5.32 to connect all the nodes of the problem. Inclusion of the arcs 7–9 and 4–5 creates loops. Hence, they are not included. The sum of the distances of the arcs in the spanning tree which is shown as in Figure 5.32 is 30.

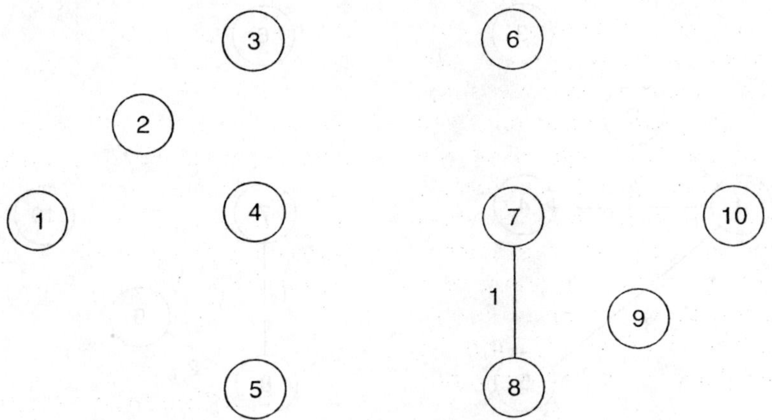

Figure 5.24 Partial spanning tree after adding arc 7–8.

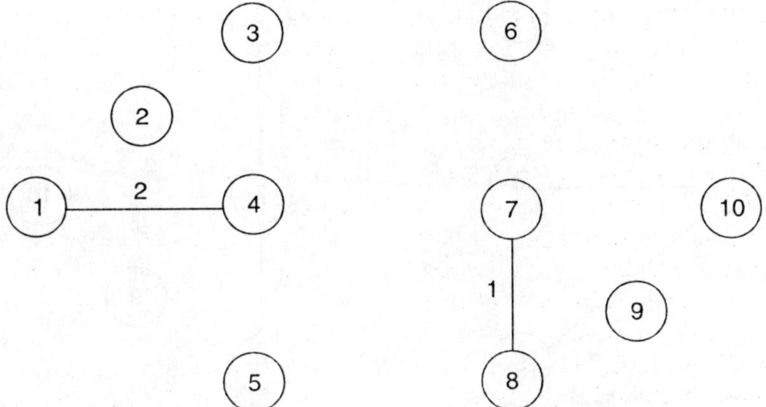

Figure 5.25 Partial spanning tree after adding arc 1–4.

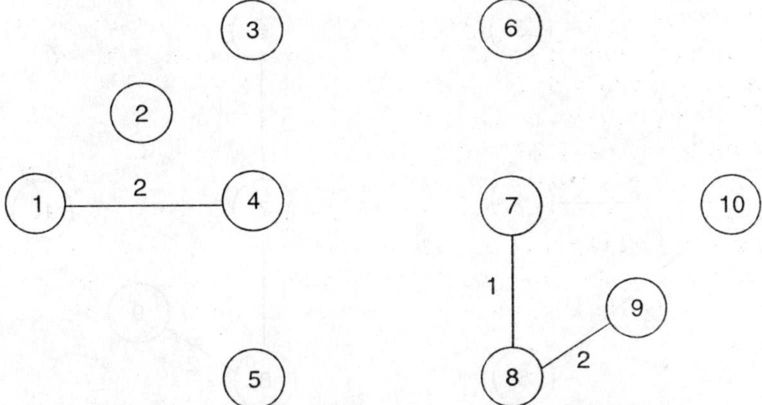

Figure 5.26 Partial spanning tree after adding arc 8–9.

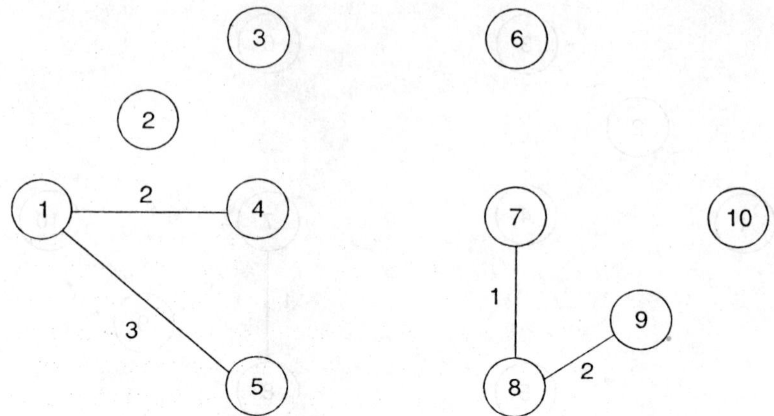

Figure 5.27 Partial spanning tree after adding arc 1–5.

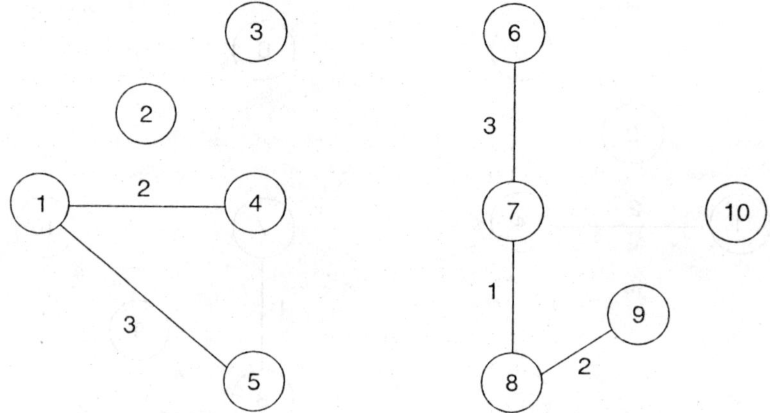

Figure 5.28 Partial spanning tree after adding arc 6–7.

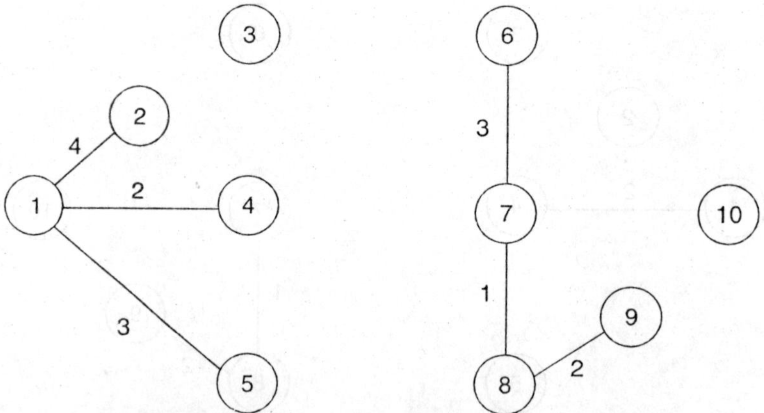

Figure 5.29 Partial spanning tree after adding arc 1–2.

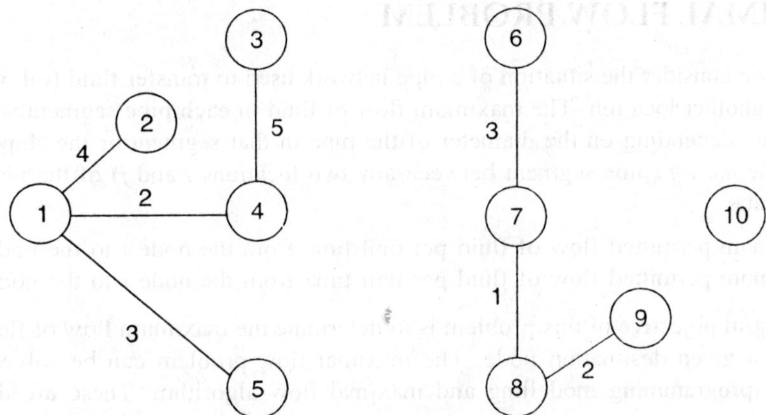

Figure 5.30 Partial spanning tree after adding arc 3–4.

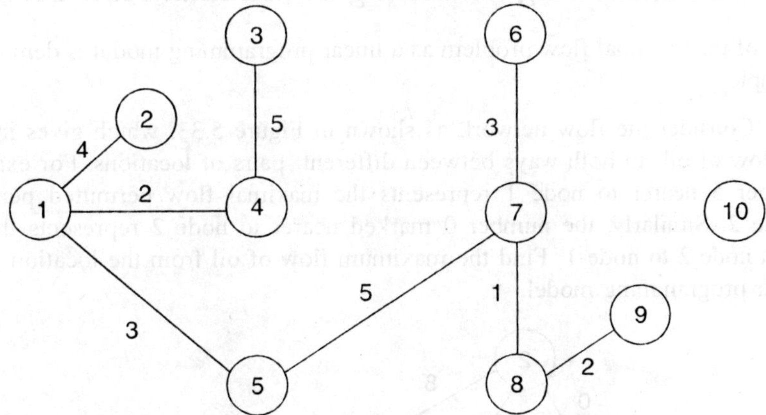

Figure 5.31 Partial spanning tree after adding arc 5–7.

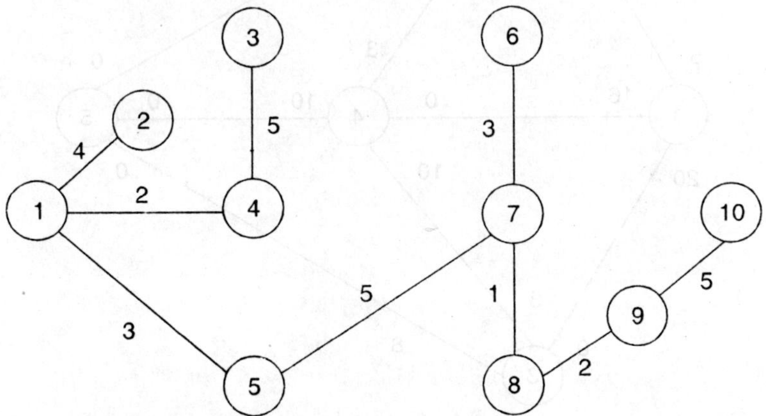

Figure 5.32 Minimum spanning tree after adding arc 9–10 (final).

5.4 MAXIMAL FLOW PROBLEM

In this section, we consider the situation of a pipe network used to transfer fluid (oil, water, etc.) from one location to another location. The maximum flow of fluid in each pipe segment will be limited to a particular value depending on the diameter of the pipe in that segment or the slope of the pipe in the segment. The arc i–j (pipe segment between any two locations i and j) of the network will have the following data:

1. Maximum permitted flow of fluid per unit time from the node i to the node j.
2. Maximum permitted flow of fluid per unit time from the node j to the node i.

A meaningful objective of this problem is to determine the maximum flow of fluid from a given source node to a given destination node. The maximal flow problem can be solved by using two methods: linear programming modelling and maximal flow algorithm. These are discussed in the following sections.

5.4.1 Linear Programming Modelling of Maximal Flow Problem

The modelling of the maximal flow problem as a linear programming model is demonstrated with the following example.

Example 5.6 Consider the flow network as shown in Figure 5.33, which gives information about capacities of flow of oil in both ways between different pairs of locations. For example, in the arc 1–2, the number 5 nearer to node 1 represents the maximal flow permitted per unit time from node 1 to node 2. Similarly, the number 0 marked nearer to node 2 represents the maximal flow permitted from node 2 to node 1. Find the maximum flow of oil from the location 1 to the location 5 using a linear programming model.

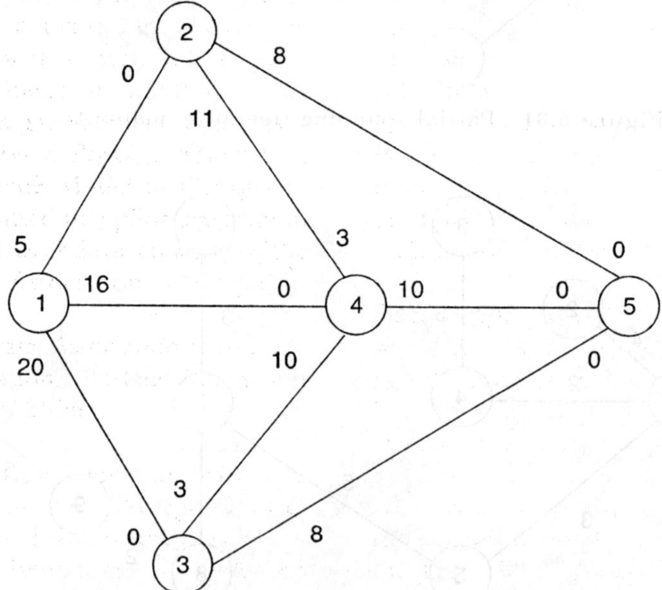

Figure 5.33 Network with flow capacities.

Solution Let us assume that the flow admitted into node 1 is equal to f. So, the flow out of node 5 is f. The different variables representing flows (forward as well as backward) on each of the arcs are summarized in Figure 5.34. In an arc (i, j), the flow capacity from the node i to node j is represented nearer to the node i and the flow capacity from the node j to node i is represented nearer to node j. For every node, the total quantity of flow into that node is equal to the total quantity of flow out of that node. On a given arc, if the flow in a particular direction is 0, there is no need to assign a variable to represent the flow in that direction.

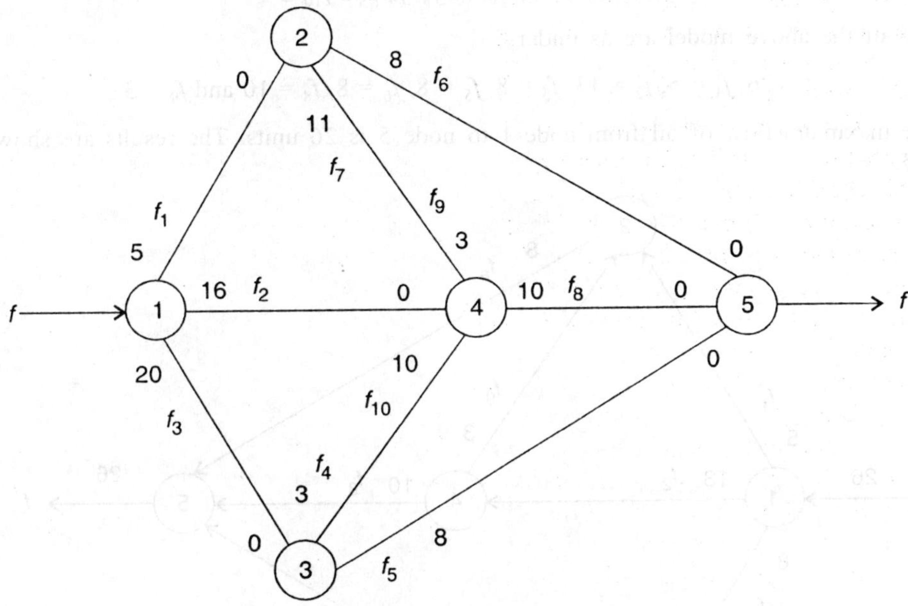

Figure 5.34 Network with flow capacities.

A linear programming model for the above problem to maximize the flow from node 1 to node 5 is presented below.

$$\text{Maximize } Z = f$$

subject to

$$f = f_1 + f_2 + f_3 \qquad \text{(at node 1)}$$
$$f_1 + f_9 = f_7 + f_6 \qquad \text{(at node 2)}$$
$$f_3 + f_{10} = f_4 + f_5 \qquad \text{(at node 3)}$$
$$f_2 + f_7 + f_4 = f_8 + f_9 + f_{10} \qquad \text{(at node 4)}$$
$$f_6 + f_8 + f_5 = f \qquad \text{(at node 5)}$$
$$f_1 \leq 5$$
$$f_2 \leq 16$$
$$f_3 \leq 20$$
$$f_4 \leq 3$$
$$f_5 \leq 8$$

$$f_6 \leq 8$$
$$f_7 \leq 11$$
$$f_8 \leq 10$$
$$f_9 \leq 3$$
$$f_{10} \leq 10$$

where

$$f_1, f_2, f_3, f_4, f_5, f_6, f_7, f_8, f_9 \text{ and } f_{10} \geq 0$$

The results of the above model are as under:

$$f = 26, f_1 = 5, f_2 = 13, f_3 = 8, f_5 = 8, f_6 = 8, f_8 = 10 \text{ and } f_9 = 3$$

Hence, the maximum flow of oil from node 1 to node 5 is 26 units. The results are shown as in Figure 5.35.

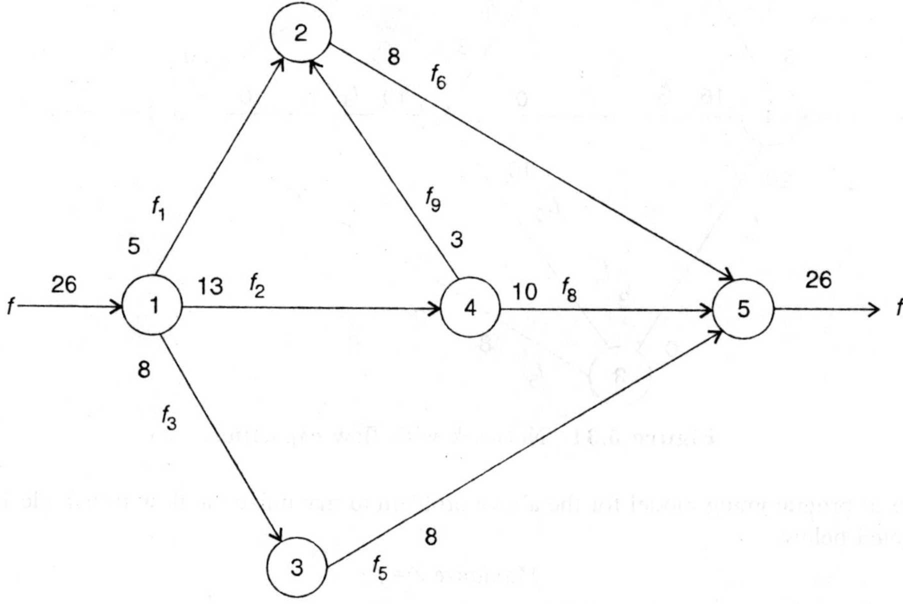

Figure 5.35 Results as per maximum flow from node 1 to node 5.

5.4.2 Maximal Flow Problem (MFP) Algorithm

The objective of this problem is to find the maximal flow from a given source (r) to a given destination (s). The algorithm to solve such problems is presented as follows:

Step 1: Form a FROM–TO capacity matrix [A] for the given flow network. A_{ij} is the flow from the node i to the node j (i = 1, 2, 3, ..., n and j = 1, 2, 3,..., n, where n is the total number of nodes in the network). Set the iteration number, k = 1 and cumulative flow, X = 0).

Step 2: Trace a path starting from the given source (r) to the given destination (s) directly or passing through some intermediate nodes which will have some feasible quantity of flow.

Step 3: Check whether such path exists. If yes, go to step 4; otherwise, go to step 7.

Step 4: Find the minimum of the capacities/remaining capacities of various links of the path traced in step 2. Let this quantity be Q_k.

$$\text{Set, } X = X + Q_k$$

(*Note*: While tracing a path in step 2, enough care should be taken to maximize Q_k.)

Step 5: Obtain the next flow matrix by performing the following steps.

(a) Subtract Q_k from all A_{ij} values corresponding to all forward links of the path traced.

(b) Add Q_k to all the A_{ij} values corresponding to all backward links of the path traced.

Step 6: Set $k = k + 1$ and go to step 2.

Step 7: Obtain a new flow matrix $[C]$ by subtracting the elements of the flow matrix $[B]$ in the last iteration from the corresponding elements of the flow matrix $[A]$ in the first iteration.

$$C_{ij} = A_{ij} - B_{ij}, \quad \text{if } A_{ij} > B_{ij} \quad \text{for all values of } i \text{ and } j;$$
$$= 0, \text{ otherwise.}$$

Step 8: (a) The cell entries of the flow matrix $[C]$ represent the flows in various arcs of the network.

(b) The maximal flow from the source node (r) to the destination node (s) is X.

(c) Map the cell entries on to the corresponding arcs of the network showing the actual flows in various arcs to achieve the maximal flow of X.

Example 5.7 Consider the pipe network shown as in Figure 5.36 showing the flow capacities between various pairs of locations in both ways. Find the maximal flow from node 1 to node 6.

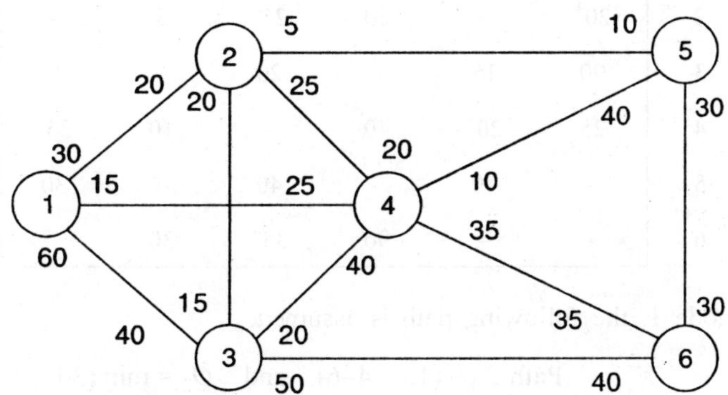

Figure 5.36 Network showing pipe capacities.

Solution

Iteration 1: The data of the Figure 5.36 is shown in matrix form as in Table 5.36.

Table 5.36 Flow Matrix of Iteration 1

		To				
	1	2	3	4	5	6
1	–	30	60^-	15	–	–
2	20	–	20	25	5	–
3	40^+	15	–	20	–	50^-
4	25	20	40	–	10	35
5	–	10	–	40	–	30
6	–	–	40^+	35	30	–

(From, rows 1–6)

To start with, starting from node 1, the following path is assumed.

$$\text{Path } 1 = 1\text{–}3\text{–}6 \quad \text{and} \quad Q_1 = \min (60 \text{ and } 50) = 50$$

Therefore, the feasible quantity of flow through the path is 50, i.e. $X = 50$.

Iteration 2: Now, assign minus sign to the cells 1–3 and 3–6 and then assign positive sign to the cells 3–1 and 6–3. Obtain Table 5.37 of iteration 2 by subtracting Q_1 from all the negatively-signed cells and adding Q_1 to all the positively-signed cells of Table 5.36.

Table 5.37 Flow Matrix of Iteration 2

		To				
	1	2	3	4	5	6
1	–	30^-	10	15	–	–
2	20^+	–	20	25^-	5	–
3	90	15	–	20	–	0
4	25	20^+	40	–	10	35^-
5	–	10	–	40	–	30
6	–	–	90	35^+	30	–

(From, rows 1–6)

Again starting from node 1, the following path is assumed.

$$\text{Path } 2 = (1\text{–}2\text{–}4\text{–}6) \quad \text{and} \quad Q_2 = \min (30, 25, 35) = 25$$

Therefore, the feasible quantity of flow through the path is 25.

$$\text{Cumulative flow, } X = X + Q_2 = 50 + 25 = 75$$

Iteration 3: Now, assign minus sign to the cells 1–2, 2–4, and 4–6 and then assign positive sign to the cells 2–1, 4–2 and 6–4. Obtain Table 5.38 of iteration 3 by subtracting Q_2 from all the negatively signed cells and adding Q_2 to all the positively signed cells of Table 5.37.

Table 5.38 Flow Matrix of Iteration 3

From \ To	1	2	3	4	5	6
1	–	5	10	15^-	–	–
2	45	–	20	0	5	–
3	90	15	–	20	–	0
4	25^+	45	40	–	10	10^-
5	–	10	–	40	–	30
6	–	–	90	60^+	30	–

Next, starting from node 1 again, the following path is assumed.

$$\text{Path } 3 = (1\text{--}4\text{--}6) \quad \text{and} \quad Q_3 = \min(15, 10) = 10$$

Therefore, the feasible quantity of flow through the path is 10.

$$\text{Cumulative flow, } X = X + Q_3 = 75 + 10 = 85$$

Iteration 4: Now, assign minus sign to the cells 1–4 and 4–6 and then assign positive sign to the cells 4–1 and 6–4. Obtain Table 5.39 of iteration 4 by subtracting Q_3 from all the negatively signed cells and adding Q_3 to all the positively signed cells of Table 5.38.

Table 5.39 Flow Matrix of Iteration 4

From \ To	1	2	3	4	5	6
1	–	5^-	10	5	–	–
2	45^+	–	20	0	5^-	–
3	90	15	–	20	–	0
4	35	45	40	–	10	0
5	–	10^+	–	40	–	30^-
6	–	–	90	70	30^+	–

Next, again starting from node 1, the following path is assumed.

$$\text{Path } 4 = (1\text{--}2\text{--}5\text{--}6) \quad \text{and} \quad Q_4 = \min(5, 5, 30) = 5$$

Therefore, the feasible quantity of flow through the path is 5.

$$\text{Cumulative flow, } X = X + Q_4 = 85 + 5 = 90$$

Iteration 5: Now, assign minus sign to the cells 1–2, 2–5 and 5–6 and then assign positive sign to the cells 2–1, 5–2 and 6–5. Obtain Table 5.40 of Iteration 5 by subtracting Q_4 from all the negatively signed cells and adding Q_4 to all the positively signed cells of Table 5.39.

Table 5.40 Flow Matrix of Iteration 5

From \ To	1	2	3	4	5	6
1	–	0	10^-	5	–	–
2	50	–	20	0	0	–
3	90^+	15	–	20^-	–	0
4	35	45	40^+	–	10^-	0
5	–	15	–	40^+	–	25^-
6	–	–	90	70	35^+	–

Again starting from node 1, the following path is assumed.

$$\text{Path 5} = (1\text{–}3\text{–}4\text{–}5\text{–}6) \quad \text{and} \quad Q_5 = \min(10, 20, 10, 25) = 10$$

Therefore, the feasible quantity of flow through the path is 10.

$$\text{Cumulative flow, } X = X + Q_5 = 90 + 10 = 100$$

Iteration 6: Now, assign minus sign to the cells 1–3, 3–4, 4–5 and 5–6 and then assign positive sign to the cells 3–1, 4–3, 5–4 and 6–5. Obtain Table 5.41 of Iteration 6 by subtracting Q_5 from all the negatively signed cells and adding Q_5 to all the positively signed cells of Table 5.40.

Table 5.41 Flow Matrix of Iteration 6

From \ To	1	2	3	4	5	6
1	–	0	0	5	–	–
2	50	–	20	0	0	–
3	100	15	–	10	–	0
4	35	45	50	–	0	0
5	–	15	–	50	–	15
6	–	–	90	70	45	–

Now, no more flow is possible from node 1 to node 6. Subtract the matrix elements of Table 5.41 from the corresponding elements of Table 5.36 and write only the positive values in Table 5.42. Table 5.42 gives the final results and the corresponding flows are mapped onto different arcs in Figure 5.37. The maximal flow from node 1 to node 6 is 100 units.

Table 5.42 Final Flow Matrix

From \ To	1	2	3	4	5	6
1	–	30	60	10	–	–
2		–	–	25	5	–
3			–	10	–	50
4				–	10	35
5					–	15
6						–

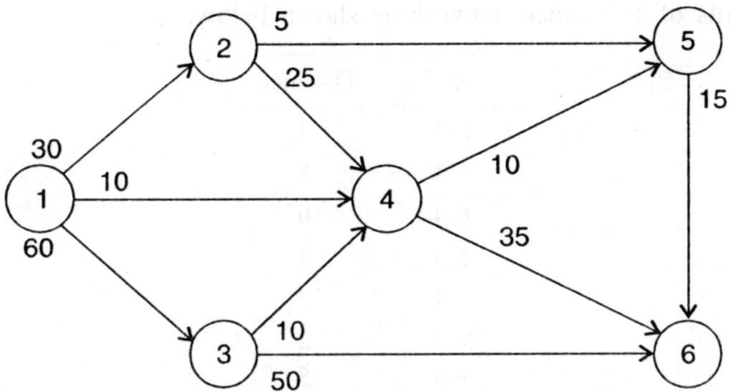

Figure 5.37 Network showing final flows in arcs with respect to maximal flow.

QUESTIONS

1. What is the shortest-path problem? Give some practical applications of the shortest-path problem.

2. What is minimum spanning problem? What are its practical applications?

3. State maximal flow problem and give its practical applications.

4. Explain the steps of the following algorithms:

    (a) Dijkstra's algorithm   (b) Floyd's algorithm

    (c) PRIM algorithm   (d) Kruskal's algorithm

5. Explain the matrix method for the maximal flow problem.

6. Consider the details of a distance network as shown below:

Arc	Distance	Arc	Distance
1–2	8	3–6	6
1–3	5	4–5	8
1–4	7	4–6	12
1–5	16	5–8	7
2–3	15	6–8	9
2–6	3	6–9	15
2–7	4	7–9	12
3–4	5	8–9	6

(a) Construct the distance network.

(b) Find the shortest path from node 1 to node 9 using the systematic method.

(c) Find the shortest path from node 1 to node 9 using Dijkstra's algorithm.

7. Consider the details of a distance network as shown below:

Arc	Distance
1–2	3
1–3	8
1–4	10
2–3	4
2–4	7
3–4	2
3–5	8
4–5	6

(a) Construct the distance network.

(b) Apply Floyd's algorithm and obtain the final matrices, D^5 and P^5.

(c) Find the shortest path and the corresponding distance for each of the following:

   (i) from node 1 to node 5

   (ii) from node 2 to node 5.

8. Consider the details of a distance network as shown below:

Arc	Distance	Arc	Distance
1–2	6	5–6	13
1–3	7	5–8	9
1–4	10	6–7	5
2–3	8	6–8	4
2–5	4	6–9	8
3–4	6	6–10	3
3–5	11	7–9	10
3–6	3	8–10	10
3–7	5	9–10	9
4–7	7		

(a) Construct the distance network.

(b) Find the minimum spanning tree using PRIM algorithm.

9. Consider the distance network which is given in Question 8 and find the minimum spanning tree using Kruskal's algorithm.

10. Consider the following details of piping network which is used to transfer oil.

   (a) Draw the flow network.

   (b) Develop a linear programming model to determine the maximum flow from the node 1 to the node 6 and also the corresponding flow pattern.

Arc i–j	Flow		Arc i–j	Flow	
	f_{ij}	f_{ji}		f_{ij}	f_{ji}
1–2	20	–	3–4	13	–
1–3	25	–	3–5	10	8
2–3	5	10	4–5	15	–
2–4	9	4	4–6	30	–
2–5	15	–	5–6	25	–

11. Consider the data of a flow network as shown below:

(a) Draw the flow network.

(b) Determine the maximum flow from node 1 to node 6 and also the corresponding flow pattern using the matrix method.

Arc i–j	Flow		Arc i–j	Flow	
	f_{ij}	f_{ji}		f_{ij}	f_{ji}
1–2	60	10	3–4	35	–
1–3	35	25	3–5	30	28
2–3	25	20	4–5	45	–
2–4	19	24	4–6	40	–
2–5	25	30	5–6	55	–

INTEGER PROGRAMMING

6

6.1 INTRODUCTION

In linear programming problems, decision variables are non-negative values which are restricted to be zero or more than zero. This demonstrates one of the basic properties of linear programming, namely, continuity, which means that fractional values of the decision variables are possible in the solution of a linear programming model. For problems like, product mix problem, nutrition problem, the assumption of continuity may be valid for some applications. In some product mix problems, for example, the production volume of different types of fertilizer in tonnages may satisfy the continuity assumption. But if the products are high valued, which means, the profit contribution per unit of them will be very high (e.g. ship, bucket wheel excavator, crane, etc.), the assumption of the continuity may lead to some practical difficulties.

Some items cannot be produced in fractions, like 101.5 cranes, 3.2 bucket wheel excavators, etc. If we round off the production volume of any one of such products to the nearest integer value, the corresponding solution would be different from the optimal solution based on the assumption of continuity. The difference would be significant if the profit per unit of each of the products in the product mix problem is very high. Hence, there is a need for integer programming methods to overcome this difficulty.

Example 6.1 Consider the following production planning situation of a company manufacturing mixies. The company plans to manufacture two types of mixies. The selling prices for these mixies are: Model A costs Rs. 1,750 per unit and Model B costs Rs. 2,000 per unit. Daily production volume of each type of these mixies is constrained by available man hours and available machine hours. The production specifications for the given problem situation are presented in Table 6.1.

Table 6.1   Example 6.1

Resource	Resource requirement/unit		Availability (in hours)
	Model A	Model B	
Man hours	4	8	32
Machine hours	6	4	36

Find the optimal production plan for the above problem. (*Note:* The values of availabilities of the resources are assumed as small numbers just to demonstrate the problem on a graph conveniently. In reality, these values will be high.)

Solution Let X_1 and X_2 be the respective number of Model A and Model B to be manufactured. An integer linear programming model for the above problem is represented as given below:

$$\text{Maximize } Z = 1750X_1 + 2000X_2$$

subject to

$$4X_1 + 8X_2 \leq 32$$

$$6X_1 + 4X_2 \leq 36$$

$$X_1 \text{ and } X_2 \geq 0 \text{ and integers}$$

The optimal linear programming solution of the above problem is given below.

$$X_1 = 5.0,\ X_2 = 1.5,\ Z(\text{optimum}) = \text{Rs. } 11,750$$

In this solution, the value of X_1 is an integer and that of X_2 is a non-integer. But the solution of this problem will be meaningful only when the values of all the decision variables are integer. A simple approach may be to round off the value of X_2 to the previous integer value of 1 to maintain feasibility. After rounding off the value of X_2, the values of the decision variables, X_1 and X_2 become 5 and 1, respectively. The corresponding total profit is Rs. 10,750 which is less than that of the optimum value of the linear programming solution.

The optimum integer solution for the given problem is as follows which is better than the rounded off solution of the linear programming problem.

$$X_1 = 4,\ X_2 = 2 \quad \text{and} \quad Z(\text{optimum}) = \text{Rs. } 11,000$$

Need for integer programming algorithms

Practical applications of integer programming have become much more common in use. Because of this fact, and the problems associated with simply rounding off linear programming solutions, there is a need for an efficient solution procedure for integer linear programming problems.

Three major algorithms have been developed to solve integer programming problems as listed below and they are explained in this chapter.

1. The *cutting plane algorithm* developed by Ralph E. Gomory.
2. The *branch-and-bound-algorithm* developed by A.H. Land and A.G. Doing.
3. The *additive algorithm* for zero-one integer programming problems developed by E. Balas.

6.2 INTEGER PROGRAMMING FORMULATIONS

Application of integer programming arises in a wide variety of managerial decision-making situations. In this section, some of them are presented with illustrative examples.

Example 6.2 (Fixed-charge problem) Consider a situation in which a fixed charge or set-up cost is incurred if a particular activity is undertaken. In such situation, the total cost of the activity will be the sum of the fixed charge for undertaking the activity plus the variable cost associated with the level of the activity.

Solution Let

X_j be the production level of the activity j.

c_j be the variable cost (cost/unit) associated with activity j.

K_j be the fixed charge associated with the activity j.

b_j be the maximum production level of the activity j.

N be the number of activities.

M be the at most value of the sum of the production levels of all the activities.

The total cost of the activity j is given by:

$$C_j(X_j) = (K_j + c_j X_j), \quad \text{if } X_j > 0$$
$$= 0, \quad \text{if } X_j = 0$$

A model for the above situation is presented below.

$$\text{Minimize } Z = \sum_{j=1}^{N} C_j(X_j) = \sum_{j=1}^{N} (K_j + c_j X_j)$$

In the above model, the objective function is nonlinear. So, it is converted into a linear objective function by using binary variables as defined below:

$$Y_j = 0, \quad \text{if } X_j = 0$$
$$= 1, \quad \text{if } X_j > 0$$

Now, a model with a linear objective function along with linear constraints for the fixed-charge problem is presented as below:

$$\text{Minimize } Z = \sum_{j=1}^{N} (c_j X_j + K_j Y_j)$$

subject to

$$\sum_{j=1}^{N} X_j = M$$
$$X_j \leq b_j Y_j$$
$$Y_j = 0 \text{ or } 1$$
$$X_j \geq 0,$$

for $j = 1, 2, 3,..., N$.

In the above model, if Y_j is equal to 1, then the value of X_j will be limited to at most b_j; otherwise, it will be made equal to 0.

Example 6.3 A textile company can use any or all of three different processes for weaving its standard white polyester fabric. Each of these production processes has a weaving machine set-up cost and per square-metre processing cost. These costs and the capacities of each of the three production processes are shown in Table 6.2.

Table 6.2 Example 6.3

Process number	Weaving machine set-up cost (Rs.)	Processing cost (Rs.)	Maximum daily capacity (sq. metre)
1	150	15	2000
2	240	10	3000
3	300	8	3500

The daily demand forecast for its white polyester fabric is 4000 sq. metre. The company's production manager wants to determine the optimal combination of the production processes and their actual daily production levels such that the total production cost is minimized.

Solution   Let X_j be the production level for process j (j = 1, 2, 3). Also let

$$Y_j = 1, \quad \text{if process } j \text{ is used, and}$$
$$Y_j = 0, \quad \text{if process } j \text{ is not used}$$

The required model is as follows:

$$\text{Minimize } Z = (15X_1 + 10X_2 + 8X_3) + (150Y_1 + 240Y_2 + 300Y_3)$$

subject to

$$X_1 + X_2 + X_3 = 4000$$
$$X_1 - 2000Y_1 \leq 0$$
$$X_2 - 3000Y_2 \leq 0$$
$$X_3 - 3500Y_3 \leq 0$$

where

$$X_j \geq 0, j = 1, 2, 3 \text{ and } Y_j = 0 \text{ or } 1, j = 1, 2, 3$$

Example 6.4 **(Covering problems)**   Consider a set of residential regions (customers) in a state which are to be covered by a set of dealers (facilities) of a particular company. A residential region is said to be covered by the dealer network of the company if there is a dealer located within say a distance of 30 km from the residential region. In this example, if the objective is to cover all the residential regions with the minimum number of dealers, then it is called *total covering problem*. Instead, if there is an upper limit on the number of dealers to be located, then it may not be possible to cover some of the residential regions. Hence, this type of problem is called *partial covering problem*. The objective of this partial covering problem is to cover as many residential regions as possible with the maximum possible number of dealers.

Solution

Zero-one programming model for total covering problem

Let M be the total number of customers, N be the total number of proposed potential sites, C_{ij} is called as covering coefficients whose value is 1, if the customer region i can be covered by the facility at the site j, and 0, otherwise. Also $Y_j = 1$, if site j is selected for assigning a facility; = 0, otherwise.

Now, a zero-one programming model for the *total covering problem* is represented as follows:

$$\text{Minimize } Z = \sum_{j=1}^{n} Y_j$$

subject to

$$\sum_{j=1}^{n} C_{ij}Y_j \geq 1, \quad \text{for } i = 1, 2, 3,..., M$$

$$Y_j = 0 \text{ or } 1, \quad \text{for } j = 1, 2, 3,..., N$$

In this model, the objective function minimizes the total number of sites selected for assigning facilities. The constraints ensure that each customer is covered by atleast one facility located at any of the selected sites.

Zero-one programming model for partial covering problem

Let M be the total number of customers, N be the total number of proposed potential sites, K be the at most number of facilities available, kk be the actual number of sites selected for assigning facilities ($kk \le K$); C_{ij} is the covering coefficients, which equals to 1 if the customer region i can be covered by the facility at the site j, and 0, otherwise. Also let,

$$Y_j = 1, \quad \text{if site } j \text{ is selected for assigning a facility;}$$
$$= 0, \quad \text{otherwise.}$$

and

$$X_i = 1, \quad \text{if the customer } i \text{ is covered by a facility;}$$
$$= 0, \quad \text{otherwise.}$$

A zero-one programming model for the *partial covering problem* is presented below.

$$\text{Maximize } Z = H \sum_{i=1}^{M} X_i - \sum_{j=1}^{N} Y_j$$

subject to

$$\sum_{j=1}^{N} Y_j \le K$$

$$\sum_{j=1}^{N} C_{ij} Y_j + (1 - X_i) \ge 1, \qquad i = 1, 2, ..., M$$

$$\sum_{j=1}^{N} C_{ij} Y_j + N(1 - X_i) \le N, \qquad i = 1, 2, ..., M$$

$$X_i = 0 \text{ or } 1, \quad i = 1, 2, ..., M \quad \text{and} \quad Y_j = 0 \text{ or } 1, \quad j = 1, 2, ..., N$$

where H is a very large value (say, $H > M$).

The objective function has the following two components:

1. Maximizing the number of customers covered and
2. Minimizing the number of sites selected for locating facilities.

The objective is to cover as many customers as possible by using the at most number of sites that are to be assigned facilities. Hence, the first component is given more weightage when compared to the second component. So, in the model, the number of customers covered is first maximized and then the number of sites selected to cover the maximum number of customers is minimized within the limit on the at most number sites to be selected. For a given problem, though the given number of sites which are to be assigned facilities is K, the actual number of sites which will be assigned facilities (kk) may be less than or equal to K.

The first constraint assures that the total number of sites which are assigned facilities is restricted to at most K. The second and the third sets of constraints jointly assure the following: If a customer is covered, then it is served by the facility from at least one site. If a customer is not covered, then it is not served by any of the facilities which are assigned to various selected sites. If the ith customer is covered, then the ith constraint in the constraint set 2 will become active and the ith constraint in the constraint set 3 will become inactive. Similarly, if the ith customer is not covered, then the *i*th constraint in the constraint set 3 will become active and the *i*th constraint in the constraint set 2 will become inactive.

In this model, the total number of zero-one variables is $M + N$ and the total number of constraints is $2M + 1$.

Example 6.5 (Modelling parallel processors under single machine scheduling) In most of the single-machine scheduling problems, it is possible to use more than one machine of the same specification to process a given set of single operation jobs. Under this situation, there will be a scope to minimize makespan measure which is considered to be very important among many measures. Here, *m* number of similar machines are continuously available to process *n* number of single operation jobs. This is an example of using parallel facilities (processors/machines).

Solution The makespan M^* is computed by the following formula:

$$M^* = \max \left\{ \frac{1}{m} \sum_{j=1}^{n} t_j, \ \max \ (t_j) \right\}$$

where t_j is the processing time of the job j, m is the number of parallel machines (processors) and n is the number of independent single operation jobs.

An *integer programming* formulation to minimize the makespan for the single machine scheduling problem with parallel processors is discussed as follows: Let m be the number of parallel processors, n be the number of independent jobs, Y be the makespan of the schedule and t_i be the processing time of the job i. Also let,

$$X_{ij} = 1, \text{ if job } i \text{ is assigned to machine } j$$
$$= 0, \text{ otherwise.}$$

A model to minimize the makespan of this problem is given below.

$$\text{Minimize } Y$$

subject to

$$Y - \sum_{i=1}^{n} t_i X_{ij} \geq 0, \qquad \text{for } j = 1, 2, 3,..., m$$

and

$$\sum_{j=1}^{m} X_{ij} = 1, \qquad \text{for } i = 1, 2, 3,..., n$$

where

$$X_{ij} = 0 \text{ or } 1, \ i = 1, 2, 3,..., n \quad \text{and} \quad j = 1, 2, 3,..., m$$

Here, the first set of constraints ensures that the makespan is at least as large as the processing time assigned to any machine. The second set of constraints ensures that each job is assigned to only one machine. This formulation contains $(m + n)$ constraints with $(mn + 1)$ variables.

6.3 THE CUTTING-PLANE ALGORITHM

An algorithm for solving fractional (pure integer) and mixed integer programming problems has been developed by **Ralph E. Gomory**.

Fractional (pure integer) algorithm

Step 1: First, relax the integer requirements.

Step 2: Solve the resulting LP problem using simplex method.

Step 3: If all the basic variables (or the required variables) have integer values, optimality of the integer programming problem is reached. So, go to step 7; otherwise go to step 4.

Step 4: Examine the constraints corresponding to the current optimal solution. Also, let m be the number of constraints, n be the number of variables (including slack, surplus and artificial variables), b_i be the right-hand side value of the ith constraint, and a_{ij} be the technological coefficients (matrix of left-hand side constants of the constraints). Then, the constraint equations are summarized as follows:

$$\sum_{j=1}^{n} a_{ij}X_j = b_i, \qquad i = 1, 2, 3,..., m$$

For each basic variable with non-integer solution in the current optimal table, find the fractional part, f_i. Therefore, $b_i = [b_i] + f_i$, where $[b_i]$ is the integer part of b_i and f_i is the fractional part of b_i.

Step 5: Choose the largest fraction among various f_i's; i.e. Max (f_i). Treat the constraint corresponding to the maximum fraction as the source row. Let the corresponding source row be as follows:

$$b_i = X_i + \sum_{j=1}^{n} a_{ij}w_j \quad \text{or} \quad X_i = b_i - \sum_{j=1}^{n} a_{ij}w_j$$

where variables X_i ($i = 1, 2, 3,..., m$) represent basic variables and variables w_j ($j = 1, 2, 3,..., n$) are the non-basic variables. This kind of assumption is for convenience only.

Some examples of diving b_i and a_{ij} into integer and fractional parts are shown as in Table 6.3.

Table 6.3 Examples of Integer and Fractional Parts

b_i or a_{ij}	$[b_i]$ or $[a_{ij}]$ (Truncated integer)	$f_i = b_i - [b_i]$ $f_{ij} = a_{ij} - [a_{ij}]$
$2\dfrac{2}{3}$	2	$\dfrac{2}{3}$
$-3\dfrac{1}{4}$	-4	$\dfrac{3}{4}$
-5	-5	0
$-\dfrac{3}{7}$	-1	$\dfrac{4}{7}$

Based on the source equation, develop an additional constraint (Gomory's constraint or fractional cut) as shown below:

$$S_i = \sum_{j=1}^{n} f_{ij}w_j - f_i \quad \text{or} \quad -f_i = S_i - \sum_{j=1}^{n} f_{ij}w_j$$

where S_i is non-negative slack variable and also an integer.

Step 6: Append the fractional cut as the last row in the latest optimal table and proceed further using dual simplex method, and find the new optimum solution. If this new optimum solution is integer then go to step 7; otherwise go to step 4.

Step 7: Print the integer solution [X_i's and Z values].

Example 6.6 Find the optimum integer solution to the following linear programming problem.

$$\text{Maximize } Z = 5X_1 + 8X_2$$

subject to

$$X_1 + 2X_2 \leq 8$$
$$4X_1 + X_2 \leq 10$$
$$X_1, X_2 \geq 0 \text{ and integers}$$

Solution The canonical form of the above problem is as follows:

$$\text{Maximize } Z = 5X_1 + 8X_2$$

subject to

$$X_1 + 2X_2 + S_1 = 8$$
$$4X_1 + X_2 + S_2 = 10$$
$$X_1, X_2, S_1 \text{ and } S_2 \geq 0 \text{ and integers}$$

The initial table is shown in Table 6.4.

Table 6.4 Initial Table

CB_i	Basic variable	C_j				Solution	Ratio
		5	8	0	0		
		X_1	X_2	S_1	S_2		
0	S_1	1	2	1	0	8	8/2 = 4*
0	S_2	4	1	0	1	10	10/1 = 10
	Z_j	0	0	0	0	0	
	$C_j - Z_j$	5	8*	0	0		

From the above table, the entering variable is X_2, since its $C_j - Z_j$ value is the maximum positive value. The minimum ratio is 4 and the corresponding variable is S_1. Therefore, S_1 leaves the basis. The resulting table is shown as in Table 6.5.

Table 6.5 Iteration 1

CB_i	Basic variable	C_j				Solution	Ratio
		5	8	0	0		
		X_1	X_2	S_1	S_2		
8	X_2	1/2	1	1/2	0	4	8
0	S_2	7/2	0	−1/2	1	6	12/7*
	Z_j	4	8	4	0	32	
	$C_j - Z_j$	1*	0	−4	0		

In Table 6.5, the maximum positive value for $C_j - Z_j$ is 1. The corresponding variable is X_1. Therefore, X_1 enters the basis. The minimum ratio is for the S_2 row. Therefore, S_2 leaves the basis. The resulting table is shown as in Table 6.6.

Table 6.6 Iteration 2 (Optimal Table)

CB_i	C_j	5	8	0	0	
	Basic variable	X_1	X_2	S_1	S_2	Solution
8	X_2	0	1	4/7	–1/7	22/7
5	X_1	1	0	–1/7	2/7	12/7
	Z_j	5	8	27/7	2/7	236/7
	$C_j - Z_j$	0	0	–27/7	–2/7	

In Table 6.6, all the values in the criterion row ($C_j - Z_j$ row) are 0 or negative. Hence, optimality for linear programming is reached. The results are as follows:

$$X_1 = \frac{12}{7}, \qquad X_2 = \frac{22}{7}, \qquad Z(\text{optimum}) = \frac{236}{7}$$

Since the values of the decision variables X_1 and X_2 are not integers, the solution is not optimal. So, to obtain integer solution for the given problem further steps are carried out.

The integer and fractional parts of the basic variables are summarized in Table 6.7.

Table 6.7 Summary of Integer and Fractional Parts

Basic variable in the optimal table	b_i	$[b_i] + f_i$
X_1	12/7	1 + (5/7)
X_2	22/7	3 + (1/7)

The fractional part, f_1 is the maximum. So, select the row X_1 as the source row for developing the first cut.

$$\frac{12}{7} = X_1 - \frac{1}{7}S_1 + \frac{2}{7}S_2$$

or

$$1 + \frac{5}{7} = X_1 + \left(-1 + \frac{6}{7}\right)S_1 + \left(0 + \frac{2}{7}\right)S_2$$

The corresponding fractional cut is

$$-\frac{5}{7} = S_3 - \frac{6}{7}S_1 - \frac{2}{7}S_2 \qquad \text{(Cut 1)}$$

This cut is appended to Table 6.6 as reproduced in Table 6.8 and further solved using dual simplex method.

Table 6.8 Table after Appending Cut 1

CB_i	Basic variable	X_1	X_2	S_1	S_2	S_3	Solution
	C_j	5	8	0	0	0	
8	X_2	0	1	4/7	−1/7	0	22/7
5	X_1	1	0	−1/7	2/7	0	12/7
0	S_3	0	0	−6/7	$\boxed{-2/7}$	1	−5/7*
	Z_j	5	8	27/7	2/7	0	236/7
	$C_j - Z_j$	0	0	−27/7	−2/7*	0	

Only the third row (containing S_3) has a negative solution value. Therefore, S_3 leaves the basis. The entering variable is determined based on Table 6.9.

Table 6.9 Determination of Entering Variable

Variable	X_1	X_2	S_1	S_2	S_3
$-(C_j - Z_j)$	0	0	27/7	2/7	0
Row S_3	0	0	−6/7	−2/7	1
Ratio (absolute value)	–	–	9/2	1	–

The smallest absolute ratio is 1 and the corresponding variable is S_2. So, the variable S_2 enters the basis. The resultant values are shown in Table 6.10.

Table 6.10 Table after Pivot Operation

CB_i	Basic variable	X_1	X_2	S_1	S_2	S_3	Solution
	C_j	5	8	0	0	0	
8	X_2	0	1	1	0	−1/2	7/2
5	X_1	1	0	−1	0	1	1
0	S_2	0	0	3	1	−7/2	5/2
	Z_j	5	8	3	0	1	33
	$C_j - Z_j$	0	0	−3	0	−1	

The solution is still non-integer. So, develop a fractional cut. The basic variables, X_2 and S_2 are not integers. The fractional parts of both of them are 1/2. The constraint X_2 is selected randomly as the source row for developing the next cut. Therefore,

$$\frac{7}{2} = X_2 + S_1 - \frac{1}{2}S_3$$

or

$$3 + \frac{1}{2} = (1 + 0)\,X_2 + (1 + 0)\,S_1 + \left(-1 + \frac{1}{2}\right)S_3$$

Therefore, the corresponding fractional cut is

$$-\frac{1}{2} = S_4 - \frac{1}{2}S_3 \qquad \text{(Cut 2)}$$

Append this constraint at the end of Table 6.10 as shown in Table 6.11.

Table 6.11   Table after Pivot Operation

CB_i	Basic variable	C_j → 5 X_1	8 X_2	0 S_1	0 S_2	0 S_3	0 S_4	Solution
8	X_2	0	1	1	0	−1/2	0	7/2
5	X_1	1	0	−1	0	1	0	1
0	S_2	0	0	3	1	−7/2	0	5/2
0	S_4	0	0	0	0	−1/2	1	−1/2*
	Z_j	5	8	3	0	1	0	33
	$C_j - Z_j$	0	0	−3	0	−1*	0	

Only row S_4 has a negative value under the solution column. Therefore, S_4 leaves the basis. The entering variable is determined based on Table 6.12.

Table 6.12   Determination of Entering Variable

Variable	X_1	X_2	S_1	S_2	S_3	S_4
$-(C_j - Z_j)$	0	0	3	0	1	0
Row S_4	0	0	0	0	−1/2	1
Ratio (absolute value)	−	−	−	−	2	−

The smallest absolute ratio (only ratio) is 2 and the corresponding variable is S_3. So, the variable S_3 enters the basis. The resultant values are shown as in Table 6.13.

Table 6.13   Table after Pivot Operation

CB_i	Basic variable	C_j → 5 X_1	8 X_2	0 S_1	0 S_2	0 S_3	0 S_4	Solution
8	X_2	0	1	1	0	0	−1	4
5	X_1	1	0	−1	0	0	2	0
0	S_2	0	0	3	1	0	−7	6
0	S_3	0	0	0	0	1	−2	1
	Z_j	5	8	3	0	0	2	32
	$C_j - Z_j$	0	0	−3	0	0	−2	

In Table 6.13, the values of all the basic variables are integers. So, the optimality is reached and the corresponding results are summarized as follows:

$$X_1 = 0, \qquad X_2 = 4 \qquad Z(\text{optimum}) = 32$$

Directions to solve mixed integer programming problems

In reality, all the decision variables of an integer programming problem need not be integers. In such problem, if the solution values to the linear programming problem are integers, the steps of adding Gomory's cut are not required. Otherwise, necessary number of iterations is to be carried out by adding different cuts until the values of the required decision variables are made integers. In each of these iterations, the source row is selected from among the rows corresponding to the decision variables whose values are restricted to integer in the given problem.

Research direction to design superior cut

The effectiveness of the integer programming procedure is determined in terms of number of cuts required to solve a given problem. Instead of simply selecting the source row based on the maximum value of f_i, one can use different approaches which will result in a reduced number of cuts required to solve a problem. This is called the *strength/effectiveness of the cut*. Researches can come out with superior cuts which will result in reduced number of cuts to solve a problem.

Example 6.7 Consider the following integer linear programming problem and solve it.

$$\text{Maximize } Z = 5X_1 + 10X_2 + 8X_3$$

subject to

$$2X_1 + 5X_2 + X_3 \leq 10$$
$$X_1 + 4X_2 + 2X_3 \leq 12$$
$$X_1, X_2 \text{ and } X_3 \geq 0$$

Solution The canonical form of the given problem is shown below.

$$\text{Maximize } Z = 5X_1 + 10X_2 + 8X_3 + 0S_1 + 0S_2$$

subject to

$$2X_1 + 5X_2 + X_3 + S_1 = 10$$
$$X_1 + 4X_2 + 2X_3 + S_2 = 12$$
$$X_1, X_2, X_3, S_1 \text{ and } S_2 \geq 0$$

The different iterations of the simplex method applied to this problem till the optimality is reached are shown in Table 6.14. The optimal results of the linear programming problem from the last iteration of the Table 6.14 are:

$$X_1 = 8/3 = 2 + 2/3, \; X_3 = 14/3 = 4 + 2/3 \text{ and } Z(\text{optimum}) = 152/3.$$

The value of the decision variables, X_1 and X_3 are not integers. Further, the fractional part of X_1 as well as X_3 is 2/3. So, the maximum fractional part is 2/3 and the row X_1 is selected for developing a Gomory's cut.

$$8/3 = X_1 + 2X_2 + 2/3S_1 - 1/3S_2$$
$$2 + 2/3 = (1 + 0)X_1 + (2 + 0)X_2 + (0 + 2/3)S_1 + (-1 + 2/3)S_2$$

The corresponding fractional cut is shown below and it is appended to the final iteration of the Table 6.14 as shown in Table 6.15.

$$-2/3 = S_3 - 2/3 S_1 - 2/3 S_2 \longrightarrow \text{Cut 1}$$

In Table 6.15, the leaving variable is S_3. The entering variable is determined as shown in Table 6.16 and it is S_1. The corresponding next iteration is presented in Table 6.17. Since, the results in Table 6.17 are integers, the optimal integer solution is reached. The optimal solution is:

$$X_1 = 2, \ X_3 = 5, \ S_1 = 1 \text{ and } Z(\text{optimum}) = 50$$

Table 6.14 Iterations of Simplex Method Applied to Example 6.7

CB_i	Basic variable	X_1	X_2	X_3	S_1	S_2	Solution	Ratio
	C_j	5	10	8	0	0		
0	S_1	2	5	1	1	0	10	2*
0	S_2	1	4	2	0	1	12	3
	Z_j	0	0	0	0	0	0	
	$C_j - Z_j$	5	10*	8	0	0		
10	X_2	2/5	1	1/5	1/5	0	2	10
0	S_2	−3/5	0	6/5	−4/5	1	4	10/3*
	Z_j	4	10	2	2	0	20	
	$C_j - Z_j$	1	0	6*	−2	0		
10	X_2	1/2	1	0	1/3	−1/6	4/3	8/3*
8	X_3	−1/2	0	1	−2/3	5/6	10/3	−
	Z_j	1	10	8	−2	5	40	
	$C_j - Z_j$	4*	0	0	2	−5		
5	X_1	1	2	0	2/3	−1/3	8/3 = 2 + 2/3	
8	X_3	0	1	1	−1/3	2/3	14/3 = 4 + 2/3	
	Z_j	5	18	8	2/3	11/3	152/3	
	$C_j - Z_j$	0	−8	0	−2/3	−11/3		

Table 6.15 Table after Appending Cut 1

CB_i	Basic variable	X_1	X_2	X_3	S_1^*	S_2	S_3	Solution
	C_j	5	10	8	0	0	0	
5	X_1	1	2	0	2/3	−1/3	0	8/3
8	X_3	0	1	1	−1/3	2/3	0	14/3
0	S_3	0	0	0	−2/3	−2/3	1	−2/3*
	Z_j	5	18	8	2/3	11/3	0	152/3
	$C_j - Z_j$	0	−8	0	−2/3	−11/3	0	

Table 6.16 Determination of Entering Variable

	X_1	X_2	X_3	S_1	S_2	S_3
$-(C_j - Z_j)$	0	8	0	2/3	11/3	0
Row S_3	0	0	0	−2/3	−2/3	1
Ratio (absolute value)	–	–	–	1*	11/2	–

Table 6.17 Table after Pivot Operation

CB_i	C_j	5	10	8	0	0	0	
	Basic variable	X_1	X_2	X_3	S_1	S_2	S_3	Solution
5	X_1	1	2	0	0	−1	1	2
8	X_3	0	1	1	0	1	−1/2	5
0	S_1	0	0	0	1	1	−3/2	1
	Z_j	5	18	8	0	3	1	50
	$C_j - Z_j$	0	−8	0	0	−3	−1	

6.4 BRANCH-AND-BOUND TECHNIQUE

If the number of decision variables in an integer programming problem is only two, a *branch-and-bound* technique can be used to find its solution graphically. Various terminologies of branch-and-bound technique are explained as under:

Branching.   If the solution to the linear programming problem contains non-integer values for some or all decision variables, then the solution space is reduced by introducing constraints with respect to any one of those decision variables. If the value of a decision variable X_1 is 2.5, then two more problems will be created by using each of the following constraints.

$$X_1 \le 2 \quad \text{and} \quad X_1 \ge 3$$

Lower bound.   This is a limit to define a lower value for the objective function at each and every node. The *lower bound* at a node is the value of the objective function corresponding to the truncated values (integer parts) of the decision variables of the problem in that node.

Upper bound.   This is a limit to define an upper value for the objective function at each and every node. The *upper bound* at a node is the value of the objective function corresponding to the linear programming solution in that node.

Fathomed subproblem/node.   A problem is said to be *fathomed* if any one of the following three conditions is true:

1. The values of the decision variables of the problem are integer.
2. The upper bound of the problem which has non-integer values for its decision variables is not greater than the current best lower bound.
3. The problem has infeasible solution.

This means that further branching from this type of fathomed nodes is not necessary.

Current best lower bound. This is the best lower bound (highest in the case of maximization problem and lowest in the case of minimization problem) among the lower bounds of all the fathomed nodes. Initially, it is assumed as infinity for the root node.

Branch-and-bound algorithm applied to maximization problem

Step 1: Solve the given linear programming problem graphically. Set, the current best lower bound, Z_B as ∞.

Step 2: Check, whether the problem has integer solution. If yes, print the current solution as the optimal solution and stop; otherwise go to step 3.

Step 3: Identify the variable X_k which has the maximum fractional part as the branching variable. (In case of tie, select the variable which has the highest objective function coefficient.)

Step 4: Create two more problems by including each of the following constraints to the current problem and solve them.

$$X_k \leq \text{Integer part of } X_k$$

$$X_k \geq \text{Next integer of } X_k$$

Step 5: If any one of the new subproblems has infeasible solution or fully integer values for the decision variables, the corresponding node is fathomed. If a new node has integer values for the decision variables, update the current best lower bound as the lower bound of that node if its lower bound is greater than the previous current best lower bound.

Step 6: Are all terminal nodes fathomed? If the answer is yes, go to step 7; otherwise, identify the node with the highest lower bound and go to step 3.

Step 7: Select the solution of the problem with respect to the fathomed node whose lower bound is equal to the current best lower bound as the optimal solution.

Example 6.8 Solve the following integer programming problem using branch-and-bound technique.

$$\text{Maximize } Z = 10X_1 + 20X_2$$

subject to

$$6X_1 + 8X_2 \leq 48$$

$$X_1 + 3X_2 \leq 12$$

$$X_1, X_2 \geq 0 \text{ and integers}$$

Solution The introduction of the non-negative constraints $X_1 \geq 0$ and $X_2 \geq 0$ will eliminate the second, third and fourth quadrants of the X_1X_2 plane as shown in Figure 6.1.

Now, from the first constraint in equation form

$$6X_1 + 8X_2 = 48$$

we get $X_2 = 6$, when $X_1 = 0$; and $X_1 = 8$, when $X_2 = 0$. Similarly from the second constraint in equation form

$$X_1 + 3X_2 = 12$$

we have $X_2 = 4$, when $X_1 = 0$; and $X_1 = 12$, when $X_2 = 0$.

Now, plot the constraints 1 and 2 as shown in Figure 6.1.

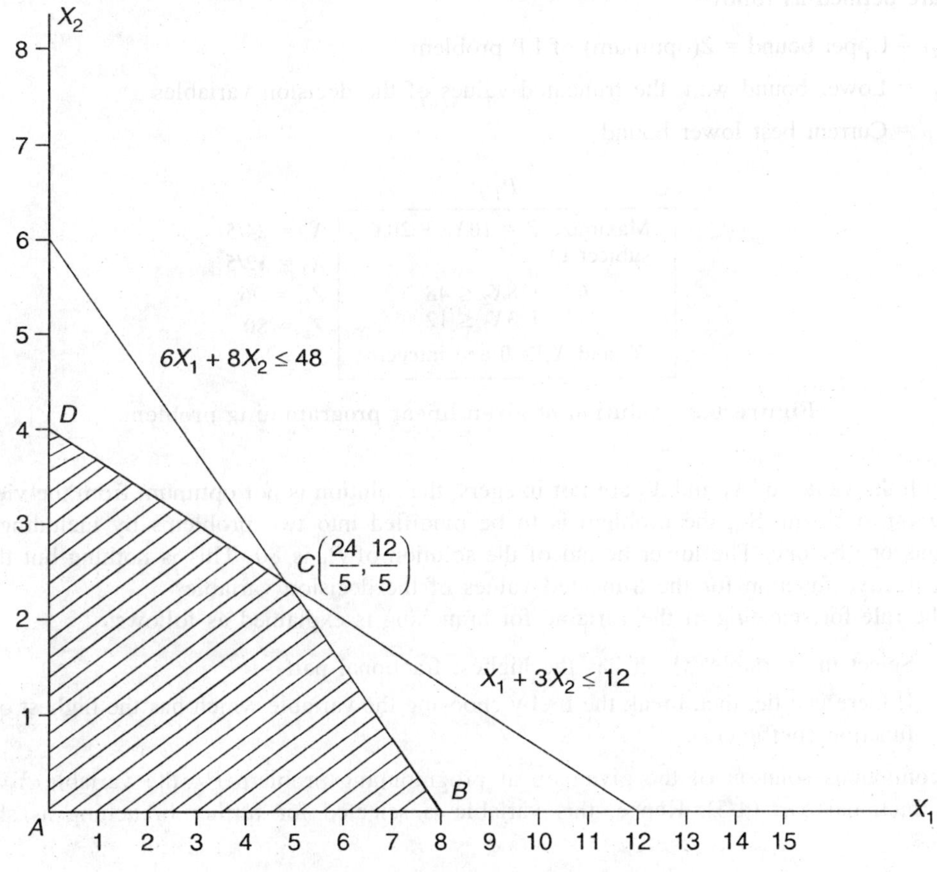

Figure 6.1 Feasible region of Example 6.8.

The closed polygon *ABCD* is the feasible region. The objective function value at each of the corner points of the closed polygon is computed as follows by substituting its coordinates in the objective function:

$$Z(A) = 10 \times 0 + 20 \times 0 = 0$$

$$Z(B) = 10 \times 8 + 20 \times 0 = 80$$

$$Z(C) = 10 \times \frac{24}{5} + 20 \times \frac{12}{5} = 96$$

$$Z(D) = 10 \times 0 + 20 \times 4 = 80$$

Since, the type of the objective function is maximization, the solution corresponding to the maximum *Z* value is to be selected as the optimum solution. The *Z* value is maximum for the corner point *C*. Hence, the corresponding solution of the continuous linear programming problem is presented below.

$$X_1 = \frac{24}{5}, \quad X_2 = \frac{12}{5}, \quad Z(\text{optimum}) = 96$$

These are jointly shown as problem P_1 in Figure 6.2. The notations for different types of lower bound are defined as follows:

Z_U = Upper bound = Z(optimum) of LP problem

Z_L = Lower bound w.r.t. the truncated values of the decision variables

Z_B = Current best lower bound

$$P_1$$

Maximize $Z = 10X_1 + 20X_2$ subject to $6X_1 + 8X_2 \le 48$ $X_1 + 3X_2 \le 12$ X_1 and $X_2 \ge 0$ and integers	$X_1 = 24/5$ $X_2 = 12/5$ $Z_U = 96$ $Z_L = 80$ $Z_B = \infty$

Figure 6.2 Solution of given linear programming problem.

Since both the values of X_1 and X_2 are not integers, the solution is not optimum from the view point of the given problem. So, the problem is to be modified into two problems by including integer constraints one by one. The lower bound of the solution of P_1 is 80. This is nothing but the value of the objective function for the truncated values of the decision variables.

The rule for selecting of the variable for branching is explained as follows:

1. Select the variable which has the highest fractional part.
2. If there is a tie, then break the tie by choosing the variable which has the highest objective function coefficient.

In the continuous solution of the given linear programming problem P_1, the variable X_1 has the highest fractional part (4/5). Hence, this variable is selected for further branching as shown in Figure 6.3.

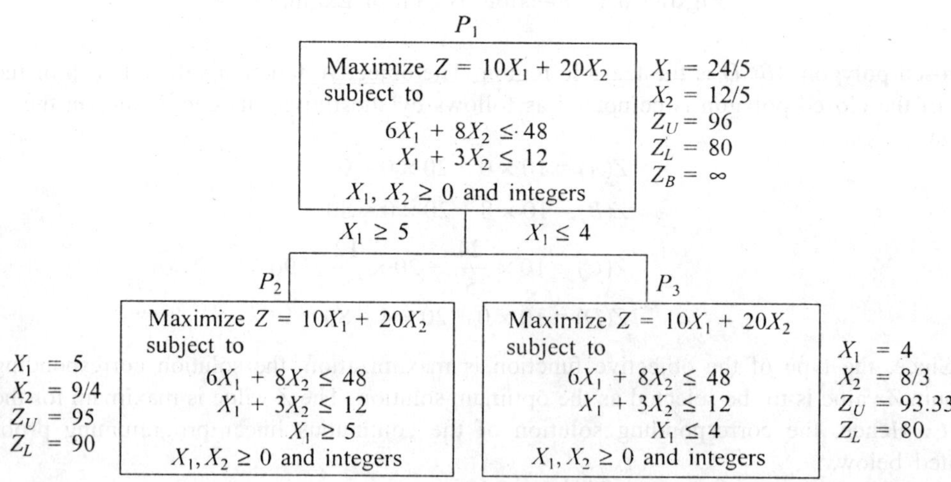

Figure 6.3 Branching from P_1.

In Figure 6.3, the problems, P_2 and P_3 are generated by adding an additional constraint. The subproblem, P_2 is created by introducing '$X_1 \geq 5$' in problem P_1, and the problem P_3 is created by introducing '$X_1 \leq 4$' in problem P_1. The corresponding effects in slicing the non-integer feasible region are shown in Figures 6.4 and 6.5, respectively. The solution for each of the subproblems, P_2 and P_3

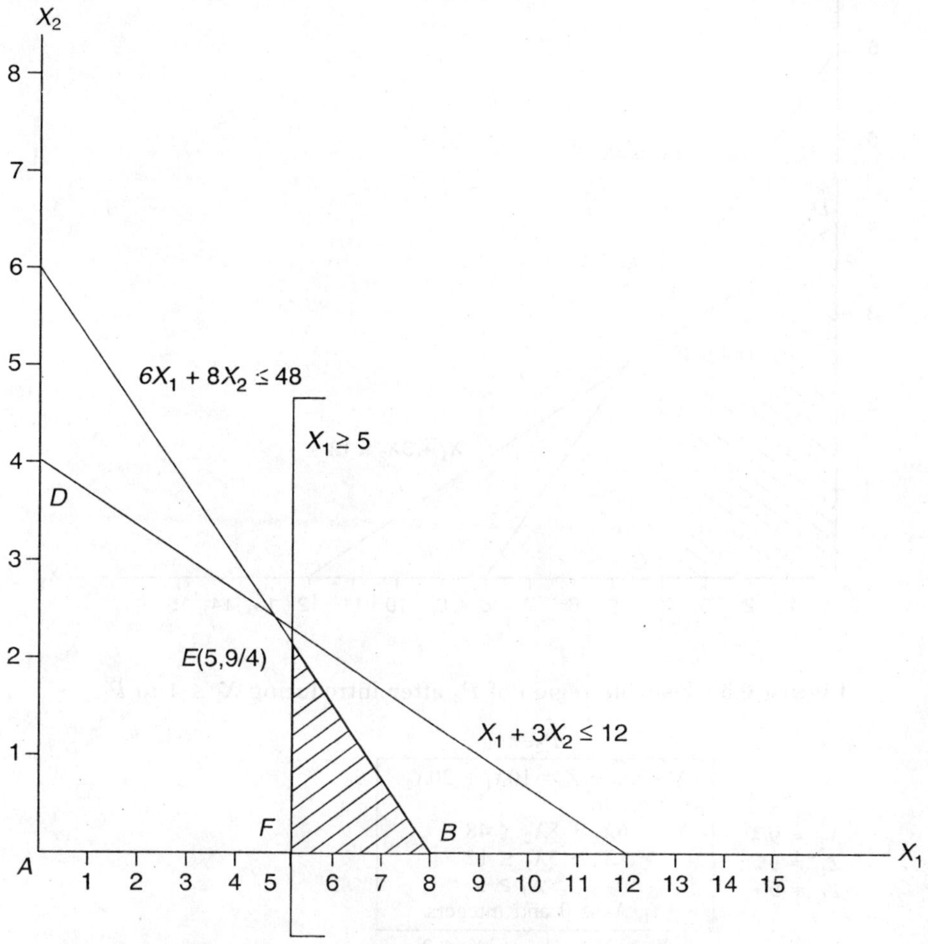

Figure 6.4   Feasible region of P_2 after introducing $X_1 \geq 5$ to P_1.

is obtained from Figures 6.4 and 6.5, respectively. These are summarized in Figure 6.3. The problem P_2 has the highest lower bound of 90 among the unfathomed terminal nodes. So, the further branching is done from this node as shown in Figure 6.6.

In Figure 6.6, the problems, P_4 and P_5 are generated by adding an additional constraint to P_2. The problem, P_4 is created by including '$X_2 \geq 3$' in problem P_2, and problem P_5 is created by including '$X_2 \leq 2$' in problem P_2. The corresponding effects in slicing the non-integer feasible region are shown in Figures 6.7 and 6.8, respectively. The solution for each of the problems P_4 and P_5 is obtained from Figures 6.7 and 6.8, respectively. The problem P_4 has infeasible solution. So, this node is fathomed. The lower bound of the node P_5 is 90. But, the solution of the node P_5 is still non-integer. Now, the

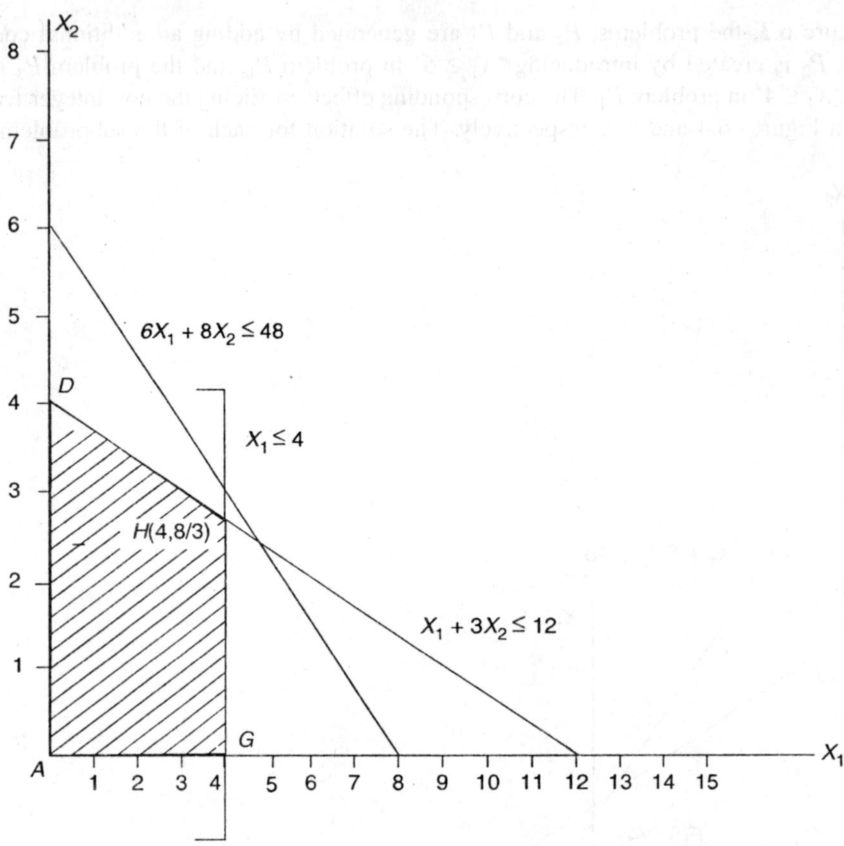

Figure 6.5 Feasible region of P_3 after introducing $X_1 \leq 4$ to P_1.

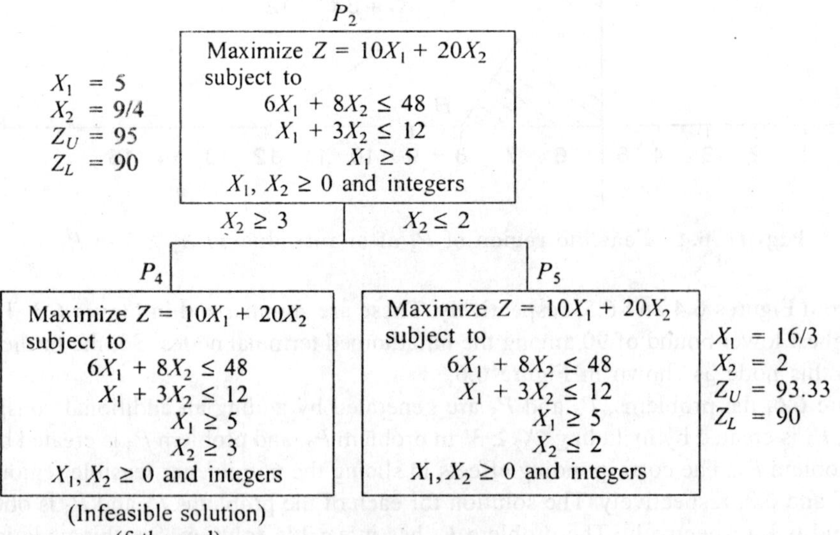

Figure 6.6 Branching from P_2.

lower bound of the node Z_L is the maximum when compared to that of all other unfathomed terminal nodes (only J) at this stage. So the further branching should be done from the node P_4 as shown in Figure 6.9.

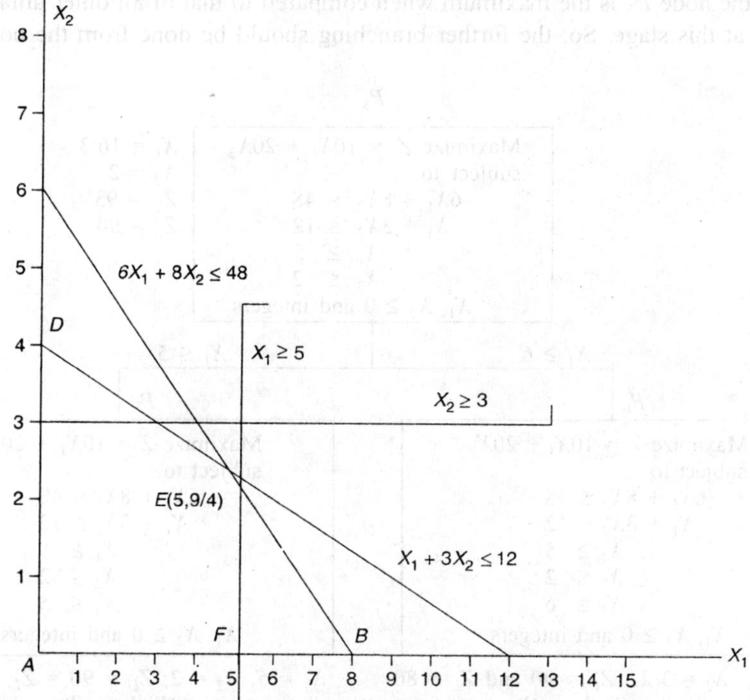

Figure 6.7 Infeasible region of P_4 after introducing $X_2 \geq 3$ to P_2.

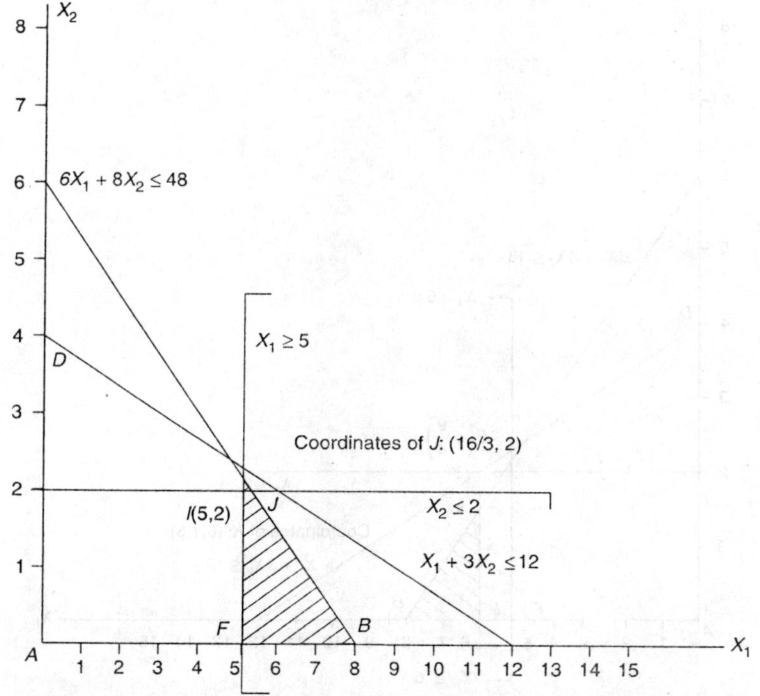

Figure 6.8 Feasible region of P_5 after introducing $X_2 \leq 2$ to P_2.

lower bound of the node P_5 is the maximum when compared to that of all other unfathomed terminal nodes (only P_3) at this stage. So, the further branching should be done from the node, P_5 as shown in Figure 6.9.

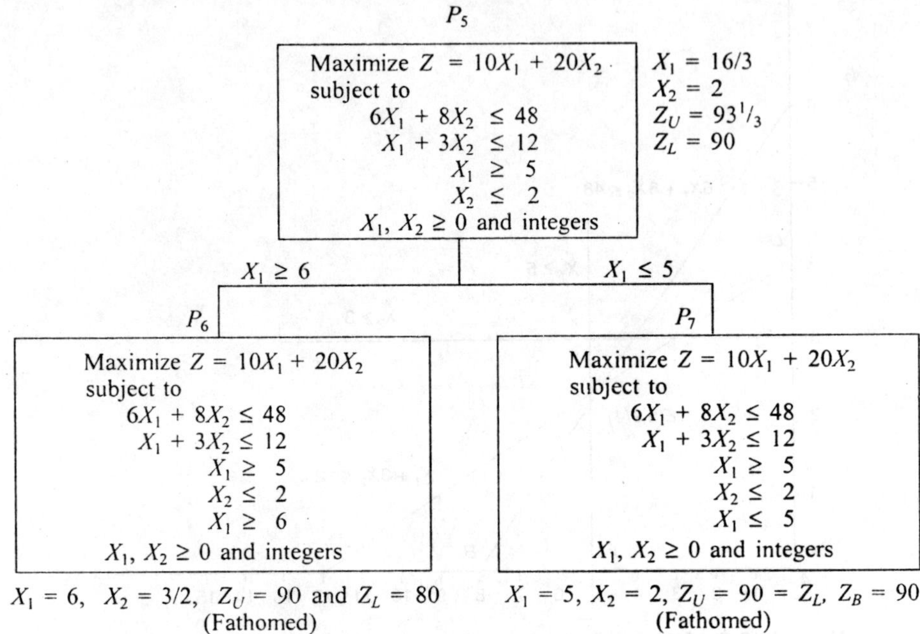

$X_1 = 6$, $X_2 = 3/2$, $Z_U = 90$ and $Z_L = 80$ (Fathomed)

$X_1 = 5$, $X_2 = 2$, $Z_U = 90 = Z_L$, $Z_B = 90$ (Fathomed)

Figure 6.9 Branching from P_5.

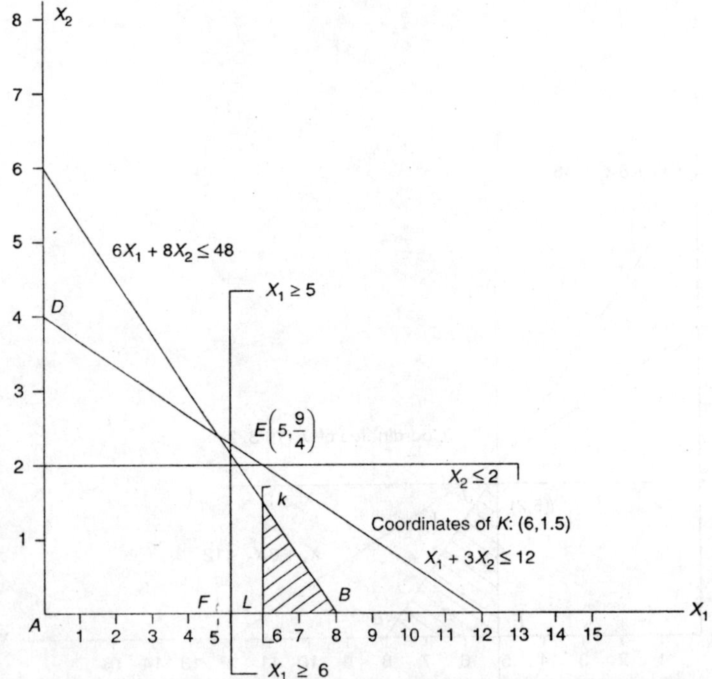

Figure 6.10 Feasible region of P_6 after introducing $X_1 \geq 6$ to P_5.

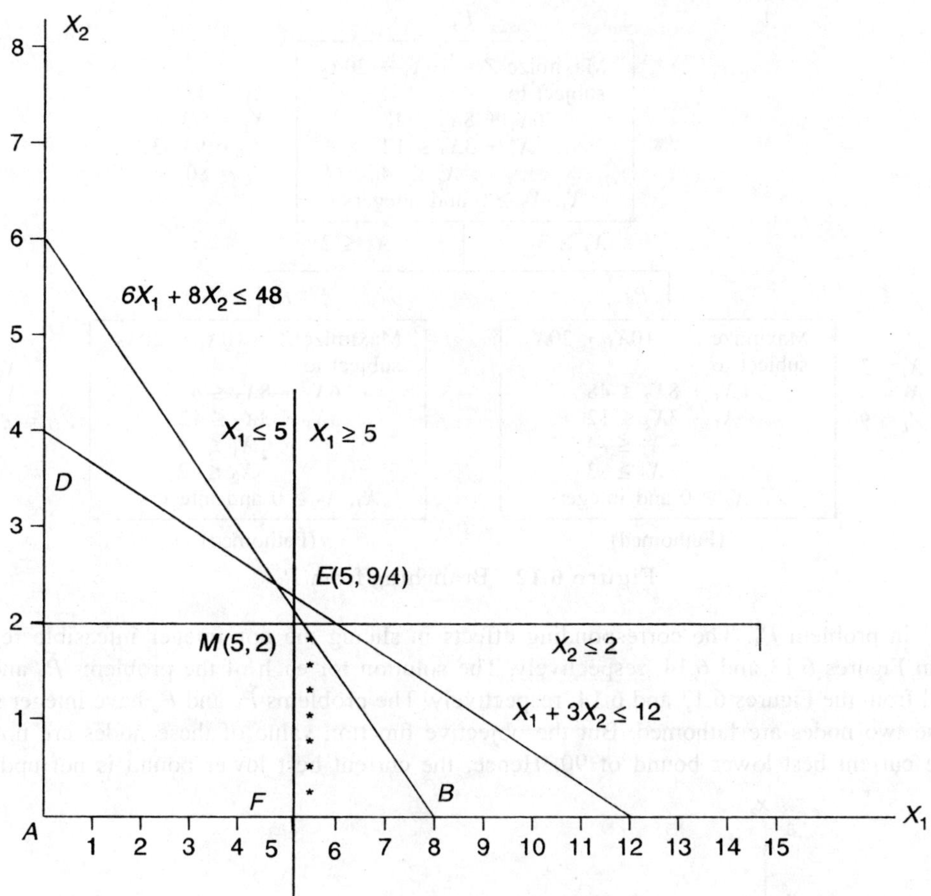

Figure 6.11 Feasible region of P_7 after introducing $X_1 \leq 5$ to P_5.
[Feasible region is the vertical line from $M(5, 2)$ to $F(5, 0)$ indicated by *s.]

In Figure 6.9, the problems, P_6 and P_7 are generated by adding an additional constraint to P_5. The problem P_6 is created by including '$X_1 \geq 6$' in the problem P_5 and problem P_7 is created by including '$X_1 \leq 5$' in problem P_5. The corresponding effects in slicing the non-integer feasible region are shown in Figures 6.10 and 6.11, respectively. The solution for each of the problems P_6 and P_7 are also obtained from these figures, respectively. The problem P_7 has integer solution. So, it is a fathomed node. Hence, the current best lower bound (Z_B) is updated to its objective function value, 90.

The solution of the node P_6 is non-integer and its lower bound and upper bound are 80 and 90, respectively. Since, the upper bound of the node P_6 is not greater than the current best lower bound of 90, the node P_6 is also fathomed and it has infeasible solution in terms of not fulfilling integer constraints for the decision variables.

Now, the only unfathomed terminal node is P_3. The further branching from this node is shown in Figure 6.12.

In Figure 6.12, the problems P_8 and P_9 are generated by adding an additional constraint to P_3. The problem P_8 is created by including '$X_2 \geq 3$' in problem P_3 and problem P_9 is created by including

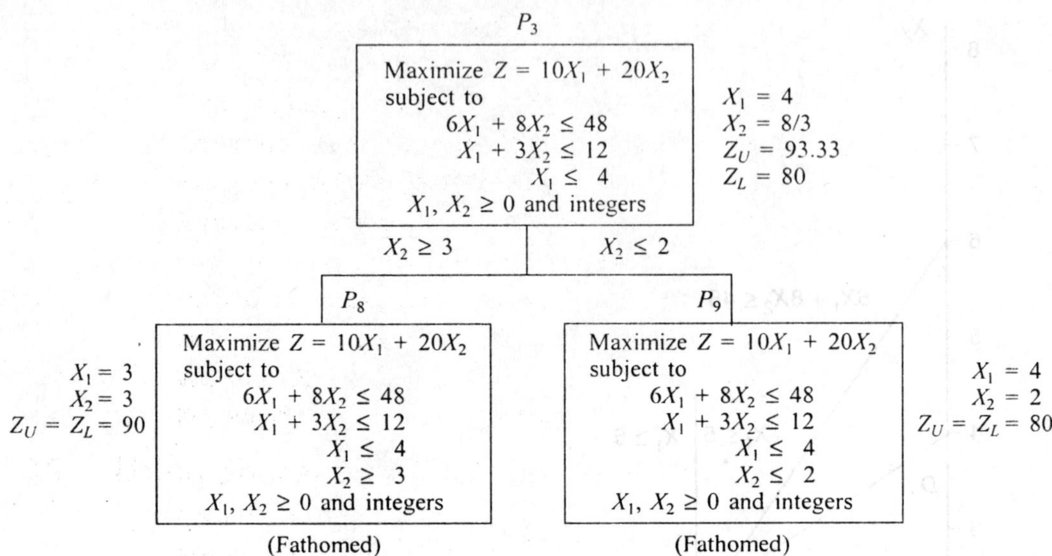

Figure 6.12 Branching from P_3.

'$X_2 \leq 2$' in problem P_3. The corresponding effects in slicing the non-integer infeasible region are shown in Figures 6.13 and 6.14, respectively. The solution for each of the problems P_8 and P_9 are obtained from the Figures 6.13 and 6.14, respectively. The problems P_8 and P_9 have integer solution. So, these two nodes are fathomed. But the objective function value of these nodes are not greater than the current best lower bound of 90. Hence, the current best lower bound is not updated.

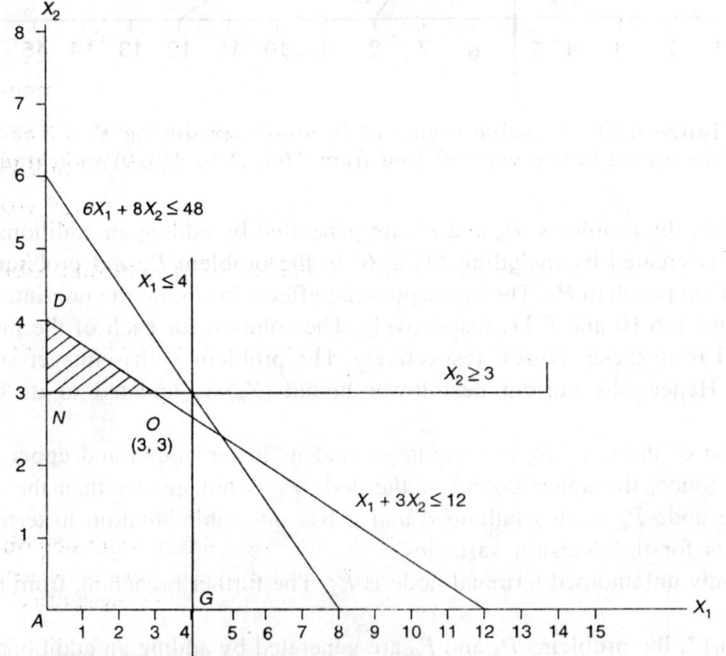

Figure 6.13 Feasible region of P_8 after introducing $X_2 \geq 3$ to P_3.

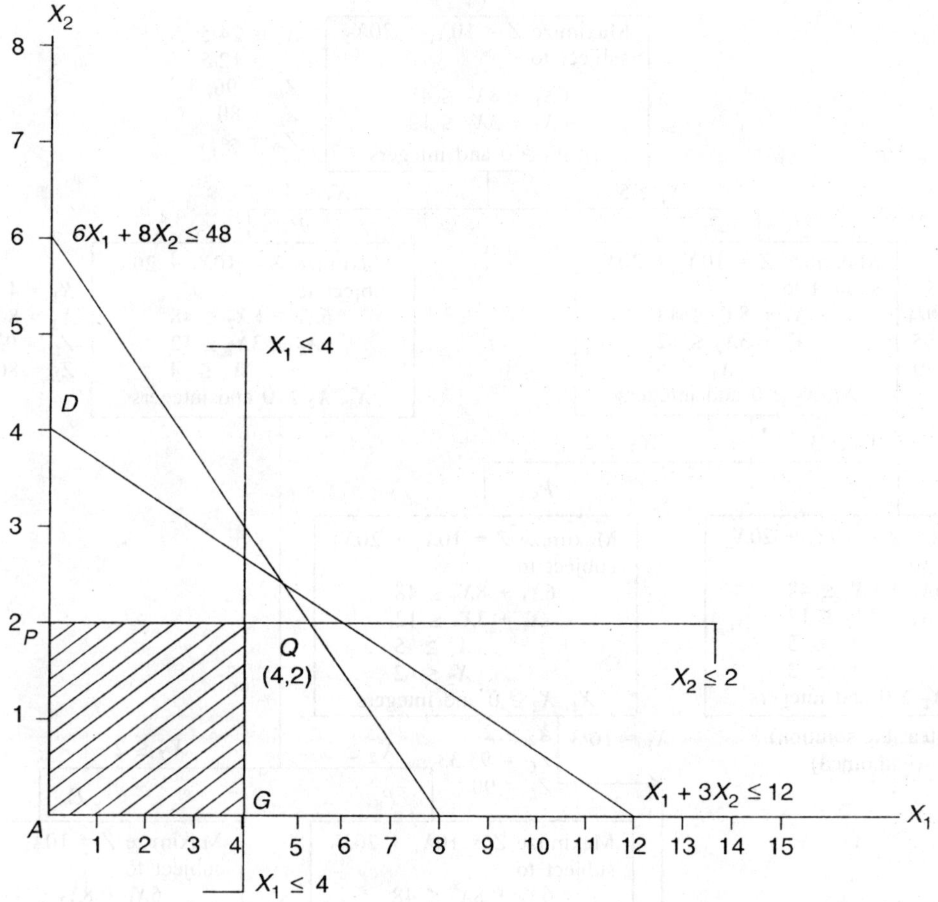

Figure 6.14 Feasible region of P_9 after introducing $X_2 \le 2$ to P_3.

Now, all the terminal nodes are fathomed. The feasible fathomed node with the current best lower bound is P_7. Hence, its solution is treated as the optimal solution as listed below. A complete branching tree is shown in Figure 6.15.

$$X_1 = 5, \qquad X_2 = 2, \qquad Z(\text{optimum}) = 90$$

Note: This problem has alternate optimum solution at P_8 with $X_1 = 3$, $X_2 = 3$, $Z(\text{optimum}) = 90$.

6.5 ZERO-ONE IMPLICIT ENUMERATION ALGORITHM

Zero-one (0-1) programming is a special kind of linear programming problem. In this type of problem, all the variables are restricted to either 0 or 1. This type of problem exists in many realistic situations like, capital budgeting problem, assignment problem, scheduling problem, portfolio problem, etc.

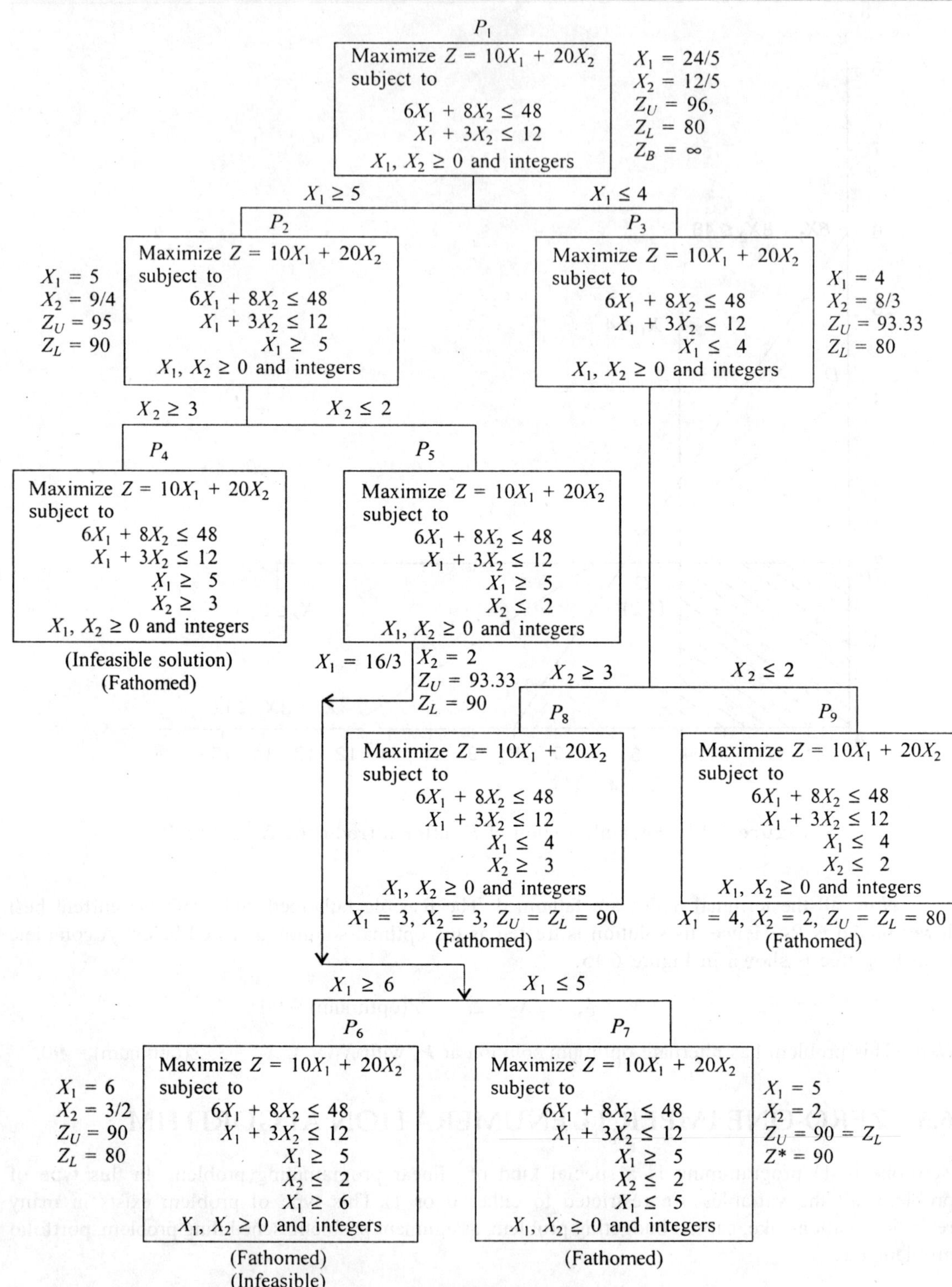

Figure 6.15 Complete tree of Example 6.8.

6.5.1   Generalized 0-1 Programming Problem

A generalized model of the 0-1 programming problem is presented below.

$$\text{Minimize } Z = \sum_{j=1}^{n} c_j X_j$$

subject to

$$\sum_{j=1}^{n} a_{ij} X_j \le b_i, \qquad i = 1, 2, 3..., m$$

$$X_j = 0 \text{ or } 1, \qquad j = 1, 2, 3..., m$$

In this model, the variable X_j is restricted to either zero or one. Hence, a specialized procedure which is known as *zero-one implicit enumeration algorithm* is required to solve this problem, and the same is demonstrated in the next section using a numerical problem. This algorithm is also known as *additive algorithm,* since it employs only additions and subtractions.

6.5.2   Zero-One Implicit Enumeration Technique

The technique for data preparation is discussed as under:

1.   Convert the problem into the minimization form with all $\le$ type constraints.
2.   If some of the coefficients in the revised objective function are negative, then using the following transformation, convert them into positive coefficients. Simultaneously, substitute the following equation in all the constraints and obtain the modified constraint set.

$Y_j = 1 - X_j,$ for all the variables with negative coefficient in the objective function
$\quad\; = X_j,$ for the remaining variables

3.   Present the problem in a convenient format as shown in Table 6.18.

Table 6.18   Format of Presenting Problem

	X_1	X_2	$\cdots$	X_n	S_1	$\cdots$	S_m	RHS
	.	.		.	.		.	Z
S_1 Row	.	.		.	.		.	.
S_2 Row	.	.		.	.		.	.
.								
.								
.								
S_m Row	.	.		.	.		.	.

A branch and bound procedure is used to solve the 0-1 programming problem. The related terminologies are presented below.

Free variable.   At any node of the tree, a binary variable is said to be a free variable if it is not selected by any of the branches leading to that node.

Partial solution.   A partial solution provides a specific binary assignment (0 or 1) for some of the variables. Let, J_t be the partial solution at the tth node (or iteration), where, $+j$ means $X_j = 1$ and $-j$ means $X_j = 0$.

The set J_t contains these elements. If the subscript of a binary variable is $+j$, then it denotes that variable is fixed (value = 1); if it is $-j$, then it denotes that variable is unfixed (value = 0). The set J_t is an ordered set which means that a new element is always augmented on the right of the partial solution.

Rules for fathomed partial solution

A partial solution is said to be fathomed if any one of the following conditions is true.

1. Further branching from the partial solution cannot lead to a better value of the objective function.
2. Further branching from the partial solution cannot lead to a feasible solution.

The general rule for generating next partial solution from a fathomed node is that if all the elements of a fathomed partial solution J_t are negative, then the enumeration is complete (stop the algorithm), otherwise, select the right-most positive element and delete all the negative elements to its right. Then, complement the selected right-most positive element with minus sign.

Tests for further branching

To branch further from a partial node (that is to select a free variable for elevating to level one), we need to do the following tests.

Let J_t be the partial solution at node t. Initially J_0 is equal to null set which means that all the variables are free and Z_t is the associated objective function value, $\bar{Z}$ is the current best upper bound of the objective function (initially $\bar{Z}$ is equal to infinity).

Test 1: For a free variable X_p, if all the technological coefficients (a_{ip}) with respect to $S_i^t < 0$ are greater than or equal to 0, then X_p cannot improve the infeasibility of the problem and must be discarded as non-promising.

Test 2: For any variable X_p, if $C_p + Z_t \geq \bar{Z}$ then X_p cannot lead to an improved solution and hence must be discarded.

Test 3: Consider the ith constraint for which $S_i^t < 0$ as shown below:

$$a_{i1}X_1 + a_{i2}X_2 + \ldots + a_{in}X_n + S_i = b_i$$

Let N_t defines the set of free variables not discarded by Test 1 and Test 2. If for at least one $S_i^t < 0$, the following condition is satisfied, then all the free variables in N_t are not promising. Under such situation, the set J_t is fathomed.

$$\sum_{j \in N_t} \min(0, a_{ij}) > S_i^t$$

Test 4: After Test 3, if the set N_t is not an empty set, find v_j^t using the following formula for all j belonging to N_t. Then identify the value of j for which v_j^t is maximum. Let it be k. The corresponding X_k is the branching variable.

$$V_j^t = \sum_{i=1}^{m} \min(0, S_i^t - a_{ij})$$

Example 6.9 Consider the capital budgeting problem where five projects are being considered for execution over the next 3 years. The expected returns for each project and the yearly expenditures (in thousands of rupees) are shown in Table 6.19. Assume that each approved project will be executed over the 3-year period. The objective is to select a combination of projects that will maximize the total returns.

Table 6.19 Example 6.9

Project	Year 1	Year 2	Year 3	Returns
		Expenditure for		
1	5	1	8	20
2	4	7	10	40
3	3	9	2	20
4	7	4	1	15
5	8	6	10	30
Maximum available funds	25	25	25	—

Formulate the problem as a zero-one integer programming problem and solve it by the additive algorithm.

Solution Let

$$Y_j = 1, \text{ if the } j\text{th project is selected.}$$
$$= 0, \text{ otherwise.}$$

Therefore, a zero-one programming model is:

$$\text{Maximize } Y_0 = 20Y_1 + 40Y_2 + 20Y_3 + 15Y_4 + 30Y_5$$

subject to

$$5Y_1 + 4Y_2 + 3Y_3 + 7Y_4 + 8Y_5 \leq 25$$
$$Y_1 + 7Y_2 + 9Y_3 + 4Y_4 + 6Y_5 \leq 25$$
$$8Y_1 + 10Y_2 + 2Y_3 + Y_4 + 10Y_5 \leq 25$$
$$Y_j = 0 \text{ or } 1, \quad j = 1, 2, 3, 4 \text{ and } 5$$

Convert the objective function into minimization type as under:

$$\text{Minimize } Z_0 = -20Y_1 - 40Y_2 - 20Y_3 - 15Y_4 - 30Y_5$$

Since all the coefficients in the objective are negative, substitute the following formula in the objective function and the constraints of the model and rearrange the terms

$$Y_j = 1 - X_j, \quad j = 1, 2, \dots, 5$$

The resulting model is presented below:

$$\text{Minimize } Z = 20X_1 + 40X_2 + 20X_3 + 15X_4 + 30X_5$$

subject to

$$-5X_1 - 4X_2 - 3X_3 - 7X_4 - 8X_5 \leq -2$$
$$-X_1 - 7X_2 - 9X_3 - 4X_4 - 6X_5 \leq -2$$
$$-8X_1 - 10X_2 - 2X_3 - X_4 - 10X_5 \leq -6$$
$$X_j = 0 \text{ or } 1, \quad j = 1, 2, 3, 4 \text{ and } 5$$

[*Note.* The sum of the constants in the objective function is omitted.]

The above model is represented in a tabular form as in Table 6.20 in which S_1, S_2 and S_3 are the slack variables of the constraints.

Table 6.20 Initial Table of Example 6.9

X_1	X_2	X_3	X_4	X_5	S_1	S_2	S_3	RHS
20	40	20	15	30	0	0	0	Z
–5	–4	–3	–7	–8	1	0	0	–2
–1	–7	–9	–4	–6	0	1	0	–2
–8	–10	–2	–1	–10	0	0	1	–6

Iteration 0:   For J_0 = [null set], $\overline{Z}$ = infinite. The corresponding tree is shown in Figure 6.16.

$$(S_1^0, S_2^0, S_3^0) = (-2, -2, -6), \quad Z_0 = 0$$

List of free variables, $N_0 = (1, 2, 3, 4, 5)$.

$$\boxed{0} \quad \begin{array}{l} Z_0 = 0 \\[4pt] \overline{Z} = \infty \end{array}$$

Figure 6.16 Root node.

Test 1: This test does not exclude any free variable.

Test 2: At this stage, the Test 2 is not applicable, since $\overline{Z}$ is infinity.

Test 3:

$$S_1: -5\ -4\ -3\ -7\ -8 = -27 < -2$$
$$S_2: -1\ -7\ -9\ -4\ -6 = -27 < -2$$
$$S_3: -8\ -10\ -2\ -1\ -10 = -31 < -6$$

So, the set N_0 cannot be discarded.

Test 4:

$$V_1^0 = 0 - 1 + 0 = -1$$
$$V_2^0 = 0 + 0 + 0 = 0$$
$$V_3^0 = 0 + 0 - 4 = -4$$
$$V_4^0 = 0 + 0 - 5 = -5$$
$$V_5^0 = 0 + 0 + 0 = 0^*$$

The maximum value of V_j^0 is 0 when j is equal to 2 and 5. Out of these two values, 5 is selected randomly as the value of k. Therefore, X_5 is the branching variable in the next iteration.

Iteration 1:   For J_1 = [5], the corresponding tree is shown in Figure 6.17.

$$(S_1^1, S_2^1, S_3^1) = (-2 + 8, -2 + 6, -6 + 10) = (6, 4, 4), \quad Z_1 = 30$$

Since it is a feasible solution, $\overline{Z} = Z_1 = 30$. Thus J_1 is fathomed.

Hence, for $J_1 = [5]$, $\bar{Z} = 30$. These are indicated in Figure 6.17.

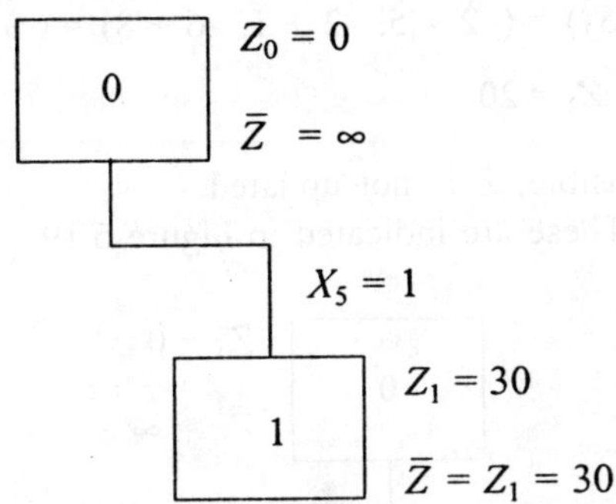

Figure 6.17   Branching tree for $J_1 = [5]$.

Iteration 2:   For $J_2 = [-5]$, the corresponding tree is shown in Figure 6.18.

$$(S_1^2, S_2^2, S_3^2) = (-2, -2, -6), \quad Z_2 = 0$$

Since the above solution is infeasible, node 2 is not fathomed. Hence, $\bar{Z}$ is not updated.

Hence for $J_2 = [-5]$, $\bar{Z} = 30$. These are indicated in Figure 6.18. Also, we have, $N_2 = (1, 2, 3, 4)$.

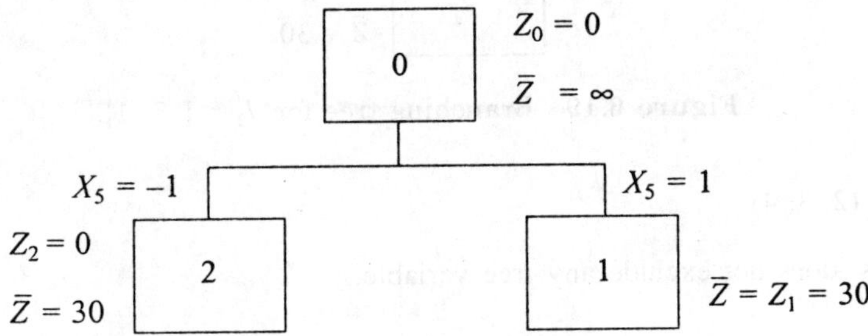

Figure 6.18   Branching tree for $J_1 = [-5]$.

Test 1:   This test does not exclude any of the variables in N_2.

Test 2:   Here $C_2 = 40$. Then $Z_2 + C_2 = 0 + 40 = 40$, which is more than $\bar{Z}(30)$. So, X_2 is excluded. Therefore, we have $N_2 = (1, 3, 4)$.

Test 3:

$$S_1: -5 - 3 - 7 = -15 < -2$$
$$S_2: -1 - 9 - 4 = -14 < -2$$
$$S_3: -8 - 2 - 1 = -11 < -6$$

So, the free variables in N_2 cannot be discarded.

Test 4:

$$V_1^2 = 0 - 1 + 0 = -1$$
$$V_3^2 = 0 + 0 - 4 = -4$$
$$V_4^2 = 0 + 0 - 5 = -5$$

Hence, k is equal to 1 and the corresponding branching variable is X_1.

Iteration 3: For $J_3 = [-5, 1]$, the corresponding tree is shown in Figure 6.19.

$$(S_1^3, S_2^3, S_3^3) = (-2 + 5, -2 + 1, -6 + 8) = (3, -1, 2),$$

$$Z_3 = 20$$

Since the above solution is infeasible, $\bar{Z}$ is not updated.

For, $J_3 = [-5, 1]$, $\bar{Z} = 30$. These are indicated in Figure 6.19.

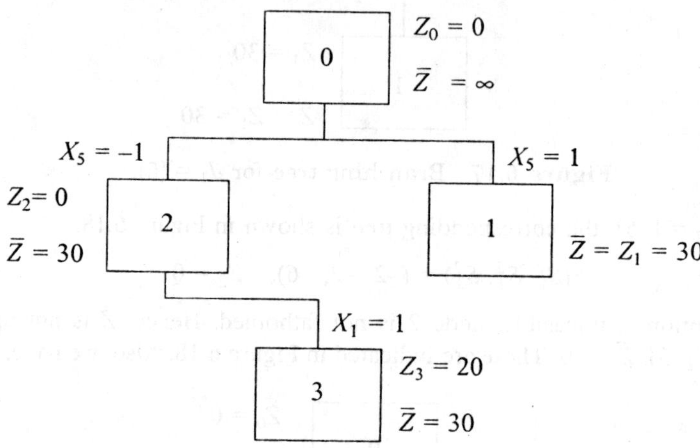

Figure 6.19 Branching tree for $J_1 = [-5, 1]$.

Also, $N_3 = (2, 3, 4)$

Test 1: This test does not exclude any free variable.

Test 2:

$$C_2 + Z_3 (40 + 20 = 60) > \bar{Z}(30). \text{ So, exclude } X_2$$

$$C_3 + Z_3 (20 + 20 = 40) > \bar{Z}(30). \text{ So, exclude } X_3$$

$$C_4 + Z_3 (15 + 20 = 35) > \bar{Z}(30). \text{ So, exclude } X_4$$

All the free variables at this stage are excluded.

Since $N_0 =$ null set, J_3 is fathomed.

Iteration 4: $J_4 = [-5, -1]$. The corresponding tree is shown in Figure 6.20

$$(S_1^4, S_2^4, S_3^4) = (-2, -2, -6), \qquad Z_4 = 0$$

Since the above solution is infeasible, $\bar{Z}$ is not updated.

$$\text{For } J_4 = [-5, -1], \qquad \bar{Z} = 30$$

These are indicated in Figure 6.20. Also, we have $N_4 = (2, 3, 4)$.

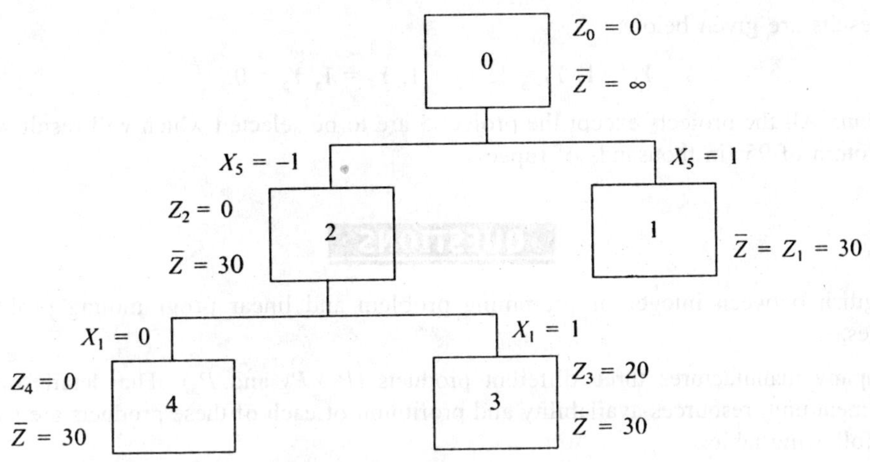

Figure 6.20 Branching tree for $J_1 = [-5, -1]$

Test 1: This test does not exclude any free variable.

Test 2:

$$C_2 + Z_4 \,(40 + 0 = 40) > \bar{Z}(30)$$
$$C_3 + Z_4 \,(20 + 0 = 20) < \bar{Z}(30)$$
$$C_4 + Z_4 \,(15 + 0 = 15) < \bar{Z}(30)$$

So, discard only X_2. Hence, $N_4 = (3, 4)$.

Test 3:

$$S_1: -10 < -2$$
$$S_2: -13 < -2$$
$$S_3: \;\; -3 > -6$$

Since for S_3 row, $\sum_{j \in N_4} \min(0, a_{ij}) > S_3^4$, discard N_4. Thus J_4 is fathomed. Since all the elements of J_4 are negative, the enumeration is complete and J_1 is the optimum fathomed node, i.e. $J_1 = [5]$. So, $X_5 = 1$ and all other variables are zero. The values for Y_1, Y_2, Y_3, Y_4 and Y_5 are derived as shown in Table 6.21.

Table 6.21 Summary of Results

j	X_j	$Y_j = 1 - X_j$	Objective function coefficient (C_j)	$C_j * Y_j$
1	0	1–0 = 1	20	20
2	0	1–0 = 1	40	40
3	0	1–0 = 1	20	20
4	0	1–0 = 1	15	15
5	1	1–1 = 0	30	0
			Total	95

The final results are given below:

$$Y_1 = 1, \ Y_2 = 1, \ Y_3 = 1, \ Y_4 = 1, \ Y_5 = 0.$$

Interpretation: All the projects except the project 5 are to be selected which will result with a total maximum return of 95 (in thousands of rupees).

QUESTIONS

1. Distinguish between integer programming problem and linear programming problem. Give examples.

2. A company manufactures three different products (P_1, P_2 and P_3). The details on resource requirement/unit, resources availability and profit/unit of each of these products are summarized in the following table:

Resource	Product			Resource availability per month
	P_1	P_2	P_3	
Man hours	2000	1500	1000	38,000
Machine hours	1000	1500	2000	33,000
Profit/unit (Rs.)	75,000	1,00,000	80,000	

Develop an integer programming model to determine the production volume of each of the products such that the total profit is maximized.

3. Develop a model for the fixed-charge problem.

4. Distinguish between total covering problem and partial covering problem. Also, discuss the related models.

5. Solve the following integer linear programming problem optimally.

$$\text{Maximize } Z = 8X_1 + 6X_2 + 10X_3$$

subject to

$$8X_1 + 4X_2 + 2X_3 \leq 155$$
$$3X_1 + 6X_2 + 12X_3 \leq 135$$
$$X_1, X_2, X_3 \geq 0 \text{ and integers}$$

6. Solve the following integer linear programming problem optimally:

$$\text{Maximize } Z = 8X_1 + 6X_2$$

subject to

$$8X_1 + 4X_2 \leq 85$$
$$3X_1 + 6X_2 \leq 95$$
$$X_1, X_2 \geq 0 \text{ and integers}$$

7. Solve the following integer linear programming problem optimally:

$$\text{Maximize } Z = 2X_1 + 5X_2$$

subject to

$$3X_1 + 6X_2 \leq 24$$
$$6X_1 + 12X_2 \leq 18$$
$$2X_1 + 8X_2 \leq 20$$
$$X_1, X_2 \geq 0 \text{ and integers}$$

8. Solve the following integer linear programming problem optimally using branch-and-bound technique.

$$\text{Maximize } Z = 6X_1 + 8X_2$$

subject to

$$4X_1 + 5X_2 \leq 22$$
$$5X_1 + 8X_2 \leq 30$$
$$X_1, X_2 \geq 0 \text{ and integers}$$

9. Solve the following integer linear programming problem optimally:

$$\text{Maximize } Z = 10X_1 + 8X_2$$

subject to

$$2X_1 + 4X_2 \leq 25$$
$$4X_1 + 6X_2 \leq 27$$
$$X_1, X_2 \geq 0 \text{ and integers}$$

10. Consider the capital budgeting problem where five projects are being considered for execution over the next 3 years. The expected returns for each project and the yearly expenditure (in thousands of rupees) are shown in the following table. Assume that each approved project will be executed over the 3-year period. The objective is to select a combination of projects that will maximize the total returns.

Project	Expenditure for			Returns
	Year 1	Year 2	Year 3	
1	6	2	6	40
2	2	5	8	25
3	5	6	3	40
4	6	3	4	20
5	8	7	5	25
Maximum available funds	20	20	20	–

Formulate the problem as a zero-one integer programming problem and solve it by the additive algorithm.

INVENTORY CONTROL

7

7.1 INTRODUCTION

Inventory is essential to provide flexibility in operating a system or organization. An inventory can be classified into raw materials inventory, work-in-process inventory and finished goods inventory. The raw material inventory removes dependency between suppliers and plants. The work-in-process inventory removes dependency between various machines of a product line. The finished goods inventory removes dependency between plants and its customers or market. The main functions of an inventory are: smoothing out irregularities in supply, minimizing the production cost and allowing organizations to cope up with perishable materials.

Some important terminologies of inventory control are discussed now.

Inventory decisions. The following two basic *inventory decisions* are generally taken by managers.

1. When to replenish the inventory of an item?
2. How much of an item to order when the inventory of that item is to be replenished?

Costs of inventory systems. The following costs are associated with the inventory system.

1. Purchase price/unit
2. Ordering cost/order
3. Carrying cost/unit/period
4. Shortage cost/unit/period.

Costs trade off. If we place frequent orders, the cost of ordering will be more, but the inventory carrying cost will be less. On the other hand, if we place less frequent orders, the ordering cost will be less, but the carrying cost will be more. In Figure 7.1, for an increase in Q (order size), the carrying cost increases and the ordering cost decreases. The total cost curve represents the sum of ordering cost and carrying cost for each order size. The order size at which the total cost is minimum is called *economic order quantity* (EOQ) or *optimal order size* (Q^*).

7.2 MODELS OF INVENTORY

There are different models of inventory. The inventory models can be classified into *deterministic* models and *probabilistic* models. The various deterministic models are:

(a) Purchase model with instantaneous replenishment and without shortages;
(b) Manufacturing model without shortages;

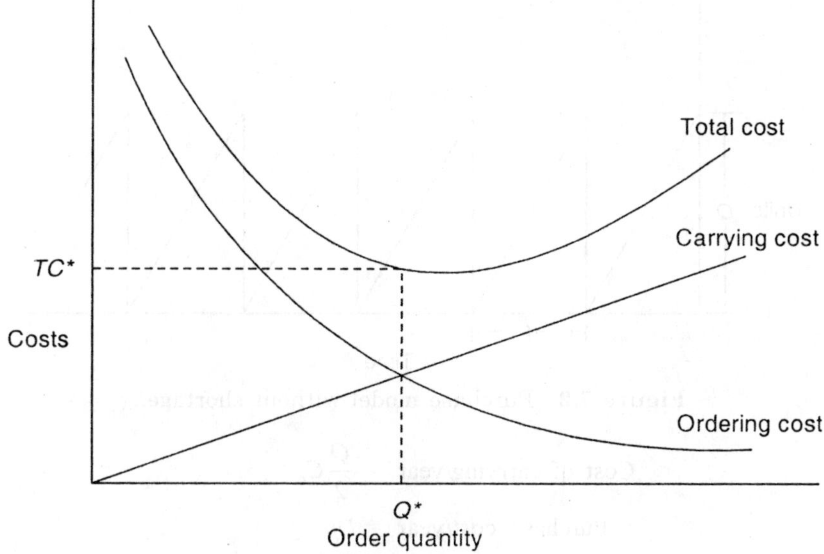

Figure 7.1 Trade-off between costs.

(c) Purchase model with instantaneous replenishment and with shortages;

(d) Manufacturing model with shortages.

These models are explained in the following sections.

7.2.1 Purchase Model with Instantaneous Replenishment and without Shortages

In this inventory model, orders of equal size are placed at periodical intervals. The items against an order are replenished instantaneously and the items are consumed at a constant rate. The purchase price per unit is same irrespective of order size.

Let us suppose, D = Annual demand in units

C_o = Ordering cost/order

C_c = Carrying cost/unit/year

p = Purchase price per unit

Q = Order size

The corresponding purchase model can be represented as shown in Figure 7.2. From the above assumptions, we have:

$$\text{The number of orders/year} = \frac{D}{Q}$$

$$\text{Average inventory} = \frac{Q}{2}$$

$$\text{Cost of ordering/year} = \frac{D}{Q} C_o$$

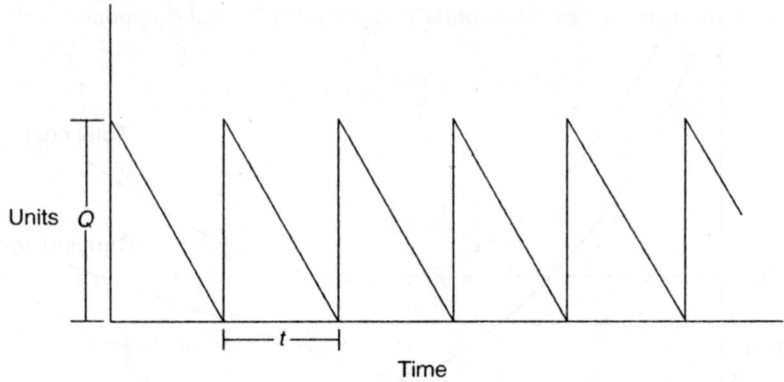

Figure 7.2 Purchase model without shortage.

$$\text{Cost of carrying/year} = \frac{Q}{2} C_c$$

$$\text{Purchase cost/year} = Dp$$

Therefore,

$$\text{Total inventory cost/year} = \frac{D}{Q} C_o + \frac{Q}{2} C_c + Dp$$

Differentiating with respect to Q yields

$$\frac{d}{dQ}(TC) = \frac{-D}{Q^2} C_o + \frac{C_c}{2}$$

Differentiating it again with respect to Q yields

$$\frac{d^2}{dQ^2}(TC) = \frac{2D}{Q^3} C_o$$

Since the second derivative is positive, the optimal value for Q is obtained by equating the first derivative to zero. Therefore,

$$\frac{-D}{Q^2} C_o + \frac{C_c}{2} = 0 \quad \text{or} \quad Q^2 = \frac{2C_o D}{C_c}$$

Hence, the optimal order size is

$$Q^* = \sqrt{\frac{2C_o D}{C_c}}$$

and

$$\text{Total number of orders per year} = \frac{D}{Q^*}$$

where

$$\text{Time between orders} = \frac{Q^*}{D}$$

***Example* 7.1** Ram Industry needs 5400 units/year of a bought-out component which will be used in its main product. The ordering cost is Rs. 250 per order and the carrying cost per unit per year is Rs. 30. Find: the economic order quantity (EOQ), the number of orders per year and the time between successive orders.

Solution

$$D = 5400 \text{ units/year}$$

$$C_o = \text{Rs. } 250/\text{order}$$

$$C_c = \text{Rs. } 30/\text{unit/year}$$

Therefore, the economic order quantity

$$\text{EOQ } (Q^*) = \sqrt{\frac{2C_o D}{C_c}} = \sqrt{\frac{2 \times 250 \times 5400}{30}} = 300 \text{ units}$$

where

$$\text{Number of orders/year} = \frac{D}{Q^*} = \frac{5400}{300} = 18$$

and

$$\text{Time between successive orders} = \frac{Q^*}{D} = \frac{300}{5400}$$

$$= 0.0556 \text{ year}$$

$$= 0.6672 \text{ month}$$

$$= 20 \text{ days (approx.)}$$

***Example* 7.2** Alpha Industry needs 15,000 units per year of a bought-out component which will be used in its main product. The ordering cost is Rs. 125 per order and the carrying cost per unit per year is 20% of the purchase price per unit. The purchase price per unit is Rs. 75. Find: economic order quantity, number of orders per year and time between successive orders.

Solution We have

$$D = 15,000 \text{ units/year}$$

$$C_o = \text{Rs. } 125/\text{order}$$

$$\text{Purchase price/unit} = \text{Rs. } 75$$

$$C_c = \text{Rs. } 75 \times 0.20$$

$$= \text{Rs. } 15/\text{unit/year}$$

Therefore, the economic order quantity is

$$\text{EOQ} = \sqrt{\frac{2C_o D}{C_c}} = \sqrt{\frac{2 \times 125 \times 15,000}{15}} = 500 \text{ units}$$

and

$$\text{Number of orders/year} = \frac{D}{Q^*} = \frac{15,000}{500} = 30$$

Time between successive orders is obtained as

$$\frac{Q^*}{D} = \frac{500}{15,000} = 0.033 \text{ year} = 0.4 \text{ month} = 12 \text{ days}$$

Example 7.3 A textile mill buys its raw material from a vendor. The annual demand of the raw material is 9000 units. The ordering cost is Rs. 100 per order and the carrying cost is 20% of the purchase price per unit per month, where the purchase price per unit is Re. 1. Find the following:

(a) Economic order quantity (EOQ)
(b) Total cost w.r.t. EOQ
(c) Number of orders per year
(d) Time between consecutive two orders

Solution   The given data are:

D = 9000 units per year
C_o = Rs. 100 per order
p = Re. 1 per unit
C_c = Re. 0.2 per unit per month = Rs. 2.4 per unit per year

(a) The economic order quantity

$$\text{EOQ}(Q^*) = \sqrt{\frac{2C_o D}{C_c}} = \sqrt{\frac{2 \times 100 \times 9000}{2.4}} = 866.03 \text{ units} = 866 \text{ units (approx.)}$$

(b) Total cost $(TC^*) = (D/Q^*)C_o + (Q^*/2)C_c + pD$
$$= (9000/866) \times 100 + (866/2) \times 2.4 + 1 \times 9000 = \text{Rs. } 11,078.46/\text{year}$$

(c) Number of orders per year = D/Q = 9000/866 = 10.39 orders per year
(d) Time between two consecutive orders = Q/D = 866/9000 = 0.0962 year = 1.15 month = 34.5 days

Example 7.4   The annual demand of an item in the stores of a foundry is 9000 units. Its annual carrying cost is 15% of the purchase price of the item per year, where the purchase price is Rs. 20 per unit. The ordering cost is Rs. 15 per order. Presently, the order size of the item is the average monthly demand of that item. Find the economic order quantity and compare its cost with the present ordering system and find the corresponding cost advantage if exists.

Solution   The given data are:

D = 9000 units per year
C_o = Rs. 15 per order
p = Rs. 20 per unit
C_c = 0.15 × 20 = Rs. 3 per unit per year

Therefore, the economic order quantity

$$\text{EOQ}(Q^*) = \sqrt{\frac{2C_o D}{C_c}} = \sqrt{\frac{2 \times 15 \times 9000}{3}} = 300 \text{ units}$$

Total cost w.r.t. to EOQ $(TC^*) = (D/Q^*)C_o + (Q^*/2)C_c + pD$

$$= (9000/300) \times 15 + (300/2) \times 3 + 20 \times 9000 = \text{Rs. } 1,80,900/\text{year}$$

As per the present ordering system, the order size $= 9000/12 = 750$ units

Total cost of the present ordering system $= (D/Q)C_o + (Q/2)C_c + pD$
$$= (9000/750) \times 15 + (750/2) \times 3 + 20 \times 9000$$
$$= \text{Rs. } 1,81,305/\text{year}$$

The total cost as per the EOQ policy is lesser when compared to the present ordering system and the corresponding cost advantage is Rs. 405 per year (i.e. Rs. 1,81,305 − Rs. 1,80,900).

Example 7.5 The purchase manager currently follows EOQ policy of ordering for an item in the stores of his company. The annual demand of the item is 1,600 units. Its carrying cost is 40% of the unit cost where the unit cost is Rs. 400. The ordering cost is Rs. 500 per order. Recently, the vendor supplying that item gives a discount of 10% in its unit cost if the order size is minimum of 500 units.

(a) Find EOQ and the corresponding total cost per year.
(b) Check whether the discount offer given by the vendor can be considered by the purchase manager.

Solution The data of the problem are:

D $= 1,600$ units per year
C_o $= \text{Rs. } 500$ per order
p $= \text{Rs. } 400$ per unit

Current ordering system:

$C_c = 0.4 \times 400 = \text{Rs. } 160$ per unit per year

Therefore, the economic order quantity

$$\text{EOQ}(Q^*) = \sqrt{\frac{2C_o D}{C_c}} = \sqrt{\frac{2 \times 500 \times 1600}{160}} = 100 \text{ units}$$

Total cost w.r.t to EOQ $(TC^*) = (D/Q^*)C_o + (Q^*/2)C_c + pD$

$$= (1600/100) \times 500 + (100/2) \times 160 + 400 \times 1600 = \text{Rs. } 6,56,000/\text{year}$$

Proposed ordering system:

Minimum order size$(Q) = 500$ units
p (after 10% discount) $= (1 \times 0.1) \times 400 = \text{Rs. } 360$
$C_c = 0.4 \times 360 = \text{Rs. } 144$ per unit per year

Total cost of the proposed ordering system $= (D/Q)C_o + (Q/2)C_c + pD$
$$= (1600/500) \times 500 + (500/2) \times 144 + 360 \times 1600$$
$$= \text{Rs. } 6,13,600 \text{ per year}$$

Since, the total cost as per the proposed ordering system (Rs. 6,13,600) is less than that of the current ordering system (Rs. 6,56,000), the purchase manager should avail the discount given by its vendor by placing orders for 500 units instead of 100 units as decided by the EOQ formula.

7.2.2 Manufacturing Model without Shortages

If a company manufactures an item which is required for its main product, then the corresponding model of inventory is called *manufacturing model*. In this model, shortages are not permitted. The rate of consumption of the item is assumed to be uniform throughout the year. The item is produced and consumed simultaneously for a portion of the cycle time. During the remaining cycle time, only the consumption of the item takes place and the cost of production per unit is same irrespective of production lot size.

Let us suppose,

r = Annual demand in units
k = Production rate of the item (total number of units produced/year)
C_o = Cost per set-up
C_c = Carrying cost/unit/year
p = Cost of production/unit
t_1 = Period of production as well as consumption of the item
t_2 = Period of consumption only
t = Cycle time (i.e. $t = t_1 + t_2$)

The operation of the manufacturing model without shortages is shown as in Figure 7.3.

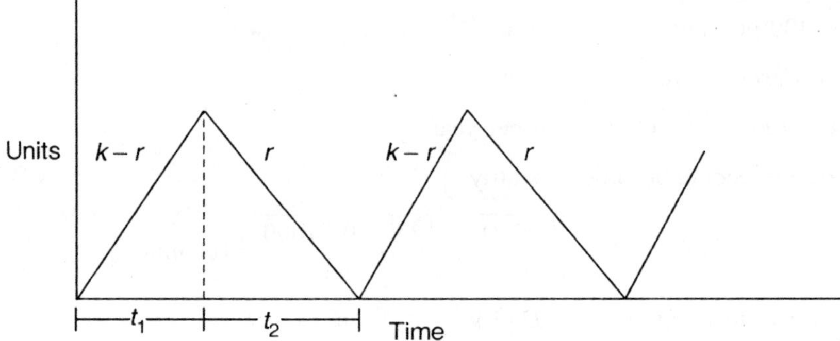

Figure 7.3 Manufacturing model without shortages.

During the period t_1, the item is produced at the rate of k units per period and simultaneously it is consumed at the rate of r units per period. During this period, the inventory is built at the rate of $k - r$ units per period. During the period t_2, the production of the item is discontinued but the consumption of the same item is continued. Hence, the inventory is decreased at the rate of r units per period during this time t_2. The various formula to be applied for this kind of situation are given below.

$$\text{Economic batch quantity (EBQ or } Q^*) = \sqrt{\frac{2C_o r}{C_c\left[1 - (r/k)\right]}}$$

$$\text{Period of production as well as consumption, } t_1^* = \frac{Q^*}{k}$$

$$\text{Period of consumption only, } t_2^* = \frac{Q^*[1 - (r/k)]}{r} = \frac{(k - r)t_1}{r}$$

$$\text{Cycle time } t = t_1^* + t_2^*$$

$$\text{Number of set-ups per year} = \frac{r}{Q^*}$$

Example 7.6 An automobile factory manufactures a particular type of gear within the factory. This gear is used in the final assembly. The particulars of this gear are: demand rate $r = 14{,}000$ units/year, production rate $k = 35{,}000$ units/year, set-up cost, $C_o = $ Rs. 500 per set-up and carrying cost, $C_c = $ Rs. 15/unit/year.

Find the economic batch quantity (EBQ) and cycle time.

Solution Applying the required formulae, we have the economic batch quantity

$$Q^* = \sqrt{\frac{2C_o r}{C_c[1 - (r/k)]}}$$

$$= \sqrt{\frac{2 \times 500 \times 14{,}000}{15[1 - (14{,}000/35{,}000)]}}$$

$$= 1247.22 \text{ units}$$

$$= 1248 \text{ (approx.)}$$

Now, the period of production as well as consumption

$$t_1^* = \frac{Q^*}{k}$$

$$= \frac{1248}{35{,}000}$$

$$= 0.0357 \text{ year}$$

$$= 0.4284 \text{ month}$$

$$= 13 \text{ days (approx.)}$$

and the period of consumption

$$t_2^* = \frac{Q^*}{r} \left(1 - \frac{r}{k}\right)$$

$$= \frac{1248}{14{,}000} \left(1 - \frac{14{,}000}{35{,}000}\right)$$

$$= 0.0535 \text{ year}$$

$$= 0.642 \text{ month}$$

$$= 20 \text{ days (approx.)}$$

Therefore, the cycle time is

$$t = t_1^* + t_2^* = 13 + 20 = 33 \text{ days}$$

Also

$$\text{The number of set-ups per year} = \frac{r}{Q^*} = \frac{14{,}000}{1248} = 11.22$$

Example 7.7 A company manufactures a low cost bearing which is used in its main product line. The demand of the bearing is 10,000 units per month and the production rate of the bearing is 25,000 units per month. The carrying cost of the bearing is Re. 0.02 per bearing per year and the set-up cost is Rs. 18 per set-up. Find the economic batch quantity (EBQ) and the cycle time (t^*)

Solution The given data are:

r = 10,000 bearings per month = 1,20,000 bearings per year
k = 25,000 bearings per month = 3,00,000 bearings per year
C_o = Rs. 18 per set-up
C_c = Re. 0.02 per bearing per year

Therefore, the economic batch quantity (EBQ) is

$$\text{EBQ}(Q^*) = \sqrt{\frac{2C_o r}{C_c[1 - (r/k)]}} = \sqrt{\frac{2 \times 18 \times 1,20,000}{0.02 \times [1 - (1,20,000/3,00,000)]}}$$

$$= 18{,}973.65 \text{ units} = 18{,}974 \text{ units (approx.)}.$$

The cycle time, $t^* = \text{EBQ}/r = 18{,}974/1{,}20{,}000 = 0.158$ year = 1.9 months = 57 days.

7.2.3 Purchase Model with Instantaneous Replenishment and with Shortages

In this model, an item on order will be received instantaneously and it is consumed at a constant rate. The purchase price per unit is same irrespective of order size. If there is no stock at the time of receiving a request for the item, it is assumed that it will be satisfied at a later date with a penalty. This is called *backordering*. The model is shown as in Figure 7.4.

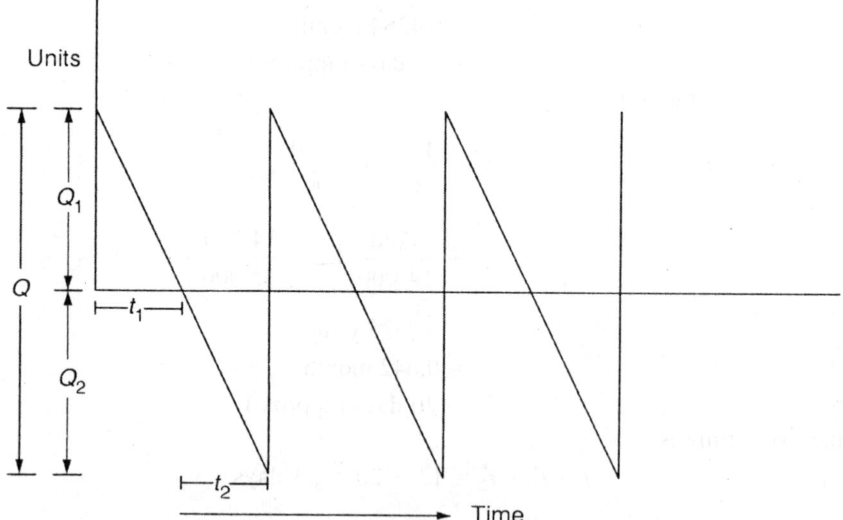

Figure 7.4 Purchase model with shortages.

The variables which are to be used in this model are:

$$D = \text{Demand/period}$$
$$C_c = \text{Carrying cost/unit/period}$$
$$C_o = \text{Ordering cost/order}$$
$$C_s = \text{Shortage cost/unit/period}$$
$$Q = \text{Order size}$$
$$Q_1 = \text{Maximum inventory}$$
$$Q_2 = \text{Maximum stock-out}$$
$$t_1 = \text{Period of positive stock}$$
$$t_2 = \text{Period of shortage}$$
$$t = \text{Cycle time } (t_1 + t_2)$$

Optimal values of the above variables are:

$$Q^* = \sqrt{\frac{2C_o D}{C_c} \frac{C_s + C_c}{C_s}}$$

$$Q_1^* = \sqrt{\frac{2C_o D}{C_c} \frac{C_s}{C_s + C_c}}$$

$$Q_2^* = Q^* - Q_1^*$$

$$t^* = \frac{Q^*}{D}$$

$$t_1^* = \frac{Q_1^*}{D}$$

$$t_2^* = \frac{Q_2^*}{D}$$

where

$$\text{Number of orders/period} = \frac{D}{Q^*}.$$

Example 7.8 The annual demand for a component is 7200 units. The carrying cost is Rs. 500/unit/year, the ordering cost is Rs. 1500 per order and the shortage cost is Rs. 2000/unit/year. Find the optimal values of economic order quantity, maximum inventory, maximum shortage quantity, cycle time (t), inventory period (t_1) and, shortage period (t_2).

Solution We have

$$D = 7200 \text{ units/year}$$
$$C_c = \text{Rs. } 500\text{/unit/year}$$
$$C_o = \text{Rs. } 1500\text{/order}$$
$$C_s = \text{Rs. } 2000\text{/unit/year}$$

Therefore,

$$\text{Economic order quantity } Q^* = \sqrt{\frac{2C_oD}{C_c}\,\frac{C_s + C_c}{C_s}}$$

$$= \sqrt{\frac{2 \times 1500 \times 7200}{500}\,\frac{2000 + 500}{2000}}$$

$$= 233 \text{ units (approx.)}$$

$$\text{Maximum inventory } Q_1^* = \sqrt{\frac{2C_oD}{C_c}\,\frac{C_s}{C_s + C_c}}$$

$$= \sqrt{\frac{2 \times 1500 \times 7200}{500}\,\frac{2000}{2000 + 500}}$$

$$= 186 \text{ units (approx.)}$$

$$\text{Maximum stock-out } Q_2^* = Q^* - Q_1^* = 233 - 186 = 47 \text{ units}$$

$$\text{Cycle time } t^* = \frac{Q^*}{D} = \frac{233}{7200} \times 365 = 12 \text{ days (approx.)}$$

$$\text{Period of positive stock } t_1^* = \frac{Q_1^*}{D} = \frac{186}{7200} \times 365 = 10 \text{ days (approx.)}$$

$$\text{Period of shortage } t_2^* = t^* - t_1^* = 12 - 10 = 2 \text{ days}$$

$$\text{Number of orders per year} = \frac{D}{Q^*} = \frac{7200}{233} = 30.9$$

Example 7.9   The demand of a bought out item in a store is 12,000 units per year. The carrying cost is Rs. 2 per unit per year and the ordering cost is Rs. 600 per order. The shortage cost is Rs. 10 per unit per year. Find the EOQ and the corresponding number of orders per year, the maximum inventory and maximum shortage quantity.

Solution   The data of the given problem are:

$$D = 12,000 \text{ units per year}$$
$$C_o = \text{Rs. } 600 \text{ per order}$$
$$C_c = \text{Rs. } 2 \text{ per unit per year}$$
$$C_s = \text{Rs. } 10 \text{ per unit per year}$$

Therefore, the economic order (EOQ) is

$$\text{EOQ}(Q^*) = \sqrt{\frac{2C_oD}{C_c}\,\frac{(C_s + C_c)}{C_s}} = \sqrt{\frac{2 \times 600 \times 12,000}{2}\,\frac{(10 + 2)}{10}}$$

$$= 2939.39 \text{ units} = 2940 \text{ units (approx.).}$$

The maximum inventory, Q_1^* is

$$Q_1^* = \sqrt{\frac{2C_oD}{C_c}\frac{C_s}{C_s+C_c}} = \sqrt{\frac{2\times600\times12,000}{2}\frac{10}{10+2}}$$

$$= 2449.49 \text{ units} = 2450 \text{ units (approx.)}$$

The maximum shortage quantity per cycle, $Q_2^* = Q^* - Q_1^* = 2940 - 2450 = 490$ units

Number of orders per year $= D/Q^* = 12,000/2940 = 4.082$ orders per year

Therefore, the shortage quantity per year $= 4.082 \times 490 = 2000.18$ units $= 2000$ units (approx.)

Example 7.10 The demand of an item in a store is 18,000 units per year. The purchase price of the item is Rs. 5 per unit and its carrying cost is Rs. 1.2 per unit per year and the ordering cost is Rs. 400 per order. The shortage cost is Rs. 5 per unit per year. Find the EOQ and the corresponding number of orders per year, the maximum inventory, maximum shortage quantity and the total cost of the system.

Solution The data of the given problem are:

$$D = 18,000 \text{ units per year}$$
$$C_o = \text{Rs. 400 per order}$$
$$p = \text{Rs. 5 per unit}$$
$$C_c = \text{Rs. 1.2 per unit per year}$$
$$C_s = \text{Rs. 5 per unit per year}$$

Therefore, the economic order (EOQ) is

$$\text{EOQ}(Q^*) = \sqrt{\frac{2C_oD}{C_c}\frac{(C_s+C_c)}{C_s}} = \sqrt{\frac{2\times400\times18,000}{1.2}\frac{(5+1.2)}{5}}$$

$$= 3857.46 \text{ units} = 3858 \text{ units (approx.)}$$

The maximum inventory, Q_1^* is

$$Q_1^* = \sqrt{\frac{2C_oD}{C_c}\frac{C_s}{C_s+C_c}} = \sqrt{\frac{2\times400\times18,000}{1.2}\frac{5}{5+1.2}}$$

$$= 3110.85 \text{ units} = 3111 \text{ units (approx.)}$$

The maximum shortage quantity per cycle, $Q_2^* = Q^* - Q_1^* = 3858 - 3111 = 747$ units

Number of orders/year $(N) = D/Q^* = 18,000/3858 = 4.666$

Shortage quantity per year $(Q_s) = N \times Q_2 = 4.666 \times 747$

$$= 3485.5 = 3486 \text{ units}$$

Total cost of the system per year,

$$TC^* = (D/Q^*)C_o + (Q^*/2)\,C_c + C_sQ_s + pD = (18,000/3858)\times400 + (3858/2)\times1.2 + 5\times3486 + 5\times18,000$$

$$= \text{Rs. } 1,11,611.05 \text{ per year}$$

7.2.4 Manufacturing Model with Shortages

In this model, an item is produced and consumed simultaneously for a portion of the cycle time. During the remaining cycle time, only the consumption of the item takes place. The cost of production per unit is the same irrespective of the production lot size. Stock-out is permitted in this model, and it is assumed that the stock-out units will be satisfied from the units which will be produced at a later date, with a penalty. This is called *backordering*. The operation of this model is shown in Figure 7.5.

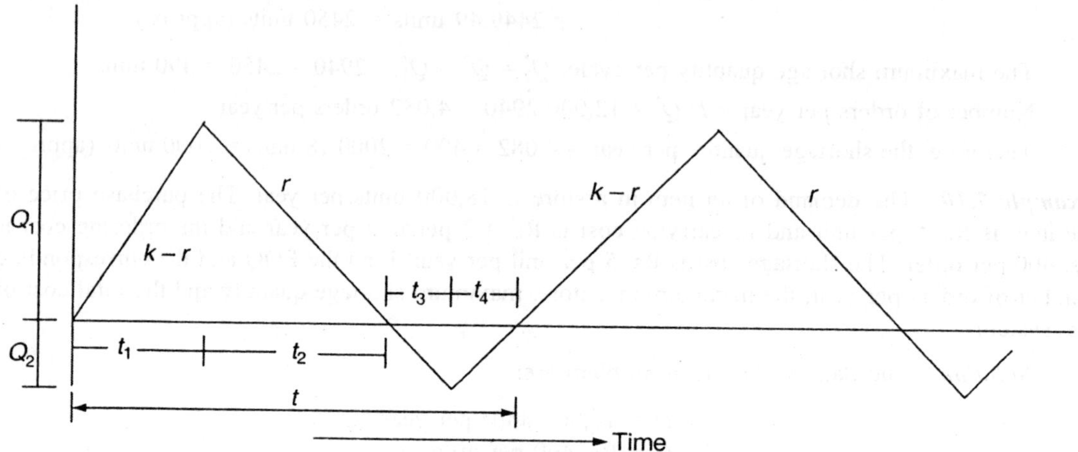

Figure 7.5   Manufacturing model with shortages.

The variables which are used in this model are given below:

r = Demand of an item/period

k = Production rate of the item (number of units produced/period)

C_o = Cost/set-up

C_c = Carrying cost/unit/period

C_s = Shortage cost/unit/period

t = Total cycle time

p = Cost of production/unit

t_1 = Period of production as well as consumption of the item satisfying period's requirement

t_2 = Period of consumption only

t_3 = Period of shortage

t_4 = Period of production as well as consumption of the item satisfying back order

The formulae for the optimal values of the above variables are presented below:

$$\text{Economic batch quantity } Q^* = \sqrt{\frac{2C_o}{C_c} \, \frac{kr}{k-r} \, \frac{C_c + C_s}{C_s}}$$

$$\text{Maximum inventory } Q_1^* = \sqrt{\frac{2C_o}{C_c} \, \frac{r(k-r)}{k} \, \frac{C_s}{C_c + C_s}}$$

$$\text{Maximum stock out } Q_2^* = \sqrt{\frac{2C_oC_c}{C_s(C_c + C_s)} \, \frac{r(k-r)}{k}}$$

$$Q_1^* = \frac{k-r}{k} Q^* - Q_2^*$$

$$t^* = \frac{Q^*}{r}$$

$$t_1^* = \frac{Q_1^*}{k-r}$$

$$t_2^* = \frac{Q_1^*}{r}$$

$$t_3^* = \frac{Q_2^*}{r}$$

$$t_4^* = \frac{Q_2^*}{k-r}$$

Example 7.11 The demand for an item is 6000 units per year. Its production rate is 1000 units per month. The carrying cost is Rs. 50/unit/year and the set-up cost is Rs. 2000 per set-up. The shortage cost is Rs. 1000 per unit per year. Find various parameters of the inventory system.

Solution Here

$$r = 6000 \text{ units/year}$$
$$k = 1000 \times 12 = 12{,}000 \text{ units/year}$$
$$C_o = \text{Rs. } 2{,}000/\text{set-up}$$
$$C_c = \text{Rs. } 50/\text{unit/year}$$
$$C_s = \text{Rs. } 1000/\text{unit/year}$$

Therefore,

$$Q^* \text{ (EBQ)} = \sqrt{\frac{2C_o}{C_c} \frac{kr}{k-r} \frac{C_c + C_s}{C_s}}$$

$$= \sqrt{\frac{2 \times 2000}{50} \frac{12{,}000 \times 6000}{12{,}000 - 6000} \frac{50 + 1000}{1000}}$$

$$= 1004 \text{ units (approx.)}$$

$$Q_2^* = \sqrt{\frac{2C_o C_c}{C_s(C_c + C_s)} \frac{r(k-r)}{k}}$$

$$= \sqrt{\frac{2 \times 2000 \times 50}{1000(50 + 1000)} \frac{6000(12{,}000 - 6000)}{12{,}000}}$$

$$= 24 \text{ units (approx.)}$$

$$Q_1^* = \frac{k-r}{k} Q^* - Q_2^*$$

$$= \frac{12{,}000 - 6000}{12{,}000} \times 1004 - 24 = 478 \text{ units}$$

$$t^* = \frac{Q^*}{r} \times 365 = \frac{1004}{6000} \times 365 = 61 \text{ days (approx.)}$$

$$t_1^* = \frac{Q_1^*}{k-r} \times 365 = \frac{478}{12,000 - 6000} \times 365 = 29 \text{ days (approx.)}$$

$$t_2^* = \frac{Q_1^*}{r} \times 365 = \frac{478}{6000} \times 365 = 29 \text{ days (approx.)}$$

$$t_3^* = \frac{Q_2^*}{r} \times 365 = \frac{24}{6000} \times 365 = 1.5 \text{ days (approx.)}$$

$$t_4^* = \frac{Q_2^*}{k-r} \times 365 = \frac{24}{12,000 - 6000} \times 365 = 1.5 \text{ days (approx.)}$$

Example 7.12 In a two wheeler manufacturing company, pistons are being fed into the main assembly line from a product line situated in the next bay. The annual demand for the pistons is 8000 units and the annual production capacity of the product line manufacturing the piston is 12,000 units. The set-up cost is Rs. 125 per set up and the carrying cost is Rs. 4 per piston per year. The shortage cost is Rs. 8 per piston per year. Find: Q^*, Q_1^*, Q_2^*, t^*, t_1^*, t_2^*, t_3^* and t_4^*.

Solution The data of the given problem are:

r = 8000 pistons per year
k = 12,000 pistons per year
C_o = Rs. 125 per set-up
C_c = Rs. 4 per piston per year
C_s = Rs. 8 per piston per year

Therefore, the economic batch quantity (EBQ) is

$$\text{EBQ}(Q^*) = \sqrt{\frac{2C_o}{C_c} \ \frac{kr}{(k-r)} \ \frac{C_c + C_s}{C_s}} = \sqrt{\frac{2 \times 125}{4} \ \frac{12,000 \times 8000}{(12,000 - 8000)} \ \frac{(4+8)}{8}}$$

$$= 1500 \text{ pistons}$$

The maximum inventory (Q_1^*) is

$$Q_1^* = \sqrt{\frac{2C_o}{C_c} \ \frac{r(k-r)}{k} \ \frac{C_s}{C_c + C_s}} = \sqrt{\frac{2 \times 125}{4} \ \frac{8000 \times (12,000 - 8000)}{12,000} \ \frac{8}{(4+8)}}$$

$$= 333.33 \text{ units} = 333 \text{ pistons (approx.)}$$

The maximum stock out (Q_2^*) is

$$Q_2^* = \sqrt{\frac{2C_o C_c}{C_s(C_c + C_s)} \ \frac{r(k-r)}{k}} = \sqrt{\frac{2 \times 125 \times 4}{8 \times (4+8)} \ \frac{8000 \times (12,000 - 8000)}{12,000}}$$

$$= 166.667 \text{ units} = 167 \text{ pistons (approx.)}$$

$t^*\ \ = Q^*/r = 1500/8000 \text{ year} = (1500/8000) \times 365 \text{ days} = 68.4 \text{ days}$

$t_1^*\ = Q_1^*/(k - r) = 333/(12,000 - 8000) \text{ year} = [333/(12,000 - 8000)] \times 365 \text{ days} = 30.4 \text{ days}$

$t_2^*\ = Q_1^*/r = 333/8000 \text{ year} = [333/8000] \times 365 \text{ days} = 15.2 \text{ days}$

$t_3^*\ = Q_2^*/r = 167/8000 \text{ year} = [167/8000] \times 365 \text{ days} = 7.6 \text{ days}$

$t_4^*\ = Q_2^*/(k - r) = 167/(12,000 - 8000) \text{ year} = [167/(12,000 - 8000)] \times 365 \text{ days} = 15.2 \text{ days}$

7.3 OPERATION OF INVENTORY SYSTEM

Consider a purchase model of inventory system as shown in Figure 7.6.

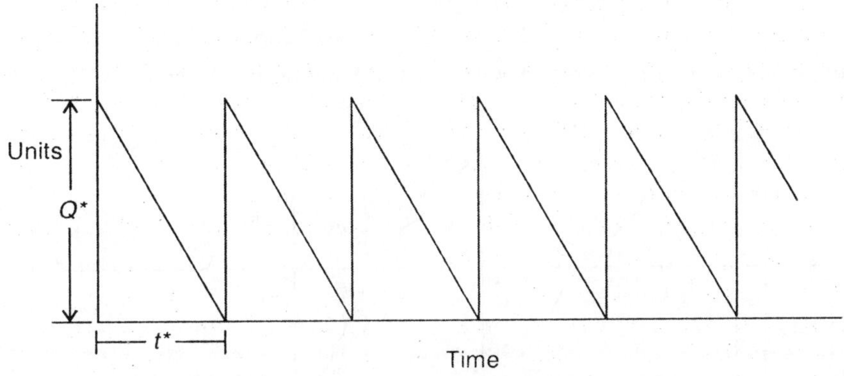

Figure 7.6 Operation of inventory system.

Here, Q^* is the economic order size and t^* is the cycle time. If we operate the system with any fluctuations in demand and lead time, we shall encounter stock-out situation very often. Even if we consider the model with constant demands and constant lead time, we shall have to place order well before the end of the cycle time, so that the items are received exactly at the end of the present cycle or at the beginning of the next cycle.

Let D_{LT} be the demand during lead time (LT). Then

$$D_{\text{LT}} = \text{demand rate (per day)} \times \text{lead time period (in days)}$$

If there is no variation in lead time and demand, then it is sufficient to have a stock of D_{LT} at the time of placing order. This is shown in Figure 7.7.

Reorder level (ROL) is the stock level at which an order is placed so that we receive the items against the order at the beginning of the next cycle. If the demand is not varying, the ROL is given by

$$\text{ROL} = D_{\text{LT}}$$

and if the demand is varying, the ROL is given by

$$\text{ROL} = D_{\text{LT}} + SS$$

where SS is the safety stock, which acts as a cushion to absorb the variation in demand. SS is defined as: $SS = K\sigma$, where σ is the standard deviation of demand and K is the standard normal statistic value for a given service level. The corresponding chart is shown in Figure 7.8.

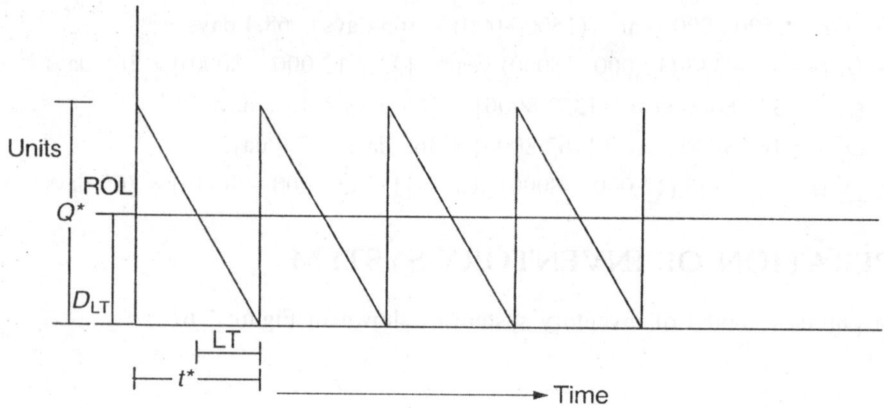

Figure 7.7 Inventory system with constant demand and constant lead time.

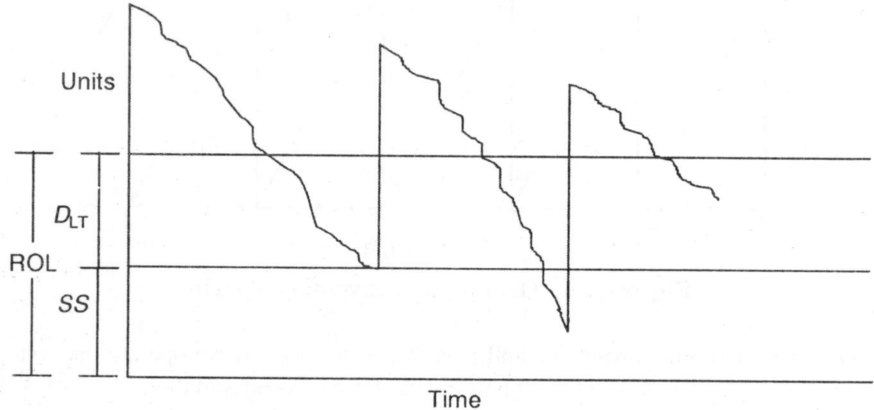

Figure 7.8 Inventory system with safety stock for variation in lead time demand.

Example 7.13 In a firm, the distribution of demand of an item during a constant lead time follows normal distribution. The standard deviation of the demand of the item is 300 units. The firm wants to have a service level of 95 per cent. Find: How much safety stock should be carried out for the item? If the demand during lead time averages 2000 units, what is the appropriate reorder level?

Solution The safety stock is given by

$$SS = K\sigma = (1.64)\,(300) = 492 \text{ units}$$

where $K = 1.64$ for 95% service level from standard normal table.

$$\text{Reorder level (ROL)} = D_{LT} + SS = 2000 + 492 = 2492 \text{ units}$$

Example 7.14 The average demand of an item is 24000 units per year. The purchase price per unit is Rs. 1.25, the ordering cost is Rs. 25 per order and the carrying cost is 6% of the unit cost. The number of working days in a year is 320 days and the lead time is 10 days. The demand follows normal distribution and the standard deviation of the demand is 100 units. Find EOQ, safety stock and reorder level by assuming a confidence level of 95%.

Solution   The data of the given problem are:

Average demand (D) = 24,000 units per year
C_o = Rs. 25 per order
p = Rs. 1.25 per unit
C_c = 0.06 × 1.25 = Re. 0.075 per unit per year
Number of working days per year = 320 days
Lead time = 10 days
Standard deviation of demand (σ) = 100 units
Confidence level $(1 - \alpha)$ = 0.95; therefore, α = 0.05
K_α = 1.64

Therefore, the economic order quantity

$$\text{EOQ}(Q^*) = \sqrt{\frac{2C_o D}{C_c}} = \sqrt{\frac{2 \times 25 \times 24,000}{0.075}} = 4000 \text{ units}$$

Demand per day = $(D$/No. of working days in one year) = 24,000/320 = 75 units per day
Lead time demand, D_{LT} = Lead time × Daily demand = 10 × 75 = 750 units
Safety Stock (SS) = $K_\alpha \sigma$ = 1.64 × 100 = 164 units
Reorder level (ROL) = D_{LT} + SS = 750 + 164 = 914 units.

7.4   QUANTITY DISCOUNT

When an item is purchased in bulk, buyers are usually given discount in the purchase price of the item. Let i be the per cent of the purchase price accounted for carrying cost/unit/period. The discount may be a step function of purchase quantity as shown below.

Quantity	Purchase price per unit
$0 \le Q_1 < b_1$	p_1
$b_1 \le Q_2 < b_2$	p_2
$b_2 \le Q_3 < b_3$	p_3
$\vdots$	$\vdots$
$b_{n-1} \le Q_n$	p_n

The procedure to compute the optimal order size for this situation is given in the following steps:

Step 1: Find the EOQ for the nth (last) price break.

$$Q_n^* = \sqrt{\frac{2C_o D}{i p_n}}$$

If it is greater than or equal to b_{n-1}, then the optimal order size Q^* is equal to Q_n^*; otherwise go to step 2.

Step 2: Find the EOQ for the $(n - 1)$th price break.

$$Q^*_{n-1} = \sqrt{\frac{2C_oD}{ip_{n-1}}}$$

If it is greater than or equal to b_{n-2}, then compute the following and select the least cost purchase quantity as the optimal order size; otherwise go to step 3.

(i) Total cost $TC(Q^*_{n-1})$.

(ii) Total cost, $TC(b_{n-1})$.

Step 3: Find the EOQ for the $(n - 2)$th price break.

$$Q^*_{n-2} = \sqrt{\frac{2C_oD}{ip_{n-2}}}$$

If it is greater than or equal to b_{n-3}, then compute the following and select the least cost purchase quantity as the optimal order size; otherwise go to step 4:

(i) Total cost $TC(Q^*_{n-2})$

(ii) Total cost $TC(b_{n-2})$

(iii) Total cost $TC(b_{n-1})$

Step 4: Continue in this manner until $Q^*_{n-i} \geq b_{n-i-1}$. Then compare total costs $TC(Q^*_{n-i})$, $TC(b_{n-i})$, $TC(b_{n-i+1}),..., TC(b_{n-1})$ corresponding to purchase quantities Q^*_{n-i}, b_{n-1}, $b_{n-i+1},..., b_{n-1}$, respectively. Finally, select the purchase quantity with respect to the minimum total cost as the optimal order size.

Example 7.15 Annual demand for an item is 6000 units. Ordering cost is Rs. 600 per order. Inventory carrying cost is 18% of the purchase price/unit/year. The price breakups are as shown below.

Quantity	Price (in Rs.) per unit
$0 \leq Q_1 < 2000$	20
$2000 \leq Q_2 < 4000$	15
$4000 \leq Q_3$	9

Find the optimal order size.

Solution Given that, $D = 6000$/year, $C_o = $ Rs. 600/order and $i = 18\%$ of the purchase price/unit/year.

Step 1: $p_3 = $ Rs. 9. Therefore,

$$Q^*_3 = \sqrt{\frac{2C_oD}{ip_3}} = \sqrt{\frac{2 \times 600 \times 6000}{0.18 \times 9}} = 2109 \text{ units (approx.)}$$

Since, $Q^*_3 < b_2$ (4000) and $Q^*_3 > b_1$ (2000), go to step 2.

Step 2: p_2 = Rs. 15. Therefore,

$$Q_2^* = \sqrt{\frac{2C_oD}{ip_2}} = \sqrt{\frac{2 \times 600 \times 6000}{0.18 \times 15}} = 1633 \text{ units (approx.)}$$

Since $Q_2^* < b_1$ (2000), go to step 3.

Step 3: p_1 = Rs. 20. Therefore,

$$Q_1^* = \sqrt{\frac{2C_oD}{ip_1}} = \sqrt{\frac{2 \times 600 \times 6000}{0.18 \times 20}} = 1415 \text{ units (approx.)}$$

Since $Q_1^* < b_1$ (2000), find the following costs and select the order size with respect to the least cost as the optimal order size.

$$TC(Q_1^*) = (20)(6000) + \frac{(600)(6000)}{1415} + \frac{(0.18)(20)(1415)}{2} = \text{Rs. } 1,25,091$$

$$TC(b_1) = (15)(6000) + \frac{(600)(6000)}{2000} + \frac{(0.18)(15)(2000)}{2} = \text{Rs. } 94,500$$

$$TC(b_2) = (9)(6000) + \frac{(600)(6000)}{4000} + \frac{(0.18)(9)(4000)}{2} = \text{Rs. } 58,140.$$

The least cost is Rs. 58,140. Hence, the optimal order size is b_2 which is equal to 4000 units.

Example 7.16   A company currently purchases one of its items for Rs. 2 per unit without quantity discount. The ordering cost is Rs. 20 per order and the carrying cost is 20% of its purchase price per unit per year. The annual demand is 2500 units. A new vendor offers quantity discount for the same item as per the following quantity discount scheme. Find the best order quantity.

Quantity	Price (in Rs.) per unit
$0 \le Q_1 < 1500$	p
$1500 \le Q_2 < 2500$	97% of p
$2500 \le Q_3$	95% of p

Solution

Ordering cost, C_o = Rs. 20 per order
Annual demand, D = 2500 units
Purchase price, p = Rs. 2 per unit without quantity discount
Price break: $2500 \le Q_3$ in which the purchase price is 95% of p.
p = 0.95 × Rs. 2 = Rs. 1.90 per unit
C_c = 0.20 × Rs. 1.90 = Re. 0.38 per unit per year

$$\text{EOQ under last price break } (Q_3^*) = \sqrt{\frac{2C_oD}{C_c}} = \sqrt{\frac{2 \times 20 \times 2500}{0.38}} = 512.99 \text{ units}$$

$$= 513 \text{ units (approx.)}$$

Since, Q_3 is in the interval $0 \leq Q_1 < 1500$, the first interval is to be examined.

Price break: $0 \leq Q_1 < 1500$ in which the purchase price is p.

p = Rs. 2

C_c = 0.20 × Rs. 2 = Re. 0.40 per unit per year

$$\text{EOQ under first price break } (Q_1^*) = \sqrt{\frac{2C_o D}{C_c}} = \sqrt{\frac{2 \times 20 \times 2500}{0.40}} = 500 \text{ units}$$

$$TC(Q_1^*) = (D/Q_1)\, C_o + (Q_1/2)\, C_c + D \times p$$
$$= (2500/500) \times 20 + (500/2) \times 0.40 + 2500 \times 2 = \text{Rs. } 5200$$

When order size is b_1 (i.e. 1500) units, price per unit, = 0.97 × Rs. 2 = Rs. 1.94/unit and C_c = 0.2 × Rs. 1.94 = Re. 0.388

Therefore, $TC(b_1 = 1500) = (D/b_1) \times C_o + (b_1/2) \times C_c + D \times p$
$$= (2500/1500) \times 20 + (1500/2) \times 0.388 + 2500 \times 1.94 = \text{Rs. } 5174.33$$

When order size is b_2 (i.e. 2500) units, price per unit, = 0.95 × Rs. 2 = Rs. 1.90/unit and C_c = 0.2 × Rs. 1.90 = Re. 0.38

Therefore, $TC(b_2 = 2500) = (D/b_2) \times C_o + (b_2/2) \times C_c + D \times p$
$$= (2500/2500) \times 20 + (2500/2) \times 0.38 + 2500 \times 1.9 = \text{Rs. } 5245$$

The minimum of $TC(Q_1)$, $TC(b_1)$ and $TC(b_2)$ is $TC(b_1)$ which is Rs. 5174.33. Hence, the best order size is 1500 units.

7.5 IMPLEMENTATION OF PURCHASE INVENTORY MODEL

The practical version of purchase model of inventory can be classified into *fixed order quantity system* (or *Q system*) and *Periodic review system* (or *P system*). (These are described in the following sections.) Any one or a combination of the following cases exists in each of the systems:

1. Constant demand and constant lead time
2. Varying demand and constant lead time
3. Constant demand and varying lead time
4. Varying demand and varying lead time.

7.5.1 Fixed Order Quantity System (Q System)

In this inventory system, whenever the stock level touches reorder level, an order is placed for a fixed quantity which is equal to EOQ. A schematic representation of this model is shown in Figure 7.9.

The average demand during the lead time (average lead time) is known as the *demand during lead time* (D_{LT}). The variation in demand during lead time (average lead time) is known as *safety stock*. The average demand during delivery delays is called *reserve stock*. The reorder level is computed as the sum of the demand during lead time (D_{LT}), the variation in demand during lead time (safety stock) and the average demand during delivery delays (reserve stock).

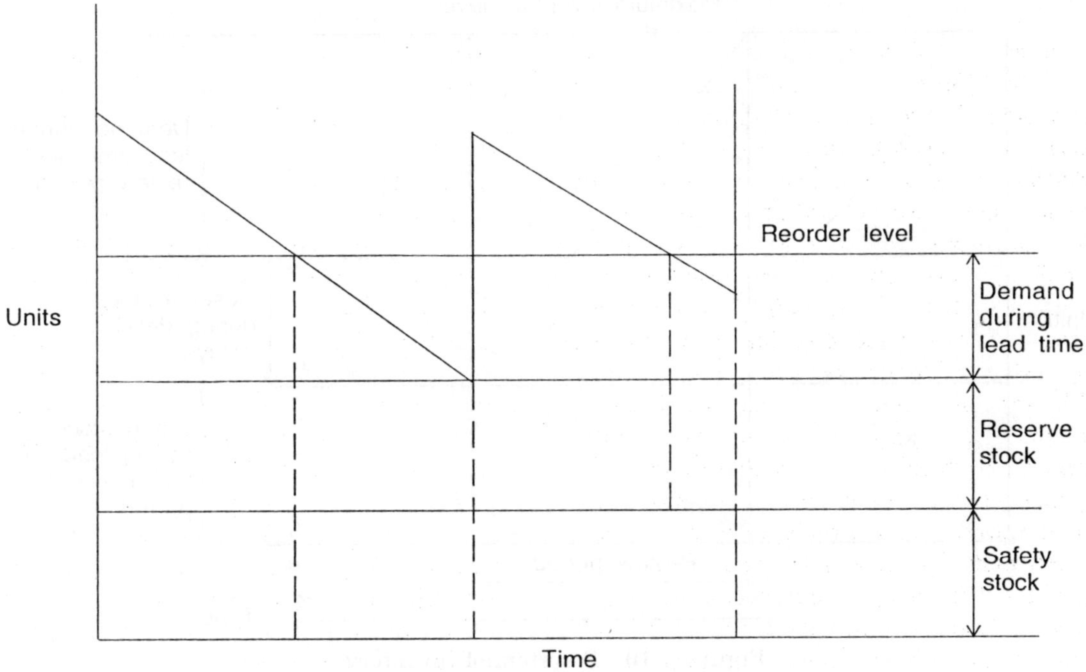

Figure 7.9   Q system of inventory.

7.5.2   Periodic Review System (P System)

In this inventory system, the stock position is reviewed once in a fixed period and an order is placed depending on the stock position, unlike a fixed quantity in the Q system of inventory. The review period is approximately equal to EOQ/D. The desired maximum inventory level is fixed as the sum of the average demand during average lead time *plus* review period, variation in demand during average lead time *plus* review period, and the average demand during delays in supply. In the beginning of each review period, the difference between the maximum inventory level and the actual stock on hand will be determined. Then an order will be placed for this quantity. A schematic representation of this model is shown in Figure 7.10.

Example 7.17   The annual demand of a product is 36,000 units. The average lead time is 3 weeks. The standard deviation of demand during the average lead-time is 150 units/week. The cost of ordering is Rs. 500 per order. The cost of purchase of the product per unit is Rs. 15. The cost of carrying per unit per year is 20% of the purchase price. The maximum delay in lead time is 1 week and the probability of this delay is 0.3. Assume a service level of 0.95.

   (a)   What is the reorder level if Q system is followed?
   (b)   What is the maximum inventory level, if P system is followed?

Solution   We have

Annual demand, $D = 36,000$ units
Ordering cost per order, $C_o = $ Rs. 500
Purchase price per unit, $p = $ Rs. 15

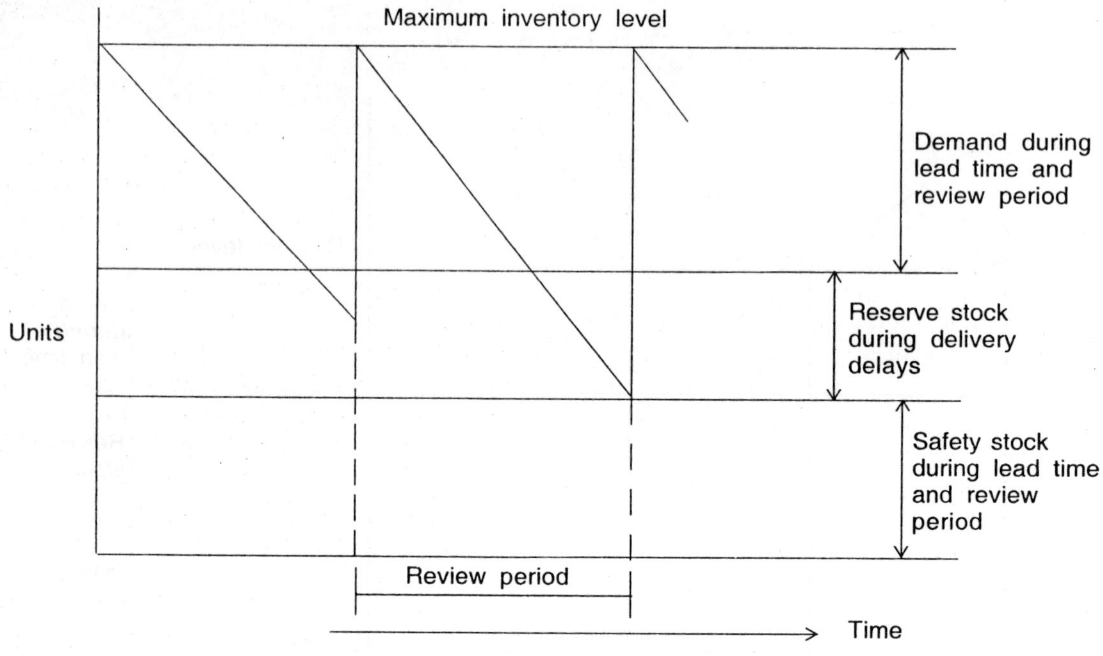

Figure 7.10 P system of inventory.

Carrying cost per unit per year = 20% of the purchase price per unit
Average lead time = 3 weeks
Standard deviation of demand during average lead time per week = 150 units
Maximum delay = 1 week
Probability of having the maximum delay = 0.3
Service level = 0.95

$$\text{Order quantity} = \sqrt{\frac{2C_oD}{\%C_c \times p}} = \sqrt{\frac{2 \times 500 \times 36{,}000}{0.20 \times 15}} = 3{,}465 \text{ units (approx.)}$$

(a) *Determination of reorder level for Q system*

$$\text{Demand during lead time, } D_{LT} = \frac{36{,}000}{52} \times 3 = 2077 \text{ units (approx.)}$$

Standard deviation in demand during lead time $(\sigma) = \sqrt{\text{Lead time}}$ (standard deviation per week)

$$= \sqrt{3} \times 150 = 260 \text{ units (approx.)}$$

and

$$\text{Safety stock during lead time } (SS) = K \times \sigma = 1.64 \times 260 = 427 \text{ units (approx.)}$$

where K is equal to 1.64 for the given service level of 0.95.

Also, average demand during delivery delays (reserve stock)

$$= \frac{D \times \text{Maximum delay}}{\text{Number of weeks/yr}} \times \text{Probability of maximum delay}$$

$$= \frac{36,000}{52} \times 1 \times 0.3$$

$$= 208 \text{ units (approx.)}$$

$$\text{Reorder level} = \text{Demand during lead time } (D_{\text{LT}})$$

$$+ \text{ Variation in demand during lead time (safety stock)}$$

$$+ \text{ Average demand during delivery delays (reserve stock)}$$

$$= 2077 + 427 + 208$$

$$= 2712 \text{ units.}$$

(b) *Determination of maximum inventory level for P system*

$$\text{Review period} = \frac{\text{EOQ}}{D} = \frac{3465}{36,000} = 0.09625 \text{ year} = 5.005 \text{ weeks}$$

The review period can be either 5 weeks or 6 weeks. When the review period (RP) is equal to 5 weeks, the total cost in calculated as:

$$\text{Total cost} = \text{Ordering cost} + \text{Carrying cost}$$

$$= \left(\frac{52}{5} \times 500 \right) + \left(\frac{36,000}{52} \times 5 \times \frac{1}{2} \times 15 \times 0.20 \right)$$

$$= \text{Rs. } 10,392.31$$

When the review period (RP) is equal to 6 weeks;

$$\text{Total cost} = \text{Ordering cost} + \text{Carrying cost}$$

$$= \left(\frac{52}{6} \times 500 \right) + \left(\frac{36,000}{52} \times 6 \times \frac{1}{2} \times 15 \times 0.20 \right)$$

$$= \text{Rs. } 10,564.10$$

The total cost is minimum when the review period is 5 weeks. Hence, the review period is selected as 5 weeks.

$$\text{Demand during lead time and review period} = \left(\frac{36,000}{52} \right) \times (3 + 5) = 5539 \text{ units (approx.)}$$

$$\text{Safety stock during lead time and review period} = (\text{LT} + \text{RP})^{1/2} \times \sigma \text{ per week} \times K$$

$$= (3 + 5)^{1/2} \times 150 \times 1.64$$

$$= 696 \text{ units (approx.)}$$

where $K = 1.64$ for the given service level of 0.95.

Average demand during delivery delays (reserve stock)

$$= \frac{D \times \text{Maximum delay}}{\text{No. of weeks/yr}} \times \text{Probability of maximum delay}$$

$$= \frac{36,000}{52} \times 1 \times 0.30$$

$$= 208 \text{ units (approx.)}$$

Maximum inventory level = Demand during lead time and review period

+ Safety stock during lead time and review period

+ Average demand during delivery delays

= 5539 + 696 + 208

= 6443 units.

7.6   MULTIPLE-ITEM MODEL WITH STORAGE LIMITATION

Consider the deterministic purchase model of inventory without shortages. In reality, there may be a situation wherein the storage space will act as a constraint for storing different items. Under such circumstance, the economic order quantities of the items which are to be stocked in the stores may have to be modified to meet the storage space limitation.

Let us suppose,

n = Number of items stocked in the stores

D_i = Demand per year of the item i

C_{oi} = Ordering cost per order of the item i

C_{ci} = Carrying cost per unit per year of the item i

Q_i = Order quantity of the item i

S_i = Space requirement per unit of the item i

K = Total space available

Based on the above definitions, the formula to find the economic order quantity of the item i is

$$Q_i^* = \sqrt{\frac{2C_{oi}D_i}{C_{ci}}}, \qquad i = 1, 2, 3,..., n$$

If the store has unlimited space, then the economic order quantities of the items need no modification. Otherwise, the best combination of the order quantities of the n items is to be determined such that the total space requirement is equal to the available space.

Let the space constraint is

$$\sum_{i=1}^{n} S_iQ_i \leq K$$

The objective function of this problem consists of two cost components, viz., ordering costs and carrying costs, and represented as:

$$Z = \sum_{i=1}^{n} \left(\frac{D_i}{Q_i}C_{oi} + \frac{Q_i}{2}C_{ci} \right)$$

A mathematical model of the proposed problem as:

$$\text{Minimize } Z = \sum_{i=1}^{n} \left(\frac{D_i}{Q_i}C_{oi} + \frac{Q_i}{2}C_{ci} \right)$$

subject to

$$\sum_{i=1}^{n} S_iQ_i \leq K$$

where, $Q_i > 0$, $i = 1, 2, 3,..., n$. In this model, the objective function finds the optimal values of Q_i such that the constraint is satisfied in equality sense. The model is translated into a *Lagrangian function*:

$$L(\mu, Q_1, Q_2, Q_3,..., Q_n) = Z(Q_1, Q_2, Q_3,..., Q_n) - \mu \left(\sum_{i=1}^{n} S_i Q_i - K \right)$$

where, μ is the *Lagrange multiplier* and it is less than zero. The partial derivatives of the Lagrangian function with respect to Q_i and μ are:

$$\frac{\delta L}{\delta Q_i} = -\frac{D_i}{Q_i^2} C_{oi} + \frac{C_{ci}}{2} - \mu S_i, \qquad i = 1, 2, 3,..., n$$

and

$$\frac{\delta L}{\delta \mu} = -\left(\sum_{i=1}^{n} S_i Q_i - K \right) = -\sum_{i=1}^{n} S_i Q_i + K$$

The second equation shows that the total space requirement for all the items is equal to K. The formula for Q_i^* is obtained by equating $\delta L/\delta Q_i$ to zero as shown below:

$$Q_i^* = \sqrt{\frac{2C_{oi}D_i}{C_{ci} - 2\mu^* S_i}}, \qquad i = 1, 2, 3,..., n$$

In the above equation, Q_i^* depends on μ^*. The optimal value for μ^* can be obtained by trial-and-error method such that the following space constraint is satisfied.

$$\sum_{i=1}^{n} S_i Q_i \leq K$$

Example 7.18 A company wants to determine the economic order size of each of the three bought-out items which are stocked in its stores. The details of the items are presented in Table 7.1. The maximum space available is 500 square metre. Find the economic order size of each of the items subject to the space constraint.

Table 7.1   Example 7.18

	Item number		
	1	2	3
Demand/year (ton)	1000	1500	750
Ordering cost/order (Rs.)	500	700	300
Carrying cost/ton/year (Rs.)	50	80	100
Space requirement/ton (sq. mt)	2	1	3

Solution The economic order quantity of the item i under the space constraint is

$$Q_i^* = \sqrt{\frac{2C_{oi}D_i}{C_{ci} - 2\mu^* S_i}}$$

The computations of total space requirement to store all the three items with respect to different values of μ are summarized in Table 7.2.

Table 7.2   Details of Computation

μ	Q_1	Q_2	Q_3	$\Sigma S_i Q_i$
0	141.4213	162.0185	67.0820	646.1071
−5.0	119.5228	152.7525	58.8348	568.3025
−10.0	105.4092	144.9137	53.0330	514.8311
−15.0	95.3463	138.1698	48.6664	474.8616
−11.0	103.1421	143.4860	52.0658	505.9676
−12.0	101.0152	142.0996	51.1496	497.5788
−11.1	102.9233	143.3455	51.9719	505.1078
−11.2	102.7060	143.2054	51.8786	504.2532
−11.3	102.4900	143.0658	51.7858	503.4032
−11.4	102.2753	142.9266	51.6934	502.5574
−11.5	102.0620	142.7877	51.6016	501.7165
−11.6	101.8501	142.6493	51.5102	500.8801
−11.7	101.6394	142.5113	51.4193	500.0480*
−11.8	101.4301	142.3736	51.3289	499.2205

In Table 7.2, initially the total space requirement is computed for different values of μ from 0 to −15 in steps of −5. Again, the procedure is repeated for μ with −11 and −12. Finally, the total space requirement is computed for different values of μ from −11.1 to −11.8 in steps of −0.1. From Table 7.2, it is clear that the total space requirement is almost equal to 500 square metre when μ is equal to −11.7. The corresponding optimal economic order quantities of the three items are presented below. (If the total space requirement is somewhat closer to 500 sq. metre, then one can determine the Q_i^* by using interpolation method with reference to its previous row or its succeeding row. But for this problem, we are able to get the results without using the interpolation method because the total space requirement is 500.048 sq. metre, when μ is equal to −11.7.

Thus,

$$Q_1^* = 101.6394 \text{ tons}, \quad Q_2^* = 142.5113 \text{ tons} \quad \text{and} \quad Q_3^* = 51.4193 \text{ tons}$$

7.7   PURCHASE MODEL OF INVENTORY FOR MULTI-ITEM WITH INVENTORY CARRYING COST CONSTRAINT

Consider the purchase model of inventory with multi-item with a constraint on the inventory carrying cost, K. The variables involved in the purchase model of inventory with inventory carrying cost constraint are listed below. In this model, shortages are not permitted.

C_{o_i} is the ordering cost of the item i
D_i is the annual demand in units of the item i
p_i is the purchase price per unit of the item i in the group of items ordered
C_{c_i} be the inventory carrying cost per unit per period of the item i
Q_i is the economic order quantity (EOQ) in units for the item i
m is the number of items in the group
K is the utmost inventory carrying cost per period

The formula for the total cost (*TC*) of this inventory system is as given below:

$$TC = \text{Total ordering cost} + \text{Total carrying cost} + \text{Total purchase cost}$$

$$= \sum_{i=1}^{m}[(D_i)/(Q_i)]C_{o_i} + \sum_{i=1}^{m}(Q_i/2)C_{c_i} + \sum_{i=1}^{m}D_i p_i$$

As per the problem statement, there is a constraint on the total carrying cost whose utmost value is K.

The Lagrangian function with the above constraint is

$$L = \sum_{i=1}^{m}[(D_i)/(Q_i)]C_{o_i} + \mu\left(\sum_{i=1}^{m}(Q_i/2)C_{c_i} - K\right) + \sum_{i=1}^{m}D_i p_i$$

where, μ is a Lagrange multiplier.

Differentiating the function L w.r.t. Q_i, we get

$$\frac{\partial L}{\partial Q_i} = [-(D_i)/(Q_i^{-2})]C_{o_i} + \mu\,(C_{c_i}/2), \text{ for } i = 1, 2, 3,..., m$$

Then by differentiating the function L by μ and equating it to 0, we get

$$\frac{\partial L}{\partial \mu} = \sum_{i=1}^{m}(Q_i/2)C_{c_i} - K = 0$$

The second equation gives the condition that the total carrying cost is restricted to utmost K. Equating the first equation to 0, the formula for Q_i is as given below:

$$Q_i = \sqrt{\frac{2C_{o_i}\,D_i}{\mu C_{c_i}}} \quad \text{for, } i = 1, 2, 3,..., m$$

subject to $\displaystyle\sum_{i=1}^{m}(Q_i/2)C_{c_i} \leq K$

Example 7.19 A company wants to determine the economic order quantity for each of its 2 items which are stored in its warehouse. The annual demand in units of the item 1 and the item 2 are 12,000 units and 18,000 units, respectively. The carrying cost per unit per year for the item 1 and the item 2 are Rs. 5 and Rs. 8, respectively. The ordering cost per order for the item 1 and the item 2 are Rs. 500 and Rs. 400, respectively. The company wants to limit the total inventory carrying cost to Rs. 4000. Find the economic order quantity of each of the items.

Solution The data of the given problem are listed below:

Annual demand of the item 1 (D_1) = Rs. 12,000 units
Annual demand of the item 2 (D_2) = Rs. 18,000 units
Carrying cost per unit per year of the item 1 (C_{c_1}) = Rs. 5
Carrying cost per unit per year of the item 2 (C_{c_2}) = Rs. 8
Ordering cost per order of the item 1 (C_{o_1}) = Rs. 500
Ordering cost per order of the item 2 (C_{o_2}) = Rs. 400
Utmost value of the total carrying cost (K) = Rs. 4000

The formula for the EOQ of the item i is

$$Q_i = \sqrt{\frac{2C_{o_i} D_i}{\mu C_{c_i}}} = \frac{1}{\sqrt{\mu}} \sqrt{\frac{2C_{o_i} D_i}{C_{c_i}}}, \, i = 1, 2, 3, \ldots, m$$

$$Q_1 = \frac{1}{\sqrt{\mu}} \sqrt{\frac{2C_{o_1} D_1}{C_{c_1}}} = \frac{1}{\sqrt{\mu}} \sqrt{\frac{2 \times 500 \times 12,000}{5}} = \frac{1}{\sqrt{\mu}} 1549.19$$

$$Q_2 = \frac{1}{\sqrt{\mu}} \sqrt{\frac{2C_{o_2} D_2}{C_{c_2}}} = \frac{1}{\sqrt{\mu}} \sqrt{\frac{2 \times 400 \times 18,000}{8}} = \frac{1}{\sqrt{\mu}} 1341.64$$

Total carrying cost = $(Q_1/2) \, C_{c_1} + (Q_2/2) \, C_{c_2} = K = 4000$
Therefore,

$$\frac{1}{\sqrt{\mu}} \left\{ \frac{1549.19}{2} \times 5 + \frac{1341.64}{2} \times 8 \right\} = 4000$$

From this equation, $\mu = 5.335562$.
Therefore,

$$Q_1 = \frac{1}{\sqrt{5.335562}} \times 1549.19 = 670.68 = 671 \text{ units}$$

$$Q_2 = \frac{1}{\sqrt{5.335562}} \times 1341.64 = 580.8263 = 581 \text{ units}$$

Total carrying cost = $(Q_1/2) \, C_{c_1} + (Q_2/2) \, C_{c_2}$

$$= (670.68/2) \times 5 + (580.8263/2) \times 8 = 4000$$

$$= K \text{ (Satisfied)}$$

7.8 EOQ MODEL FOR MUTI-ITEM JOINT REPLENISHMENT

In practice, the order processing department of an organization may process the purchases of multi-item jointly to replenish them. Under such situation, the ordering cost is to be redefined. It has a fixed component which is the cost to the organization for processing the common activities involved in purchasing all the items in the group of items that is being purchased jointly and a variable cost which is specific to each item in the group. This variable cost/marginal cost of processing some specific purchase activities of an item which are not part of the common activities for the entire group of items depends on that item. Hence, the variable cost of an item is called as *marginal cost* or *item dependent cost* of that item.

In this book, the purchase model of inventory for multi-item joint replenishment without shortages and manufacturing model of inventory for multi-item joint replenishment without shortages are presented.

7.8.1 Purchase Model of Inventory for Multi-Item Joint Replenishment without Shortages

The variables involved in the purchase model of inventory for multi-item joint replenishment without shortages are listed below. In this model, shortages are not permitted.

C_o is the fixed ordering cost in rupees for a group of items
C_{o_i} is the marginal cost of ordering for the item i
D_i is the annual demand in units of the item i
Dr is the annual demand in rupees for the group of items ordered
dr_i is the annual demand in rupees for the item i in the group of items ordered
p_i is the purchase price per unit of the item i in the group of items ordered
I is the inventory carrying cost in percentage of the unit cost (For example, 10% is to be written as 0.10)
Qr is the economic order quantity (EOQ) in rupees for the group of items ordered
qr_i is the economic order quantity (EOQ) in rupees for the item i
Q_i is the economic order quantity (EOQ) in units for the item i
N is the number of order cycles per year
m is the number of items in the group
T is the time between consecutive orders

The formula for the total cost (TC) of this inventory system is as given below:

$$TC = \text{Total ordering cost} + \text{Total carrying cost} + \text{Total purchase cost}$$

$$= [Dr)/(Qr)]\left\{ C_o + \sum_{i=1}^{m} C_{o_i} \right\} + (Qr/2)I + \sum_{i=1}^{m} p_i D_i$$

Differentiating the function TC w.r.t. Qr, we get

$$\frac{d}{dQr}(TC) = [-(Dr)/(Qr^{-2})]\left\{ C_o + \sum_{i=1}^{m} C_{o_i} \right\} + I/2$$

Again, differentiating the function TC w.r.t. Qr, we get

$$\frac{d^2}{dQr^2}(TC) = [(Dr)/(Qr^{-3})]\left\{ C_o + \sum_{i=1}^{m} C_{o_i} \right\}$$

As per the principles of minima, the second derivative is positive. Hence, equating the first derivative to 0, we get,

$$Qr = \sqrt{\frac{2\left(C_o + \sum_{i=1}^{m} C_{o_i} \right) Dr}{I}}$$

The formulas to get the EOQ in rupees for the item i (qr_i) and the EOQ in units for the item i (Q_i) are given below:

$$qr_i = Qr \times (dr_i/Dr), \qquad \text{for } i = 1, 2, 3, \ldots, m$$
$$Q_i = qr_i/p_i, \qquad \text{for } i = 1, 2, 3, \ldots, m$$

Further, the formulas to obtain N and T are as given below:

$$N = Dr/Qr$$
$$T = Qr/Dr$$

N is computed for the group of items ordered. But, the number of intervals for the individual item may differ from N. There are many procedures to obtain the individual interval multiples, n_i, $i = 1, 2, 3,..., m$. In this book, an algorithm called *Silver's Algorithm* is presented to find such n_i values.

Silver's Algorithm

Step 1: Find the ratios C_{oi}/dr_i, for $i = 1, 2, 3,..., m$ and find their minimum. Let the item corresponding to this minimum be k.

Step 2: Set the interval multiple for the item k to 1 (i.e. $n_k = 1$).

Step 3: Compute the interval multiples, n_i for the remaining items, $i = 1, 2, 3,..., m$ and $i \neq k$ using the following formula and round off each of them to the nearest integer (*Note:* If n_i is less than 1, then round off it to 1 irrespective of its value).

$$n_i = \sqrt{\frac{C_{o_i}}{dr_i}\frac{dr_k}{C_o + C_{o_k}}} \quad \text{for, } i = 1, 2, 3,..., m \text{ and } i \neq k$$

Example 7.20 A company orders for three items jointly. The details of the situation are given in Table 7.3. The fixed ordering cost is Rs. 200. The inventory carrying charge is 12% of the purchase prices of items in the group. Find the EOQ in units as well as the interval multiple of each of the items.

Table 7.3 Data of Example 7.20

Item i	Annual demand in units	Price/unit p_i (Rs.)	Marginal cost of ordering item i (Rs.)
1	12,000	20	30
2	1000	5	40
3	24,000	15	25

Solution The additional details which are required to solve the problem are shown in Table 7.4.

Table 7.4 Data of Example 7.20 with Additional Details

Item i	Annual demand in units (D_i)	Price/unit p_i (Rs.)	Annual demand in rupees (dr_i)	Marginal cost of ordering item i in rupees (C_{o_i})
1	12,000	20	2,40,000	30
2	1000	5	5000	40
3	24,000	15	3,60,000	25
Total			$Dr = 6,05,000$	$\Sigma C_{o_i} = 95$

Fixed ordering cost (C_o) = Rs. 200
Sum of marginal costs of ordering the items (ΣC_{o_i}) = Rs. 95
Annual demand in rupees of all items in the group (Dr) = Rs. 6,05,000
Inventory carrying charge percentage in decimal (I) = 0.12

Therefore, the EOQ in rupees of all the items put together in the group (Qr) is computed as shown below.

$$Qr = \sqrt{\dfrac{2\left(C_o + \sum_{i=1}^{3} C_{o_i}\right)Dr}{I}}$$

$$= \sqrt{\dfrac{2 \times (200 + 95) \times 6,05,000}{0.12}}$$

$$= \text{Rs. } 54,539.74$$

From the value of Qr, the value of qr_i and Q_i for each of the items are computed as shown below.

$$qr_1 = \dfrac{dr_1}{Dr} \times Qr = (2,40,000/6,05,000) \times 54,539.74 = \text{Rs. } 21,635.60$$

$$qr_2 = \dfrac{dr_2}{Dr} \times Qr = (5000/6,05,000) \times 54,539.74 = \text{Rs. } 450.74$$

$$qr_3 = \dfrac{dr_3}{Dr} \times Qr = (3,60,000/6,05,000) \times 54,539.74 = \text{Rs. } 32,453.40$$

Similarly, the EOQ in units of each item i is computed using the formula

$Q_i = qr_i/p_i$, for $i = 1, 2, 3, \ldots, m$

$Q_1 = qr_1/p_1 = 21,635.60/20 = 1081.78 = 1082$ units (approx.)

$Q_2 = qr_2/p_2 = 450.74/5 = 90.148 = 90$ units (approx.)

$Q_3 = qr_3/p_3 = \text{Rs. } 32,453.4/15 = 2163.56 = 2164$ units (approx.)

$N = Dr/Qr = 6,05,000/54,539.74 = 11.09$

$T = 1/N = 1/11.09 = 0.09$ year

The calculations for determining the interval multiples $(n_i,\ i = 1, 2, 3)$ of the individual items as per Silver's algorithm are summarized in Table 7.5. The formula for the interval multiple, n_i is as given below.

$$n_i = \sqrt{\dfrac{C_{o_i}}{dr_i}\dfrac{dr_k}{C_o + C_{o_k}}} \quad \text{for, } i = 1, 2, 3, \ldots, m \text{ and } i \neq k$$

From Table 7.5, one can see that the interval multiples for the item 1, item 2 and item 3 are 1, 4 and 1, respectively.

Table 7.5   Determination of Interval Multiples of Individual Items

Item i	dr_i	C_{o_i}	C_{o_i}/dr_i	$dr_k/(C_o + C_{o_k})$ where, $k = 3$	n_i	n_i^*
1	2,40,000	30	0.000125	1600	0.4472	1
2	5000	40	0.008000	1600	3.5778	4
3	3,60,000	25	0.000069	–	–	1

7.8.2   Manufacturing Model of Inventory with Multi-Item Joint Replenishment without Shortages

The variables used in the manufacturing model of inventory with muti-item joint replenishment are listed below. In this model, shortages are not permitted.

C_o is the fixed set-up cost in rupees for a group of items
C_{o_i} is the marginal cost of set-up for the item i
D_i is the annual demand in units of the item i
D_r is the annual demand in rupees for the group of items manufactured
dr_i is the annual demand in rupees for the item i in the group of items manufactured
K_i is the annual production in units of the item i
p_i is the purchase price per unit of the item i in the group of items manufactured
I is the inventory carrying cost in percentage of the unit cost (For example, 10% is to be written as 0.10)
Qr is the economic batch quantity (EBQ) in rupees for the group of items manufactured
qr_i is the economic batch quantity (EBQ) in rupees of the item i in the group of items manufactured
Q_i is the economic batch quantity (EBQ) in units of the item i
N is the number of batch intervals per year
m is the number of items in the group
T is the time between consecutive batches of production

The formula to find the economic batch quantity (EBQ) for the group of items in rupees (Qr) is given below.

$$Qr = \sqrt{\frac{2\left(C_o + \sum_{i=1}^{m} C_{o_i}\right) Dr}{I\left(1 - \dfrac{\sum_{i=1}^{m} D_i}{\sum_{i=1}^{m} K_i}\right)}}$$

The formula to obtain the economic batch quantity for the item i is

$$qr_i = Qr\,(dr_i/Dr), \quad i = 1, 2, 3,\ldots, m$$

$$\text{Total set-up cost} = (Dr/Qr)\left(C_0 + \sum_{i=1}^{m} C_{0_i}\right)$$

$$\text{Total inventory carrying cost} = (Qr/2)I$$

$$\text{Total cost of the inventory system} = (Dr/Qr)\left(C_0 + \sum_{i=1}^{m} C_{0_i}\right) + (Qr/2)I.$$

Example 7.21 A company manufactures 5 items which are in turn fed into its main assembly line. The company wants to jointly replenish these five items. The details of the situation are given in Table 7.6. The fixed set-up cost is Rs. 75. The inventory carrying charge is 18% of the costs of items in the group. Find the EBQ in units of each item (Q_i), the number of set-ups per year (N), time between consecutive set-ups (T) and the total cost (TC).

Table 7.6 Data of Example 7.21

Item i	Annual demand in units	Price/unit p_i (Rs.)	Marginal cost of set-up of item i (Rs.)	Annual production quantity in units
1	5000	8	7	10,000
2	4000	6	5	9000
3	6000	9	6	12,000
4	8000	5	8	18,000
5	7000	10	4	21,000

Solution The additional details which are required to solve the problem are shown in Table 7.7.

Table 7.7 Data of Example 7.21

Item i	Annual demand in units (D_i)	Price/unit p_i (Rs.)	Annual demand in rupees (dr_i) (Rs.)	Marginal cost of set-up of item i (Rs.) (C_{0_i})	Annual production quantity in units (K_i)
1	5000	8	40,000	7	10,000
2	4000	6	24,000	5	9000
3	6000	9	54,000	6	12,000
4	8000	5	40,000	8	18,000
5	7000	10	70,000	4	21,000
	$D_i = 30,000$		$Dr = 2,28,000$	$\Sigma C_{0_i} = 30$	$\Sigma K_i = 70,000$

Fixed set-up cost, $C_0 = $ Rs. 75

Sum of marginal costs of set-ups of the items $(\Sigma C_{0_i}) = $ Rs. 30

Annual demand in rupees of all items in the group $(Dr) = $ Rs. 2,28,000

Inventory carrying charge percentage in decimal $(I) = 0.18$

Therefore, the EBQ (i.e. Qr) in rupees of the items in the group is

$$Qr = \sqrt{\dfrac{2\left(C_o + \displaystyle\sum_{i=1}^{m} C_{o_i}\right)Dr}{I\left(1 - \dfrac{\displaystyle\sum_{i=1}^{m} D_i}{\displaystyle\sum_{i=1}^{m} K_i}\right)}}$$

$$= \sqrt{\dfrac{2 \times (75 + 30) \times 2,28,000}{0.18 \times (1 - 30,000/70,000)}}$$

$$= \text{Rs. } 21,575.45$$

From the value of Qr, the value of qr_i and Q_i for each of the items are computed as shown below.

$$qr_1 = \dfrac{dr_1}{Dr} \times Qr = (40,000/2,28,000) \times 21,575.45 = \text{Rs. } 3785.17$$

$$qr_2 = \dfrac{dr_2}{Dr} \times Qr = (24,000/2,28,000) \times 21,575.45 = \text{Rs. } 2271.10$$

$$qr_3 = \dfrac{dr_3}{Dr} \times Qr = (54,000/2,28,000) \times 21,575.45 = \text{Rs. } 5109.98$$

$$qr_4 = \dfrac{dr_4}{Dr} \times Qr = (40,000/2,28,000) \times 21,575.45 = \text{Rs. } 3785.17$$

$$qr_5 = \dfrac{dr_5}{Dr} \times Qr = (70,000/2,28,000) \times 21,575.45 = \text{Rs. } 6624.04$$

Similarly, the EBQ in units of each item i is computed using the formula

$Q_i = qr_i/p_i$, for $i = 1, 2, 3,\ldots, m$

$Q_1 = qr_1/p_1 = 3785.17/8 = 473.15 = 473$ units (approx.)

$Q_2 = qr_2/p_2 = 2271.10/6 = 378.52 = 379$ units (approx.)

$Q_3 = qr_3/p_3 = 5109.98/9 = 567.78 = 568$ units (approx.)

$Q_4 = qr_4/p_4 = 3785.17/5 = 757.03 = 757$ units (approx.)

$Q_5 = qr_5/p_5 = 6624.04/10 = 662.40 = 662$ units (approx.)

$N = Dr/Qr = 2,28,000/21,575.45 = 10.568$ set-ups

$T = 1/N = 1/10.568 = 0.0946$ year

$$\text{Total set-up cost} = (Dr/Qr) \times \left(C_o + \sum_{i=1}^{5} C_{o_i}\right)$$

$$= (2,28,000/21,575.45) \times (75 + 30) = \text{Rs. } 1109.60$$

Total carrying cost $= (Qr/2)I = (21,575.45/2) \times 0.18 = \text{Rs. } 1941.79$

Total cost of the inventory system $= \text{Rs. } 1109.60 + \text{Rs. } 1941.79 = \text{Rs. } 3051.39$

7.9 EOQ FOR THE PURCHASE MODEL OF INVENTORY FOR MUTI-ITEM JOINT REPLENISHMENT WITH SPACE CONSTRAINT

The EOQ model for multi-item joint replenishment has been already presented in Section 7.8. In this section, the model with space constraint is presented. In practice, there will be a limitation on the storage space. So, the EOQ of each item should be decided by taking the available storage space into consideration.

The variables involved in the purchase model of inventory for multi-item joint replenishment with space constraint are listed below. In this model, shortages are not permitted.

C_o is the fixed ordering cost in rupees for a group of items

C_{o_i} is the marginal cost of ordering the item i

D_i is the annual demand in units of the item i

Dr is the annual demand in rupees for the group of items ordered

dr_i is the annual demand in rupees for the item i in the group of items ordered

p_i is the purchase price per unit of the item i in the group of items ordered

I is the inventory carrying cost in percentage of the unit cost (For example, 10% is to be written as 0.10)

Qr is the economic order quantity (EOQ) in rupees for the group of items ordered

qr_i is the economic order quantity (EOQ) in rupees for the item i

Q_i is the economic order quantity (EOQ) in units for the item i

N is the number of order cycles per year

m is the number of items in the group

T is the time between consecutive orders

s_i is the required space per unit of the item i

S is the utmost space available to store all the items

The formula for the total cost (TC) of this inventory system is as given below.

TC = Total ordering cost + Total carrying cost + Total purchase cost

$$= (Dr/Qr)\left(C_o + \sum_{i=1}^{m} C_{o_i}\right) + (Qr/2)I + \sum_{i=1}^{m} p_i D_i$$

The objective is to determine Qr which results in minimum total cost subject to fulfilling the space constraint.

$$\text{Minimize } Z = (Dr/Qr)\left(C_o + \sum_{i=1}^{m} C_{o_i}\right) + (Qr/2)I + \sum_{i=1}^{m} p_i D_i$$

subject to

$$\sum_{i=1}^{m} \frac{Qr(dr_i/Dr)}{p_i} s_i = S$$

where,

$$Qr(dr_i/Dr)/p_i = Q_i, \ i = 1, 2, 3,\ldots, m$$

The Lagrangian function for this model is as given below.

$$L = (Dr/Qr)\left(C_o + \sum_{i=1}^{m} C_{o_i}\right) + (Qr/2)I + \sum_{i=1}^{m} p_i D_i - \mu\left[\sum_{i=1}^{m}\{Qr(dr_i/Dr)/p_i\}s_i - S\right]$$

where, μ is a Lagrange multiplier and it is less than 0.

Differentiating the Lagrangian function L w.r.t. Qr, we get

$$\frac{dL}{dQr} = -(Dr/Qr^{-2})\left(C_o + \sum_{i=1}^{m} C_{o_i}\right) + I/2 - \mu\left[\sum_{i=1}^{m}\{(dr_i/Dr)/p_i\}s_i\right]$$

$$\frac{dL}{d\mu} = \sum_{i=1}^{m}\{Qr(dr_i/Dr)/p_i\}s_i - S$$

The equation w.r.t. the second partial derivative ensures that the total required space is equal to the utmost space available as indicated below.

$$\sum_{i=1}^{m}\{Qr(dr_i/Dr)/p_i\}s_i = S$$

From the equation w.r.t. the first partial derivative, we get

$$Qr = \sqrt{\frac{2\left(C_o + \sum_{i=1}^{m} C_{o_i}\right)Dr}{I - \mu\sum_{i=1}^{m}\{(dr_i/Dr)/p_i\}s_i}}$$

The formulas to get the EOQ in rupees for the item i (qr_i), EOQ in units for the item i (Q_i), N and T are given below.

$qr_i = Qr \times (dr_i/Dr)$, for $i = 1, 2, 3,\ldots, m$

$Q_i = qr_i/p_i$, for $i = 1, 2, 3,\ldots, m$

$N = Dr/Qr$

$T = 1/N$

N is computed for the group of items ordered. But, the number of intervals for the individual item may differ from N. There are many procedures to obtain the individual interval multiples, n_i, $i = 1, 2, 3,\ldots, m$. If n_i is less than 1, then round off it to 1 irrespective of its value. In this book, an algorithm called Silver's algorithm is presented to find such n_i values. The steps of this algorithm are already presented in Section 7.8.

The formula for the interval multiples of the items is

$$n_i = \sqrt{\frac{C_{o_i}}{dr_i}\frac{dr_k}{C_o + C_{o_k}}} \quad \text{for,} \quad i = 1, 2, 3,\ldots, m \quad \text{and} \quad i \neq k$$

where, k is the item number for which C_o/dr_i is the minimum and the value of n_k is 1.

Example 7.22 A company orders for three items jointly. The details of the situation are given in Table 7.8. The fixed ordering cost is Rs. 250. The inventory carrying charge is 15% of the purchase prices of items in the group. The utmost space available is 3000 cubic metre. Find the economic order quantity for each of the items. Also, find the interval multiples of the items.

Table 7.8 Data of Example 7.22

Item i	Annual demand in units	Price/unit p_i (Rs.)	Marginal cost of ordering item i (Rs.)	Space/unit cubic meter
1	8000	25	40	3
2	2000	10	25	2
3	9000	20	50	5

Solution The additional details which are required to solve the problem are shown in Table 7.9.

Table 7.9 Data of Example 7.22 with Additional Details

Item i	Annual demand in units (D_i)	Price/unit p_i (Rs.)	Annual demand in rupees (dr_i)	Marginal cost of ordering item i in rupees (C_{o_i})	Space per unit
1	8000	25	2,00,000	40	3
2	2000	10	20,000	25	2
3	9000	20	1,80,000	50	5
Total			Dr = 4,00,000	ΣC_{o_i} = 115	

Fixed ordering cost (C_o) = Rs. 250
Sum of marginal costs of ordering the items (ΣC_{o_i}) = Rs. 115
Annual demand in rupees of all items in the group (Dr) = Rs. 4,00,000
Inventory carrying charge percentage in decimal (I) = 0.15
Utmost storage space available (S) = 3000 cubic metre

Therefore, the EOQ in rupees of all the items put together in the group (Qr), the EOQ in units for each of the items and the total space requirement for all the items are computed for different μ as shown in Table 7.10 till the total space requirement is just equal to the utmost space available. The formulas to get Qr, qr_i, and Q_i are as given below.

$$Qr = \sqrt{\frac{2\left(C_o + \sum_{i=1}^{3} C_{o_i}\right) Dr}{I - \mu \sum_{i=1}^{3} \left\{(dr_i/Dr)/p_i\right\} s_i}}$$

$qr_i = Qr(dr_i/Dr)$, $i = 1, 2$ and 3

$Q_i = qr_i/p_i$, $i = 1, 2$ and 3

Table 7.10 Calculations of Qr, qr_i, Q_i and Total Space Required

μ	Qr	qr_1	qr_2	qr_3	Q_1	Q_2	Q_3	Total space required
0.00	44121.05	22060.52	2206.05	19854.47	882.42	220.61	992.7235	8052.091
−1.00	29634.36	14817.18	1481.72	13335.46	592.69	148.17	666.7732	5408.271
−2.00	23811.56	11905.78	1190.58	10715.20	476.23	119.06	535.7601	4345.609
−3.00	20460.64	10230.32	1023.03	9207.29	409.21	102.30	460.3645	3734.068
−4.00	18215.88	9107.94	910.79	8197.15	364.32	91.08	409.8572	3324.398
−5.00	16577.80	8288.90	828.89	7460.01	331.56	82.89	373.0006	3025.449
−6.00	15314.64	7657.32	765.73	6891.59	306.29	76.57	344.5794	2794.922
−5.01	16563.58	8281.79	828.18	7453.61	331.27	82.82	372.6806	3022.854
−5.02	16549.40	8274.70	827.47	7447.23	330.99	82.75	372.3615	3020.266
−5.03	16535.25	8267.63	826.76	7440.87	330.71	82.68	372.0432	3017.684
−5.04	16521.14	8260.57	826.06	7434.52	330.42	82.61	371.7258	3015.109
−5.05	16507.07	8253.54	825.35	7428.18	330.14	82.54	371.4091	3012.540
−5.06	16493.03	8246.52	824.65	7421.87	329.86	82.47	371.0932	3009.978
−5.07	16479.03	8239.52	823.95	7415.56	329.58	82.40	370.7782	3007.423
−5.08	16465.06	8232.53	823.25	7409.28	329.30	82.33	370.4639	3004.874
−5.09	16451.13	8225.57	822.56	7403.01	329.02	82.26	370.1505	3002.332
−5.10	16437.24	8218.62	821.86	7396.76	328.75	82.19	369.8378	2999.796

From the last column of Table 7.10, it is clear that the total space required goes on decreasing from 8052.091 cubic metre and it becomes less than 3,000 cubic metre when the value of μ is −6. So, for the values of μ in between −5 and −6 in steps of −0.01, the different calculations are done and presented in the later half of the Table 7.10. In this range of values of μ, when μ is −5.1, the total space required is 2999.796 which is very nearer to the utmost value of the available space 3000 cubic metre. The corresponding economic order quantities of the items are given below.

Qr = Rs. 16,437.24

Q_1 = 328.75 units = 329 units (approx.)

Q_2 = 82.19 units = 82 units (approx.)

Q_3 = 369.8378 units = 370 units (approx.)

The number of orders per year is computed as shown below.

$N = Dr/Qr = 4,00,000/16,437.24 = 24.335$ orders

$T = 1/N = 1/24.335 = 0.0411$ year

The calculations for determining the interval multiples of the individual items as per Silver's algorithm are summarized in Table 7.11.

The formula for the interval multiples of the items is

$$n_i = \sqrt{\frac{C_{o_i}}{dr_i} \frac{dr_k}{C_o + C_{o_k}}} \quad \text{for, } i = 1, 2, 3 \text{ and } i \neq k$$

If n_i is less than 1, then round off it to 1 irrespective of its value.

From Table 7.11, one can see that the interval multiples for the item 1, item 2 and item 3 are 1, 1 and 1, respectively.

Table 7.11 Determination of Interval Multiples of Individual Items

Item i	dr_i	C_{o_i}	C_{o_i}/dr_i	$dr_k/(C_o + C_{o_k})$ where, $k = 1$	n_i	n_i^*
1	2,00,000	40	0.000200	–	–	1
2	20,000	25	0.001250	547.9452	0.82761	1
3	1,80,000	50	0.000278	547.9452	0.39014	1

7.10 DETERMINATION OF STOCK LEVEL OF PERISHABLE ITEMS UNDER PROBABILISTIC CONDITION

In some reality, the demand of an item may not be constant and it will follow some probability distribution. Consider the case of a perishable item, namely newspaper. Assume that the demand of the newspaper does not follow a fixed pattern. But a probability distribution can be established for the same. The probability distribution may be a discrete distribution, or a uniform distribution, or a normal distribution.

For example, for each unsold newspaper, there will be a penalty which is given by the formula:

Marginal cost of surplus/unit, S_1 = Purchase price/unit – salvage value/unit

Similarly for each shortage unit, there will be a penalty which is given by the formula:

Marginal cost of shortage/unit, S_2 = Selling price/unit – Purchase price/unit

Let the generalized probability distribution of the demand of the item be a discrete distribution as shown in Table 7.12.

Table 7.12 Generalized Probability Distribution of Demand

Observation, i	1	2	3	4	. . .	i	. . .	n
Demand, D_i	D_1	D_2	D_3	D_4	. . .	D_i	. . .	D_n
Probability, p_i	p_1	p_2	p_3	p_4	. . .	p_i	. . .	p_n

Under such situation, the optimal order size D_i^* is determined by using the following relation:

$$P_{i-1} < \frac{S_2}{S_1 + S_2} < P_i$$

where, P_i is the cumulative probability of having demand up to D_i^*.

Example 7.23 The daily demand of sweet bread at a kiosk follows a discrete distribution as given in Table 7.13.

Table 7.13 Example 7.23

Serial No.	1	2	3	4	5	6	7	8	9	10	11
Demand, D_i	25	26	27	28	29	30	31	32	33	34	35
Probability, p_i	0.2	0.11	0.10	0.09	0.08	0.12	0.14	0.05	0.04	0.04	0.03

The purchase price of the sweet bread is Rs. 8 per packet. The selling price is Rs. 11 per packet. If the bread packets are not sold within the day of purchase, they are sold at Rs. 4 per packet to hotels for secondary use. Find the optimal order size of the sweet bread.

Solution The following are given:

$$\text{Purchase price/packet} = \text{Rs. } 8$$
$$\text{Selling price/packet} = \text{Rs. } 11$$
$$\text{Salvage value/packet} = \text{Rs. } 4$$
$$\text{Marginal cost of surplus, } S_1 = 8 - 4 = \text{Rs. } 4$$
$$\text{Marginal cost of shortages, } S_2 = 11 - 8 = \text{Rs. } 3$$

Therefore

$$\text{Cumulative probability, } P = \frac{S_2}{S_1 + S_2} = \frac{3}{4 + 3} = 0.43$$

Based on Table 7.13, the cumulative probability distribution of the demand is presented in Table 7.14.

Table 7.14 Cumulative Distribution of Sweet Bread

Serial No.	1	2	3	4	5	6	7	8	9	10	11
Demand, D_i	25	26	27	28	29	30	31	32	33	34	35
Probability, p_i	0.2	0.11	0.10	0.09	0.08	0.12	0.14	0.05	0.04	0.04	0.03
Cumulative probability, P_i	0.2	0.31	0.41	0.50	0.58	0.70	0.84	0.89	0.93	0.97	1.00

From Table 7.14, one can check the relation:

$$P_3 < \frac{S_2}{S_1 + S_2} < P_4 \quad \text{or} \quad 0.41 < 0.43 < 0.50$$

Therefore, the optimal order size is D_4, which is equal to 28 bread packets.

Example 7.24 In a railway private canteen, the daily demand for packed meals follows uniform distribution as presented below:

$$p(x) = \frac{1}{250 - 150}, \qquad 150 \le x \le 250$$

The cost of production per packed meals is Rs. 9. The selling price is Rs. 15 per packed meals. The surplus packets on each day are sold at Rs. 7 in a nearby public place. Find the optimal number of packets of meals to be prepared on each day.

Solution   We are given

$$\text{Cost of production/packed meals} = \text{Rs. } 9$$
$$\text{Selling price/packed meals} = \text{Rs. } 15$$
$$\text{Salvage value/packed meals} = \text{Rs. } 7$$
$$\text{Marginal cost of surplus, } S_1 = 9 - 7 = \text{Rs. } 2$$
$$\text{Marginal cost of shortages, } S_2 = 15 - 9 = \text{Rs. } 6$$

Therefore,

$$\text{Cumulative probability, } P = \frac{S_2}{S_1 + S_2} = \frac{6}{2 + 6} = 0.75$$

Let Q^* be the optimal production size. The cumulative probability distribution function of the uniform distribution when x is equal to Q^* is:

$$\int_{150}^{Q^*} \frac{1}{250 - 150}\, dx = \frac{1}{250 - 150}\,[x]_{150}^{Q^*} = \frac{1}{100}(Q^* - 150)$$

Now, equating the above expression to $S_2/(S_1 + S_2)$ and solving for Q^*, we get

$$\frac{1}{100}\left(Q^* - 150\right) = \frac{S_2}{S_1 + S_2} = 0.75$$

or

$$Q^* = 225$$

Therefore, the optimal production size is 225 packed meals.

Example 7.25   Alpha fish stall is planning for its optimal purchase quantity of a costly variety of fish. The daily demand of the fish follows normal distribution with a mean of 500 kg and standard deviation of 50 kg. The purchase price of the fish is Rs. 120/kg. The selling price is Rs. 180/kg. If the fish is not sold on the day of the purchase, it is sold to a dry fish manufacturing firm at Rs. 100/kg. Find the optimal daily purchase quantity of the fish.

Solution   We have

$$\text{Purchase price of fish/kg} = \text{Rs. } 120$$
$$\text{Selling price of fish/kg} = \text{Rs. } 180$$
$$\text{Salvage value of fish/kg} = \text{Rs. } 100$$
$$\text{Marginal cost of surplus, } S_1 = 120 - 100 = \text{Rs. } 20$$
$$\text{Marginal cost of shortages, } S_2 = 180 - 120 = \text{Rs. } 60$$

So,

$$\text{Cumulative probability, } P = \frac{S_2}{S_1 + S_2} = 60/(20 + 60) = 0.75$$

The daily demand of the fish follows normal distribution with the following mean and standard deviation:

$$\text{Mean demand of fish, } \mu = 500 \text{ kg}$$
$$\text{Standard deviation of the demand of fish, } \sigma = 50 \text{ kg}$$

Let, Q^* be the optimal daily purchase quantity of fish. The standard normal statistic Z for the demand is

$$Z = \frac{Q^* - \mu}{\sigma} = \frac{Q^* - 500}{50}$$

For a cumulative probability of 0.75, the value of Z is 0.675. So, we get

$$\frac{Q^* - 500}{50} = 0.675 \quad \text{or} \quad Q^* = 533.75 \text{ kg}$$

Therefore, the optimal daily order size of the fish is 534 kg (approx.).

QUESTIONS

1. What are the reasons for stocking items in inventory?
2. List and explain different types of costs in inventory system.
3. Name the types of models of inventory system and explain them in detail.
4. Derive the EOQ formula for the purchase model without shortages.
5. Distinguish between P and Q systems of inventory.
6. Beta industry estimates that it will sell 24,000 units of its product for the forthcoming year. The ordering cost is Rs. 150 per order and the carrying cost per unit per year is 20% of the purchase price per unit. The purchase price per unit is Rs. 50. Find: the economic order size, the number of orders per year and the time between successive orders.
7. For a product to be manufactured within the company, the details are as follows: $r = 36,000$ units/year; $k = 72,000$ units/year; $C_o =$ Rs. 250 per set-up; $C_c =$ Rs. 25/unit/year; Find the EBQ and cycle time.
8. The annual demand for an automobile component is 36,000 units. The carrying cost is Re. 0.50/unit/year, the ordering cost is Rs. 25.00 per order and the shortage cost is Rs. 15.00/unit/year. Find the optimal values of the following:

    (a) Economic ordering quantity
    (b) Maximum inventory
    (c) Maximum shortage quantity
    (d) Cycle time
    (e) Inventory period (t_1)
    (f) Shortage period (t_2).

9. The demand for an item is 24,000 per year. Its production rate is 4000 per month. The carrying cost is Re. 0.25/unit/month and the set-up cost is Rs. 800 per set-up. The shortage cost is Rs. 15/unit/year. Find various parameters of the inventory system.
10. A firm has a demand distribution during a constant lead time with a standard deviation of 400 units. The firm wants to provide 95 per cent service.

    (a) How much safety stock should be carried?
    (b) If the demand during lead time averages 1500 units, what is the appropriate reorder level?
11. Annual demand for an item is 5400 units. Ordering cost is Rs. 600 per order. Inventory carrying cost is 30% of the purchase price/unit/year. The price breaks are shown as:

Quantity	Price (Rs.)
$0 \leq Q_1 < 2400$	12
$2400 \leq Q_2 < 3000$	10
$3000 \leq Q_3$	8

Find the optimal order size. If the order cost is changed to Rs. 300 per order, find the optimal order size.

12. The annual demand of a product is 48,000 units. The average lead time is 3 weeks. The standard deviation of demand during the average lead time is 100 units/week. The cost of ordering is Rs. 500 per order. The cost of purchase of the product per unit is Rs. 15. The cost of carrying per unit per year is 20% of the purchase price. The maximum delay in lead time is 2 weeks and the probability of this delay is 0.30. Assume a service level of 0.90.

(a) If Q system is followed, find the reorder level.
(b) If P system is followed, find the maximum inventory level.

13. A company wants to determine the economic order size of each of the three bought-out items which are stocked in its stores. The details of the items are presented in the following table. The maximum space available is 1000 square metre. Find the economic order size of each of the items subject to the space constraint.

	Item number		
	1	2	3
Demand/year (ton)	2000	2500	1750
Ordering cost/order (Rs.)	600	900	500
Carrying cost/ton/year (Rs.)	60	90	110
Space requirement/ton (sq. mt)	3	2	4

14. A company wants to determine the economic order quantity for each of its 2 items which are stored in its warehouse. The annual demand in units of the item 1, and the item 2 are 8000 units and 10,000 units, respectively. The carrying cost per unit per year for the item 1 and the item 2 are Rs. 8 and Rs. 12, respectively. The ordering cost per order for the item 1 and the item 2 are Rs. 800 and Rs. 500, respectively. The company wants to limit the total inventory carrying cost to Rs. 6000. Find the economic order quantity of each of the items.

15. A company orders for three items jointly. The details of the situation are given below. The fixed ordering cost is Rs. 300. The inventory carrying charge is 10% of the purchase prices of items in the group. Find the EOQ in units as well as the interval multiple of each of the items.

Item i	Annual demand in units	Price/unit p_i (Rs.)	Marginal cost of ordering item i (Rs.)
1	15,000	25	45
2	2000	8	50
3	20,000	15	40

16. A company manufactures 4 items which are in the main assembly line produced within that company. The company wants to jointly replenish these four items. The details of the situation

are given below. The fixed set-up cost is Rs. 125. The inventory carrying charge is 12% of the costs of items in the group. Find the EBQ in units of each item (Q_i), and the total cost (*TC*).

Item i	Annual demand in units	Price/unit p_i (Rs.)	Marginal cost of set-up of item i (Rs.)	Annual production quantity in units
1	7000	7	20	14,000
2	5000	9	15	10,000
3	8000	12	30	15,000
4	9000	10	10	12,000

17. A company orders for three items jointly. The details of the situation are given below. The fixed ordering cost is Rs. 350. The inventory charge is 10% of the purchase prices of items in the group. The utmost space available is 4000 cubic metre. Find the economic order quantity for each of the items. Also, find the interval multiples of the items.

Item i	Annual demand in units	Price/unit p_i (Rs.)	Marginal cost of ordering item i (Rs.)	Space/unit (cubic metre)
1	9000	30	60	5
2	4000	15	40	3
3	6000	25	45	6

18. The daily demand of sandwich bread at a kiosk follows a discrete distribution as given in the following table.

Discrete Distribution of Sweet Bread

Serial No.	1	2	3	4	5	6	7	8	9	10	11
Demand, D_i	35	36	37	38	39	40	41	42	43	44	45
Probability, p_i	0.10	0.11	0.20	0.07	0.10	0.14	0.12	0.04	0.05	0.05	0.02

The purchase price of the bread is Rs. 10 per packet. The selling price is Rs. 14 per packet. If the bread packets are not sold on the day of purchase, they are sold at Rs. 7 per packet to hotels for secondary use. Find the optimal order size of the sandwich bread.

19. In a private canteen, the daily demand for packed meals follows uniform distribution as presented below.

$$p(x) = \frac{1}{450 - 230}, \quad 230 \le x \le 450$$

The cost of production per packet of meals is Rs. 8. The selling price is Rs. 16 per packet. The surplus packets on each day are sold at Rs. 6 per packet in a nearby public place. Find the optimal number of packets of meals to be prepared on each day.

20. A fish stall is planning for its optimal purchase quantity of a costly variety of fish. The daily demand of the fish follows normal distribution with a mean of 800 kg and standard deviation of 75 kg. The purchase price of the fish is Rs. 150 per kg. The selling price is Rs. 200 per kg. If the fish is not sold on the day of the purchase, it is sold to a dry fish manufacturing firm at Rs. 110 per kg. Find the optimal daily purchase quantity of the fish.

DYNAMIC PROGRAMMING 8

8.1 INTRODUCTION

Dynamic programming is a special kind of optimization technique which subdivides an original problem into as many number of subproblems as the variables, solves each subproblem individually and then obtains the solution of the original problem by integrating the solutions of the subproblems. It is a systematic, complete enumeration technique.

The terminologies used in the dynamic programming problem are presented below:

Stage i. Each subproblem of the original problem is known as a *stage i*.

Alternative, m_i. In a given stage i, there may be more than one choice of carrying out a task. Each choice is known as an *alternative, m_i*.

State variable, x_i. A possible value of a resource within its permitted range at a given stage i is known as *state variable, x_i*.

Recursive function, $f_i(x_i)$. A function which links the measure of performance of interest of the current stage with the cumulative measure of performance of the previous stages/succeeding stages as a function of the state variable of the current stage is known as the *recursive function* of the current stage. Let

$$f_1(x_1) = \max \, [R(m_1)]$$

$$f_i(x_i) = \max \, \{R(m_i) + f_{i-1}[x_i - c(m_i)]\}, \, i = 2, 3,..., n$$

for possible m_i where, n is the total number of stages, $R(m_i)$ is the measure of performance (like, return) due to alternative m_i of the stage i, $c(m_i)$ is the cost/resource required for the alternative m_i of the stage i and $f_i(x_i)$ is the value of the measure of performance up to the current stage i from the stage 1, if the amount of resource allocated up to the current stage is x_i when forward recursion is used.

Best recursive function value. In a given stage i, the lowest (minimization problems)/highest (maximization problems) value of the recursive function for a given value of x_i is known as the *best recursive function* value.

Best alternative in a given stage i. In a given stage i, the alternative corresponding to the best recursive function value for a given value of x_i is known as the *best alternative* for that value of x_i.

Backward recursive function. Here computation begins from the last stage/subproblem, and this stage will be numbered as stage 1, while the first subproblem will be numbered as the last stage. Since the recursion proceeds in a backward direction, this type of recursive function is known as *backward recursive function.*

Forward recursive function. While defining the stages of the original problem, the first subproblem will be numbered as stage 1 and last subproblem will be numbered as the last stage. Then, the recursive function will be defined as per this assumption. This type of recursive function is known as *forward recursive function.*

8.2 APPLICATION OF DYNAMIC PROGRAMMING

The dynamic programming can be applied to many real-life situations. A sample list of applications of the dynamic programming is given below. The details of these problems are explained while solving them.

1. Capital budgeting problem
2. Reliability improvement problem
3. Stage-coach problem (shortest-path problem)
4. Cargo loading problem
5. Minimizing total tardiness in single machine scheduling problem
6. Optimal subdividing problem
7. Linear programming problem

They are discussed in the following sections:

8.2.1 Capital Budgeting Problem

A *capital budgeting problem* is a problem in which a given amount of capital is allocated to a set of plants by selecting the most promising alternative for each selected plant such that the total revenue of the organization is maximized. This is demonstrated using a numerical problem.

Example 8.1 An organization is planning to diversify its business with a maximum outlay of Rs. 5 crores. It has identified three different locations to instal plants. The organization can invest in one or more of these plants subject to the availability of the fund. The different possible alternatives and their investment (in crores of rupees) and present worth of returns during the useful life (in crores of rupees) of each of these plants are summarized in Table 8.1. The first row of Table 8.1 has zero cost and zero return for all the plants. Hence, it is known as *do-nothing* alternative. Find the optimal allocation of the capital to different plants which will maximize the corresponding sum of the present worth of returns.

Table 8.1 Example 8.1

Alternative	Plant 1		Plant 2		Plant 3	
	Cost	Return	Cost	Return	Cost	Return
1	0	0	0	0	0	0
2	1	15	2	14	1	3
3	2	18	3	18	2	7
4	4	28	4	21	–	–

Solution Maximum capital amount, C = Rs. 5 crores. Each plant is treated as a stage. So, the number of stages is equal to 3. The plants 1, 2 and 3 are defined as stage 1, stage 2 and stage 3, respectively. So, the forward recursive function is used for this problem.

Stage 1. The recursive function for a given combination of the state variable, x_1 and alternative, m_1 in the first stage is presented below. The corresponding returns are summarized in Table 8.2. For each value of the state variable, the best return and the corresponding alternative are presented in the last two columns, respectively.

$$f_1(x_1) = R(m_1)$$

Table 8.2 Calculations for Stage 1 (Plant 1)

State variable	Alternative m_1									
	1		2		3		4			
	C	R	C	R	C	R	C	R		
x_1	0	0	1	15	2	18	4	28	$f_1^*(x_1)$	m_1^*
0	0		–		–		–		0	1
1	0		15		–		–		15	2
2	0		15		18		–		18	3
3	0		15		18		–		18	3
4	0		15		18		28		28	4
5	0		15		18		28		28	4

C = cost, R = return.

Stage 2. The recursive function $f_2(x_2)$ for a given combination of the state variable, x_2 and alternative, m_2 in the second stage is given by

$$f_2(x_2) = R(m_2) + f_1[x_2 - C(m_2)]$$

The corresponding returns are summarized in Table 8.3. For each value of the state variable, the best return and the corresponding alternative(s) are presented in the last two columns, respectively.

Table 8.3 Calculations for Stage 2 (Plant 2)

State variable	Alternative m_2									
	1		2		3		4			
	C	R	C	R	C	R	C	R		
x_2	0	0	2	14	3	18	4	21	$f_2(x_2)^*$	m_2^*
0	0		–		–		–		0	1
1	0 + 15 = 15		–		–		–		15	1
2	0 + 18 = 18		14 + 0 = 14		–		–		18	1
3	0 + 18 = 18		14 + 15 = 29		18 + 0 = 18		–		29	2
4	0 + 28 = 28		14 + 18 = 32		18 + 15 = 33		21 + 0 = 21		33	3
5	0 + 28 = 28		14 + 18 = 32		18 + 18 = 36		21 + 15 = 36		36	3 and 4

Stage 3. The recursive function $f_3(x_3)$ for different combinations of the state variable, x_3 and alternative, m_3 in the third stage is:

$$f_3(x_3) = R(m_3) + f_2[x_3 - C(m_3)]$$

The corresponding returns are summarized in Table 8.4. For each value of the state variable, the best return and the corresponding alternative(s) are presented in the last two columns, respectively.

Table 8.4 Calculations for Stage 3 (Plant 3)

State variable	Alternative m_3							
	1		2		3			
	C	R	C	R	C	R	$f_3(x_3)^*$	m_3^*
x_3	0	0	1	3	2	7		
5	$0 + 36 = 36$		$3 + 33 = 36$		$7 + 29 = 36$		36	1, 2 and 3

The final results of the original problem is traced as in Table 8.5. From this table, one can visualize the fact that the original problem has four alternate optimal solutions.

Table 8.5 Final Results

Stage 3		Stage 2		Stage 1		Optimal alternatives stage		
C^*	m_3^*	C^*	m_2^*	C^*	m_1^*	1	2	3
5	1	$5 - 0 = 5$ — 3		$5 - 3 = 2$	3	3 — 3	—	1
		4		$5 - 4 = 1$	2	2 — 4	—	1
5	2	$5 - 1 = 4$	3	$4 - 3 = 1$	2	2 — 3	—	2
5	3	$5 - 2 = 3$	2	$3 - 2 = 1$	2	2 — 2	—	3

8.2.2 Reliability Improvement Problem

Generally, electronic equipments are made up of several components in series or parallel. Assuming that the components are connected in series, if there is a failure of a component in the series, it will make the equipment inoperative. The reliability of the equipment can be increased by providing optimal number of standby units to each of the components in the series such that the total reliability of the equipment is maximized subject to a cost constraint. Application of dynamic programming technique to this problem is illustrated in Example 8.2.

Example 8.2 An electronic item has three components in series. (The reliability of the system is equal to the product of the reliabilities of the three components, i.e. $R = r_1 r_2 r_3$. It is a known fact that the reliability of the system can be improved by providing standby units at extra cost.) The details of costs and reliabilities for different number of standby units for each of the components of the system are summarized in Table 8.6.

Table 8.6 Example 8.2

No. of standby units	Component 1		Component 2		Component 3	
	Cost (Rs.)	Reliability	Cost (Rs.)	Reliability	Cost (Rs.)	Reliability
1	1	0.75	3	0.84	2	0.80
2	2	0.88	4	0.94	3	0.91
3	4	0.94	6	0.97	5	0.96

The total capital budgeted for this purpose is Rs. 9. Determine the optimal number of standby units for each of the components of the system such that the total reliability of the system is maximized.

Solution Maximum capital budgeted/unit, K = Rs. 9. Each component is treated as a stage. So, the number of stages is equal to 3. The components 3, 2 and 1 are defined as stage 1, stage 2 and stage 3, respectively. Hence, the backward recursive function is used for this problem.

Range for state variable x_1 at Stage 1. The minimum amount of money required to have at least one standby unit for Component 3 is Rs. 2. Therefore, the lower limit of the state variable x_1 = Rs. 2. Similarly, a sum of Rs. 4 (Rs. 3 + Re. 1) is required to have at least one standby unit in Stage 2 and Stage 3 (i.e. the sum of the cost of one standby unit for each of the Component 2 and Component 1). Therefore, the upper limit of the state variable x_1 = 9 – 4 = Rs. 5.

Based on these guidelines, the effective range of the state variable x_1 is:

$$2 \leq x_1 \leq 5$$

Range for state variable x_2 at Stage 2. The minimum amount of money required to have at least one standby unit in each of the stages up to the current stage is Rs. 5 (the sum of the cost of one standby unit for each of the Component 3 and Component 2). Therefore, the lower limit of the state variable x_2 = Rs. 5. Similarly, the amount required to have at least one standby unit in stage 3 is Re. 1 (i.e. the cost of one standby unit for Component 1). Therefore, the upper limit of the state variable x_2 = 9 – 1 = Rs. 8.

Based on these guidelines, the effective range of the state variable x_2 is:

$$5 \leq x_2 \leq 8$$

Range for state variable x_3 at Stage 3. The minimum amount of money required to have at least one standby unit in each of the stages up to the current stage is Rs. 6 (the sum of the cost of one standby unit for each of the Component 3, Component 2 and Component 1). Therefore, the lower limit of the state variable x_3 = Rs. 6. Also, the upper limit of the state variable x_3 = Rs. 9. The effective range of the state variable x_3 is:

$$6 \leq x_3 \leq 9$$

Stage 1. The recursive function $f_1(x_1)$ for a given combination of the state variable, x_1 and alternative, m_1 in the first stage is:

$$f_1(x_1) = r(m_1)$$

The corresponding reliabilities are shown in Table 8.7. For each value of the state variable x_1, the best reliability and the corresponding alternative are presented in the last two columns, respectively.

Table 8.7   Calculations for Stage 1 (Component 3)

State variable	Alternative m_1							
	1		2		3			
	C	R	C	R	C	R		
x_1	2	0.8	3	0.91	5	0.96	$f_1(x_1)^*$	m_1^*
2		0.8	–		–		0.8	1
3		0.8	0.91		–		0.91	2
4		0.8	0.91		–		0.91	2
5		0.8	0.91		0.96		0.96	3

C = cost, R = reliability.

Stage 2.   The recursive function $f_2(x_2)$ for a given combination of the state variable, x_2 and alternative, m_2 in the second stage is:

$$f_2(x_2) = r(m_2) \times f_1[x_2 - C(m_2)]$$

The corresponding reliabilities are summarized in Table 8.8. For each value of the state variable, the best reliability and the corresponding alternative are presented in the last two columns, respectively.

Table 8.8   Calculations for Stage 2 (Component 2)

State variable	Alternative m_2							
	1		2		3			
	C	R	C	R	C	R		
x_2	3	0.84	4	0.94	6	0.97	$f_2(x_2)^*$	m_2^*
5	$0.84 \times 0.8 = 0.672$		–		–		0.672	1
6	$0.84 \times 0.91 = 0.764$		$0.94 \times 0.8 = 0.752$		–		0.764	1
7	$0.84 \times 0.91 = 0.764$		$0.94 \times 0.91 = 0.855$		–		0.855	2
8	$0.84 \times 0.96 = 0.806$		$0.94 \times 0.91 = 0.855$		$0.97 \times 0.8 = 0.776$		0.855	2

Stage 3.   The recursive function for a given combination of the state variable, x_3 and alternative, m_3 in the third stage is presented below:

$$f_3(x_3) = r(m_3) \times f_2[x_3 - C(m_3)]$$

The corresponding reliabilities are summarized in Table 8.9. For each value of the state variable, the best reliability and the corresponding alternative are presented in the last two columns, respectively.

Table 8.9 Calculations for Stage 3 (Component 1)

State variable x_3	Alternative m_3						$f_3(x_3)^*$	m_3^*
	1		**2**		**3**			
	C R		C R		C R			
	1 0.75		2 0.88		4 0.94			
6	$0.75 \times 0.672 = 0.504$		—		—		0.504	1
7	$0.75 \times 0.764 = 0.573$		$0.88 \times 0.672 = 0.591$		—		0.591	2
8	$0.75 \times 0.855 = 0.641$		$0.88 \times 0.764 = 0.672$		—		0.672	2
9	$0.75 \times 0.855 = 0.641$		$0.88 \times 0.855 = 0.752$		$0.94 \times 0.672 = 0.632$		0.752	2

The final result of the original problem is traced as shown in Table 8.10. From the table it is clear that each of the three components requires two standby units to have a maximum reliability of 0.752 with an additional total cost of Rs. 9.

Table 8.10 Final Results

Stage						Stage		
3		**2**		**1**				
C^*	m_3^*	C^*	m_2^*	C^*	m_1^*	**1**	**2**	**3**
9	2	$9 - 2 = 7$	2	$7 - 4 = 3$	2	$m_1^* = 2$	$m_2^* = 2$	$m_3^* = 2$

8.2.3 Stage-coach Problem (Shortest-path Problem)

Stage-coach problem is a shortest-path problem in which the objective is to find the shortest distance and the corresponding path from a given source node to a given destination node in a given distance network. Application of dynamic programming technique to this problem is illustrated using Example 8.3.

Example 8.3 A distance network consists of eleven nodes which are distributed as shown in Figure 8.1. Find the shortest path from node 1 to node 11 and also the corresponding distances.

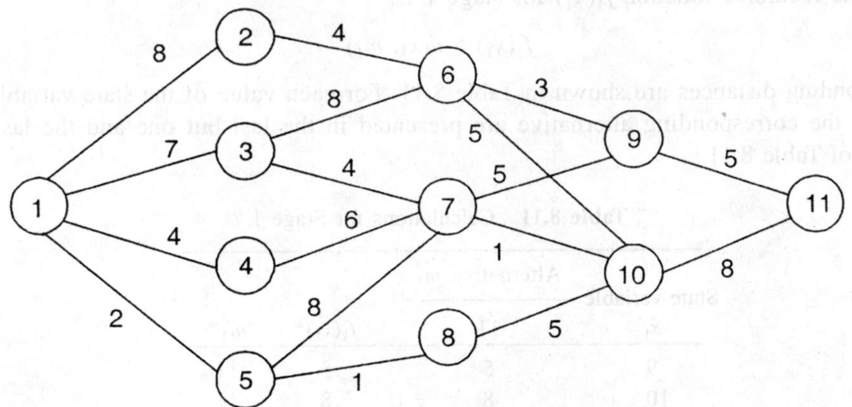

Figure 8.1 Distance network.

Solution Each pair of adjacent vertical columns of nodes is treated as a stage. As shown in Figure 8.2, there are four stages in this problem. Since the stages are defined from right to left, backward recursive function is to be used.

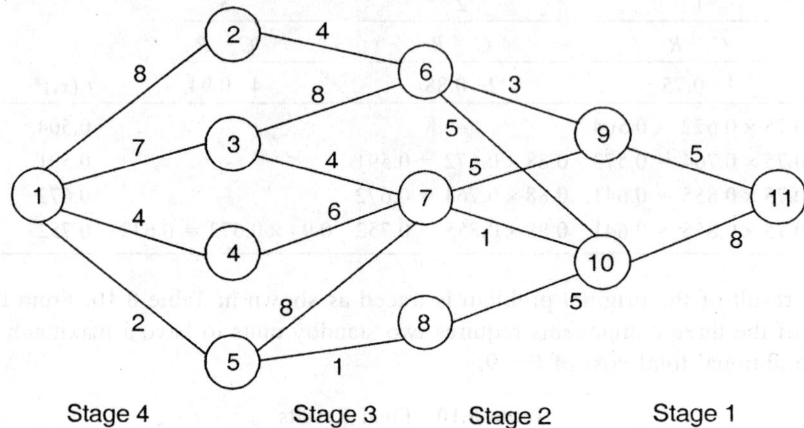

Figure 8.2 Distance network with stages.

In stage 1, the possible alternative is only one, i.e. node 11. In the same stage (stage 1), the possible state variables are nodes 9 and 10.

The *recursive function* $f_1(x_1)$ for a given combination of the state variable, x_1 and alternative, m_1 in the first stage is:

$$f_1(x_1) = d(x_1, m_1)$$

where $d(x_1, m_1)$ is the distance between node x_1 and node m_1,

The *recursive function* for a given combination of the state variable, x_i and alternative, m_i in the stage i for i more than 1 is presented below:

$$f_i(x_i) = d(x_i, m_i) + f_{i-1}(x_{i-1} = m_i)$$

where $d(x_i, m_i)$ be the distance between the nodes x_i and m_i, $f_i(x_i)$ be the shortest distance from node x_i in the current stage i to the destination node in stage 1 (node 11 in this example).

Stage 1. The recursive function $f_1(x_1)$ for stage 1 is:

$$f_1(x_1) = d(x_1, m_1)$$

The corresponding distances are shown in Table 8.11. For each value of the state variable, the best distance and the corresponding alternative are presented in the last but one and the last columns, respectively of Table 8.11.

Table 8.11 Calculations for Stage 1

State variable	Alternative m_1		
x_1	11	$f_1(x_1)^*$	$m_1{}^*$
9	5	5	11
10	8	8	11

Stage 2. The recursive function $f_2(x_2)$ for a given combination of the state variable, x_2 and alternative, m_2 in the second stage is:

$$f_2(x_2) = d(x_2, m_2) + f_1(x_1 = m_2)$$

The corresponding distances are summarized in Table 8.12. For each value of the state variable, the best distance and the corresponding alternative are presented in the last two columns of Table 8.12, respectively.

Table 8.12 Calculations for Stage 2

State variable x_2	Alternative m_2		$f_2(x_2)^*$	m_2^*
	9	10		
6	3 + 5 = 8	5 + 8 = 13	8	9
7	5 + 5 = 10	1 + 8 = 9	9	10
8	–	5 + 8 = 13	13	10

Stage 3. The recursive function $f_3(x_3)$ for a given combination of the state variable, x_3 and alternative, m_3 is:

$$f_3(x_3) = d(x_3, m_3) + f_2(x_2 = m_3)$$

The corresponding distances are summarized in Table 8.13. For each value of the state variable, the best distance and the corresponding alternative are presented in the last but one and the last columns of Table 8.13, respectively.

Table 8.13 Calculations for Stage 3

State variable x_3	Alternative m_3			$f_3(x_3)^*$	m_3^*
	6	7	8		
2	4 + 8 = 12	–	–	12	6
3	8 + 8 = 16	4 + 9 = 13	–	13	7
4	–	6 + 9 = 15	–	15	7
5	–	8 + 9 = 17	1 + 13 = 14	14	8

Stage 4. The recursive function $f_4(x_4)$ for a given combination of the state variable, x_4 and alternative, m_4 in the fourth stage is:

$$f_4(x_4) = d(x_4, m_4) + f_3(x_3 = m_4)$$

The corresponding distances are summarized in Table 8.14. For each value of the state variable, the best distance and the corresponding alternative are presented in the last two columns of Table 8.14, respectively.

Table 8.14 Calculations for Stage 4

State variable x_4	Alternative m_4				$f_4(x_4)^*$	m_4^*
	2	3	4	5		
1	8 + 12 = 20	7 + 13 = 20	4 + 15 = 19	2 + 14 = 16	16	5

The final results of the original problem are traced in Tables 8.14 to Table 8.11 backwards. Therefore, the shortest path is 1–5–8–10–11. Hence, the corresponding shortest distance = 16 units.

8.2.4 Cargo Loading Problem

Cargo loading problem is an optimization problem in which a logistic company is left with the option of loading a desirable combination of items in a cargo subject to its weight or volume or both constraints. In this process, the return to the company is to be maximized. The application of dynamic programming to the cargo loading problem which has only the weight constraint is illustrated.

Example 8.4 Alpha logistic company has to load a cargo out of four items whose details are shown in Table 8.15. The maximum weight of the cargo is 7 tons. Find the optimal cargo loading using dynamic programming method such that the total return is maximized.

Table 8.15 Data of Example 8.4

Item i	1	2	3	4
Weight, w_i/unit (in tons)	2	1	4	3
Return, r_i/unit (in rupees)	1000	400	2100	1400

Solution In this problem, each item is treated as a stage starting from stage 1 to stage 4 for the item 1 to item 4, respectively. The maximum weight of the cargo is 7 tons. So, the weights allocated to the alternatives in each of the stages are zero and the multiples of the unit weight (less than or equal to 7) of the item corresponding to that stage. In stage 1, there are four alternatives and the weights allocated to those alternatives are 0, 2, 4 and 6; in stage 2, there are ten alternatives and the weights allocated to those alternatives are 0, 1, 2, 3, 4, 5, 6, 7, 8 and 9; the stage 3 has two alternatives and the weights allocated to them are 0 and 4; the stage 4 has three alternatives and the weights allocated to those alternatives are 0, 3 and 6. The range of values of the state variable in each stage is from 0 to 7 with an increment of 1.

Stage 1. The recursive function of this stage is shown below and the calculations are presented in Table 8.16.

$$f_1(x_1) = (\text{Allocated weight}/2) \times 1000$$

Table 8.16 Calculations for Stage 1

State variable x_1	Alternative m_1				f_1^*	m_1^*
	1	2	3	4		
	Allocated weight					
	0	2	4	6		
0	0	–	–	–	0	1
1	0	–	–	–	0	1
2	0	1000	–	–	1000	2
3	0	1000	–	–	1000	2
4	0	1000	2000	–	2000	3
5	0	1000	2000	–	2000	3
6	0	1000	2000	3000	3000	4
7	0	1000	2000	3000	3000	4

Stage 2. The recursive function of this stage is shown below and the calculations are presented in Table 8.17.

$$f_2(x_2) = \text{Allocated weight} \times 400 + f_1(x_2 - \text{Allocated weight})$$

Table 8.17 Calculations for Stage 2

State variable	Alternative m_2									
	1	2	3	4	5	6	7	8		
	Allocated weight									
x_2	0	1	2	3	4	5	6	7	f_2^*	m_2
0	0	–	–	–	–	–	–	–	0	1
1	0+0=0	400	–	–	–	–	–	–	400	2
2	0+1000 =1000	400+0 =400	–	–	–	–	–	–	1000	1
3	0+1000 =1000	400+1000 =1400	800+0 =800	–	–	–	–	–	1400	2
4	0+2000 =2000	400+1000 =1400	800+1000 =1800	1200+0 =1200	1600	–	–	–	2000	1
5	0+2000 =2000	400+2000 =2400	800+1000 =1800	1200+1000 =2200	1600+0 =1600	2000	–	–	2400	2
6	0+3000 =3000	400+2000 =2400	800+2000 =2800	1200+1000 =2200	1600+1000 =2600	2000+0 =2000	2400	–	3000	1
7	0+3000 =3000	400+3000 =3400	800+2000 =2800	1200+2000 =3200	1600+1000 =2600	2000+1000 =3000	2400+0 =2400	2800	3400	2

Stage 3. The recursive function of this stage is shown below and the calculations are presented in Table 8.18.

$$f_3(x_3) = (\text{Allocated weight}/4) \times 2100 + f_2(x_3 - \text{Allocated weight})$$

Table 8.18 Calculations for Stage 3

State variable	Alternative m_3			
	1	2		
	Allocated weight			
x_3	0	4	f_3^*	m_3^*
0	0	–	0	1
1	0 + 400 = 400	–	400	1
2	0 + 1000 = 1000	–	1000	1
3	0 + 1400 = 1400	–	1400	1
4	0 + 2000 = 2000	2100	2100	2
5	0 + 2400 = 2400	2100 + 400 = 2500	2500	2
6	0 + 3000 = 3000	2100 + 1000 = 3100	3100	2
7	0 + 3400 = 3400	2100 + 1400 = 3500	3500	2

Stage 4. The recursive function of this stage is shown below and the calculations are presented in Table 8.19.

$$f_4(x_4) = (\text{Allocated weight}/3) \times 1400 + f_3(x_4 - \text{Allocated weight})$$

Table 8.19 Calculations for Stage 4

State variable x_4	Alternative m_4			f_4^*	m_4^*
	1	2	3		
	Allocated weight				
	0	3	6		
0	0	–	–	0	1
1	$0 + 400 = 400$	–	–	400	1
2	$0 + 1000 = 1000$	–	–	1000	1
3	$0 + 1400 = 1400$	1400	–	1400	1, 2
4	$0 + 2100 = 2100$	$1400 + 400 = 1800$	–	2100	1
5	$0 + 2500 = 2500$	$1400 + 1000 = 2400$	–	2500	1
6	$0 + 3100 = 3100$	$1400 + 1400 = 2800$	2800	3100	1
7	$0 + 3500 = 3500$	$1400 + 2100 = 3500$	$2800 + 400 = 3200$	3500	1, 2

Tracing the solution from the stage 4 gives two solutions as shown below and each of them gives the optimal return of Rs. 3500.

$$m_4: \boxed{1} \rightarrow 7 - 0 = 7 \rightarrow m_3: \boxed{2} \rightarrow 7 - 4 = 3 \rightarrow m_2: \boxed{2} \rightarrow 3 - 1 = 2 \rightarrow m_1: \boxed{2}$$

$$m_4: \boxed{2} \rightarrow 7 - 3 = 4 \rightarrow m_3: \boxed{2} \rightarrow 4 - 4 = 0 \rightarrow m_2: \boxed{1} \rightarrow 0 - 0 = 0 \rightarrow m_1: \boxed{1}$$

The details of the weights of different items in the cargo are summarized in Table 8.20.

Table 8.20 Summary of Weights of Items in the Cargo (Two Solutions)

| Solution number | Weight (tons) | | | | Total return (Rs.) |
	Item 1	Item 2	Item 3	Item 4	
1	2	1	4	0	3500
2	0	0	4	3	3500

8.2.5 Minimizing Total Tardiness in Single Machine Scheduling Problem

Single machine scheduling problem consists of n independent jobs which require processing in the same single machine.

Let, n be the total number of independent jobs

    t_j be the processing time of the job j

    d_j be the due date of the job j

    C_j be the completion time of the job j

    T_j be the tardiness of the job j

$$T_j = C_j - d_j, \text{ if } C_j > d_j$$
$$= 0, \text{ otherwise.}$$

There are many measures of performance which are to be optimized in single machine scheduling problem. The dominant one is the minimization of total tardiness. There will be $n!$ sequences. The computation of the total tardiness for a sample sequence is demonstrated using the data shown in Table 8.21.

Table 8.21   Data for Sample Single Machine Scheduling Problem

Job j	1	2	3	4
t_j	6	8	3	9
d_j	10	16	8	18

Consider a sample sequence: 3-2-1-4. The calculations to determine the total tardiness for this sequence are summarized in Table 8.22.

Table 8.22   Summary of Calculations for Total Tardiness

Job j	3	2	1	4
t_j	3	8	6	9
C_j	3	11	17	26
d_j	8	16	10	18
T_j	0	0	7	8

$$\text{Total tardiness} = \sum_{j=1}^{4} T_j = 0 + 0 + 7 + 8 = 15$$

The application of dynamic programming to determine the sequence(s) which minimizes/minimize the total tardiness in the single machine scheduling problem is illustrated in this section.

Example 8.5   Consider the single machine scheduling problem which is shown in Table 8.23 in which each of the 5 jobs can be fully processed in the same single machine. The other data in this table are the processing time and the due date for each of the jobs. Find the best sequence which minimizes the total tardiness using dynamic programming method. ($T_j = C_j - d_j$, if $C_j > d_j$; otherwise, it is 0, where C_j is the completion time of the job j).

Table 8.23 Data for Example 8.5

Job j	1	2	3	4	5
t_j	5	8	6	7	9
d_j	15	25	15	10	20

Solution

Let, C_j be the completion time of the job j

T_j be the tardiness of the job j [Also, it is known as $T_j(C_j)$]

T_j $= C_j - d_j$, if $C_j > d_j$

   $= 0$, otherwise

σ be the set of scheduled jobs at the end of the sequence

σ^1 be the set of unscheduled jobs

q_σ be the sum of the processing times of unscheduled jobs in σ^1 $\left(q_\sigma = \sum_{j \,\varepsilon\, \sigma^1} t_j \right)$

$S(\sigma)$ be the sum of the tardiness values of the jobs in σ

$$S(\sigma) = \underset{j\,\varepsilon\,\sigma}{\text{Min}} \ [T_j\,(q_\sigma + t_j) + S(\sigma - \{j\})]$$

= Min [Tardiness of the job j in σ which is scheduled immediately after q_σ + Sum of the tardiness values of other jobs in σ which are scheduled after the job j]

where, $S(\phi) = 0$

The jobs in σ cannot be scheduled before q_σ. If the job j is scheduled immediately after q_σ, then the remaining jobs in σ [i.e. $(\sigma - \{j\})$] are to be scheduled after completion of the job j.

As per dynamic programming concept, this problem is divided into five stages, where each stage corresponds to a position in the sequence. That is the stage 1 corresponds to the fifth position [5] in the sequence, the stage 2 corresponds to the fourth position [4] in the sequence, the stage 3 corresponds to the third position [3] in the sequence, the stage 4 corresponds to the second position [2] in the sequence and the stage 5 corresponds to the first position [1] in the sequence. The jobs in different σ at each stage are given in Table 8.24.

Table 8.24 Jobs in Different σ at Different Stages

Stage	Jobs in different σ
1	{1} {2} {3} {4} {5}
2	{1, 2} {1, 3} {1, 4} {1, 5} {2, 3}, {2, 4} {2, 5} {3, 4} {3, 5} {4, 5}
3	{1, 2, 3} {1, 2, 4} {1, 2, 5} {1, 3, 4} {1, 3, 5} {1, 4, 5} {2, 3, 4} {2, 3, 5} {2, 4, 5} {3, 4, 5}
4	{1, 2, 3, 4} {1, 2, 3, 5} {1, 2, 4, 5} {1, 3, 4, 5} {2, 3, 4, 5}
5	{1, 2, 3, 4, 5}

The calculations of different stages are shown from Table 8.25 to Table 8.29.

Table 8.25 Calculations for Stage 1

σ	{1}	{2}	{3}	{4}	{5}
q_σ	30	27	29	28	26
$j \,\varepsilon\, \sigma$	{1}	{2}	{3}	{4}	{5}
$T_j(q_\sigma + t_j)$	20*	10	20	25	15
$S(\sigma - \{j\})$	0	0	0	0	0
$S(\sigma)$	20	10	20	25	15

*$T_j(q_\sigma + t_j)$ when j is $1 = C_1 - d_1 = (q_1 + t_1) - d_1 = (30 + 5) - 15 = 20$
[If $q_\sigma + t_j \le d_j$, then $T_j(q_\sigma + t_j) = 0$]

Table 8.26 Calculations for Stage 2

σ	{1, 2}	{1, 3}	{1, 4}	{1, 5}	{2, 3}	{2, 4}	{2, 5}	{3, 4}	{3, 5}	{4, 5}
q_σ	22	24	23	21	21	20	18	22	20	19
$j \,\varepsilon\, \sigma$	[1] 2	[1] 3	[1] 4	[1] 5	2 [3]	2 [4]	2 [5]	[3] 4	[3] 5	[4] 5
$T_j(q_\sigma + t_j)$	12 5	14 15	13 20	11 10	4 12	3 17	1 7	13 19	11 9	16 8
$S(\sigma - \{j\})$	10 20	20 20	25 20	15 20	20 10	25 10	15 10	25 20	15 20	15 25
Min. $S(\sigma)$	22 –	34 –	38 –	26 –	– 22	– 27	16 –	38 –	26 –	31 –

Note: For $\sigma = \{1, 2\}$, $S(\sigma - \{j\})$ for $j = 1$ is equal to the minimum of $S(\sigma = 2)$ in stage 1 where, σ contains the remaining job(s) of σ in the current stage under consideration, that is $\{2\}$. Similarly, for different j in different σ of the current stage, $S(\sigma - \{j\})$ can be obtained by referring to $S(\sigma)$ in stage 1.

Table 8.27 Calculations for Stage 3

σ	{1, 2, 3}	{1, 2, 4}	{1, 2, 5}	{1, 3, 4}	{1, 3, 5}	{1, 4, 5}	{2, 3, 4}	{2, 3, 5}	{2, 4, 5}	{3, 4, 5}
q_σ	16	15	13	17	15	14	14	12	11	13
$j \,\varepsilon\, \sigma$	[1] 2 3	[1] 2 4	[1] 2 5	1 [3] 4	[1] 3 5	[1] 4 5	2 [3] 4	2 [3] 5	2 [4] 5	[3] 4 5
$T_j(q_\sigma + t_j)$	6 0 7	5 0 12	3 0 2	7 8 11	5 6 4	4 11 3	0 5 11	0 3 1	0 8 0	4 10 2
$S(\sigma - \{j\})$	22 34 22	27 38 22	16 26 22	38 38 34	26 26 34	31 26 38	38 27 22	26 16 22	31 16 27	31 26 38
Min. $S(\sigma)$	28 – –	32 – –	19 – –	45 – 45	31 – –	35 – –	– 32 –	– 19 –	– 24 –	35 – –

Note: For $\sigma = \{1, 2, 3\}$ in stage 3, $S(\sigma - \{j\})$ for $j = 1$ is equal to the minimum of $S(\sigma = 2, 3)$ in stage 2 where, σ contains the remaining jobs of σ in the current stage under consideration, that is $\{2, 3\}$.

Table 8.28 Calculations for Stage 4

σ	{1, 2, 3, 4}	{1, 2, 3, 5}	{1, 2, 4, 5}	{1, 3, 4, 5}	{2, 3, 4, 5}
q_σ	9	7	6	8	5
$j \,\varepsilon\, \sigma$	[1] 2 [3] 4	[1] 2 [3] 5	1 2 [4] 5	[1] [3] 4 5	2 3 [4] 5
$T_j(q_\sigma + t_j)$	0 0 0 6	0 0 0 0	0 0 3 0	0 0 5 0	0 0 2 0
$S(\sigma - \{j\})$	32 45 32 28	19 31 19 28	24 35 19 32	35 35 31 45	35 24 19 32
Min. $S(\sigma)$	32 – 32 –	19 – 19 –	– – 22 –	35 35 – –	– – 21 –

Note: For $\sigma = \{1, 2, 3, 5\}$ in stage 4, $S(\sigma - \{j\})$ for $j = 2$ is equal to the minimum of $S(\sigma = 1, 3, 5)$ in stage 3, where, σ contains the remaining jobs of σ in the current stage under consideration, that is $\{1, 3, 5\}$.

Table 8.29 Calculations for Stage 5

σ				{1, 2, 3, 4, 5}	
q_σ				0	
$j \varepsilon \sigma$	1	2	3	$\boxed{4}$	5
$T_j(q_\sigma + t_j)$	0	0	0	0	0
$S(\sigma - \{j\})$	21	35	22	19	32
Min. $S(\sigma)$	–	–	–	19	–

Note: For $\sigma = \{1, 2, 3, 4, 5\}$ in stage 5, $S(\sigma - \{j\})$ for $j = 5$ is equal to the minimum of $S(\sigma = 1, 2, 3, 4)$ in stage 4 where, σ contains the remaining jobs of σ in the current stage under consideration, that is $\{1, 2, 3, 4\}$.

Tracing the solution from the stage 5 gives two sequences, 4-1-3-2-5 and 4-3-1-2-5. These sequences yield the same minimum total tardiness of 19 units.

In the stage 5 (Table 8.29), the minimum $S(\sigma)$ is 19 and the corresponding $j \varepsilon \sigma$ is 4. It is assigned to the first position in the sequence, that is [1] = 4. After excluding the job 4 from σ at this stage $\{1, 2, 3, 4, 5\}$, we get σ for the stage 4 as $\{1, 2, 3, 5\}$. In stage 4 (Table 8.28), under this σ, the job 1 as well as job 3 is having the least $S(\sigma)$ of 19. If we select the job 1 for assigning to the position 2 in the sequence, then [2] = 1. After excluding the job 1 from the σ at this stage, we get σ for the stage 3 as $\{2, 3, 5\}$. In stage 3 (Table 8.27), under this σ, job 3 is having the least $S(\sigma)$ of 19. This job 3 is assigned to the position 3 in the sequence, then [3] = 3. After excluding the job 3 from the σ at this stage, we get σ for the stage 2 as $\{2, 5\}$. In stage 2 (Table 8.26), under this σ, job 2 is having the least $S(\sigma)$ of 16. This job 2 is assigned to the position 4 in the sequence, that is [4] = 2. So, the remaining job in σ is assigned to the position 5, that is [5] = 5. So the resultant sequence is 4-1-3-2-5. Similarly, by working backward from the stage 4 where tie occurs under $\sigma = \{1, 2, 3, 5\}$, we get another sequence as 4-3-1-2-5. The total tardiness for each of these two sequences is 19 units.

8.2.6   Optimal Subdividing Problem

Consider a cable of length k units. The objective is to subdivide the cable into n parts each having a lengt. p_i, where i varies from 1 to n such that the product of the lengths of the parts is maximized.

A mathematical model for the above situation is presented below:

$$\text{Maximize } Z = p_1 p_2 p_3 \cdots p_i \cdots p_n$$

subject to

$$p_1 + p_2 + p_3 + \cdots + p_i + \cdots + p_n = k$$

$$p_i > 0, \qquad i = 1, 2, 3, \ldots, n$$

To solve this problem using dynamic programming technique, we suppose the length of the cable $= k$ units and number of parts to be cut out of the cable of length $k = n$, Also, let, the part i of the cable corresponds to the stage i of the dynamic programming problem, p_i be the length of the part in the stage i, where $i = 1, 2, 3, \ldots, n$, y_i be the sum of the lengths of the parts 1, 2,... and part i of the cable.

Since, y_i at the stage i, for i varying from 1 to n, is a non-negative variable, there will be infinite number of alternatives in each stage.

Recursive function. We have

$$f_1(y_1) = y_1 \qquad \text{and} \qquad f_i(y_i) = \max \left[p_i \, f_{i-1}(y_i - p_i) \right], \quad i = 2, 3, 4, \ldots, n$$

In the second function shown above, $y_i - p_i$ is the total length of the cable which accounts for the stages from 1 to $i - 1$. Substitute k in the first and the second functions as shown below:

$$f_1(k) = k \quad \text{and} \quad f_2(k) = \max \left[p_2 \, f_1(k - p_2) \right] = \max \left[p_2(k - p_2) \right]$$

By the principle of maxima, one can verify that the function $p_2(k - p_2)$ yields the maximum value when p_2 is equal to $k/2$. Therefore,

$$f_2(k) = \frac{k}{2} \left(k - \frac{k}{2} \right) = \left(\frac{k}{2} \right)^2$$

Similarly, one can prove the following:

$$f_3(k) = \left(\frac{k}{3} \right)^3$$

$$\vdots \qquad \vdots$$

$$f_n(k) = \left(\frac{k}{n} \right)^n$$

Thus the optimal length of each part of the cable is:

$$p_1 = p_2 = p_3 = \cdots = p_n = \frac{k}{n}$$

Hence, the corresponding value of the objective function $= (k/n)^n$.

Example 8.6 Solve the following model of the optimal subdividing of a cable of length 10 units into three parts such that the product of their lengths is maximized, using dynamic programming technique.

$$\text{Maximize } z = p_1 \times p_2 \times p_3$$

subject to

$$p_1 + p_2 + p_3 = 10$$

$$p_1, p_2 \text{ and } p_3 > 0$$

Solution. Length of the cable = 10 units. Number of parts to be cut out of the cable of length $10 = 3$, let the part i of the cable corresponds to the stage i of the dynamic programming problem, p_i be the length of the part in the stage i, where $i = 1, 2,$ and 3. y_i be the sum of the lengths of the part 1, and up to part i of the cable. Since, y_i at the stage i, when i varying from 1 to 3, is a non-negative variable, there will be infinite number of alternatives in each stage.

Recursive function is obtained as under: We have

$$f_1(y_1) = y_1 \qquad \text{and} \qquad f_i(y_i) = \max \left[p_i \, f_{i-1}(y_i - p_i) \right], \quad i = 2 \text{ and } 3$$

In the second function shown above, $y_i - p_i$ is the total length of the cable which accounts for the stages from 1 to $i - 1$. Substitute y_1 and y_i, for i varying from 2 to 3 with 10 in the first and the second functions, respectively, as shown below.

$$f_1(10) = 10 \quad \text{and} \quad f_2(10) = \max \left[p_2 \, f_1(10 - p_2) \right] = \max \left[p_2 \, (10 - p_2) \right]$$

By the principle of maxima, one can verify that the function $p_2(10 - p_2)$ yields the maximum value when p_2 is equal to 10/2. Therefore,

$$f_2(10) = \frac{10}{2}\left(10 - \frac{10}{2}\right) = \left(\frac{10}{2}\right)^2$$

Similarly, one can prove that

$$f_3(10) = \left(\frac{10}{3}\right)^3$$

Therefore, the optimal length of each part of the cable:

$$p_1 = p_2 = p_3 = \frac{10}{3}$$

Hence, the corresponding value of the objective function $= (10/3)^3$.

8.2.7 Solution of Linear Programming Problem through Dynamic Programming

Let us take the generalized linear programming problem for the product-mix problem in which X_j is the volume of production of the product j.

$$\text{Maximize } Z = c_1X_1 + c_2X_2 + \cdots + c_jX_j + \cdots + c_nX_n$$

subject to

$$a_{11}X_1 + a_{12}X_2 + \cdots + a_{1j}X_j + \cdots + a_{1n}X_n \leq b_1$$
$$a_{21}X_1 + a_{22}X_2 + \cdots + a_{2j}X_j + \cdots + a_{2n}X_n \leq b_2$$
$$\vdots \qquad \vdots \qquad \vdots \qquad \vdots \qquad \vdots$$
$$a_{i1}X_1 + a_{i2}X_2 + \cdots + a_{ij}X_j + \cdots + a_{in}X_n \leq b_i$$
$$\vdots \qquad \vdots \qquad \vdots \qquad \vdots \qquad \vdots$$
$$a_{m1}X_1 + a_{m2}X_2 + \cdots + a_{mj}X_j + \cdots + a_{mn}X_n \leq b_m$$
$$X_j \geq 0, \qquad j = 1, 2, 3,\ldots, n$$

In this linear programming problem, the product j is treated as stage j, where j varies from 1 to n. So, the total number of stages of the problem is equal to n. At stage j, a value (production volume) of the decision variable X_j is known as an *alternative*.

Since, in the LPP, X_j is a continuous variable, for j varying from 1 to n, there will be infinite number of alternatives in each of the stages.

The method of backward recursion is used to solve this problem. Product 1 is treated as Stage 1, Product 2 is treated as stage 2 and so on. Since the backward recursion is to be used, the working of the problem will commence from stage n.

Let us suppose that b_{ij} be the state of the system with respect to Constraint i in Stage j. (i.e. the amount of resource i allocated to the activities of the current stage and its succeeding stages). So, b_{1j}, b_{2j}, b_{3j},..., b_{ij}, and b_{mj} are the states of the system at stage j with respect to the resources, 1, 2, 3,..., i, and m, respectively. $f_j(b_{1j}, b_{2j}, b_{3j},..., b_{ij},..., b_{mj})$ be the optimum objective function value at stage j.

Now, the objective function for stage n is

$$f_n(b_{1n}, b_{2n}, b_{3n},..., b_{in},..., b_{mn}) = \max_{\substack{0 \le a_{in} X_n \le b_{in} \\ \text{for } i = 1,2,...,m}} (c_n X_n)$$

and the objective function for stage j is

$$f_j(b_{1j}, b_{2j}, b_{3j},..., b_{ij},..., b_{mj}) = \max_{\substack{0 \le a_{ij} X_j \le b_{ij} \\ \text{for } i = 1,2,...,m \\ \text{and } j = 1,2,...,n-1}} \left[\begin{array}{l} c_j X_j + f_{j+1}[(b_{1j} - a_{1j} X_j), (b_{2j} - a_{2j} X_j),..., \\ \quad (b_{ij} - a_{ij} X_j),..., (b_{mj} - a_{mj} X_j)] \end{array} \right]$$

with

$$0 \le b_{ij} \le b_i, \qquad i = 1, 2,..., m \quad \text{and} \quad j = 1, 2,..., n$$

In the above function, the quantity $(b_{ij} - a_{ij} X_j)$ is the sum of the resource i allocated to all the succeeding stages (i.e. from stage $j + 1$ through stage n) with respect to the current stage.

Example 8.7 Solve the following LPP using dynamic programming technique:

$$\text{Maximize } Z = 10X_1 + 30X_2$$

subject to

$$3X_1 + 6X_2 \le 168$$

$$12X_2 \le 240$$

$$X_1 \text{ and } X_2 \ge 0$$

Solution The number of decision variables in the given problem is equal to 2. So, there will be two stages (i.e. stage 1 is assigned to the decision variable X_1 and stage 2 is assigned to the decision variable X_2). Since backward recursion is used to solve the problem, stage 2 is to be considered first. The sets of states of different stages are summarized in Table 8.30.

Table 8.30 Sets of States of Different Stages

Stage j	Decision variable	Set of states
2	X_2	$\{b_{12}, b_{22}\}$
1	X_1	$\{b_{11}, b_{21}\}$

Recursive function for stage 2 with respect to X_2 is based on the backward recursion. Therefore,

$$f_2(b_{12}, b_{22}) = \max_{\substack{0 \le 6X_2 \le b_{12} \\ 0 \le 12X_2 \le b_{22}}} 30X_2$$

To maintain feasibility, X_2 should be the minimum of $b_{12}/6$ and $b_{22}/12$, the above objective function is modified as follows:

$$f_2(b_{12}, b_{22}) = 30 \min \left(\frac{b_{12}}{6}, \frac{b_{22}}{12} \right) \quad \text{and} \quad X_2^* = \min \left(\frac{b_{12}}{6}, \frac{b_{22}}{12} \right)$$

Recursive function for stage 1 with respect to X_1 is

$$f_1(b_{11}, b_{21}) = \max_{0 \le 3X_1 \le b_{11}} [10X_1 + f_2(b_{11} - 3X_1, b_{21})]$$

$$= \max_{0 \le 3X_1 \le b_{11}} \left[10X_1 + 30 \min \left(\frac{b_{11} - 3X_1}{6}, \frac{b_{21}}{12}\right)\right]$$

The stage 1 is the last in the series of backward recursion. Therefore, $b_{11} = 168$, and $b_{21} = 240$. Determine the upper limit for X_1^*, we have

$$f_1\left(\frac{X_1}{b_{11}}, b_{21}\right) = \max \left[f_1\left(\frac{X_1}{168}, 240\right)\right]$$

$$= \max \left[10X_1 + 30 \min \left(\frac{168 - 3X_1}{6}, \frac{240}{12}\right)\right]$$

Now, to which in the ranges of X_1 is defined, $(168 - 3X_1)/6$ can be as high as 20 or as low as 0. So, equate it to 20 as well as to 0 and solve for X_1, as explained below:

$$\frac{168 - 3X_1}{6} = 20, \quad \text{or} \quad X_1 = 16$$

$$\frac{168 - 3X_1}{6} = 0, \quad \text{or} \quad X_1 = 56$$

The ranges for X_1 are follows:

$$0 \le X_1 \le 16 \quad \text{and} \quad 16 \le X_1 \le 56$$

Now, $f_1(X_1/b_{11}, b_{21})$ is rewritten as:

$$f_1\left(\frac{X_1}{b_{11}}, b_{21}\right) = \max \begin{bmatrix} 10X_1 + 30 \min \left(\dfrac{168 - 3X_1}{6}, 20\right), & 0 \le X_1 \le 16 \\[2ex] 10X_1 + 30 \min \left(\dfrac{168 - 3X_1}{6}, 20\right), & 16 \le X_1 \le 56 \end{bmatrix}$$

$$f_1\left(\frac{X_1}{168}, 240\right) = \max \begin{bmatrix} 10X_1 + 30 \times 20 & 0 \le X_1 \le 16 \\[2ex] 10X_1 + 30 \times \dfrac{168 - 3X_1}{6}, & 16 \le X_1 \le 56 \end{bmatrix}$$

$$= \max \begin{bmatrix} 10X_1 + 600, & 0 \le X_1 \le 16 \\[1ex] 10X_1 + 840 - 15X_1, & 16 \le X_1 \le 56 \end{bmatrix}$$

$$= \max \begin{bmatrix} 10X_1 + 600, & 0 \le X_1 \le 16 \\[1ex] 10X_1 + 840 - 15X_1, & 16 \le X_1 \le 56 \end{bmatrix}$$

To maximize each of the above cases, substitute 16 for X_1. Now, we get

$$f_1\left(\frac{X_1}{168}, 240\right) = \max (760, 760) = 760$$

Therefore,

$$X_1^* = 16 \quad \text{and} \quad f_1\left(\frac{X_1}{168}, 240\right) = 760$$

For tracing the value of X_2^*, we have

$$b_{12} = b_{11} - 3X_1 = 168 - 3 \times 16 = 120$$
$$b_{22} = b_{21} - 0 = 240 - 0 = 240$$

Therefore,

$$X_2^* = \min\left(\frac{b_{12}}{6}, \frac{b_{22}}{12}\right) = \min\left(\frac{120}{6}, \frac{240}{12}\right) = \min (20, 20) = 20$$

The optimal results are:

$$X_1^* = 16, \ X_2^* = 20, \ Z(\text{optimum}) = 760.$$

QUESTIONS

1. Define dynamic programming problem. List and explain the terminologies of dynamic programming problem. What are the application areas of dynamic programming?

2. An organization is planning to diversify its business with a maximum outlay of Rs. 4 crores. It has identified three different locations to instal plants. The organization can invest in one or more of these plants subject to the availability of the fund. The different possible alternatives and their investment (in crores of rupees) and present worth of returns during the useful life (in crores of rupees) of each of these plants are summarized in the following table. The first row of the table has zero cost and zero return for all the plants. Hence, it is known as *do-nothing* alternative. Find the optimal allocation of the capital to different plants which will maximize the corresponding sum of the present worth of returns.

Alternatives	Plant 1		Plant 2		Plant 3	
	Cost	Return	Cost	Return	Cost	Return
1	0	0	0	0	0	0
2	1	12	2	16	2	9
3	2	15	3	20	3	12
4	3	19	4	25	–	–

3. An electronic item has three components in series. So, the reliability of the system is equal to the product of the reliabilities of the three components ($R = r_1 r_2 r_3$). It is a known fact that the reliability of the system can be improved by providing standby units at extra cost. The details of costs and reliabilities for different number of standby units for each of the components of the system are summarized in the following table.

No. of standby units	Component 1		Component 2		Component 3	
	Cost (Rs.)	Reliability	Cost (Rs.)	Reliability	Cost (Rs.)	Reliability
1	1	0.70	3	0.85	2	0.85
2	2	0.85	4	0.95	3	0.92
3	3	0.95	6	0.98	5	0.97

The total capital budgeted for this purpose is Rs. 8. Determine the optimal number of standby units for each of the components of the system such that the total reliability of the system is maximized.

4. A distance network consists of eleven nodes which are distributed as shown in the following table. Find the shortest path from node 1 to node 11 and the corresponding distance.

Arc	Distance	Arc	Distance
1–2	8	5–8	12
1–3	7	5–9	7
1–4	1	6–9	9
2–5	5	7–9	6
3–5	9	7–10	13
3–6	2	8–11	4
3–7	8	9–11	2
4–7	10	10–11	15

5. Alpha logistic company has to load a cargo out of three items whose details are shown below. The maximum weight of the cargo is 10 tons. Find the optimal cargo loading using dynamic programming method such that the total return is maximized.

Item i	1	2	3
Weight, w_i/unit (in tons)	2	3	1
Return, r_i/unit (in rupees)	500	900	300

6. Consider the single machine scheduling problem which is shown below in which each of the 5 jobs can be fully processed in the same single machine. The other data in this table are the processing time and the due date for each of the jobs. Find the best sequence which minimizes the total tardiness using dynamic programming method. (*Note:* $T_j = C_j - d_j$, if $C_j > d_j$; otherwise, it is 0, where C_j is the completion time of the job j).

Job j	1	2	3	4	5
t_j	5	8	7	8	10
d_j	12	14	12	9	20

7. Consider a cable of length k units. The objective is to subdivide this cable into n parts each having a length p_i, where i varies from 1 to n such that the product of the lengths of the parts is maximized.

A mathematical model for the above situation is presented below:

$$\text{Maximize } Z = p_1 p_2 p_3 \cdots p_i \cdots p_n$$

subject to

$$p_1 + p_2 + p_3 + \cdots + p_i + \cdots + p_n = k$$
$$p_i > 0, \; i = 1, 2, 3,\ldots, n$$

Solve the problem using dynamic programming technique.

8. Solve the following model of the optimal subdividing of a cable of length 20 units into four parts such that the product of their lengths is maximized using dynamic programming technique.

$$\text{Maximize } Z = p_1 p_2 p_3 p_4$$

subject to

$$p_1 + p_2 + p_3 + p_4 = 20$$
$$p_1, p_2, p_3, p_4 > 0$$

9. Solve the following linear programming problem using dynamic programming technique.

$$\text{Maximize } Z = 30X_1 + 15X_2$$

subject to

$$6X_1 + 8X_2 \leq 180$$
$$15X_2 \leq 210$$
$$X_1 \text{ and } X_2 \geq 0$$

10. Solve the following linear programming problem using dynamic programming technique.

$$\text{Maximize } Z = 30X_1 + 15X_2 + 6X_3$$

subject to

$$6X_1 + 8X_2 + 9X_3 \leq 210$$
$$12X_2 + 6X_3 \leq 180$$
$$X_1, X_2 \text{ and } X_3 \geq 0$$

QUEUEING THEORY 9

9.1 INTRODUCTION

In many real-world applications such as railways and airlines reservation counters, bank counters, gasoline stations, etc. incoming customers become part of the respective queueing system. In fact, waiting for service has become an integral part of our daily life, albeit at a considerable cost most of the times. However, the adverse impact of the queueing up phenomena can be brought down to a minimum by applying various queueing models.

In general, the queueing system consists of one or more queues and one or more servers, and operates under a set of procedures. Let us consider the reservation counter of an airlines where customers from different parts of the world/country arrive and wait at the reservation counter. Depending on the server status, the incoming customer either waits at the queue or gets the turn to be served. If the server in the reservation counter is free at the time of arrival of a customer, the customer can directly enter into the counter for getting service and then leave the system. In this process, over a period of time, the system may experience 'customer waiting' and/or 'server idle time'. In any service system/manufacturing system involving queueing situation, the objective is to design the system in such a manner that the average waiting time of the customers is minimized and the percentage utilization of the server is maintained above a desired level.

The various types of queueing system in many service/manufacturing situations are described in Table 9.1.

Table 9.1 Application Areas of Queueing System

Example	Members of queue	Server(s)
Bank counter	Account holders	Counter clerk
Toll gate	Vehicles	Toll collectors
Ration shop	Ration card holders	Shop clerk
Main frame computer centre	Programs	Computer
Library	Students	Counter clerk
Traffic signal	Vehicles	Signal point
Final inspection station of T.V. assembly line	Assembled T.V. sets	Inspector
Airport runways	Planes	Runways
Telephone booth	Customers	Telephones
Maintenance shop	Breakdown machines	Mechanics

In addition to these examples, many subsystems of production, finance, personnel and marketing functions of an organization can be modelled as queueing systems for management decision making.

9.2 TERMINOLOGIES OF QUEUEING SYSTEM

A schematic representation of a simple queueing system that consists of a queue and a service station is shown in Figure 9.1.

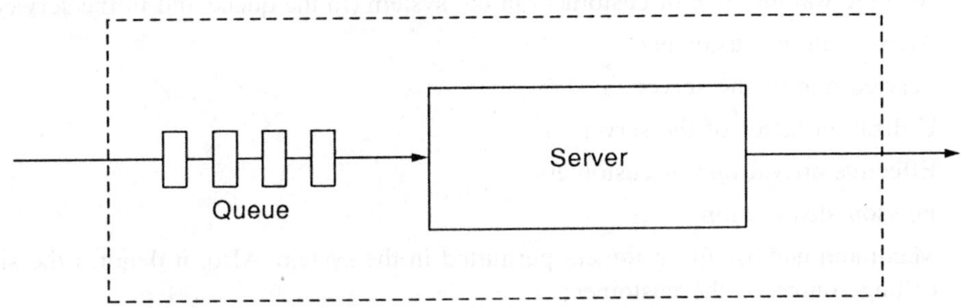

Figure 9.1 Queueing system.

Customers who come to the system to get the required service will directly enter the service station without waiting in the queue if the server is free at that point of time. Otherwise, they will wait in the queue till the server becomes free. But in reality, there may be variations of the system as given below.

1. The number of queues may be more than one. If there is a queue for male as well as for female customers, then generally, alternate mode of selecting customers from each queue is followed.
2. The number of servers may be more than one. This is an example of *parallel counters* for providing service.
3. Sometimes, the service may be provided in multistages in sequential order. This type of system is known as *queues in tandem*.

Bulk arrival. Generally, it is assumed that the customers arrive into the system one by one. But, in some reality, customers may arrive in groups. Such arrival is called as *bulk arrival*.

Jockeying. If there is more than one queue, the customers from one queue will be tempted to join another queue because of its smaller size. This behaviour of the customers is known as *queue jockeying*.

Balking. If the queue length appears very large to a customer he/she may not join the queue. This property is known as *balking* of customers.

Reneging. Sometimes, a customer who is already in the queue will leave the queue in anticipation of longer waiting time. This kind of departure from the queue without receiving the service is known as *reneging*.

The list of variables which is used in queueing models is presented below:

n Number of customers in the system

C Number of servers in the system

$p_n(t)$ Probability of having n customers in the system at time t

p_n Steady-state probability of having n customers in the system

p_0 Probability of having 0 customer in the system

L_q — Average number of customers waiting in the queue

L_s — Average number of customers waiting in the system (in the queue and in the service station)

W_q — Average waiting time of customers in the queue

W_s — Average waiting time of customers in the system (in the queue and in the service station)

δ — Arrival rate of customers

μ — Service rate of the server

ϕ — Utilization factor of the server

δ_{eff} — Effective arrival rate of customers

M — Poisson distribution

N — Maximum number of customers permitted in the system. Also, it denotes the size of the calling source of the customers

GD — General discipline for service. This may be first-in-first-serve (FIFS), last-in-first-serve (LIFS), random order (RO), etc.

9.3 EMPIRICAL QUEUEING MODELS

The basic queueing models can be classified into six categories using Kendall notation which in turn uses six parameters to define a model such as $(P/Q/R):(X/Y/Z)$. The parameters of this notation are:

P = Arrival rate distribution

Q = Service rate distribution

R = Number of servers

X = Service discipline

Y = Maximum number of customers permitted in the system

Z = Size of the calling source of the customers

The basic six queueing models as per this classification are as shown in Table 9.2.

Table 9.2 Classification of Queueing Models

Kendall notation	Parameter	Description
$(M/M/1):(GD/\infty/\infty)$	P	Poisson arrival rate
	Q	Poisson service rate
	R	Single server
	X	General discipline
	Y	Infinite number of customers is permitted in the system
	Z	Size of the calling source is infinite
$(M/M/C):(GD/\infty/\infty)$	P	Poisson arrival rate
	Q	Poisson service rate
	R	Multiservers
	X	General discipline
	Y	Infinite number of customers is permitted in the system
	Z	Size of the calling source is infinite

Table 9.2 Classification of Queueing Models *(Contd.)*

Kendall notation	Parameter	Description
$(M/M/1):(GD/N/\infty)$	P	Poisson arrival rate
	Q	Poisson service rate
	R	Single server
	X	General discipline
	Y	Finite number of customers is permitted in the system
	Z	Size of the calling source is infinite
$(M/M/C):(GD/N/\infty)$	P	Poisson arrival rate
	Q	Poisson service rate
	R	Multiservers
	X	General discipline
	Y	Finite number of customers is permitted in the system
	Z	Size of the calling source is infinite
$(M/M/1):(GD/N/N)$	P	Poisson arrival rate
	Q	Poisson service rate
	R	Single server
	X	General discipline
	Y	Finite number of customers is permitted in the system
	Z	Size of the calling source is finite
$(M/M/C):(GD/N/N)$	P	Poisson arrival rate
	Q	Poisson service rate
	R	Multiservers
	X	General discipline
	Y	Finite number of customers is permitted in the system
	Z	Size of the calling source is finite

9.3.1 $(M/M/1):(GD/\infty/\infty)$ Model

The parameters of this model are given as follows:

   (i) Arrival rate follows Poisson distribution.
  (ii) Service rate follows Poisson distribution.
 (iii) Number of servers is one.
 (iv) Service discipline is general discipline.
  (v) Maximum number of customers permitted in the system is infinite.
 (vi) Size of the calling source is infinite.

The steady-state formula to obtain the probability of having n customers in the system p_n, and the formulas for p_0, L_s, L_q, W_s and W_q are presented below:

$$p_n = (1 - \phi)\phi^n, \quad n = 0, 1, 2, 3, ..., \infty \qquad \text{where, } \phi = \frac{\delta}{\mu} < 1$$

$$L_s = \frac{\phi}{1 - \phi}$$

$$L_q = L_s - \frac{\delta}{\mu} = \frac{\phi^2}{1 - \phi}$$

$$W_s = \frac{L_s}{\delta} = \frac{1}{(1-\phi)\mu} = \frac{1}{\mu-\delta}$$

$$W_q = \frac{L_q}{\delta} = \frac{\phi}{\mu-\delta}$$

Example 9.1 The arrival rate of customers at a banking counter follows Poisson distribution with a mean of 45 per hour. The service rate of the counter clerk also follows Poisson distribution with a mean of 60 per hour.

(a) What is the probability of having 0 customer in the system (p_0)?
(b) What is the probability of having 5 customers in the system (p_5)?
(c) What is the probability of having 10 customers in the system (p_{10})?
(d) Find L_s, L_q, W_s and W_q.

Solution We have the following data:

$$\text{Arrival rate, } \delta = 45 \text{ per hour}$$
$$\text{Service rate, } \mu = 60 \text{ per hour}$$

$$\text{Utilization factor, } \phi = \frac{\delta}{\mu} = \frac{45}{60} = 0.75$$

(a) $p_0 = 1 - \phi = 1 - 0.75 = 0.25$

(b) $p_5 = (1 - \phi)\phi^5 = (1 - 0.75)0.75^5 = 0.0593$

(c) $p_{10} = (1 - \phi)\phi^{10} = (1 - 0.75)0.75^{10} = 0.0141$

(d) $L_s = \dfrac{\phi}{1-\phi} = \dfrac{0.75}{1-0.75} = 3 \text{ customers}$

$$L_q = \frac{\phi^2}{1-\phi} = \frac{0.75^2}{1-0.75} = 2.25 \text{ customers}$$

$$W_s = \frac{1}{\mu-\delta} = \frac{1}{60-45} = 0.067 \text{ hour}$$

$$W_q = \frac{\phi}{\mu-\delta} = \frac{0.75}{60-45} = 0.05 \text{ hour.}$$

Example 9.2 Vehicles pass through a toll gate at a rate of 90 per hour. The average time to pass through the gate is 36 seconds. The arrival rate and service rate follow Poisson distribution. There is a complaint that the vehicles wait for long duration. The authorities are willing to instal one more gate to reduce the average time to pass through the toll gate to 30 seconds if the idle time of the toll gate is less than 10% and the average queue length at the gate is more than 5 vehicles. Check whether the installation of the second gate is justified.

Solution Arrival rate of vehicles at the toll gate, $\delta = 90$ per hour and the time taken to pass through the gate = 36 seconds. Therefore,

$$\text{The service rate, } \mu = \frac{1}{36} \times 3600 = 100 \text{ vehicles per hour}$$

$$\text{Utilization factor, } \phi = \frac{\delta}{\mu} = \frac{90}{100} = 0.9$$

(a) Waiting number of vehicles in the queue,

$$L_q = \frac{\phi^2}{1 - \phi} = \frac{0.9^2}{1 - 0.9} = 8.1 \text{ vehicles}$$

(b) Revised time taken to pass through the gate = 30 seconds. Therefore,

$$\text{The service rate, } \mu = \frac{1}{30} \times 3600 = 120 \text{ vehicles per hour}$$

$$\text{Utilization factor, } \phi = \frac{\delta}{\mu} = \frac{90}{120} = 0.75$$

$$\text{Percentage idle time of the gate} = 1 - \phi = 1 - 0.75 = 0.25 = 25\%$$

From the above results, it is clear that the average waiting number of vehicles in the queue is more than 5 but the idle time of the toll gate is not less than 10%. Hence, the installation of another gate is not justified.

Example 9.3   The arrival rate of customers at the single window booking counter of a two wheeler agency follows Poisson distribution and the service time follows exponential (negative) distribution and hence, the service rate also follows Poisson distribution. The arrival rate and the service rate are 25 customers per hour and 35 customers per hour, respectively. Find the following:

(a) Utilization of the booking clerk.
(b) Average number of waiting customers in the queue.
(c) Average number of waiting customers in the system.
(d) Average waiting time per customer in the queue.
(e) Average waiting time per customer in the system.

Solution

The arrival rate of customers at the booking counter, $\delta = 25$ customers per hour
The service rate of the booking clerk, $\mu = 35$ customers per hour
This problem comes under $(M/M/1) : (GD/\infty/\infty)$.

(a) Utilization of the booking clerk, $\phi = \dfrac{\delta}{\mu} = \dfrac{25}{35} = 0.71429$

(b) Average waiting number of customers in the queue, $L_q = \dfrac{\phi^2}{(1 - \phi)} = \dfrac{0.71429^2}{1 - 0.71429}$

$$= 1.7857 \text{ customers}$$

(c) Average waiting number of customers in the system, $L_s = \dfrac{\phi}{(1 - \phi)} = \dfrac{0.71429}{(1 - 0.71429)}$

$$= 2.5 \text{ customers}$$

(d) Average waiting time per customer in the queue, $W_q = \dfrac{\phi}{(\mu - \delta)} = \dfrac{0.71429}{(35 - 25)}$

$$= 0.07143 \text{ hour}$$

$$= 4.286 \text{ minutes}$$

(e) Average waiting time per customer in the system, $W_s = \dfrac{1}{(\mu - \delta)} = \dfrac{1}{(35 - 25)}$

$$= 0.1 \text{ hour} = 6 \text{ minutes.}$$

Example 9.4 A harbour has single dock to unload the containers from the incoming ships. The arrival rate of ships at the harbour follows Poisson distribution and the unloading time for the ships follows exponential (negative) distribution and hence, the service rate also follows Poisson distribution. The arrival rate and the service rate are 8 ships per week and 14 ships per week, respectively. Find the following:

(a) Utilization of the dock.
(b) Average number of waiting ships in the queue.
(c) Average number of waiting ships in the system.
(d) Average waiting time per ship in the queue.
(e) Average waiting time per ship in the system.

Solution

The arrival rate of ships at the harbour, $\delta = 8$ ships per week
The service rate of the dock, $\mu = 14$ ships per week
This problem comes under $(M/M/1) : (GD/\infty/\infty)$.

(a) Utilization of the dock, $\phi = \dfrac{\delta}{\mu} = \dfrac{8}{14} = 0.57143$

(b) Average waiting number of ships in the queue, $L_q = \dfrac{\phi^2}{(1 - \phi)} = \dfrac{0.57143^2}{(1 - 0.57143)}$

$$= 0.7619 \text{ ship}$$

(c) Average waiting number of ships in the system, $L_s = \dfrac{\phi}{(1 - \phi)} = \dfrac{0.57143}{(1 - 0.57143)}$

$$= 1.3333 \text{ ships}$$

(d) Average waiting time per ship in the queue, $W_q = \dfrac{\phi}{(\mu - \delta)} = \dfrac{0.57143}{(14 - 8)}$

$$= 0.09524 \text{ week} = 0.6667 \text{ day}$$

(e) Average waiting time per ship in the system, $W_s = \dfrac{1}{(\mu - \delta)} = \dfrac{1}{(14 - 8)}$

$$= 0.16667 \text{ week} = 1.1667 \text{ days}$$

9.3.2 (*M/M/C*):(GD/∞/∞) Model

The parameters of this model are as follows:

  (i) Arrival rate follows Poisson distribution.
  (ii) Service rate follows Poisson distribution.
  (iii) Number of servers is C.
  (iv) Service discipline is general discipline.
  (v) Maximum number of customers permitted in the system is infinite.
  (vi) Size of the calling source is infinite.

The steady-state formula to obtain the probability of having n customers in the system p_n, and the formula for p_0, L_s, L_q, W_q and W_s are presented below.

$$p_n = \frac{\phi^n}{n!} \, p_0, \qquad 0 \le n \le C$$

$$= \frac{\phi^n}{C^{n-C} C!} \, p_0, \qquad n > C$$

where,

$$\frac{\phi}{C} < 1 \quad \text{or} \quad \frac{\delta}{\mu C} < 1$$

$$p_0 = \left\{ \sum_{n=0}^{C-1} \frac{\phi^n}{n!} + \frac{\phi^C}{C![1 - (\phi/C)]} \right\}^{-1}$$

$$L_q = \frac{\phi^{C+1}}{(C-1)!\,(C-\phi)^2} \, p_0 = \frac{C\phi p_C}{(C-\phi)^2}$$

$$L_s = L_q + \phi$$

$$W_q = \frac{L_q}{\delta}$$

$$W_s = W_q + \frac{1}{\mu}$$

Now the formulae for p_0 and L_q under special conditions are:

$$p_0 \approx 1 - \phi, \quad L_q \approx \frac{\phi^{C+1}}{C^2}, \qquad \text{where } \phi \ll 1$$

and

$$p_0 \approx \frac{(C-\phi)(C-1)!}{C^C} \quad \text{and} \quad L_q = \frac{\phi}{C-\phi}, \qquad \text{where } \frac{\phi}{C} \approx 1.$$

Example 9.5 At a central warehouse, vehicles arrive at the rate of 18 per hour and they arrival rate follows Poisson distribution. The unloading time of the vehicles follows exponential distribution and the unloading rate is 6 vehicles per hour. There are 4 unloading crews. Find the following:

(a) p_0 and p_3

(b) L_q, L_s, W_q and W_s

Solution   We have

$$\text{Arrival rate, } \delta = 18 \text{ per hour}$$
$$\text{Unloading rate, } \mu = 6 \text{ per hour}$$
$$\text{Number of unloading crews, } C = 4$$

and

$$\phi = \frac{\delta}{\mu} = \frac{18}{6} = 3$$

(a) Therefore, p_0 is computed as:

$$p_0 = \left\{ \sum_{n=0}^{C-1} \frac{\phi^n}{n!} + \frac{\phi^C}{C![1 - (\phi/C)]} \right\}^{-1}$$

$$= \left\{ \sum_{n=0}^{3} \frac{3^n}{n!} + \frac{3^4}{4![1 - (3/4)]} \right\}^{-1}$$

$$= \left\{ \frac{3^0}{0!} + \frac{3^1}{1!} + \frac{3^2}{2!} + \frac{3^3}{3!} + \frac{3^4}{4![1 - (3/4)]} \right\}^{-1}$$

$$= 0.0377$$

Now to compute p_3, we have

$$p_n = \frac{\phi^n}{n!} p_0, \ 0 \le n \le C$$

Therefore,

$$p_3 = \frac{\phi^3}{3!} p_0 = \frac{3^3}{6} \times 0.0377 = 0.1697$$

(b) L_q, L_s, W_q and W_s are computed as under:

$$L_q = \frac{\phi^{C+1}}{(C-1)! \times (C - \phi)^2} p_0 = \frac{3^5}{3! \times 1} \times 0.0377 = 1.53 \approx 2 \text{ vehicles}$$

$$L_s = L_q + \phi = 1.53 + 3 = 4.53 \approx 5 \text{ vehicles}$$

$$W_q = \frac{L_q}{\delta} = \frac{1.53}{18} = 0.085 \text{ hour} = 5.1 \text{ minutes}$$

$$W_s = W_q + \frac{1}{\mu} = 0.085 + \frac{1}{6} = 0.252 \text{ hour} = 15.12 \text{ minutes.}$$

Example 9.6   The arrival rate of breakdown machines at a maintenance shop follows Poisson distribution with a mean of 4 per hour. The service rate of machines by a maintenance mechanic also follows Poisson distribution with a mean of 3 per hour. The downtime cost per hour of a breakdown machine is Rs. 200. The labour rate per hour is Rs. 50. Determine the optimal number of maintenance mechanics to be employed to repair the machines such that the total cost is minimized.

Solution The following are given:

$$\text{Arrival rate, } \delta = 4 \text{ per hour}$$
$$\text{Service rate, } \mu = 3 \text{ per hour}$$
$$\text{Downtime cost of machine} = \text{Rs. 200 per hour}$$
$$\text{Labour cost} = \text{Rs. 50 per hour}$$

Since, $\delta > \mu$, a minimum of 2 mechanics will be required. Therefore,

$$\phi = \frac{\delta}{\mu} = \frac{4}{3} = 1.33$$

Whatever may be the number of mechanics, the breakdown machines have to wait for the same amount of time in the service station. So, the value of L_q is alone used to compute the cost of downtime of machines.

Case 1: When $C = 2$

$$p_0 = \left\{ \sum_{n=0}^{C-1} \frac{\phi^n}{n!} + \frac{\phi^C}{C![1-(\phi/C)]} \right\}^{-1} = \left\{ \sum_{n=0}^{2-1} \frac{1.33^n}{n!} + \frac{1.33^2}{2![1-(1.33/2)]} \right\}^{-1}$$

$$= \left\{ \frac{1.33^0}{0!} + \frac{1.33^1}{1!} + \frac{1.33^2}{2![1-(1.33/2)]} \right\}^{-1}$$

$$= 0.2$$

and

$$L_q = \frac{\phi^{C+1}}{(C-1)!(C-\phi)^2} p_0 = \frac{1.33^3}{1 \times (2-1.33)^2} \times 0.2 = 1.048 \approx 1.05$$

Therefore,

$$\text{Downtime cost per hour} = L_q \times 200 = 1.05 \times 200 = \text{Rs. 210}$$
$$\text{Wage paid to 2 mechanics/hour} = 2 \times 50 = \text{Rs. 100}$$
$$\text{Total cost} = 210 + 100 = \text{Rs. 310}$$

Case 2: When $C = 3$

$$p_0 = \left[\sum_{n=0}^{3-1} \frac{\phi^n}{n!} + \frac{\phi^3}{3!(1-\phi/3)} \right]^{-1}$$

$$= \left\{ \frac{1.33^0}{0!} + \frac{1.33^1}{1!} + \frac{1.33^2}{2!} + \frac{1.33^3}{3![1-(1.33/3)]} \right\}^{-1}$$

$$= 0.255$$

and

$$L_q = \frac{\phi^{C+1}}{(C-1)!(C-\phi)^2} p_0 = \frac{1.33^4}{2! \times (3-1.33)^2} \times 0.255 = 0.143$$

Therefore,

$$\text{Downtime cost per hour} = L_q \times 200 = 0.143 \times 200 = \text{Rs. } 28.60$$
$$\text{Wage paid to 3 mechanics/hour} = 3 \times 50 = \text{Rs. } 150.00$$
$$\text{Total cost} = 28.60 + 150.00 = \text{Rs. } 178.60$$

Case 3: When $C = 4$, the total wage for 4 mechanics itself is Rs. 200. Hence, the total cost with respect to 4 mechanics will be greater than Rs. 200. The minimum total cost is Rs. 178.60, when the number of maintenance mechanics is 3. Therefore, the optimal number of mechanics to be employed is 3.

Example 9.7   There are three clerks in the loan section of a bank to process the initial queries of customers. The arrival rate of customers follows Poisson distribution and it is 20 per hour. The service rate also follows Poisson distribution and it is 9 customers per hour. Find the following:

(a) Average waiting number of customers in the queue as well as in the system.
(b) Average waiting time per customer in the queue as well as in the system.

Solution

The arrival rate of customers in the loan section of the bank, $\delta = 20$ customers per hour
The service rate, $\mu = 9$ customers per hour
Number of clerks, $C = 3$
This problem comes under $(M/M/C) : (GD/\infty/\infty)$.

Therefore, $\phi = \dfrac{\delta}{\mu} = \dfrac{20}{9} = 2.222222$

p_0 is computed as:

$$p_0 = \left(\sum_{n=0}^{C-1} \frac{\phi^n}{n!} + \frac{\phi^C}{C![1-(\varphi/C)]} \right)^{-1}$$

$$= \left(\sum_{n=0}^{2} \frac{2.222222^n}{n!} + \frac{2.222222^3}{3![1-(2.222222/3)]} \right)^{-1}$$

$$= 0.0784558$$

(a) Average waiting number of customers in the queue, L_q is computed as:

$$L_q = \frac{\phi^{C+1}}{(C-1)! \times (C-\phi)^2} \, p_0 = \frac{2.222222^4}{(3-1)! \times (3-2.222222)^2} = \times \, 0.0784558 = 1.581372 \text{ customers}$$

Therefore, the average waiting number of customers in the system, $L_s = L_q + \phi$

$$= 1.581372 + 2.222222$$
$$= 3.803594 \text{ customers}$$

(b) Average waiting time per customer in the queue, $W_q = \dfrac{L_q}{\delta} = \dfrac{1.581372}{20}$

$$= 0.0790686 \text{ hour} = 4.744 \text{ minutes}$$

Average waiting time per customer in the system, $W_s = W_q + \dfrac{1}{\mu} = 0.0790686 + \dfrac{1}{9}$

$$= 0.1901797 \text{ hour} = 11.41 \text{ minutes.}$$

Example 9.8 There are four booking counters in a railway station. The arrival rate of customers follows Poisson distribution and it is 30 per hour. The service rate also follows Poisson distribution and it is 10 customers per hour. Find the following:

(a) Average waiting number of customers in the queue as well as in the system.
(b) Average waiting time per customer in the queue as well as in the system.

Solution

The arrival rate of customers at booking counters of a railway station, $\delta = 30$ customers per hour
The service rate, $\mu = 10$ customers per hour
Number of booking counters, $C = 4$
This problem comes under $(M/M/C) : (GD/\infty/\infty)$.

Therefore, $\phi = \dfrac{\delta}{\mu} = \dfrac{30}{10} = 3$

p_0 is computed as:

$$p_0 = \left(\sum_{n=0}^{C-1} \frac{\phi^n}{n!} + \frac{\phi^C}{C![1-(\phi/C)]} \right)^{-1}$$

$$= \left(\sum_{n=0}^{3} \frac{3^n}{n!} + \frac{3^4}{4![1-(3/4)]} \right)^{-1}$$

$$= 0.03773585$$

(a) Average waiting number of customers in the queue, L_q is computed as:

$$L_q = \frac{\phi^{C+1}}{(C-1)! \times (C-\phi)^2} \, p_0 = \frac{3^5}{(4-1)! \times (4-3)^2} \times 0.03773585 = 1.5283 \text{ customers}$$

Therefore, the average waiting number of customers in the system, $L_s = L_q + \phi$

$$= 1.5283 + 3$$

$$= 4.5283 \text{ customers}$$

(b) Average waiting time per customer in the queue, $W_q = \dfrac{L_q}{\delta} = \dfrac{1.5283}{30}$

$$= 0.0509434 \text{ hour} = 3.0566 \text{ minutes}$$

Average waiting time per customer in the system, $W_s = W_q + \dfrac{1}{\mu} = 0.0509434 + \dfrac{1}{10}$

$$= 0.1509434 \text{ hour} = 9.0566 \text{ minutes}$$

9.3.3 $(M/M/1):(GD/N/\infty)$ Model

The parameters of this model are defined below:

(i) Arrival rate follows Poisson distribution.
(ii) Service rate follows Poisson distribution.

(iii) Number of servers is one.
(iv) Service discipline is general discipline.
(v) Maximum number of customers permitted in the system is N.
(vi) Size of the calling source is infinite.

The steady-state formula to obtain the probability of having n customers in the system p_n, and the formulas for p_0, L_s, L_q, W_s and W_q are presented below.

$$p_n = \frac{1 - \phi}{1 - \phi^{n+1}} \phi^n, \qquad \phi \neq 1 \text{ and } n = 0, 1, 2, 3, ..., N$$

$$= \frac{1}{N + 1}, \qquad \phi = 1$$

$$L_s = \frac{\phi\left[1 - (N + 1)\phi^N + N\phi^{N+1}\right]}{(1 - \phi)(1 - \phi^{N+1})}, \qquad \phi \neq 1$$

$$= \frac{N}{2}, \qquad \phi = 1$$

$$\delta_{\text{eff}} = \delta(1 - p_N) = \mu(L_s - L_q)$$

$$L_q = L_s - \frac{\delta_{\text{eff}}}{\mu} = L_s - \frac{\delta(1 - p_N)}{\mu}$$

$$W_q = \frac{L_q}{\delta_{\text{eff}}} = \frac{L_q}{\delta(1 - p_N)}$$

$$W_s = W_q + \frac{1}{\mu} = \frac{L_s}{\delta_{\text{eff}}} = \frac{L_s}{\delta(1 - p_N)}.$$

Example 9.9 Cars arrive at a drive-in restaurant with a mean arrival rate of 24 cars per hour and the service rate of the cars is 20 cars per hour. The arrival rate and the service rate follow Poisson distribution. The number of parking space for cars is only 4. Find the standard results of this system.

Solution Here

$$\text{Arrival rate, } \delta = 24 \text{ cars per hour}$$
$$\text{Service rate, } \mu = 20 \text{ cars per hour}$$
$$N = 4$$

$$\phi = \frac{\delta}{\mu} = \frac{24}{20} = 1.2$$

Therefore, we get

$$L_s = \frac{\phi\left[1 - (N + 1)\phi^N + N\phi^{N+1}\right]}{(1 - \phi)(1 - \phi^{N+1})} = \frac{1.2\left[1 - (4 + 1)1.2^4 + 4 \times 1.2^5\right]}{(1 - 1.2)(1 - 1.2^5)} = 2.36 \text{ cars}$$

and

$$p_N = \frac{1 - \phi}{1 - \phi^{N+1}} \phi^N = \frac{1 - 1.2}{1 - 1.2^5} \times 1.2^4 = 0.2787$$

The other results are:

$$\delta_{\text{eff}} = \delta(1 - p_N) = 24(1 - 0.2787) = 17.3112 \text{ per hour}$$

$$L_q = L_s - \frac{\delta_{\text{eff}}}{\mu} = 2.36 - \frac{17.3112}{20} = 1.494 \text{ cars}$$

$$W_q = \frac{L_q}{\delta_{\text{eff}}} = \frac{1.494}{17.3112} = 0.0863 \text{ hour} = 5.2 \text{ minutes}$$

$$W_s = \frac{L_s}{\delta_{\text{eff}}} = \frac{2.36}{17.3112} = 0.1363 \text{ hour} = 8.2 \text{ minutes.}$$

Example 9.10 Assembled television sets are inspected for volume control in an assembly line. The arrival rate of the television sets follows Poisson distribution and it is 21 sets per hour. The inspection rate also follows Poisson distribution and it is 25 sets per hour. In front of the inspection station, the waiting space is sufficient for a maximum of 6 television sets. Find the following:

(a) Average waiting number of television sets in the queue in front of the inspection station as well as in the system.
(b) Average waiting time per television set in the queue in front of the inspection station as well as in the system.

Solution

The arrival rate of television sets in front of the inspection station, $\delta = 21$ sets per hour
The service rate of the inspection station, $\mu = 25$ sets per hour
Number of inspection station, $C = 1$
Number of waiting space, $N = 6$
This problem comes under $(M/M/1) : (GD/N/\infty)$.

Therefore, $\phi = \dfrac{\delta}{\mu} = \dfrac{21}{25} = 0.84$ which is not equal to 1.

(a) Computation of L_s and L_q:

$$L_s = \frac{\phi[1-(N+1)\,\phi^N + N\phi^{N+1}]}{(1-\phi)(1-\phi^{N+1})} = \frac{0.84\,[1-(6+1)\times 0.84^6 + 6\times 0.84^7]}{(1-0.84)\times(1-0.84^7)} = 2.319648 \text{ TV sets}$$

$$p_N = \frac{(1-\phi)}{(1-\phi^{N+1})}\,\phi^N = \frac{(1-0.84)}{(1-0.84^7)} \times 0.84^6 = 0.07973743$$

$$\delta_{\text{eff}} = \delta(1 - p_N) = 21 \times (1 - 0.07973743) = 19.3255 \text{ TV sets per hour}$$

$$L_q = L_s - \frac{\delta_{\text{eff}}}{\mu} = 2.319648 - \frac{19.3255}{25} = 1.546627 \text{ TV sets}$$

(b) Computation of W_q and W_s

Average waiting time per TV set in the queue, $W_q = \dfrac{L_q}{\delta_{\text{eff}}} = \dfrac{1.546627}{19.3255}$

$$= 0.08 \text{ hour} = 4.8 \text{ minutes}$$

Average waiting time per TV set in the system, $W_s = W_q + \dfrac{1}{\mu} = 0.08 + \dfrac{1}{25}$

$$= 0.12 \text{ hour} = 7.2 \text{ minutes.}$$

Example 9.11 A weighing station has single weighing bridge. The arrival rate of the vehicles coming to the weighing station follows Poisson distribution and it is 45 vehicles per hour. The service rate also follows Poisson distribution and it is 55 vehicles per hour. In front of the weighing bridge, the waiting space is sufficient for a maximum of 10 vehicles. Find the following:

(a) Average waiting number of vehicles in the queue in front of the weighing bridge as well as in the weighing station.
(b) Average waiting time per vehicle in front of the weighing bridge as well as in the weighing station.

Solution

The arrival rate of vehicles at the weighing station, $\delta = 45$ vehicles per hour
The service rate of the weighing bridge, $\mu = 55$ vehicles per hour
Number of inspection station, $C = 1$
Number of waiting space, $N = 10$
This problem comes under $(M/M/1) : (GD/N/\infty)$.

Therefore, $\phi = \dfrac{\delta}{\mu} = \dfrac{45}{55} = 0.8181818$

(a) Computation of L_s and L_q:

$$L_s = \frac{\phi[1-(N+1)\,\phi^N + N\phi^{N+1}]}{(1-\phi)(1-\phi^{N+1})} = \frac{0.8181818\,[1-(10+1)\times 0.8181818^{10} + 10\times 0.8181818^{11}]}{(1-0.8181818)\times(1-0.8181818^{11})}$$

$$= 3.1406 \text{ vehicles per hour}$$

$$p_N = \frac{(1-\phi)}{(1-\phi^{N+1})}\,\phi^N = \frac{(1-0.8181818)}{(1-0.8181818^{11})}\times 0.8181818^{10} = 0.0274625$$

$$\delta_{\text{eff}} = \delta(1 - p_N) = 45\times(1 - 0.0274625) = 43.76419 \text{ vehicles per hour}$$

$$L_q = L_s - \frac{\delta_{\text{eff}}}{\mu} = 3.1406 - \frac{43.76419}{55} = 2.34489 \text{ vehicles per hour}$$

(b) Computation of W_q and W_s

Average waiting time per vehicle in the queue, $W_q = \dfrac{L_q}{\delta_{\text{eff}}} = \dfrac{2.34489}{43.76419}$

$$= 0.05358 \text{ hour}$$

$$= 3.2148 \text{ minutes}$$

Average waiting time per vehicle in the system, $W_s = W_q + \dfrac{1}{\mu} = 0.05358 + \dfrac{1}{55}$

$$= 0.071762 \text{ hour}$$

$$= 4.305721 \text{ minutes.}$$

9.3.4 $(M/M/C):(GD/N/\infty)$ Model (for $C \leq N$)

The parameters of this model are defined as follows:

    (i) Arrival rate follows Poisson distribution.
    (ii) Service rate follows Poisson distribution.
    (iii) Number of servers is C.
    (iv) Service discipline is general discipline.
    (v) Maximum number of customers permitted in the system is N.
    (vi) Size of the calling source is infinite.

In this model, the following assumptions are made:

$$\delta_n = \delta, \qquad 0 \leq n \leq N$$
$$= 0, \qquad n \geq N$$

and

$$\mu_n = n\mu, \qquad 0 \leq n \leq C$$
$$= C\mu, \qquad C \leq n \leq N$$

The steady-state formula to obtain the probability of having n customers in the system p_n, and the formulas for p_0, L_s, L_q, W_s and W_q are presented below:

$$p_n = \frac{\phi^n}{n!}\, p_0, \qquad 0 \leq n \leq C$$

$$= \frac{\phi^n}{C!\,C^{n-C}}\, p_0, \qquad C \leq n \leq N$$

$$p_0 = \left\{ \sum_{n=0}^{C-1} \frac{\phi^n}{n!} + \frac{\phi^C [1 - (\phi/C)]^{N-C+1}}{C![1 - (\phi/C)]} \right\}^{-1} \qquad \text{when } \frac{\phi}{C} \neq 1$$

$$= \left[\sum_{n=0}^{C-1} \frac{\phi^n}{n!} + \frac{\phi^C}{C!}(N - C + 1) \right]^{-1} \qquad \text{when } \frac{\phi}{C} = 1$$

$$L_q = p_0 \frac{\phi^{C+1}}{(C-1)!\,(C-\phi)^2} \left[1 - \left(\frac{\phi}{C}\right)^{N-C} - (N-C)\left(\frac{\phi}{C}\right)^{N-C}\left(1 - \frac{\phi}{C}\right) \right] \qquad \text{for } \frac{\phi}{C} \neq 1$$

$$= p_0 \frac{\phi^C (N-C)(N-C+1)}{2C!} \qquad \text{for } \frac{\phi}{C} = 1$$

and

$$L_s = L_q + (C - \bar{C}) = L_q + \frac{\delta_{\text{eff}}}{\mu}$$

where

$$\bar{C} = \sum_{n=0}^{C} (C - n)p_n$$

which is the expected number of idle servers. Therefore,

$$\delta_{eff} = \delta(1 - p_N) = \mu(C - \bar{C})$$

$$W_q = \frac{L_q}{\delta_{eff}}$$

$$W_s = \frac{L_s}{\delta_{eff}}.$$

Example 9.12 In a harbour, ships arrive with a mean rate of 18 per week. The harbour has 4 docks to handle unloading and loading of ships. The service rate of individual dock is 6 per week. The arrival rate and the service rate follow Poisson distribution. At a point in time, the maximum number of ships permitted in the harbour is 6. Find p_0, L_q, L_s, W_q and W_s.

Solution We have

$$\text{Arrival rate of ships, } \delta = 18 \text{ per week}$$
$$\text{Service rate of individual dock, } \mu = 6 \quad \text{per week}$$

Therefore,

$$\phi = \frac{\delta}{\mu} = \frac{18}{6} = 3$$

Now, applying the required formulae, we calculate the given quantities as follows:

$$p_0 = \left\{ \sum_{n=0}^{C-1} \frac{\phi^n}{n!} + \frac{\phi^C[1 - (\phi/C)]^{N-C+1}}{C![1 - (\phi/C)]} \right\}^{-1} \qquad \text{when } \frac{\phi}{C} \neq 1$$

$$= \left\{ \sum_{n=0}^{4-1} \frac{3^n}{n!} + \frac{3^4[1 - (3/4)]^{6-4+1}}{4![1 - (3/4)]} \right\}^{-1}$$

$$= \left\{ \frac{3^0}{0!} + \frac{3^1}{1!} + \frac{3^2}{2!} + \frac{3^3}{3!} + \frac{3^4[1 - (3/4)]^{6-4+1}}{4![1 - (3/4)]} \right\}^{-1}$$

$$= 0.0757$$

$$L_q = p_0 \frac{\phi^{C+1}}{(C - 1)!(C - \phi)^2} \left[1 - \left(\frac{\phi}{C}\right)^{N-C} - (N - C)\left(\frac{\phi}{C}\right)^{N-C}\left(1 - \frac{\phi}{C}\right) \right]$$

$$= 0.0757 \frac{3^5}{3!(1)^2} \left[1 - \left(\frac{3}{4}\right)^{6-4} - (6 - 4) \times \left(\frac{3}{4}\right)^{6-4}\left(1 - \frac{3}{4}\right) \right]$$

$$= 0.479 \text{ ship}$$

$$\bar{C} = \sum_{n=0}^{4} (4 - n)p_n$$

$$= (4 - 0)p_0 + (4 - 1)p_1 + (4 - 2)p_2 + (4 - 3)p_3 + (4 - 4)p_4$$

$$= 4p_0 + 3\frac{\phi^1}{1!}\,p_0 + 2\frac{\phi^2}{2!}\,p_0 + 1\frac{\phi^3}{3!}\,p_0 + 0$$

$$= p_0\left(4 + 3\frac{\phi^1}{1!} + 2\frac{\phi^2}{2!} + 1\frac{\phi^3}{3!}\right)$$

$$= 0.0757\left(4 + 3\frac{3!}{1!} + 2\frac{3^2}{2!} + 1\frac{3^3}{3!}\right)$$

$$= 2$$

Therefore,

$$p_N = \frac{\phi^N}{C!\,C^{N-C}}\,p_0, \quad \text{for } n = N$$

and

$$p_N = \frac{3^6}{4!\,4^{6-4}} \times 0.0757 = 0.1437$$

The other results are:

$$\delta_{\text{eff}} = \delta(1 - p_N) = 18(1 - 0.1437) = 15.41 \text{ ships per week}$$

$$L_s = L_q + \frac{\delta_{\text{eff}}}{\mu} = 0.479 + \frac{15.41}{6} = 3.05 \text{ ships}$$

$$W_q = \frac{L_q}{\delta_{\text{eff}}} = \frac{0.479}{15.41} = 0.031 \text{ week} = 0.217 \text{ day}$$

$$W_s = \frac{L_s}{\delta_{\text{eff}}} = \frac{3.05}{15.41} = 0.198 \text{ week} = 1.386 \text{ day}$$

Example 9.13 There are three docks in a harbour. The arrival rate of ships follows Poisson distribution and it is 36 ships per month. The service rate (loading and unloading of containers) also follows Poisson distribution and it is 13 ships per month. The waiting space in the harbour can accommodate a maximum of 10 ships. Find the following:

(a) Average waiting number of ships in the queue as well as in the system.
(b) Average waiting time per ship in the queue as well as in the system.

Solution

The arrival rate of ships, $\delta = 36$ ships per month
The service rate, $\mu = 13$ ships per month
Number of docks, $C = 3$
The number of waiting spaces for ships, $N = 10$
This problem comes under $(M/M/C) : (GD/N/\infty)$.

Therefore, $\phi = \dfrac{\delta}{\mu} = \dfrac{36}{13} = 2.76923$

p_0 is computed as:

$$p_0 = \left(\sum_{n=0}^{C-1} \frac{\phi^n}{n!} + \frac{\phi^C [1 - (\phi/C)]^{N-C+1}}{C! [1 - (\phi/C)]} \right)^{-1} \text{ for } \frac{\phi}{C} \neq 1$$

$$= \left\{ \left(\sum_{n=0}^{2} \frac{2.76923^n}{n!} \right) + \frac{2.76923^3 [1 - (2.76923/3)]^{10-3+1}}{3! [1 - (2.76923/3)]} \right\}^{-1}$$

$$= 0.102352$$

(a) Average waiting number of ships in the queue, L_q is computed as:

$$L_q = \frac{\phi^{C+1}}{(C-1)! \times (C-\phi)^2} \left(1 - \left\{ \frac{\phi}{C} \right\}^{N-C} - (N-C) \left\{ \frac{\phi}{C} \right\}^{N-C} \left\{ 1 - \frac{\phi}{C} \right\} \right) p_0$$

$$= \frac{2.76923^4}{2! \times (3 - 2.76923)^2} \left(1 - \left\{ \frac{2.76923}{3} \right\}^7 - 7 \times \left\{ \frac{2.76923}{3} \right\}^7 \times \left\{ 1 - \frac{2.76923}{3} \right\} \right) \times 0.102352$$

$$= 6.86516 \text{ ships}$$

$$p_N = \frac{\phi^N}{C! \, C^{N-C}} p_0 = \frac{2.76923^{10}}{3! \times 3^7} \times 0.102352 = 0.206865$$

$$\delta_{\text{eff}} = \delta(1 - p_N) = 36 \times (1 - 0.206865) = 28.55286 \text{ ships/month}$$

Therefore, the average waiting number of ships in the system, $L_s = L_q + \dfrac{\delta_{\text{eff}}}{\mu}$

$$= 6.86516 + \frac{28.55286}{13}$$

$$= 9.06153 \text{ ships}$$

(b) Average waiting time per ship in the queue, $W_q = \dfrac{L_q}{\delta_{\text{eff}}} = \dfrac{6.86516}{28.55286}$

$$= 0.240437 \text{ month}$$

$$= 7.213 \text{ days}$$

Average waiting time per ship in the system, $W_s = \dfrac{L_s}{\delta_{\text{eff}}} = \dfrac{9.06153}{28.55286}$

$$= 0.31736 \text{ month}$$

$$= 9.52 \text{ days.}$$

Example 9.14 There are five mechanics in an automobile workshop to do minor checkup for automobiles. The arrival rate of cars follows Poisson distribution and it is 15 cars per hour. The service rate also follows Poisson distribution and it is 4 cars per hour. The waiting space in the workshop can accommodate a maximum of 6 cars.

(a) Average waiting number of cars in the queue as well as in the system.
(b) Average waiting time per car in the queue as well as in the system.

Solution The arrival rate of cars, $\delta = 15$ cars per hour and the service rate of the individual mechanic, $\mu = 4$ cars per hour. The number of mechanics, C is 5. The number of waiting space for the cars, N is 6. So, this problem comes under $(M/M/C) : (GD/N/\infty)$.

Therefore, $\phi = \dfrac{\delta}{\mu} = \dfrac{15}{4} = 3.75$

p_0 is computed as:

$$p_0 = \left(\sum_{n=0}^{C-1} \frac{\phi^n}{n!} + \frac{\phi^C[1-(\phi/C)]^{N-C+1}}{C![1-(\phi/C)]} \right)^{-1} \qquad \text{when } \frac{\phi}{C} \neq 1$$

$$= \left\{ \left(\sum_{n=0}^{4} \frac{3.75^n}{n!} \right) + \frac{3.75^5 \times [1-(3.75/5)]^{6-5+1}}{5![1-(3.75/5)]} \right\}^{-1}$$

$$= 0.03467$$

(a) Average waiting number of cars in the queue, L_q is computed as:

$$L_q = \frac{\phi^{C+1}}{(C-1)! \times (C-\phi)^2} \left(1 - \left\{ \frac{\phi}{C} \right\}^{N-C} - (N-C) \left\{ \frac{\phi}{C} \right\}^{N-C} \left\{ 1 - \frac{\phi}{C} \right\} \right) p_0, \text{ when } \frac{\phi}{C} \neq 1$$

$$= \frac{3.75^6}{4! \times (5-3.75)^2} \left(1 - \left\{ \frac{3.75}{5} \right\}^1 - 1 \times \left\{ \frac{3.75}{5} \right\}^1 \times \left\{ 1 - \frac{3.75}{5} \right\} \right) \times 0.03467$$

$$= 0.16069 \text{ car}$$

$$p_N = \frac{\phi^N}{C! \, C^{N-C}} \, p_0 = \frac{3.75^6}{5! \times 5^1} \times 0.03467 = 0.16069$$

$$\delta_{\text{eff}} = \delta(1 - p_N) = 15 \times (1 - 0.16069) = 12.58964 \text{ cars per hour}$$

Therefore, the average waiting number of cars in the system, $L_s = L_q + \dfrac{\delta_{\text{eff}}}{\mu}$

$$= 0.16069 + \frac{12.58964}{4}$$

$$= 3.3081 \text{ cars}$$

(b) Average waiting time per car in the queue, $W_q = \dfrac{L_q}{\delta_{\text{eff}}} = \dfrac{0.16069}{12.58964}$

$$= 0.0127637 \text{ hour} = 0.7658 \text{ minute}$$

Average waiting time per car in the system, W_s $\quad = \dfrac{L_s}{\delta_{\text{eff}}} = \dfrac{3.3081}{12.58964}$

$$= 0.2627637 \text{ hour}$$

$$= 15.766 \text{ minutes.}$$

9.3.5 $(M/M/C) : (GD/N/N)$ Model (for $C < N$)

The parameters of this model are defined below:

    (i) Arrival rate follows Poisson distribution.
    (ii) Service rate follows Poisson distribution.
    (iii) Number of servers is C.
    (iv) Service discipline is general discipline.
    (v) Maximum number of customers permitted in the system is N.
    (vi) Size of the calling source is N.

In this model, the following assumptions are made.

$$\delta_n = (N - n)\delta, \qquad 0 \le n \le N$$
$$= 0, \qquad\qquad n \ge N$$

and

$$\mu_n = n\mu, \qquad\qquad 0 \le n \le C$$
$$= C\mu, \qquad\qquad C \le n \le N$$
$$= 0, \qquad\qquad\; n \ge N$$

The steady-state formula to obtain the probability of having n customers in the system p_n, and the formulas for p_0 are:

$$p_n = {}^{N}C_n\, \phi^n\, p_0, \qquad\qquad 0 \le n \le C$$

$$= {}^{N}C_n\, \frac{n!\phi^n}{C!C^{n-C}}\, p_0, \qquad C \le n \le N$$

$$p_0 = \left[\sum_{n=0}^{C} {}^{N}C_n\, \phi^n + \sum_{n=C+1}^{N} {}^{N}C_n\, \frac{n!\phi^n}{C!C^{n-C}} \right]^{-1}$$

We also have formulas for L_q and L_s as:

$$L_q = \sum_{n=C+1}^{N} (n - C)p_n$$

and

$$L_s = L_q + (C - \overline{C}\,) = L_q + \frac{\delta_{\text{eff}}}{\mu}$$

where

$$\overline{C} = \sum_{n=0}^{C} (C - n)\, p_n \quad \text{and} \quad \delta_{\text{eff}} = \mu(C - \overline{C}\,) = \delta(N - L_s)$$

The formulas for W_q and W_s are:

$$W_q = \frac{L_q}{\delta_{\text{eff}}} \qquad W_s = \frac{L_s}{\delta_{\text{eff}}}.$$

Example 9.15   In the machine shop of a small-scale industry, machines breakdown with a mean rate of 2 per hour. The maintenance shop of the industry has 2 mechanics who can attend the breakdown machines individually. The service rate of each of the mechanics is 1.5 machines per hour. Initially, there are 5 working machines in the machine shop. Find p_0, L_q, L_s, W_q and W_s.

Solution Breakdown rate of machines, $\delta = 2$ per hour. Service rate of individual mechanic, $\mu = 1.5$ machines per hour. Total number of machines in the shop, $N = 5$. Therefore,

$$\phi = \frac{\delta}{\mu} = \frac{2}{1.5} = 1.33$$

So, we get

$$p_0 = \left(\sum_{n=0}^{C} {}^{N}C_n\, \phi^n + \sum_{n=C+1}^{N} {}^{N}C_n\, \frac{n!\,\phi^n}{C!\,C^{n-C}} \right)^{-1}$$

$$= \left(\sum_{n=0}^{2} {}^{5}C_n\, (1.33)^n + \sum_{n=2+1}^{5} {}^{5}C_n\, \frac{n!\,(1.33)^n}{2!\,2^{n-2}} \right)^{-1}$$

$$= 0.0072$$

and

$$L_q = \sum_{n=C+1}^{N} (n - C)p_n$$

$$= \sum_{n=C+1}^{N} (n - C)\, {}^{N}C_n\, \frac{n!\,\phi^n}{C!\,C^{n-C}}\, p_0$$

$$= \sum_{n=2+1}^{5} (n - 2)\, {}^{5}C_n\, \frac{n!\,(1.33)^n}{2!\,2^{n-2}} \times 0.0072$$

$$= 1.604 \text{ machines}$$

where

$$\bar{C} = \sum_{n=0}^{C} (C - n)p_n = \sum_{n=0}^{2} (2 - n)\, p_n = 2p_0 + p_1 = 0.06228$$

The other results are:

$$\delta_{\text{eff}} = \mu(C - \bar{C}) = 1.5(2 - 0.06228) = 2.907 \text{ per hour}$$

$$L_s = L_q + (C - \bar{C}) = 1.604 + (2 - 0.06228) = 3.54172 \text{ machines}$$

$$W_q = \frac{L_q}{\delta_{\text{eff}}} = \frac{1.604}{2.907} = 0.552 \text{ hour}$$

$$W_s = \frac{L_s}{\delta_{\text{eff}}} = \frac{3.54172}{2.907} = 1.2183 \text{ hour.}$$

Example 9.16 In a hospital there are two physicians to treat the in-patients. The number of beds for the in-patients in the hospital is 10. The arrival rate of patients at the in-patient section is 4 per hour and the service rate of the individual physician is 3 patients per hour. The arrival rate as well as service rate follow Poisson distribution. Find the following:

(a) The probability that all the physicians are free (p_0).
(b) Average waiting number of patients in the queue.
(c) Average waiting number of patients in the system.

(d) Average waiting time per patient in the queue.

(e) Average waiting time per patient in the system.

Solution Arrival rate, $\delta = 4$ patients per hour. Service rate of individual physician, $\mu = 3$ patients per hour. The total number of physicians, $C = 2$. The size of the calling population (N) is equal to the number of beds which is 10 and hence, the number of waiting spaces for the in-patients, N is 10. So, this problem comes under $(M/M/C) : (GD/N/N)$. Therefore, the utilization factor, ϕ is

$$\phi = \frac{\delta}{\mu} = \frac{4}{3} = 1.333333$$

(a) The probability that all the physicians will be free is equal to p_0 which is as computed below:

$$p_0 = \left(\sum_{n=0}^{C} NC_n\, \phi^n + \sum_{n=C+1}^{N} NC_n\, \frac{n!\,\phi^n}{C!\,C^{n-C}} \right)^{-1}$$

$$= \left(\sum_{n=0}^{2} 10C_n\, 1.333333^n + \sum_{n=3}^{10} 10C_n\, \frac{n! \times 1.333333^n}{2!\,2^{n-2}} \right)^{-1}$$

$$= 0.000001772877.$$

(b) The average waiting number of patients in the queue, L_q is:

$$L_q = \sum_{n=C+1}^{N} (n-C)p_n = 1 \times p_3 + 2 \times p_4 + 3 \times p_5 + 4 \times p_6 + 5 \times p_7 + 6 \times p_8 + 7 \times p_9 + 8 \times p_{10}$$

$$= 6.5 \text{ patients}$$

(c) The average waiting number of patients in the system, L_s is computed as shown below:

$$\bar{C} = \sum_{n=0}^{C} (C-n)p_n = 2 \times p_0 + 1 \times p_1 + 0 \times p_2 = 0.000027184$$

$$\delta_{\text{eff}} = \mu(C - \bar{C}) = 3 \times (2 - 0.000027184) = 5.999919 \text{ patients/hour}$$

$$L_s = L_q + \frac{\delta_{\text{eff}}}{\mu} = 6.5 + \frac{5.999919}{3} = 8.5 \text{ patients}$$

(d) The average waiting time per patient in the queue, W_q is computed as shown below:

$$W_q = \frac{L_q}{\delta_{\text{eff}}} = \frac{6.5}{5.999919} = 1.083356 \text{ hours}$$

(e) The average waiting time per patient in the system, W_s is:

$$W_s = \frac{L_s}{\delta_{\text{eff}}} = \frac{8.5}{5.999919} = 1.416689 \text{ hours}$$

Example 9.17 In a machine shop, there are 3 tool grinding machines. The shop consists of 8 lathes. The lathe operators come to these grinding machines for tool grinding as and when necessary. The arrival rate of the lathe operators for tool grinding is 3 per hour. The service rate of individual grinding machine is 1.5 operators per hour. The arrival rate as well as service rate follow Poisson distribution. Find the following:

(a) The probability that all the grinding machines will be free (p_0).
(b) Average waiting number of lathe operators in the queue.
(c) Average waiting number of lathe operators in the system.
(d) Average waiting time per lathe operator in the queue.
(e) Average waiting time per lathe operator in the system.

Solution Arrival rate, $\delta = 3$ lathe operators per hour. Service rate of individual grinding machine, $\mu = 1.5$ lathe operators per hour. The total number of tool grinding machines, $C = 3$. The size of the calling population (N) is equal to the number of lathe operators which is 8 and hence, the number of waiting spaces (N) is 8. So, this problem comes under $(M/M/C) : (GD/N/N)$. Therefore, the utilization factor, ϕ is

$$\phi = \frac{\delta}{\mu} = \frac{3}{1.5} = 2$$

(a) The probability that all the tool grinding machines will be free is equal to p_0 which is as computed below:

$$p_0 = \left(\sum_{n=0}^{C} NC_n \, \phi^n + \sum_{n=C+1}^{N} NC_n \, \frac{n! \, \phi^n}{C! \, C^{n-C}} \right)^{-1}$$

$$= \left(\sum_{n=0}^{3} 8C_n \, 2^n + \sum_{n=4}^{8} 8C_n \, \frac{n! \times 2^n}{3! \, 3^{n-3}} \right)^{-1}$$

$$= 0.00003153$$

(b) The average waiting number of lathe operators in the queue, L_q is:

$$L_q = \sum_{n=C+1}^{N} (n - C)p_n = 1 \times p_4 + 2 \times p_5 + 3 \times p_6 + 4 \times p_7 + 5 \times p_8$$

$$= 3.506952 \text{ lathe operators.}$$

(c) The average waiting number of lathe operators in the system, L_s is computed as shown below:

$$\bar{C} = \sum_{n=0}^{C} (C - n)p_n = 3 \times p_0 + 2 \times p_1 + 1 \times p_2 + 0 \times p_3 = 0.00463492 \text{ grinding machine}$$

$$\delta_{\text{eff}} = \mu(C - \bar{C}) = 1.5 \times (3 - 0.00463492) = 4.493048 \text{ lathe operators per hour}$$

$$L_s = L_q + \frac{\delta_{\text{eff}}}{\mu} = 3.506952 + \frac{4.493048}{1.5} = 6.502318 \text{ lathe operators}$$

(d) The average waiting time per lathe operator in the queue, W_q is computed as shown below:

$$W_q = \frac{L_q}{\delta_{\text{eff}}} = \frac{3.506952}{4.493048} = 0.7805286 \text{ hour}$$

(e) The average waiting time per lathe operator in the system, W_s is:

$$W_s = \frac{L_s}{\delta_{\text{eff}}} = \frac{6.502318}{4.493048} = 1.447195 \text{ hours.}$$

9.3.6 $(M/M/1):(GD/N/N)$ Model (for $N > 1$)

The parameters of this model are as follows:

(i) Arrival rate follows Poisson distribution.
(ii) Service rate follows Poisson distribution.
(iii) Number of servers is 1.
(iv) Service discipline is general discipline.
(v) Maximum number of customers permitted in the system is N.
(vi) Size of the calling source is N.

In this model, the following assumptions are made:

$$\delta_n = (N - n)\delta, \qquad 0 \leq n \leq N$$
$$= 0, \qquad n \geq N$$

and

$$\mu_n = n\mu, \qquad 0 \leq n \leq 1$$
$$= \mu, \qquad 1 \leq n \leq N$$
$$= 0, \qquad n \geq N$$

The steady-state formula to obtain the probability of having n customers in the system p_n, and the formulas for p_0, L_s, L_q, W_s and W_q are presented below:

$$p_n = {}^{N}C_n \, \phi^n \, p_0, \qquad 0 \leq n \leq C$$

$$= {}^{N}C_n \, \frac{n! \, \phi^n}{C! \, C^{n-C}} \, p_0, \quad C \leq n \leq N$$

Therefore,

$$p_0 = \left(\sum_{n=0}^{C} {}^{N}C_n \phi^n + \sum_{n=C+1}^{N} {}^{N}C_n \frac{n!\phi^n}{C!C^{n-C}} \right)^{-1}$$

The other results are:

$$L_q = N - \left(1 + \frac{1}{\phi}\right)(1 - p_0)$$

$$L_s = N - \frac{1 - p_0}{\phi}$$

$$W_q = \frac{L_q}{\delta_{eff}}$$

$$W_s = \frac{L_s}{\delta_{eff}}$$

where

$$\delta_{eff} = \mu(1 - p_0).$$

Example 9.18 In the machine shop of a small-scale industry, machines breakdown with a mean rate of 2 per hour. The maintenance shop of the industry has only one mechanic who can attend to the breakdown machines. The service rate of the mechanic is 1.5 machines per hour. Initially, there are 5 working machines in the machine shop. Find p_0, L_q, L_s, W_q and W_s.

Solution Breakdown rate of machines, $\delta = 2$ per hour. Service rate of individual mechanic, $\mu = 1.5$ machines per hour. Total number of machines in the shop, $N = 5$. Therefore, the utilization factor is:

$$\phi = \frac{\delta}{\mu} = \frac{2}{1.5} = 1.33$$

The other results are as follows:

$$p_0 = \left(\sum_{n=0}^{C} {}^{N}C_n \phi^n + \sum_{n=C+1}^{N} {}^{N}C_n \frac{n!\phi^n}{C!C^{n-C}} \right)^{-1}$$

$$= \left(\sum_{n=0}^{1} {}^{5}C_n 1.33^n + \sum_{n=1+1}^{5} {}^{5}C_n \frac{n! \, 1.33^n}{1! \, 1^{n-1}} \right)^{-1}$$

$$= 0.000944$$

$$L_q = N - \left(1 + \frac{1}{\phi}\right)(1 - p_0) = 5 - \left(1 + \frac{1}{1.33}\right)(1 - 0.000944) = 3.25 \text{ machines}$$

$$L_s = N - \frac{1 - p_0}{\phi} = 5 - \frac{1 - 0.000944}{1.33} = 4.25 \text{ machines}$$

$$\delta_{eff} = \mu(1 - p_0) = 1.5(1 - 0.000944) = 1.4986 \text{ machines per hour}$$

$$W_q = \frac{L_q}{\delta_{eff}} = \frac{3.25}{1.4986} = 2.169 \text{ hours}$$

$$W_s = \frac{L_s}{\delta_{eff}} = \frac{4.25}{1.4986} = 2.836 \text{ hours.}$$

Example 9.19 *(Only for classroom discussion or assignment purpose)*

In a textile mill, there are 25 looms. The breakdown rate of these looms is 4 looms per hour. There is only one crew to attend to the breakdowns of these looms. The service rate of the maintenance crew is 3 looms per hour. Both the arrival rate and service rate follow Poisson distribution. Find p_0, L_q, L_s, W_q and W_s.

Solution Breakdown rate of looms, $\delta = 4$ looms per hour and the service rate of the maintenance crew, $\mu = 3$ looms per hour. The number of maintenance crew, C is 1. The total number looms in the mill (size of the calling population), $N = 25$ and hence, the number of waiting spaces (N) is 25. So, this problem comes under $(M/M/1) : (GD/N/N)$. Therefore, the utilization factor, ϕ is

$$\phi = \frac{\delta}{\mu} = \frac{4}{3} = 1.333333$$

$$p_0 = \left(\sum_{n=0}^{C} NC_n \, \phi^n + \sum_{n=C+1}^{N} NC_n \, \frac{n! \phi^n}{C! C^{n-C}} \right)^{-1}$$

$$= \left(\sum_{n=0}^{1} 25C_n \, 1.333333^n + \sum_{n=2}^{25} 25C_n \, \frac{n! \times 1.333333^n}{1! 1^{n-1}} \right)^{-1} = 2.291736 \times 10^{-29}$$

$$L_q = N - \left(1 + \frac{1}{\phi}\right)(1 - p_0) = 25 - \left(1 + \frac{1}{1.333333}\right) \times (1 - 2.291736 \times 10^{-29}) = 23.25 \text{ looms}$$

$$L_s = N - \frac{1 - p_0}{\phi} = 25 - \frac{1 - 2.291736 \times 10^{-29}}{1.333333} = 24.25 \text{ looms}$$

$$\delta_{eff} = \mu(1 - p_0) = 3 \times (1 - 2.291736 \times 10^{-29}) = 3 \text{ looms per hour}$$

$$W_q = \frac{L_q}{\delta_{eff}} = \frac{23.25}{3} = 7.75 \text{ hours}$$

$$W_s = \frac{L_s}{\delta_{eff}} = \frac{24.25}{3} = 8.083 \text{ hours.}$$

Example 9.20 A research investigator has 6 research associates working under him. The arrival rate of research associates to the room of the research investigator for discussion is 2.5 research associates per hour and the service rate of the research investigator is 1.5 research associates per hour. Both the arrival rate and the service rate follows poisson distribution. Find the following:

(a) The probability that the research investigator is free (p_0).
(b) Average waiting number of research associates in the queue.
(c) Average waiting number of research associates in the system.
(d) Average waiting time per research associate in the queue.
(e) Average waiting time per research associate in the system.

Solution Arrival rate, $\delta = 2.5$ research associates per hour and the service rate of the research investigator, $\mu = 1.5$ research associates per hour. The number of research associates is the size of the calling population (N) which is 6 and hence, the number of waiting spaces (N) is also 6. So, this problem comes under $(M/M/1) : (GD/N/N)$. Therefore, the utilization factor, φ is:

$$\phi = \frac{\delta}{\mu} = \frac{2.5}{1.5} = 1.66667$$

(a) The probability that the research investigator is free is equal to p_0 which is as computed below:

$$p_0 = \left(\sum_{n=0}^{C} NC_n \, \phi^n + \sum_{n=C+1}^{N} NC_n \, \frac{n! \, \phi^n}{C! \, C^{n-C}} \right)^{-1}$$

$$= \left(\sum_{n=0}^{1} 6C_n \, 1.66667^n + \sum_{n=2}^{6} 6C_n \, \frac{n! \times 1.66667^n}{1! \, 1^{n-1}} \right)^{-1} = 0.000035563$$

(b) The average waiting number of research associates in the queue, L_q is:

$$L_q = N - \left(1 + \frac{1}{\phi} \right)(1 - p_0) = 6 - \left(1 + \frac{1}{1.66667} \right) \times (1 - 0.000035563) = 4.4 \text{ research associates}$$

(c) The average waiting number of research associates in the system, L_s is:

$$L_s = N - \frac{1 - p_0}{\phi} = 6 - \frac{1 - 0.000035563}{1.66667} = 5.4 \text{ research associates}$$

(d) The average waiting time per research associate in the queue, W_q is computed as shown below:

$$\delta_{\text{eff}} = \mu(1 - p_0) = 1.5 \times (1 - 0.000035563) = 1.499947 \text{ research associates per hour}$$

$$W_q = \frac{L_q}{\delta_{\text{eff}}} = \frac{4.4}{1.499947} = 2.933476 \text{ hours}$$

(e) The average waiting time per research associate in the system, W_s is:

$$W_s = \frac{L_s}{\delta_{\text{eff}}} = \frac{5.4}{1.499947} = 3.6 \text{ hours.}$$

9.4 SIMULATION

Simulation is an experiment conducted on a model of some system to collect necessary information on the behaviour of that system. Some of the systems on which the simulation experiment can be carried out are: reservation counter of a transport corporation, bank counters, inspection station of an assembly line, automobile assembly line, computer network, toll gate, and berths in a harbour.

9.4.1 Need for Simulation

Consider an example of the queueing system, namely the reservation system of a transport corporation. The elements of the system are booking counter (server) and waiting customers (queue). Generally, the arrival rate of the customers follows Poisson distribution and the service time follows exponential distribution. If the reservation system has the above combination of distributions for its elements, then the queueing model, namely $(M/M/1){:}(GD/\infty/\infty)$ can be used to find the standard results. But in reality, the following combinations of distributions may exist.

1. Arrival rate does not follow Poisson distribution but the service time follows exponential distribution.

2. Arrival rate follows Poisson distribution but the service time does not follow exponential distribution.
3. Arrival rate does not follow Poisson distribution and the service time also does not follow exponential distribution.

In each of the above cases, the non-standard distribution for the queueing system may be discrete distribution, uniform distribution, normal distribution, etc. Under such situation, the standard model $(M/M/1){:}(GD/\infty/\infty)$ cannot be used. The last resort to find the solution for the non-standard queueing model is to use simulation. Similarly, if there is a deviation from other standard models of queueing system, one can use simulation. In practice, a system with any kind of assumptions can be tackled using simulation.

To quote a few of them, the variations of the queueing system—bulk arrival, jockeying, balking, reneging, priority are provided. Under these variations, simulation is the only method to find solution for a given problem.

Model. *Model* is an abstraction of a reality and it is used for experimentation. For example, the maintenance shop of a manufacturing industry; the shop can be modelled as a queueing system with maintenance mechanics as servers and the breakdown machines as customers. Later, this model can be experimented to obtain the standard results as:

- Percentage utilization of servers,
- Average waiting number of breakdown machines,
- Average waiting time per breakdown machine.

9.4.2 Types of Simulation

There are four types of simulation, discussed as follows:

Identity simulation. Observance of the behaviour of the real system under as many operational configurations as possible to get an insight into system behaviour is called *identity simulation*. But the identity simulation is very expensive, seldom feasible, time consuming and permits little control over parameters that affect the response of the system.

Quasi-identity simulation. In quasi-identity simulation, aspects of the real-world system are preserved while some elements, whose presence would make the identity simulation impossible, are excluded. Simulation of an air attack over a country by its enemy aircraft is an example of quasi-identity simulation.

Laboratory simulation. This is a cheaper method than the other two methods described earlier. Consider an example of an operational gaming in which the business system is represented by its entities such as production, personnel, finance, marketing, logistics and its players (executives) within its competing environment. In this type of simulation, transactions are carried over a specified period and then the operating performance of the business is reviewed.

Computer simulation. If we remove people from the laboratory simulation with a well designed computer logic and retain other aspects, then it is called as 'computer simulation'.

The following are the advantages of simulation.

1. The time for experimentation can be compressed.
2. The system performance can be studied under all possible conditions.
3. The success or failure of a system can be tested using a conceptual system.

Since the use of simulation is a costly exercise, it should be used as the last resort.

9.4.3   Major Steps of Simulation

The major steps of simulation are presented in Figure 9.2 in the form of a flow diagram.

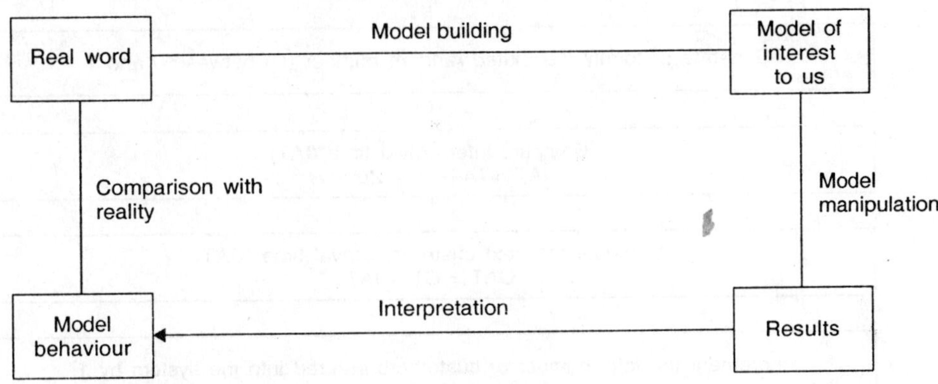

Figure 9.2   Major steps of simulation experiment.

9.4.4   Simulation using High-level Languages

There are many specialized simulation languages like GPSS, SIMSCRIPT, GASP, etc. In this chapter, a brief account of GPSS is presented. Before discussing GPSS, guidelines for simulation using high-level languages (like Fortran, C, Basic, etc.) are presented in the next section which help the readers to have better grip over the simulation concept.

Simulation method can be classified into *discrete-event simulation* and *continuous simulation*. Discrete model is applied when changes occur in a system at given instants. For example, the average waiting time for a customer in a queue changes when somebody leaves or enters the queue. These entry and exit occur at discrete points, hence the name discrete-event simulation.

The following terminologies of the single-server model are explained first to understand the concept of discrete-event simulation.

- Next customer arrival time (CAT)
- Service completion time (SCT)
- End of simulation (ES). (This is also known as *simulation run time* of the simulation.)
- Current time/clock time (CT)
- Next most imminent event (NMIE)

At any point of time, the CAT, SCT and ES will have some specific values. In discrete-event simulation, the progress of simulation experiment is made by updating CT to the time of next most imminent event which is the minimum of the values of CAT, SCT and ES.

In continuous simulation, the current time (CT) is incremented by an equal amount. The equal increment in the CT is decided based on the sensitivity of the system performance.

Single-server queueing system.   A flow chart for the single server queueing system is presented in Figure 9.3.

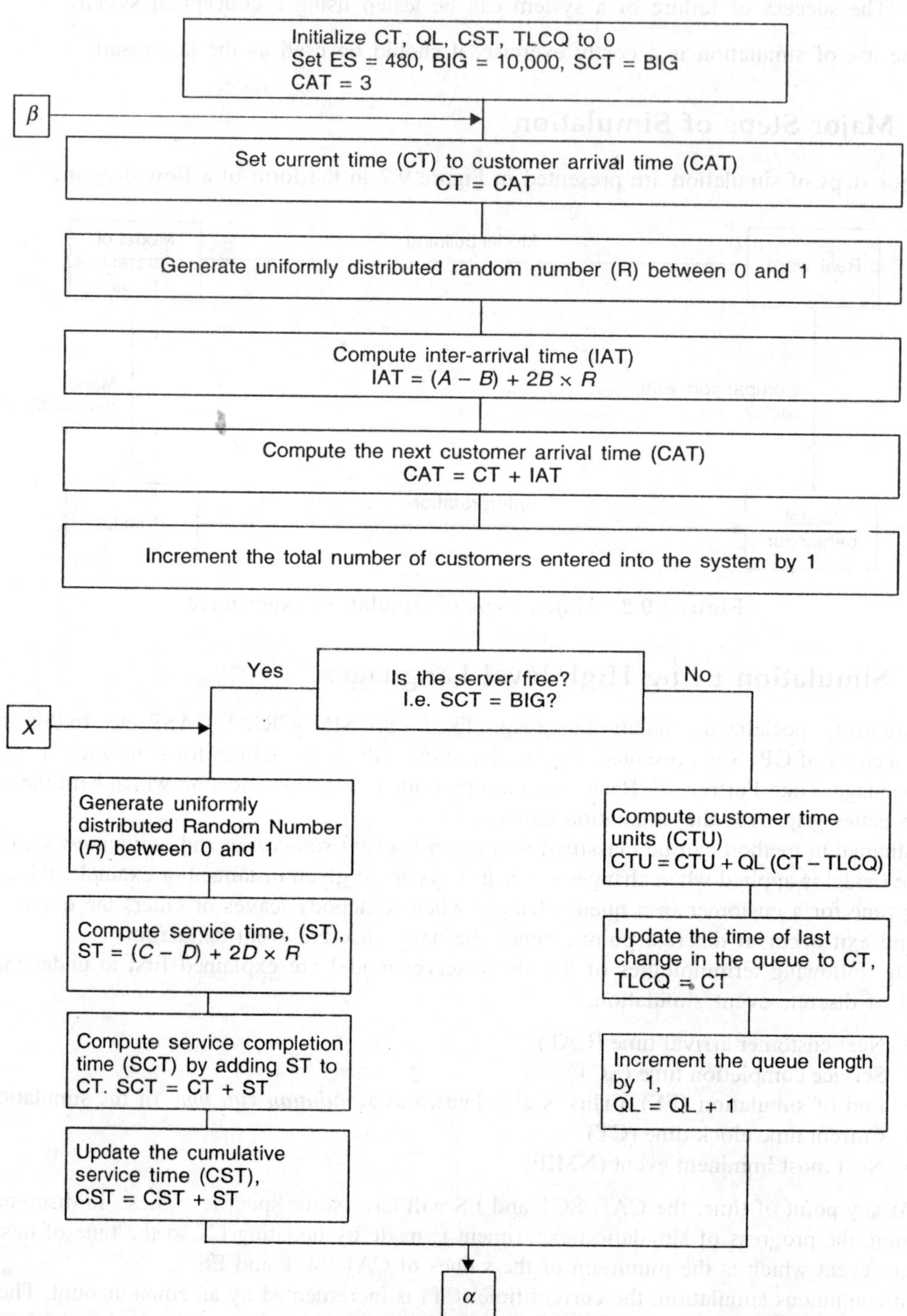

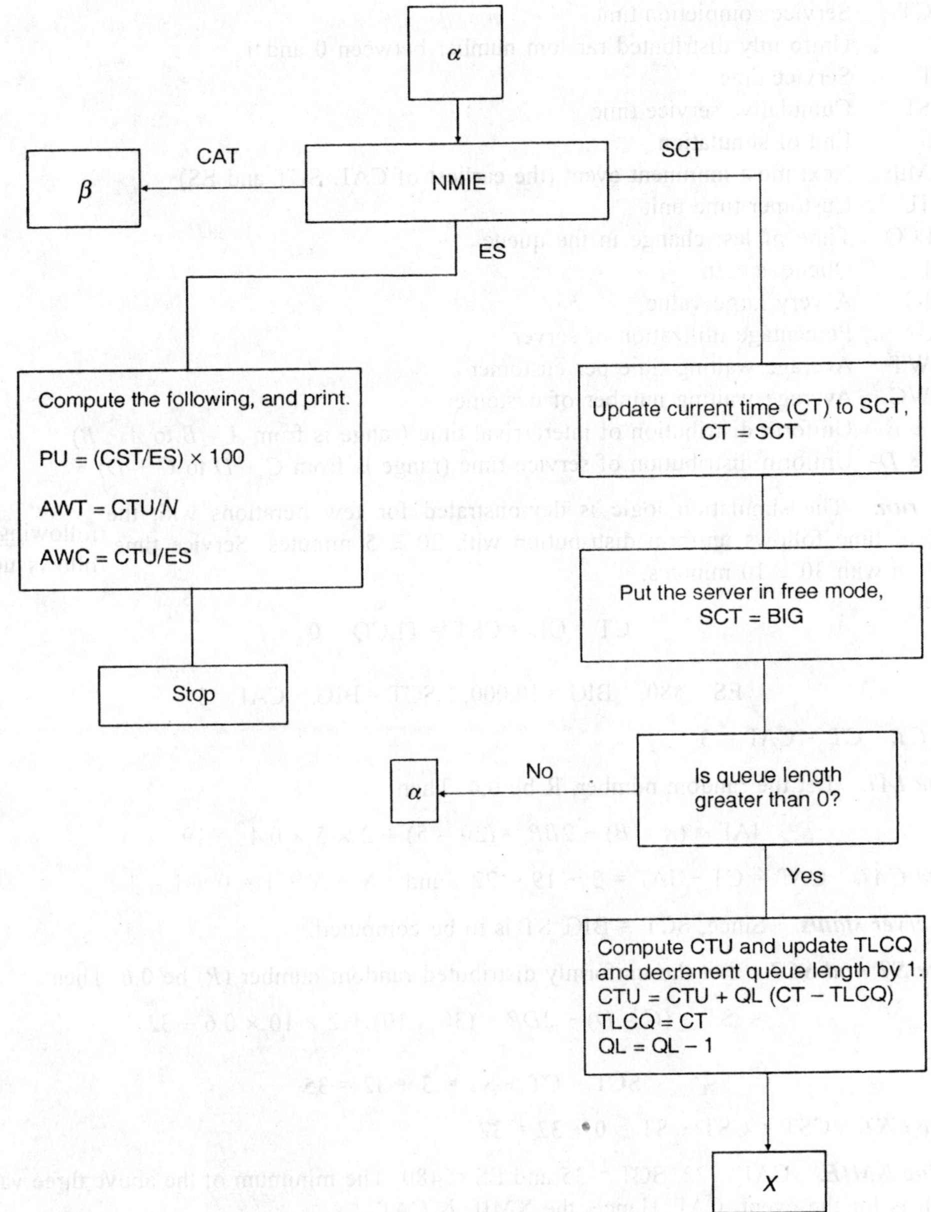

Figure 9.3 Flow chart for single-server model.

The terminologies used in the figure are listed below:

CT   Current time
CAT   Customer arrival time
IAT   Inter-arrival time
N   Total number of customers entered into the system

SCT — Service completion time

R — Uniformly distributed random number between 0 and ·1

ST — Service time

CST — Cumulative service time

ES — End of simulation

NMIE — Next most imminent event (the earliest of CAT, SCT and ES)

CTU — Customer time unit

TLCQ — Time of last change in the queue

QL — Queue length

BIG — A very large value

PU — Percentage utilization of server

AWT — Average waiting time per customer

AWC — Average waiting number of customer

$A \pm B$ — Uniform distribution of interarrival time (range is from $A - B$ to $A + B$)

$C \pm D$ — Uniform distribution of service time (range is from $C - D$ to $C + D$).

Sample run. The simulation logic is demonstrated for few iterations with the following data. Interarrival time follows uniform distribution with 20 ± 5 minutes. Service time follows uniform distribution with 30 ± 10 minutes.

Set

$$CT = QL = CST = TLCQ = 0$$

and

$$ES = 480, \quad BIG = 10{,}000, \quad SCT = BIG, \quad CAT = 3$$

Update CT.   CT = CAT = 3

Compute IAT.   Let the random number, R be 0.4. Then

$$IAT = (A - B) + 2BR = (20 - 5) + 2 \times 5 \times 0.4 = 19$$

Compute CAT.   CAT = CT + IAT = 3 + 19 = 22   and $N = N + 1 = 0 + 1 = 1$

Check server status.   Since, SCT = BIG, ST is to be computed.

Compute ST and SCT.   Let the uniformly distributed random number (R) be 0.6. Then

$$ST = (C - D) + 2DR = (30 - 10) + 2 \times 10 \times 0.6 = 32$$

and

$$SCT = CT + ST = 3 + 32 = 35$$

Compute CST.   CST = CST + ST = 0 + 32 = 32

Determine NMIE.   CAT = 22, SCT = 35 and ES = 480. The minimum of the above three values is 22 which is for the event, CAT. Hence, the NMIE is CAT.

Update CT.   CT = CAT = 22.

Compute CAT.   Let, $R = 0.99$. Then

$$IAT = (A - B) + 2BR = (20 - 5) + 2 \times 5 \times 0.99 = 25 \text{ (approx.)}$$
$$CAT = CT + IAT = 22 + 25 = 47$$
$$N = N + 1 = 1 + 1 = 2$$

Check server status.   Since SCT is not equal to BIG, add the new customer in the queue.

Add customer in the queue. We get

$$CTU = CTU + QL(CT - TLCQ) = 0 + 0(22 - 0) = 0$$
$$TLCQ = CT = 22$$
$$QL = QL + 1 = 0 + 1 = 1$$

Determine NMIE. CAT = 47, SCT = 35 and ES = 480. The minimum of the above three values is 35 which is for the event, SCT. Hence, the NMIE is SCT.

Update CT. CT = SCT = 35

Set server in free mode. SCT = BIG

Check queue content. Since, QL > 0, fetch the first customer from the queue for service.

Updating queue details and picking customer from queue. We get

$$CTU = CTU + QL(CT - TLCQ) = 0 + 1(35 - 22) = 13$$
$$TLCQ = 35$$
$$QL = QL - 1 = 0$$

Compute SCT. Let, $R = 0.3$. Then

$$ST = (C - D) + 2DR = (30 - 10) + 2 \times 10 \times 0.3 = 26$$
$$SCT = CT + ST = 35 + 26 = 61$$

Compute CST. CST = CST + ST = 32 + 26 = 58.

Determine NMIE. CAT = 47, SCT = 61 and ES = 480. The minimum of the three values is 47 which is for the event, CAT. Hence, the NMIE is CAT.

The above results along with some more calculations are summarized in Table 9.3. Similarly, the steps are to be carried out till the NMIE is ES. Finally, the results, PU, AWT and AWC can be computed using the given formula.

Now, the concept of CTU calculation is demonstrated in Figure 9.4 with the help of Table 9.3. The area of each rectangle in Figure 9.4 is given within itself. The sum of the rectangles in Figure 9.4 gives the total value for CTU. From this value, the AWT and AWC can be calculated by using the formulae:

$$AWT = \frac{CTU}{N} \quad \text{and} \quad AWC = \frac{CTU}{ES}$$

Figure 9.4 Schematic view of customer-time-unit (CTU) based on Table 9.3.

Table 9.3 Summary of Calculation for Single Server Model

Iteration	CT	R	IAT	CAT	N	SCT?	R	ST	SCT	CST	CTU	TLCQ	QL	NMIE	CT	SCT	QL > 0?
1	3	0.4	19	22	1	BIG	0.6	32	35	32	–	–	–	CAT = 22* SCT = 35 ES = 480	–	–	–
2	22	0.99	25	47	2	35	–	–	–	–	0	22	1	CAT = 47 SCT = 35* ES = 480	35	BIG	YES
											13	35	0				
3	35	–	–	–	–	BIG	0.3	26	61	58	–	–	–	CAT = 47* SCT = 61 ES = 480	–	–	–
4	47	0.7	22	69	3	61	–	–	–	–	13	47	1	CAT = 69 SCT = 61* ES = 480	61	BIG	YES
											27	61	0				
5	61	–	–	–	–	BIG	0.8	36	97	94	–	–	–	CAT = 69* SCT = 97 ES = 480	–	–	–
6	69	0.9	24	93	4	97	–	–	–	–	27	69	1	CAT = 93* SCT = 97 ES = 480	–	–	–
7	93	0.99	25	118	5	97	–	–	–	–	51	93	2	CAT = 118 SCT = 97* ES = 480	97	BIG	YES
											59	97	1				
8	97	–	–	–	–	BIG	0.2	24	121	118	–	–	–	CAT = 118* SCT = 121 ES = 480	–	–	–
9	118	0.5	20	138	6	121	–	–	–	–	80	118	2	CAT = 138 SCT = 121* ES = 480	121	BIG	YES
											86	121	1				
10	121	–	–	–	–	BIG	0.7	34	155	152	–	–	–	CAT = 138* SCT = 155 ES = 480	–	–	–

Single-stage parallel servers queueing system. A logical flow chart for the single stage parallel servers queueing system is presented in Figure 9.5 and the corresponding programming flow chart is shown in Figure 9.6. The terminologies used in these figures are listed below:

C	Number of parallel servers
CT	Current time

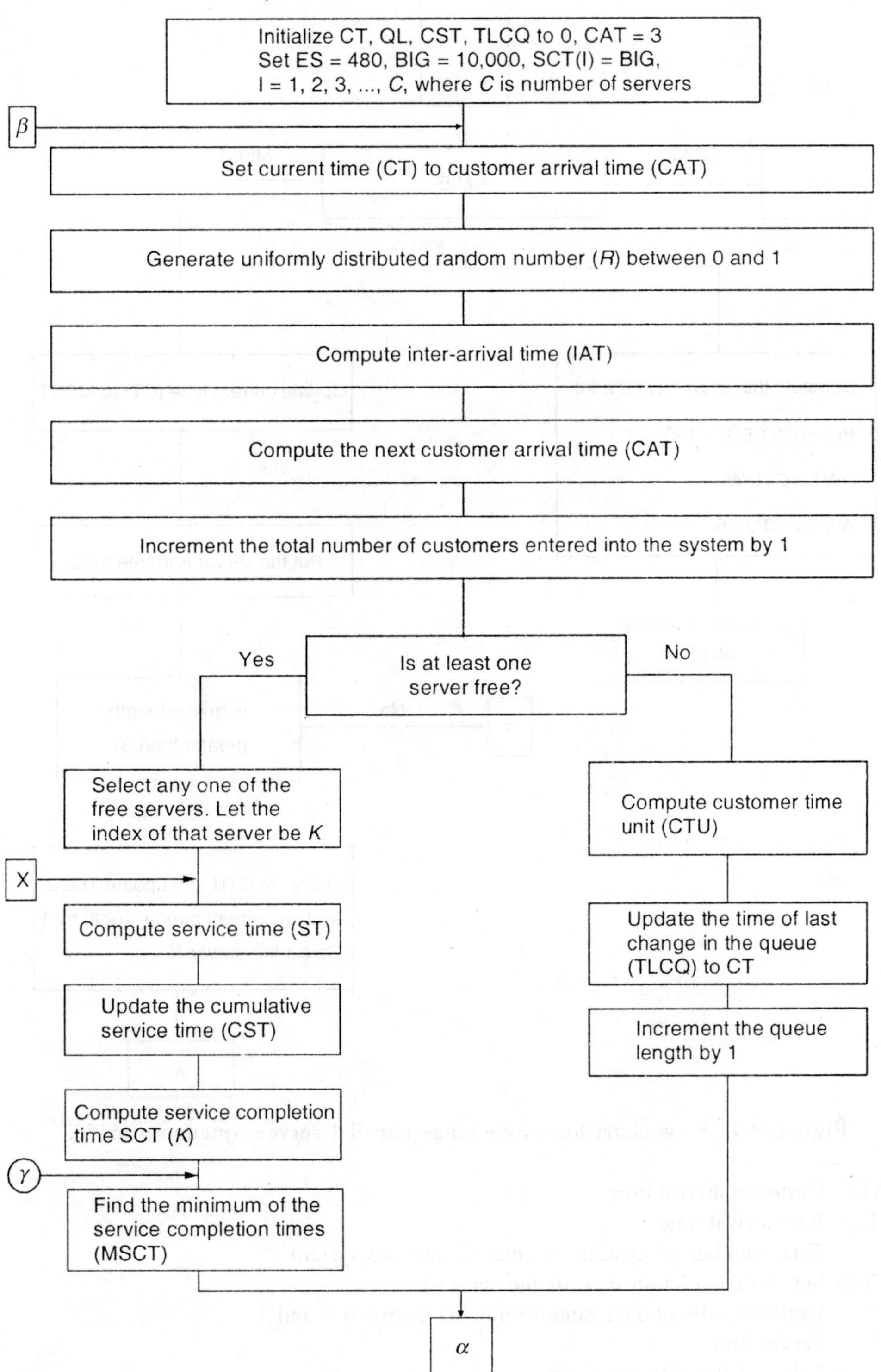
Initialize CT, QL, CST, TLCQ to 0, CAT = 3
Set ES = 480, BIG = 10,000, SCT(I) = BIG,
I = 1, 2, 3, ..., C, where C is number of servers
β
Set current time (CT) to customer arrival time (CAT)
Generate uniformly distributed random number (R) between 0 and 1
Compute inter-arrival time (IAT)
Compute the next customer arrival time (CAT)
Increment the total number of customers entered into the system by 1
Yes
Is at least one server free?
No
Select any one of the free servers. Let the index of that server be K
X
Compute customer time unit (CTU)
Compute service time (ST)
Update the time of last change in the queue (TLCQ) to CT
Update the cumulative service time (CST)
Increment the queue length by 1
Compute service completion time SCT (K)
γ
Find the minimum of the service completion times (MSCT)
α

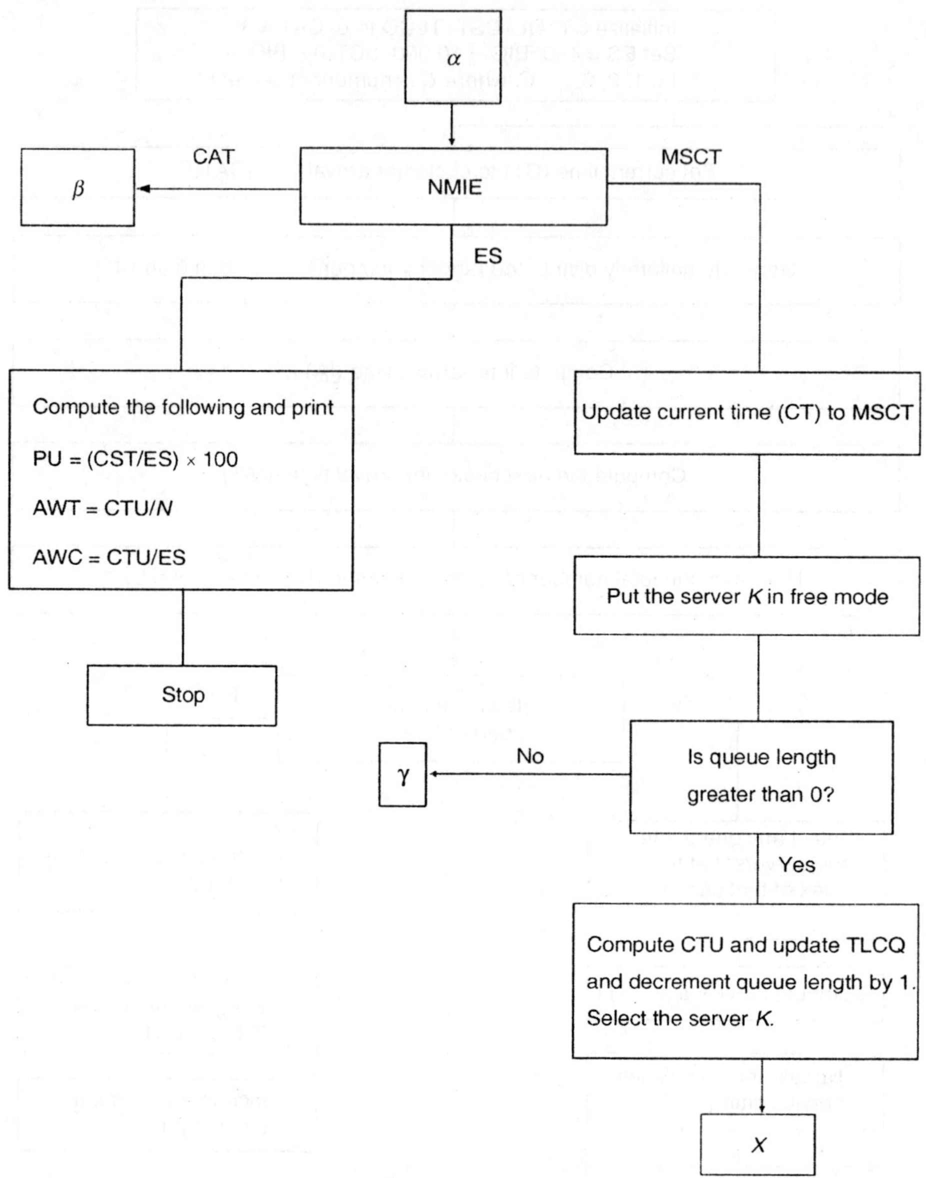

Figure 9.5 Flow chart for single-stage parallel servers queueing model.

CAT   Customer arrival time
IAT   Inter-arrival time
N Total number of customers entered into the system
SCT(I) Service completion time of the server I
R Uniformly distributed random number between 0 and 1
ST   Service time
K Index of the selected server

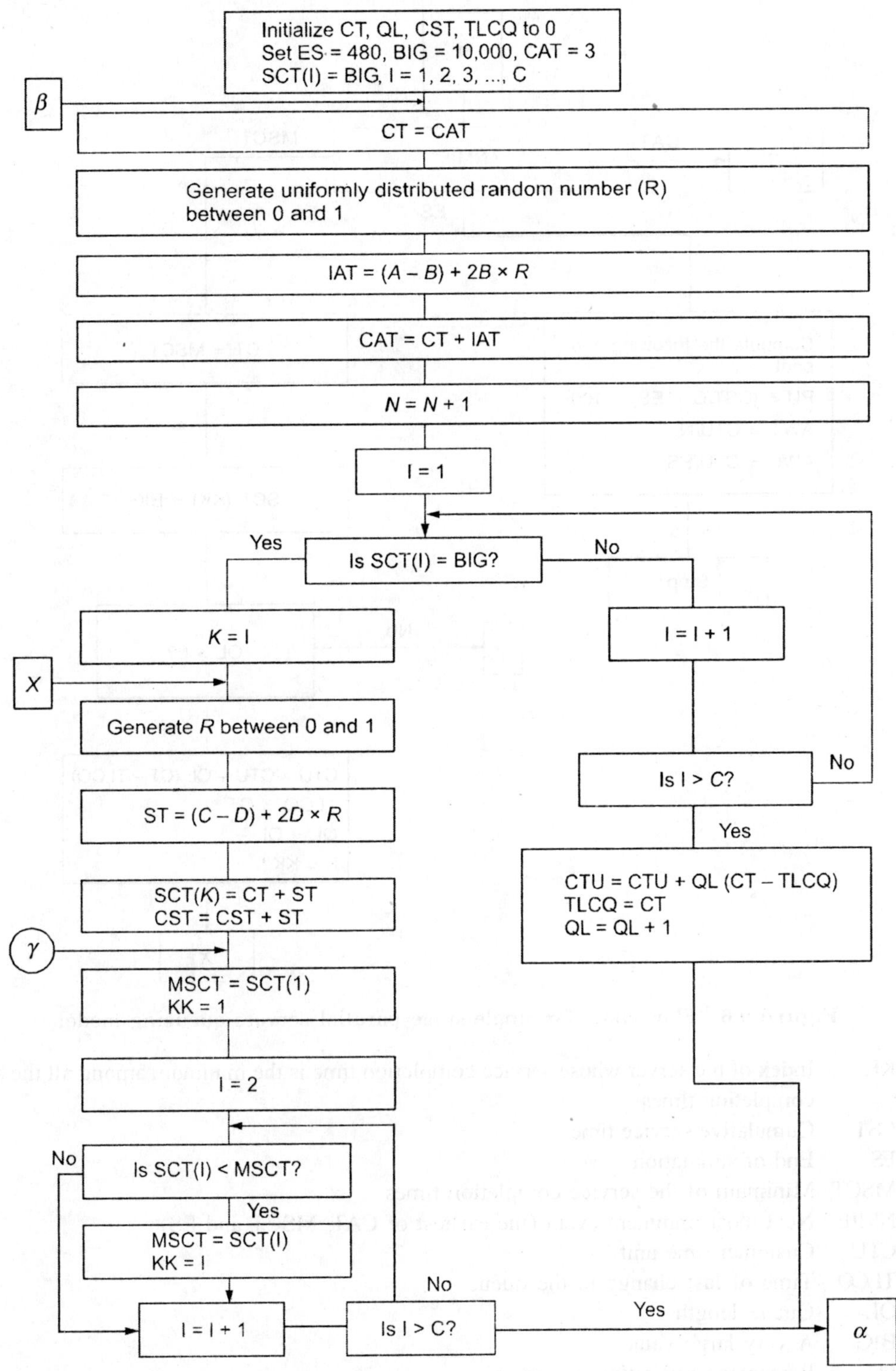

Initialize CT, QL, CST, TLCQ to 0
Set ES = 480, BIG = 10,000, CAT = 3
SCT(I) = BIG, I = 1, 2, 3, ..., C
β
CT = CAT
Generate uniformly distributed random number (R) between 0 and 1
IAT = (A − B) + 2B × R
CAT = CT + IAT
N = N + 1
I = 1
Is SCT(I) = BIG?
Yes
No
K = I
I = I + 1
X
Generate R between 0 and 1
Is I > C?
No
ST = (C − D) + 2D × R
Yes
SCT(K) = CT + ST
CST = CST + ST
CTU = CTU + QL (CT − TLCQ)
TLCQ = CT
QL = QL + 1
γ
MSCT = SCT(1)
KK = 1
I = 2
Is SCT(I) < MSCT?
No
Yes
MSCT = SCT(I)
KK = I
I = I + 1
Is I > C?
No
Yes
α

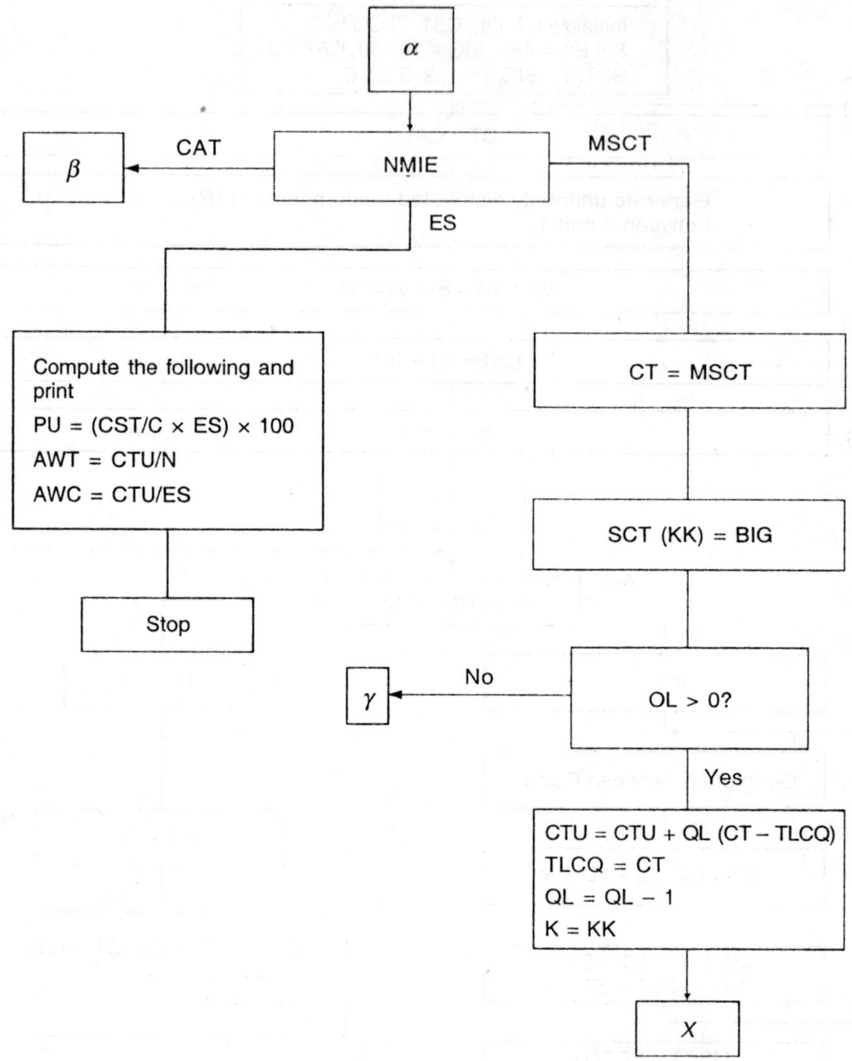

Figure 9.6 Flow chart for single-stage parallel servers queueing model.

KK	Index of the server whose service completion time is the minimum among all the service completion times
CST	Cumulative service time
ES	End of simulation
MSCT	Minimum of the service completion times
NMIE	Next most imminent event (the earliest of CAT, MSCT and ES)
CTU	Customer time unit
TLCQ	Time of last change in the queue
QL	Queue length
BIG	A very large value
PU	Percentage utilization of server

AWT Average waiting time per customer
AWC Average waiting number of customer
$A \pm B$ Uniform distribution of inter-arrival time (range is from $A - B$ to $A + B$)
$C \pm D$ Uniform distribution of service time (range is from $C - D$ to $C + D$).

9.4.5 General Purpose Simulation System (GPSS)

GPSS is a simulation language mainly to deal with non-standard queueing system. In reality, one may come across many sub-systems of business organizations or public systems, which can be modelled as queueing systems.

In the previous section, an introduction and guidelines were presented for simulating a single-server queueing model and parallel-server queueing model using a high-level language. But the difficulty of using high-level languages is that they take longer development time for many real-life systems. Hence, a simplified approach to simulation using GPSS is presented.

Based on the flow charts presented in the previous section, one can visualize the fact that the queueing simulation logic consists of some standard modules. GPSS simulation language is aimed at providing subroutines for each and every module of any queueing system. These subroutines are called as *blocks* in GPSS. The generalized sequence of items in each block is as represented below.

Block number Location Operation Operands (A, B, C, D, E)

A sample set of basic blocks is listed in Table 9.4.

Table 9.4 Basic Set of GPSS Blocks

Name of the Block	Purpose
GENERATE	It introduces customers (transactions) into the system (i.e. computation of CAT).
QUEUE and DEPART	The QUEUE block adds customers into the queue and DEPART block removes customers from the queue, and the usage of these two Blocks updates the following: CTU, TLCQ and QL.
SEIZE and RELEASE	These two blocks pertain to single-server model. The SEIZE block checks the status of the server and if the server is free, the incoming customer or the first customer in the queue or the customer with higher priority is selected for service. The RELEASE block makes the server free as soon as the service on the present customer is over. These two blocks jointly update the following: ST, SCT and CST.
STORAGE, ENTER and LEAVE	These three blocks are jointly used under parallel servers situation. The STORAGE block defines the number of servers (C). The ENTER block is equivalent to the SEIZE block in single-server model, and the LEAVE block is equivalent to the RELEASE block in the single-server model. These blocks will update the following: ST, SCT(I), CST and MSCT.
TERMINATE	This block removes the customers (transactions) from the system.
TRANSFER Blocks	There are different types of TRANSFER blocks. These blocks are used to transfer transactions (customers) to non-sequential blocks based on certain conditions.

The blocks listed in the table are being discussed now.

GENERATE block. This block can be thought of a door through which the customers (transactions) are allowed into the system. A GPSS program can have any number of GENERATE blocks. The symbol for the GENERATE block is as shown in Figure 9.7. The syntax for the GENERATE block is presented below:

GENERATE A, B, C, D, E

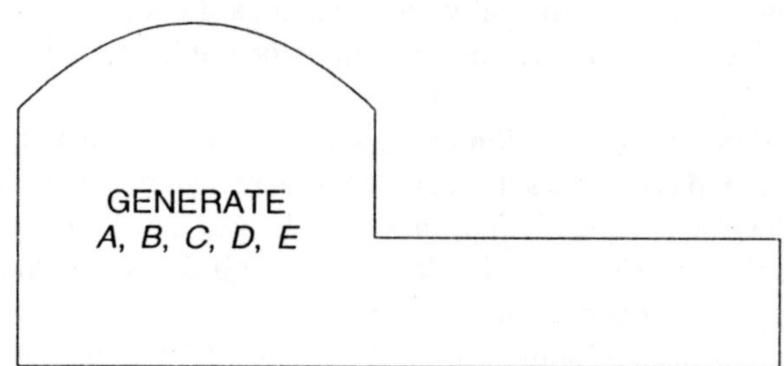

Figure 9.7   Symbol for GENERATE block.

This block consists of five operands, namely A, B, C, D and E. The details of these operands are presented in Table 9.5.

Table 9.5   Details of Operands of GENERATE Block

Operand	Significance	Default value
A	Average inter-arrival time	Zero
B	Half of the range over which the inter-arrival time is uniformly distributed	Zero
C	Offset interval	This operand is not used
D	Limit count (maximum number of transactions allowed into the system)	Infinite
E	Priority level (higher the number, higher the priority)	Zero

The usage of the GENERATE block is illustrated with some sample situations as in Table 9.6.

Table 9.6   Illustrations for GENERATE Block

Example	Explanation
GENERATE 40, 5, 7, 500, 1	Mean inter-arrival time = 40 Half of the range = 5 Off-set interval = 7 Maximum number of transactions allowed = 500 Priority of the customers = 1
GENERATE 30, 8, , , 2	Mean inter-arrival time = 30 Half of the range = 8 Off-set interval = 0 Maximum number of transactions allowed = infinite Priority of the customers = 2

Table 9.6   Illustrations for GENERATE Block (*Contd.*)

GENERATE 40, 10	Mean inter-arrival time = 40 Half of the range = 10 Off-set interval = 0 Maximum number of transactions allowed = infinite Priority of the customers = 0
GENERATE 40, , 7, , 1	Mean inter-arrival time = 40 Half of the range = 0 Off-set interval = 7 Maximum number of transactions allowed = infinite Priority of the customers = 1

QUEUE and DEPART blocks.   QUEUE block allows the transaction that is entering in it. Later, it checks the server status. If the server is busy, the transaction is held in this block itself; otherwise, the DEPART block moves the transaction into the next SEIZE/ENTER block. In this process, the necessary data for computing the statistics on the queue (e.g. CTU, TLCQ, QL, etc.) will be updated.

The symbol for QUEUE and DEPART blocks are shown in Figures 9.8(a) and 9.8(b), respectively. The details of the operands of the above two blocks are summarized in Table 9.7.

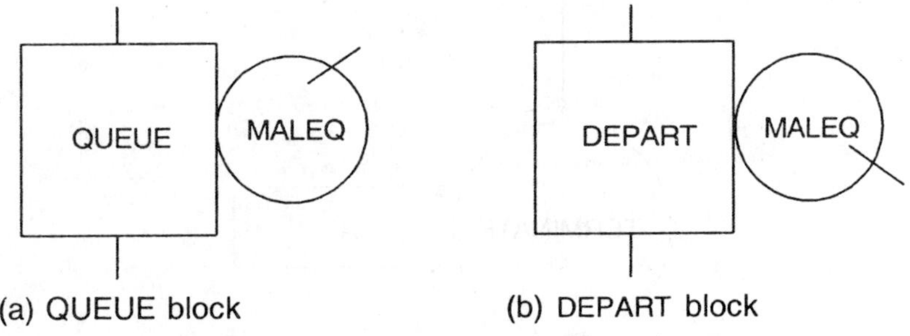

Figure 9.8 Symbols for QUEUE and DEPART blocks.

Table 9.7 Details of Operands of QUEUE and DEPART Blocks

Block	Operand	Significance	Default value
QUEUE	A	Name of the queue	Error
	B	Size of the bulk arrival (quantum of increase in the queue content) Syntax: QUEUE A, B	1
DEPART	A	Name of the queue	Error
	B	Size of the bulk of transactions moving out of the queue (quantum of decrease in the queue content). Syntax: DEPART A, B	1

The usages of the QUEUE and DEPART blocks are illustrated with some sample situations as in Table 9.8.

Table 9.8 Illustrations for QUEUE and DEPART Blocks

Example	Explanation
QUEUE MALEQ, 3	Name of the queue = MALEQ Increment in the queue length = 3
QUEUE FMALQ	Name of the queue = FMALQ Increment in the queue length = 1
DEPART MALEQ, 2	Name of the queue = MALEQ Decrement in the queue length = 2
DEPART FMALQ	Name of the queue = FMALQ Decrement in the queue length = 1

TERMINATE block.   Whenever a transaction moves into a TERMINATE block, it is removed from the system. The TERMINATE block always accepts incoming transactions. A GPSS program can have any number of TERMINATE blocks.

The symbol which is used to represent the TERMINATE block is shown in Figure 9.9.

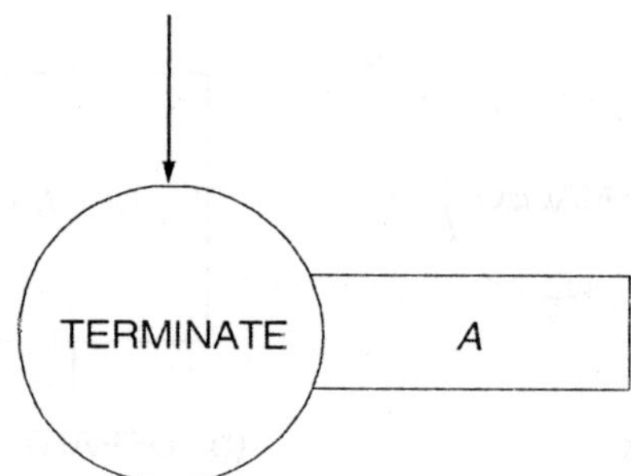

Figure 9.9 Symbol of TERMINATE block.

The TERMINATE block has only 'A' operand and it represents the value by which the termination counter is to be decreased whenever a transaction passes through it. Also, this is called as *termination counter decrement*. The default value of this operand is 1. The syntax of this block is:

TERMINATE *A*

In addition to removing transactions from the model, the TERMINATION counter helps control the simulation run time with the help of START command as shown below. The following GPSS programming segment is known as *timer segment*. The START command is to initiate the simulation run and its argument (operand) carries the initial value of the termination counter. The simulation run will be stopped whenever the value of this counter is reduced to zero.

GENERATE 480

TERMINATE 1

START 1

As per the above timer segment, at exactly 480 minutes, a transaction will be introduced into the model. When it passes through the TERMINATE block, the termination counter (argument of START) will be decreased by the value of the TERMINATE block (1). As a result, the termination counter value (argument of START) will be reduced to zero and hence the simulation will be shutdown at 480 minutes.

The various ways of controlling the simulation run time using the timer segment are summarized in Table 9.9.

Table 9.9 Ways of Controlling Simulation Run Time

Timer segment	Simulation run time (in time unit)
GENERATE 480 TERMINATE 1 START 2	960
GENERATE 240 TERMINATE 1 START 2	480
GENERATE 1000 TERMINATE 1 START 3	3000
GENERATE 1 TERMINATE 1 START 1000	1000

SEIZE and RELEASE blocks.   SEIZE and RELEASE blocks are used to fetch/engage and then release/disengage a server (facility), respectively.

The entry of the incoming transaction into the SEIZE block is only conditional. This is equivalent to asking: Is SCT = BIG? If the facility is free, then the incoming transaction will move into the SEIZE block and immediately it will enter into ADVANCE block where the amount of service time and related details are updated. So, the stay-over time of the transaction in the SEIZE block is just a point of time. If the facility is not free, the transaction will be held in the previous block and it waits for its turn. The details of waiting will be updated in the QUEUE and DEPART pair of blocks.

The RELEASE block sets the facility free and it acts as a bridge to transfer the transaction to some block beyond its location. So, the transactions which are attempting to enter the RELEASE block are always accepted into it and immediately they are moved into some succeeding blocks.

The symbols used for SEIZE and RELEASE blocks are shown in Figures 9.10(a) and 9.10(b), respectively.

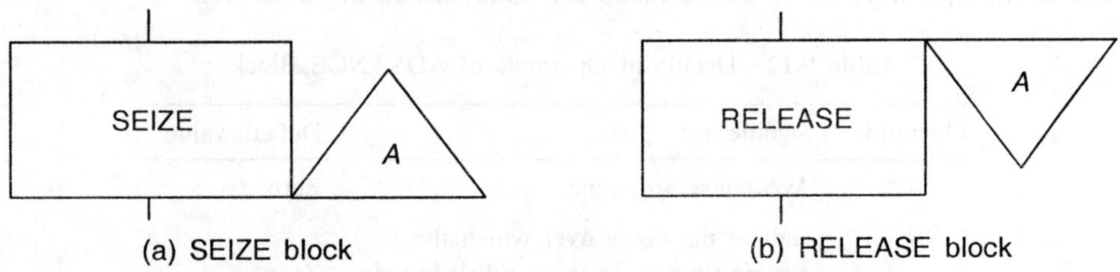

(a) SEIZE block   (b) RELEASE block

Figure 9.10 Symbols for SEIZE and RELEASE blocks.

The details of the operands of the above blocks are summarized in Table 9.10.

Table 9.10 Details of Operands of SEIZE and RELEASE Blocks

Block	Operand	Significance	Default value
SEIZE	A	Name of the facility to be seized	Error
RELEASE	A	Name of the facility to be released	Error

The facility name is identified by an alphanumeric or a numeric variable. If it is alphanumeric, it should be 3 to 5 alphanumeric characters such that the first-three be alphabetic; otherwise, a whole number may be used to represent the facility. The usages of the SEIZE and RELEASE blocks are illustrated with some sample situations in Table 9.11.

Table 9.11 Illustrations for SEIZE and RELEASE Blocks

Example
SEIZE   CLERK
RELEASE CLERK
SEIZE   CLK10
RELEASE CLK10
SEIZE   CLK
RELEASE CLK
SEIZE 25
RELEASE 25

ADVANCE block.   ADVANCE block computes the service time and updates related details for computing percentage utilization of the facility. This block always accepts incoming transactions. The symbol used for the ADVANCE block is shown in Figure 9.11.

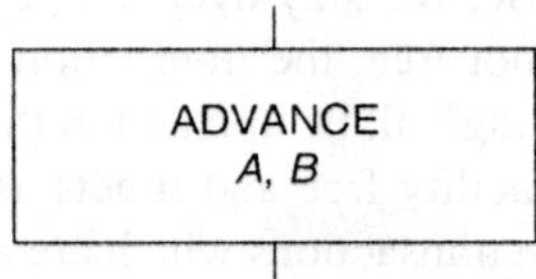

Figure 9.11   Symbol of ADVANCE block.

The details of the operands of the above block are summarized in Table 9.12.

Table 9.12   Details of Operands of ADVANCE Block

Operand	Significance	Default value
A	Average service time	Zero
B	Half of the range over which the service time is uniformly distributed	Zero

The usage of the ADVANCE block is illustrated with some sample situations as in Table 9.13.

Table 9.13 Examples of ADVANCE Block

Example	Explanation
ADVANCE 20, 2	Service time follows uniform distribution with 20 ± 2 units of time.
ADVANCE 50	Service time is deterministic with 50 units of time.

Example 9.21   The customer arrival time and service time at a single-window banking counter follow uniform distribution. The inter-arrival time is 10 ± 2 minutes and the service time is 13 ± 7 minutes. Develop a GPSS block diagram and a program to simulate the system for 8 hours.

Solution   The table of definition for this problem is shown in Table 9.14.

Table 9.14 Table of Definition

Main Model Segment

    Queue: Customer queue is named as LINE
    Facility: Facility is named as CLERK

Timer Segment

    GENERATE block's A operand = 480 minutes
    TERMINATE block's A operand = 1
    Termination counter value = 1

The GPSS block diagram for Example 9.21 is shown in Figure 9.12 and the GPSS program is shown in Figure 9.13. The results of this model can be seen in the standard format of the GPSS output file.

ENTER, LEAVE and STORAGE blocks.   Consider the case of a banking system with two or more counters providing service in parallel. Under this situation, ENTER, LEAVE and STORAGE blocks are jointly used to manage data on facilities. The ENTER block and LEAVE block of the parallel servers model are similar to the SEIZE block and RELEASE block of the single server model, respectively. The STORAGE block defines the number of the facility and the name of the facility which is used in the ENTER and LEAVE blocks.

The block diagrams of the ENTER block and the LEAVE block are shown in Figures 9.14(a) and 9.14(b), respectively.

The details of the operands of the above blocks are summarized in Table 9.15.

Table 9.15 Details of Operands of ENTER and LEAVE Blocks

Block	Operand	Significance	Default value
ENTER	*A*	Name of the facility to be seized	Error
LEAVE	*A*	Name of the facility to be released	Error

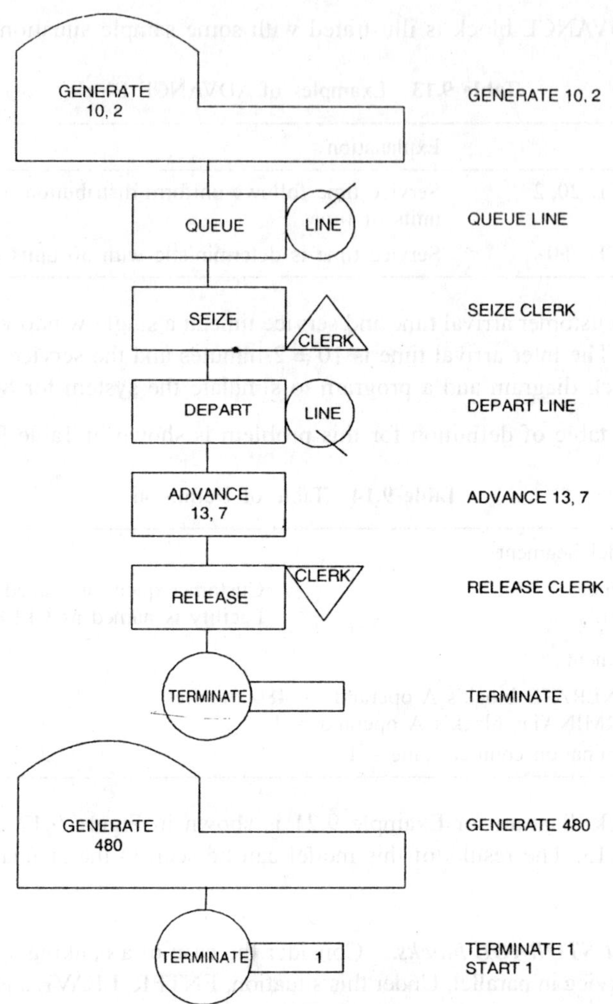

Figure 9.12 GPSS block diagram for Example 9.9.

```
Block No.  Loc.    Operation A,B,C,D,E
           *       Main Model Segment
  1                GENERATE 10,2
  2                QUEUE LINE
  3                SEIZE CLERK
  4                DEPART LINE
  5                ADVANCE 13,7
  6                RELEASE  CLERK
  7                TERMINATE
           *       Timer Segment
  8                GENERATE 480
  9                TERMINATE 1
                   START 1
```

Figure 9.13 GPSS program for Example 9.9.

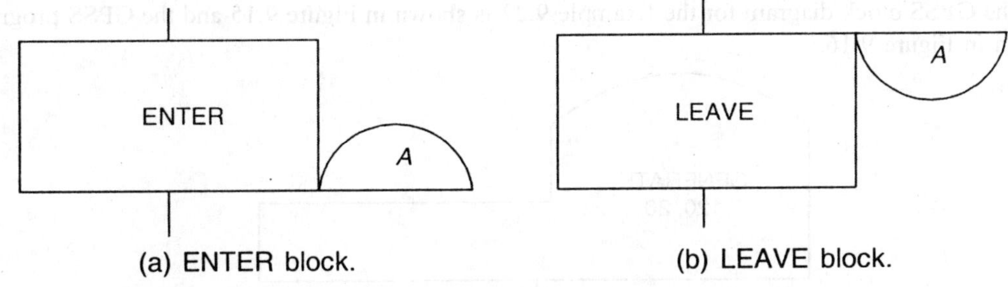

(a) ENTER block.   (b) LEAVE block.

Figure 9.14 Symbols for ENTER and LEAVE blocks.

The facility name is identified by an alphanumeric or a numeric variable. If it is alphanumeric, it should be 3 to 5 alphanumeric characters such that the first-three be alphabetic; otherwise, some whole number may be used to represent the facility.

The details of the STORAGE block are as explained below. In the STORAGE block, the name of the facility is defined in the location area. The operand A of the STORAGE block defines the size of the storage, namely the number of parallel facilities used in the ENTER and the LEAVE blocks.

A sample usage of these three blocks is presented below:

1	CLERK	STORAGE	3
—	—	—	—
—	—	—	—
—	—	—	—
8		ENTER	CLERK
—	—	—	—
—	—	—	—
15		LEAVE	CLERK
—	—	—	—
—	—	—	—

In the above programming segment, the name of the parallel facility is CLERK and its size is 3.

Example 9.22   A toll gate in a highway consists of 5 lanes. The inter-arrival time of the vehicles at the toll gate follows uniform distribution with 120 ± 20 seconds. The service time also follows uniform distribution with 30 ± 10 seconds. Draw a GPSS block diagram and prepare a program to simulate the system for 10 hours.

Solution   The table of definition of this problem is given in Table 9.16.

Table 9.16   Table of Definition

Main Model Segment	
Queue:	Vehicle queue is named as VEHQ
Facility:	Facility is named as TGATE
Timer Segment	
GENERATE block's *A* operand = 36,000 seconds	
TERMINATE block's *A* operand = 1	
Termination counter value = 1	

The GPSS block diagram for the Example 9.22 is shown in Figure 9.15 and the GPSS program is shown in Figure 9.16.

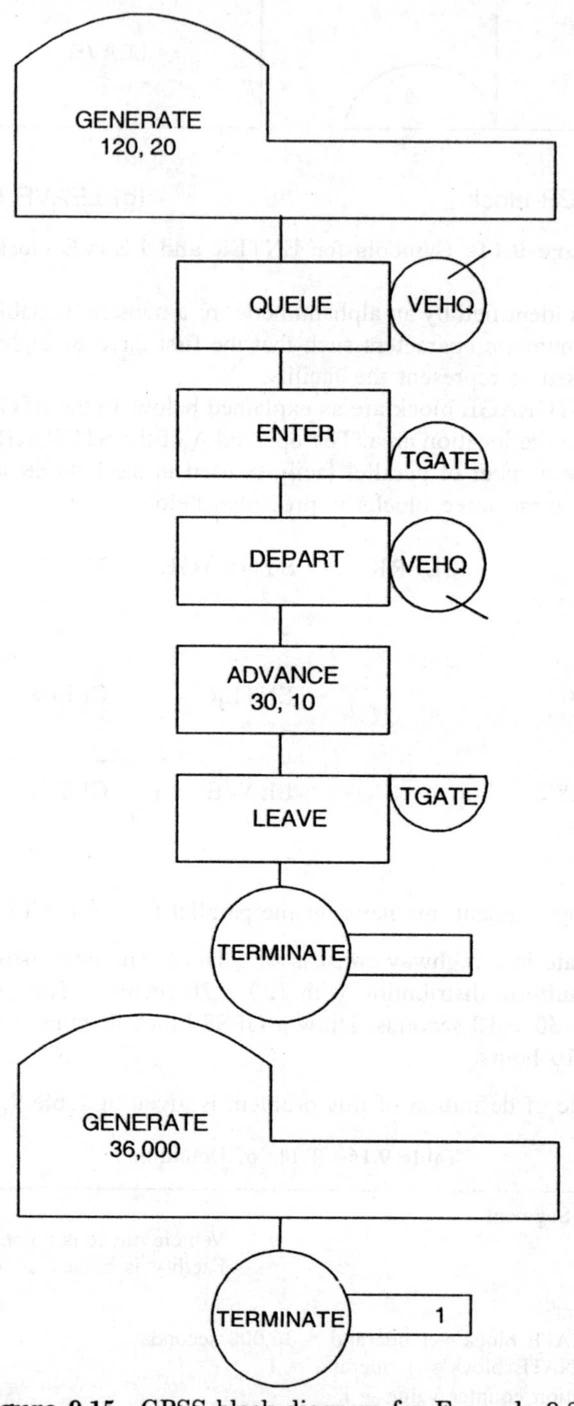

Figure 9.15 GPSS block diagram for Example 9.22.

```
Block No.  Loc.     Operation     A, B, C, D, E
             *       Main Model    Segment
   1       TGATE     STORAGE       5
   2                 GENERATE      120,20
   3                 QUEUE         VEHQ
   4                 ENTER         TGATE
   5                 DEPART        VEHQ
   6                 ADVANCE       30,10
   7                 LEAVE         TGATE
   8                 TERMINATE

             *       Timer Segment

   9                 GENERATE      36000
  10                 TERMINATE     1
                     START         1
```

Figure 9.16 GPSS Program for Example 9.22.

Sampling probability distribution

If the inter-arrival time/service time follows some discrete/continuous distributions other than the uniform distribution, then FUNCTION block is used to describe the arrival/service distribution. A sample usage of the FUNCTION block is shown below.

```
     1       SDIST   FUNCTION   R3, D3
```

In the above statement, the first number, i.e. 1 represents the statement number. SDIST is the name of the function. The operand A is placed with R_3. This means that the function uses the third random number generator (the GPSS has 8 random number generators). The operand B is placed with D_3. This means that the function, SDIST is a discrete distribution and it has 3 values for the random variable. The distribution can be either discrete or continuous (D_n is used for discrete distribution and C_n is used for continuous distribution, where n is the number of values for the random variable). The cumulative probability of each of the values of the random variable along with that value of the random variable is to be written continuously separated by slash symbol (/) immediately in the next line with respect to the FUNCTION block.

Example 9.23 The interarrival time of customers follows a probability distribution as shown in Table 9.17. Write the GPSS block for the same.

Table 9.17 Probability Distribution of Inter-arrival Time

Inter-arrival time (minutes)	Probability
1	0.10
2	0.20
3	0.25
4	0.25
5	0.10
6	0.10

Solution Let the function for the inter-arrival time of the customers be ITIME. There are 6 values for the random variable ITIME and their cumulative probability values are shown in Table 9.18.

Table 9.18 Cumulative Probability Distribution of Inter-arrival Time

Inter-arrival time (in min)	Probability	Cumulative probability	Cumulative probability used in the model*
1	0.10	0.10	0.09
2	0.20	0.30	0.29
3	0.25	0.55	0.54
4	0.25	0.80	0.79
5	0.10	0.90	0.89
6	0.10	1.00	0.99

*The beginning of the first interval of the cumulative probability is 0.

The representation of the function in GPSS is shown below:

$$1 \quad \text{ITIME} \quad \text{FUNCTION} \quad R_3, D_6$$

$$0.09,1/0.29, 2/0.54, 3/0.79, 4/0.89, 5/0.99, 6$$

In the above statement, ITIME represents the name of the function. The operand A of the FUNCTION block is R_3 which means that the third random number generator is used. The operand B of the function is D_6 which means that there are 6 values for the random variable ITIME. The values are placed in the next line with respect to the FUNCTION block along with the respective cumulative probabilities (cumulative probability, first value of the random variable/cumulative probability, second value of the random variable/...).

Example 9.24   At a milling shop which has 6 similar machines, jobs are arriving as per the discrete distribution shown in Table 9.17. The machining time of the jobs follows uniform distribution with 5 ± 3 minutes. Develop a GPSS block diagram to simulate the above system for 8 hours and also write the corresponding GPSS program.

Solution   The table of definition for this problem is as follows (Table 9.19).

Table 9.19   Table of Definition

Main Model Segment	
Queue:	Jobs queue is named as JOBQ
Facility:	Facility is named as MACHE
Timer Segment	
GENERATE block's *A* operand = 480	
TERMINATE block's *A* operand = 1	
Termination counter value = 1	

The cumulative probability distribution of inter-arrival time (ITIME) is shown in Table 9.18. The GPSS block diagram for the example problem is shown in Figure 9.17 and the GPSS program is shown in Figure 9.18.

Guidelines to handle continuous distribution

While sampling any continuous distribution (viz., normal distribution, exponential distribution, etc.), the values of the random variable and the corresponding probabilities of the respective distribution

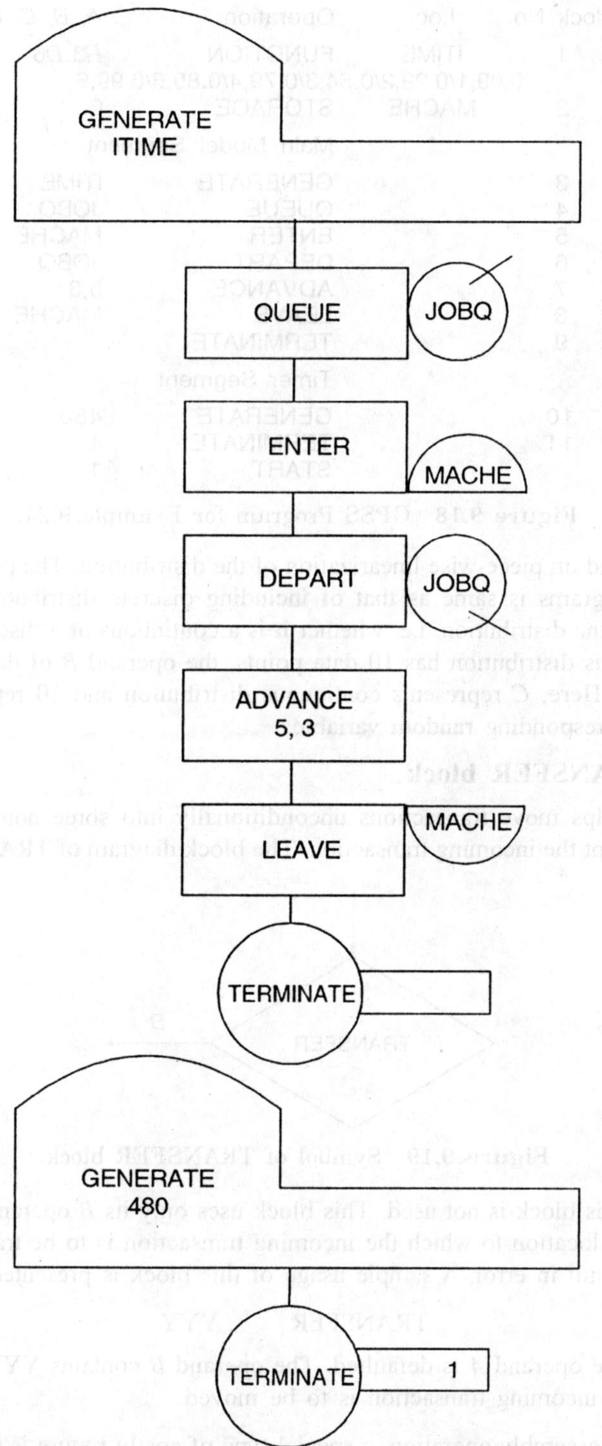

Figure 9.17 GPSS block diagram for Example 9.24.

Block No.	Loc.	Operation	A, B, C, D, E
1	ITIME	FUNCTION	R3,D6
	0.09,1/0.29,2/0.54,3/0.79,4/0.89,5/0.99,6		
2	MACHE	STORAGE	6
	*	Main Model Segment	
3		GENERATE	ITIME
4		QUEUE	JOBQ
5		ENTER	MACHE
6		DEPART	JOBQ
7		ADVANCE	5,3
8		LEAVE	MACHE
9		TERMINATE	
	*	Timer Segment	
10		GENERATE	480
11		TERMINATE	1
		START	1

Figure 9.18 GPSS Program for Example 9.24.

are to be obtained based on piece-wise linearization of the distribution. The procedure to include this function in GPSS programs is same as that of including discrete distributions except the way of defining the nature of the distribution, i.e. whether it is a continuous or a discrete distribution. If, for example, the continuous distribution has 10 data points, the operand B of the FUNCTION block is to be defined as C_{10}. Here, C represents continuous distribution and 10 represents the number of data points of the corresponding random variable.

Unconditional TRANSFER block

TRANSFER block helps move transactions unconditionally into some non-sequential block. This block will always accept the incoming transaction. The block diagram of TRANSFER block is shown in Figure 9.19.

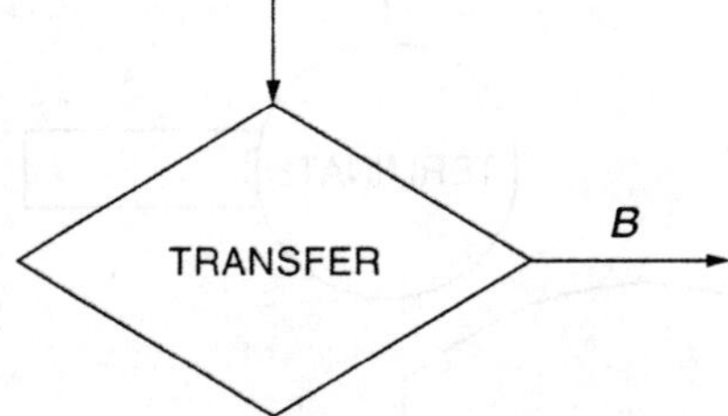

Figure 9.19 Symbol of TRANSFER block.

Operand A of this block is not used. This block uses only its B operand. This operand carries the label of the block location to which the incoming transaction is to be transferred. Defaulting of the operand B will result in error. A sample usage of this block is presented below.

TRANSFER   , YYY

In the above block, the operand A is defaulted. The operand B contains YYY and it represents the location to which the incoming transaction is to be moved.

Example 9.25 In an assembly operation, a special type of costly fixture is used. An assembler will assemble the product on a fixture and then push it into an oven for drying. After drying, the product is knocked out of the fixture and returned to the assembler. The assembly time follows uniform distribution with 15 ± 3 minutes and the oven-drying time also follows uniform distribution with 3 ± 1 minutes. Assume that 5 assemblers are in duty to assemble the product in parallel and then they use the oven. Develop a GPSS block diagram to simulate the above situation for 8 hours to get standard results and also write the corresponding GPSS program.

Solution   The table of definition of this problem will be as shown in Table 9.20.

Table 9.20   Table of Definition

Main Model Segment	
Queue:	Products queue is named as PRODQ
Facility:	Facility is named as OVEN
Timer Segment	
GENERATE block's *A* operand = 480 minutes	
TERMINATE block's *A* operand = 1	
Termination counter value = 1	

The GPSS block diagram for Example 9.25 is shown in Figure 9.20 and the GPSS program is shown in Figure 9.21. In the main model segment, the five assemblers are introduced simultaneously using the limit count (operand *D* of the GENERATE block).

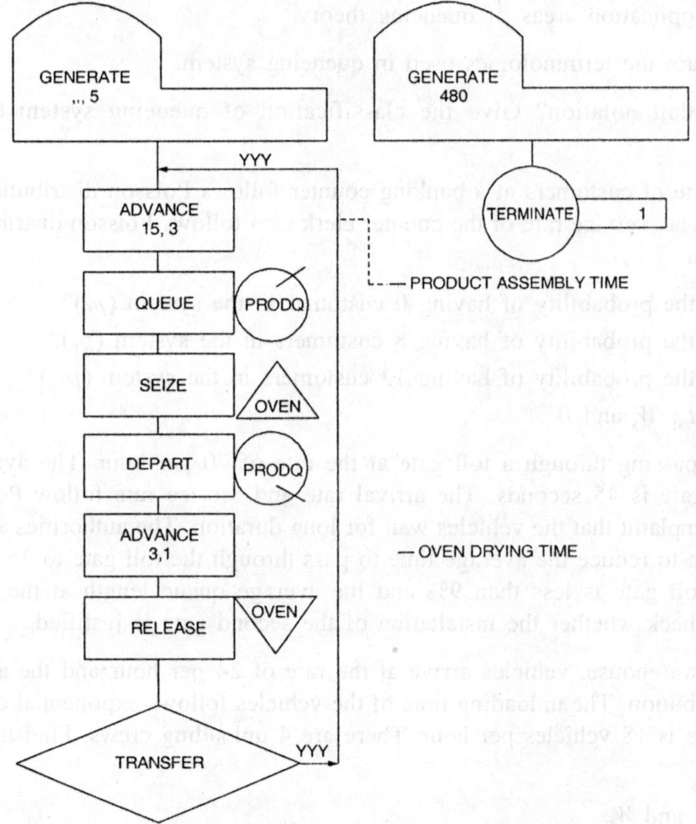

Figure 9.20 GPSS block diagram for Example 9.25.

	Block No.	Loc.	Operation	A, B, C, D, E
*			Main Model Segment	
	1		GENERATE	, , , 5
	2	YYY	ADVANCE	15,3
	3		QUEUE	PRODQ
	4		SEIZE	OVEN
	5		DEPART	PRODQ
	6		ADVANCE	3,1
	7		RELEASE	OVEN
	8		TRANSFER	,YYY
*			Timer Segment	
	9		GENERATE	480
	10		TERMINATE	1
			START	1

Figure 9.21 GPSS Program for Example 9.25.

QUESTIONS

1. Discuss the application areas of queueing theory.

2. List and explain the terminologies used in queueing system.

3. What is Kendall notation? Give the classification of queueing system based on Kendall notation.

4. The arrival rate of customers at a banking counter follows Poisson distribution with a mean of 30 per hour. The service rate of the counter clerk also follows Poisson distribution with a mean of 45 per hour.

   (a) What is the probability of having 0 customer in the system (p_0)?
   (b) What is the probability of having 8 customers in the system (p_8)?
   (c) What is the probability of having 12 customers in the system (p_{12})?
   (d) Find L_s, L_q, W_s and W_q.

5. Vehicles are passing through a toll gate at the rate of 70 per hour. The average time to pass through the gate is 45 seconds. The arrival rate and service rate follow Poisson distribution. There is a complaint that the vehicles wait for long duration. The authorities are willing to instal one more gate to reduce the average time to pass through the toll gate to 35 seconds if the idle time of the toll gate is less than 9% and the average queue length at the gate is more than 8 vehicles. Check whether the installation of the second gate is justified.

6. At a central warehouse, vehicles arrive at the rate of 24 per hour and the arrival rate follows Poisson distribution. The unloading time of the vehicles follows exponential distribution and the unloading rate is 18 vehicles per hour. There are 4 unloading crews. Find the following:

   (a) p_0 and p_3
   (b) L_q, L_s, W_q and W_s.

7. The arrival rate of breakdown machines at a maintenance shop follows Poisson distribution with a mean of 6 per hour. The service rate of machines by a maintenance mechanic also follows Poisson distribution with a mean of 4 per hour. The downtime cost per hour of a breakdown machine is Rs. 300. The labour hour rate is Rs. 60. Determine the optimal number of maintenance mechanics to be employed to repair the machines such that the total cost is minimized.

8. Cars arrive at a drive-in restaurant with a mean arrival rate of 30 cars per hour and the service rate of the cars is 22 per hour. The arrival rate and the service rate follow Poisson distribution. The number of parking space for cars is only 5. Find the standard results of this system.

9. In a harbour, ships arrive with a mean rate of 24 per week. The harbour has 3 docks to handle unloading and loading of ships. The service rate of individual dock is 12 per week. The arrival rate and the service rate follow Poisson distribution. At any point of time, the maximum number of ships permitted in the harbour is 8. Find p_0, L_q, L_s, W_q and W_s.

10. In the machine shop of a small-scale industry, machines break down with a mean rate of 4 per hour. The maintenance shop of the industry has 3 mechanics who can attend to the break down machines individually. The service rate of each of the mechanics is 2 machines per hour. Initially, there are 10 working machines in the machine shop. Find p_0, L_q, L_s, W_q and W_s.

11. In the machine shop of a small-scale industry, machines break down with a mean rate of 4 per hour. The maintenance shop of the industry has only one mechanic who can attend to the break down machines. The service rate of the mechanic is 2 machines per hour. Initially, there are 7 working machines in the machine shop. Find p_0, L_q, L_s, W_q and W_s.

12. Define simulation and its advantages. Also, discuss various application areas of simulation.

13. What are the types of simulation? Explain them in brief.

14. Discuss the steps of simulation.

15. Give a logical flow chart for single-server simulation model and parallel-server simulation model.

16. Explain the operands of the following GPSS blocks.

    GENERATE   QUEUE   DEPART   SEIZE   RELEASE   ADVANCE   TERMINATE
    ENTER   LEAVE   STORAGE   TRANSFER

17. The customer arrival time and service time at a single-window reservation counter follow uniform distribution. The inter-arrival time is 20 ± 2 minutes and the service time is 25 ± 7 minutes. Develop a GPSS block diagram and a program to simulate the system for 8 hours.

18. A checkpost in a highway consists of 6 lanes. The inter-arrival time of the vehicles at the checkpost follows uniform distribution with 150 ± 30 seconds. The service time also follows uniform distribution with 20 ± 10 seconds. Draw a GPSS block diagram and prepare a program to simulate the system for 20 hours.

19. The inter-arrival time of customers follows a discrete distribution as shown in the following table:

Probability Distribution of Inter-arrival Time

Inter-arrival time (in min)	Probability
1	0.20
2	0.25
3	0.35
4	0.10
5	0.08
6	0.02

Write the GPSS block for the same.

20. At a machine shop which has 5 similar machines, jobs arrive as per the discrete distribution shown in the following table:

Probability Distribution of Inter-arrival Time

Inter-arrival (in min)	Probability
1	0.05
2	0.15
3	0.20
4	0.30
5	0.20
6	0.10

The machining time of the jobs follows uniform distribution with 6 ± 2 minutes. Develop a GPSS block diagram to simulate the above system for 10 hours and also write the corresponding GPSS program.

21. Illustrate the use of TRANSFER block with a suitable example.

PROJECT MANAGEMENT 10

10.1 INTRODUCTION

A *project* consists of interrelated activities which are to be executed in a certain order before the entire task is completed. The activities are interrelated in a logical sequence which is known as *precedence relationship*. The work on a project cannot be started until all its immediate preceding activities that involve planning, procuring the inputs, etc., are completed. For example, excavation for foundation should be done only after site preparation; whitewashing should be done after plastering. Some of the typical projects are as follows:

- Construction of a house
- Commissioning of a factory
- Construction of a ship
- Fabrication of a steam boiler
- Construction of a bridge
- Construction of a dam
- Commissioning of a power plant
- Shutdown maintenance of major equipments/plants
- State-level professional course admission process
- New-product launching
- Launching a new weapon system
- Conducting national election
- Research to develop a new technology
- Construction of railway coaches.

Project management is generally applied for constructing items of public utilities, large industrial projects, organizing mega events, etc. Project management is considered to be an important area in production scheduling, mainly because many of the industrial activities can also be viewed as project management problems, for example, fabrication of boilers, construction of railway coaches, launching satellites, product launching, organizing R&D activities, etc.

From the examples, one can recognize the fact that many of the projects are taken up repeatedly either by the same organization or by different organizations. Though they are repeated in nature, each project is unique by itself. In the case of the new product launching of an organization, the first launching is generally organized in a big city, and thereafter in other cities as per priorities fixed by the marketing department of the organization. The project schedule which is prepared for the first city cannot be applied to other cities without any modification because the time estimates of various activities of the project (product launching) will be different for each city due to environmental conditions. In some cases, addition or deletion of some activities will take place. So, if a project is repeated at a different place/different time, then a detailed planning effort is required.

Project is represented in the form of a network for the purpose of analytical treatment to get solutions for scheduling and controlling its activities. A network consists of a set of arcs which are connected meaningfully through a set of nodes. The precedence relationship among various activities of a project can be conveniently represented using a network. So, the collection of precedence relationships among various activities of a project is known as *project network*.

There are two methods of representing any project in the network form: *activities on arrows diagram* (AOA diagram) and *activities on nodes diagram* (AON diagram). The AOA diagram is commonly used in the project management.

Critical path method (CPM) and *project evaluation and review technique* (PERT), are two main basic techniques used in project management. CPM was developed by E.I. Du Pont de Nemours & Company as an application to construction projects and was later extended by Mauchly Associates. PERT was developed by a consulting firm for scheduling the research and development activities for the polaris missile program of US Navy. In CPM, the activities timings are deterministic in nature. But in PERT, each activity will have three time estimates, viz., *optimistic time, most likely time* and *pessimistic time*. There are some more advanced topics in project management which are: crashing of a project network, resource levelling and resource allocation.

Example 10.1 A construction company has listed down various activities that are involved in constructing a community hall. These are summarized along with immediate predecessor(s) details in Table 10.1.

Table 10.1 Details of Activities and Immediate Predecessor(s)

Activity	Description	Immediate predecessor(s)
A	Plan approval	–
B	Site preparation	–
C	Arranging foundation materials	A
D	Excavation for foundation	B
E	Carpentry work for door and window main supporting frames	A
F	Laying foundation	C, D
G	Raising wall from foundation to window base	F
H	Raising wall from window base to lintel level	E, G
I	Roofing	H
J	Electrical wiring and fitting	I
K	Plastering	J, L
L	Making doors and windows and fitting them	A
M	Whitewashing	K
N	Clearing the site before handing over	M

Draw a project network for the above project.

Solution The project network summarizing the precedence relationships of various activities of constructing the community hall is shown in Figure 10.1. In this figure, the activities A and B are concurrent activities. Activities C, E and L can be started only after completing A. Activity D follows B, and activity F can be started only when C and D are completed. One can use similar logic to infer the remaining precedence relationships in Figure 10.1.

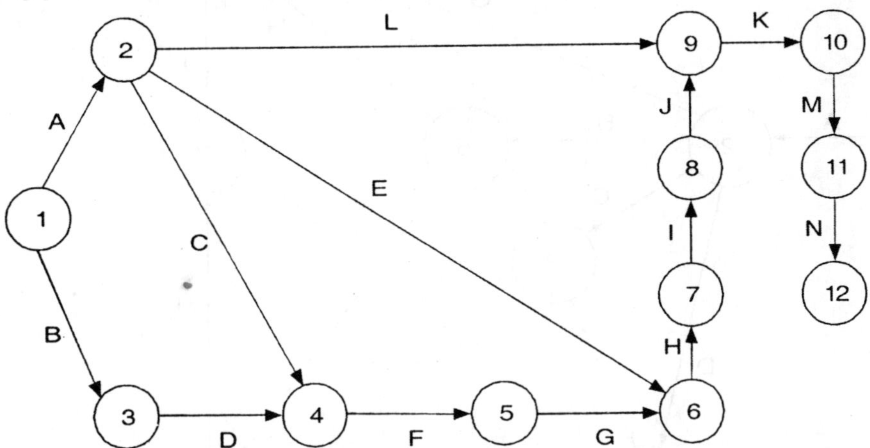

Figure 10.1 Project network of community hall construction.

Example 10.2 Nowadays, in all professional institutes, students selection is done through a written test, group discussion and interview. Prior to the written test, there are so many activities, which are to be performed starting from deciding the date of conducting all other events. Though the system looks simple, the precedence relationships among the activities necessitate coordination of these activities in completing the whole task of admitting students to any professional course. A delay at any stage would lead to ineffective operation of the system. Hence, this problem may be treated as a project consisting of activities with precedence relationships as shown in Table 10.2.*

Table 10.2 Precedence Relationships of Professional Course Admission Process

Activity	Description	Immediate predecessor(s)
A	Decide the date, time of the entrance test, group discussion and interview	—
B	Decide the examination venue(s)	A
C	Call for application through advertisement	A
D	Print application form	A
E	Despatch applications	D, C
F	Receive and process applications and send hall tickets	E, B
G	Set question paper	A
H	Arrange invigilators and examiner	B
I	Make seating arrangements	F
J	Conduct examination	G, H, I
K	Valuation and announcement of entrance test result	J
L	Conduct group discussion and interview	K
M	Publish admission result	L

Construct a critical path network (CPM) for this problem.

*Rajkumar and Panneerselvam (1992).

Solution The CPM network for the above problem is shown in Figure 10.2.

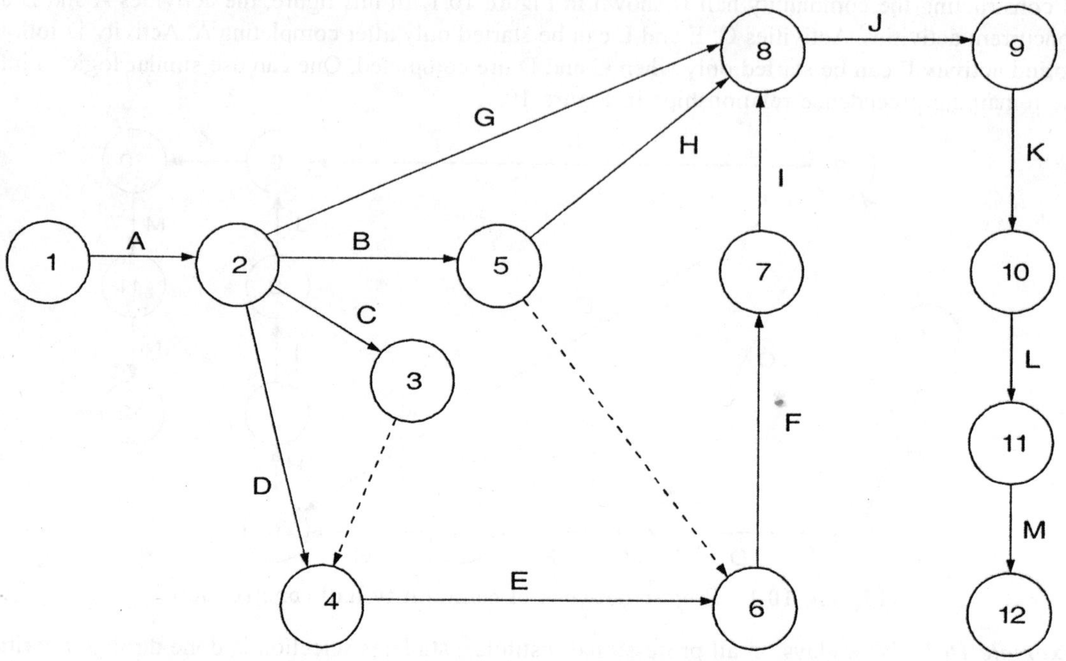

Figure 10.2 Network of professional course admission process.

The figure shows how the precedence relations between various activities of the admission project are defined. In this figure, activity A is the starting activity. Activities, B, C, D and G can be taken up only after A is completed. Activity E can be started after completing D and C. The precedence relationship between all other activities are defined in this manner. Activity M is the final activity representing the end of the selection process.

10.2   PHASES OF PROJECT MANAGEMENT

Project management has three phases: planning, scheduling and controlling. The planning phase has the following steps:

1. Dividing the project into distinct activities.
2. Estimating time requirement for each activity.
3. Establishing precedence relationships among the activities.
4. Construction of the arrow diagram (network).

The scheduling phase determines the start and end times of each and every activity. These can be summarized in the form of a time chart/Gantt chart. For each non-critical activity, the amount of slack time must be shown on the same time chart. This will be useful at the time of adjusting non-critical activities for resource levelling or resource allocation.

The control phase uses the arrow diagram and time chart for continuous monitoring and progress reporting. In this phase, the network will be updated if there is any variation in the proposed schedule.

10.3 GUIDELINES FOR NETWORK CONSTRUCTION

The terminologies which are used in network construction are explained as follows:

Node. Generally, a *node* represents the starting or ending of an activity.

Branch/arc. A *branch* represents the actual activity which consumes some kind of resource.

Precedence relations of activities. For any activity, the *precedence relations* provide the information about: the activities that precede it, the activities that follow it, and the activities that may be concurrent with it.

Network construction requires a detailed list of individual activities of a project, estimates of activity durations and specifications of precedence relationships among various activities of the project.

Rules for network construction

The following are the primary rules for constructing AOA diagram.

1. The starting event and ending event of an activity are called *tail event* and *head event*, respectively.
2. The network should have a unique starting node (tail event).
3. The network should have a unique completion node (head event).
4. No activity should be represented by more than one arc in the network.
5. No two activities should have the same starting node and the same ending node.
6. Dummy activity is an imaginary activity indicating precedence relationship only. Duration of a dummy activity is zero.

10.4 CRITICAL PATH METHOD (CPM)

As stated earlier, CPM deals with project management involving deterministic time estimates. In this section, the concept of CPM is demonstrated through an example.

Example 10.3 Consider Table 10.3 summarizing the details of a project involving 14 activities.

Table 10.3 Project Details

Activity	Immediate predecessor(s)	Duration (months)
A	–	2
B	–	6
C	–	4
D	B	3
E	A	6
F	A	8
G	B	3
H	C, D	7
I	C, D	2
J	E	5
K	F, G, H	4
L	F, G, H	3
M	I	13
N	J, K	7

(a) Construct the CPM network.

(b) Determine the critical path and project completion time.

(c) Compute total floats and free floats for non-critical activities.

Solution (a) The CPM network is shown in Figure 10.3.

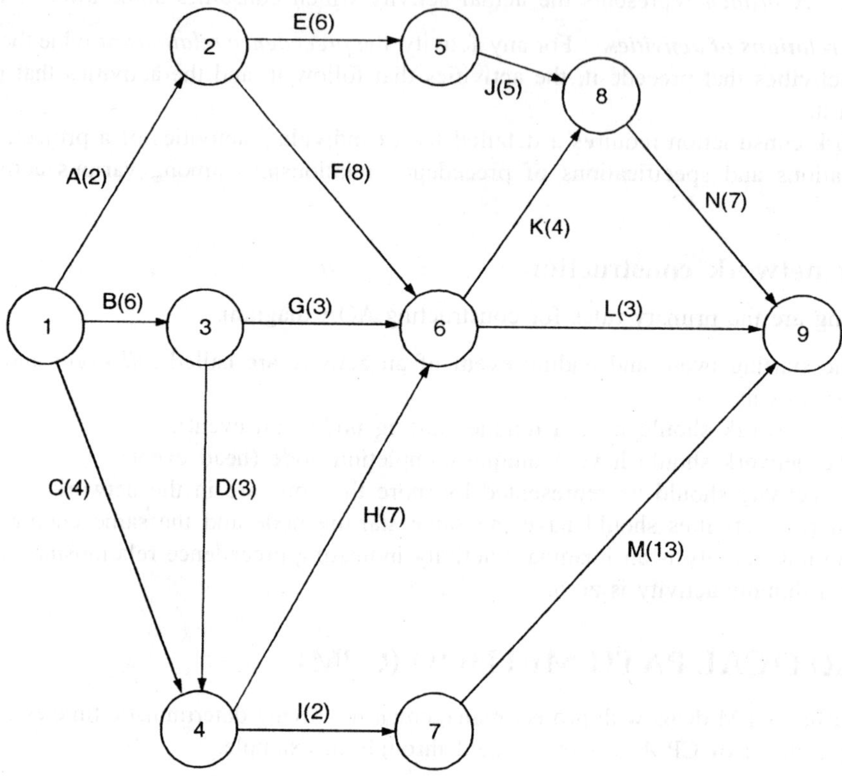

Figure 10.3 CPM network for Example 10.3.

(b) The *critical path* of a project network is the longest path in the network. This can be identified by simply listing out all the possible paths from the start node of the project (node 1) to the end node of the project (node 9) and then selecting the path with the maximum sum of activity times on that path.

This method has several drawbacks. In a large network, one may commit mistake in listing all the paths. Moreover, this method will not provide necessary details, such as total floats and free floats for further analysis. Hence, a different approach is to be used to identify the critical path. This consists of two phases: Phase 1 determines *earliest start times* (ES) of all the nodes. This is called *forward pass*; Phase 2 determines *latest completion times* (LC) of various nodes. This is called *backward pass*.

Let D_{ij} be the duration of the activity (i, j). ES_j be the earliest start times of all the activities which are emanating from node j (this is shown in the lower square which is by side of node j). LC_j be the latest completion times of all the activities which are ending at node j (this is shown in the upper square which is by the side of node j).

Determination of earliest start times (ES_j).   During forward pass, use the following formula to compute earliest start times for all nodes:

$$ES_j = \max_i \ (ES_i + D_{ij})$$

The calculations of ES_j are summarized below:

Node 1: For node 1, $ES_1 = 0$
Node 2: $ES_2 = ES_1 + D_{1,2} = 0 + 2 = 2$
Node 3: $ES_3 = ES_1 + D_{1,3} = 0 + 6 = 6$
Node 4: $ES_4 = \max_{i=1,3} \ (ES_i + D_{i4})$

$$= \max \ (ES_1 + D_{1,4}, \ ES_3 + D_{3,4})$$
$$= \max \ (0 + 4, \ 6 + 3) = 9$$

Node 5: $ES_5 = ES_2 + D_{2,5} = 2 + 6 = 8$
Node 6: $ES_6 = \max_{i=2,3,4} \ (ES_i + D_{i,6})$

$$= \max \ (ES_2 + D_{2,6}, \ ES_3 + D_{3,6}, \ ES_4 + D_{4,6})$$
$$= \max \ (2 + 8, \ 6 + 3, \ 9 + 7)$$
$$= \max \ (10, 9, 16)$$
$$= 16$$

Similarly, the ES_j values for all other nodes are computed and summarized in Figure 10.4.

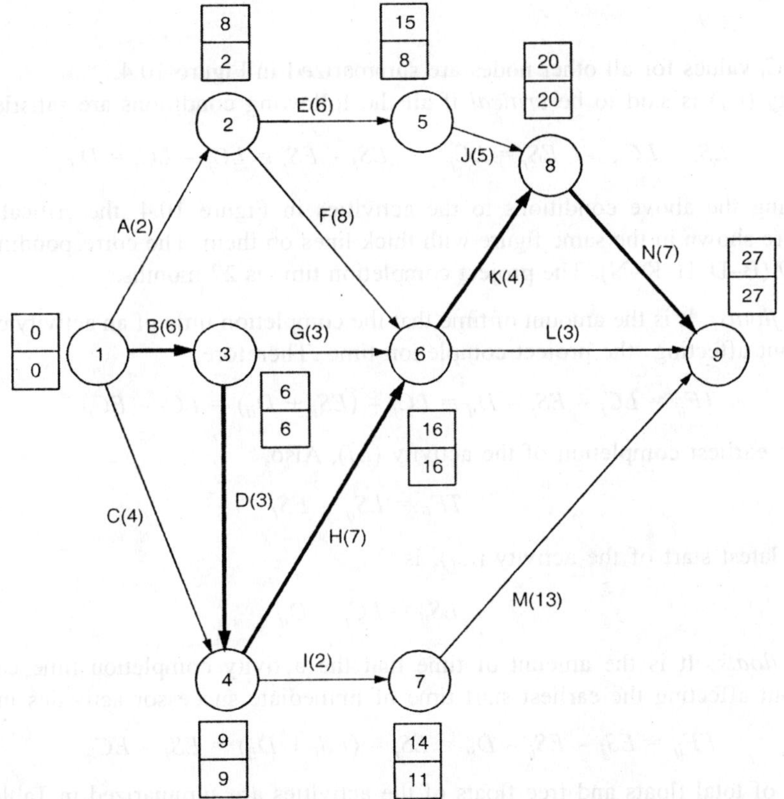

Figure 10.4 Network with critical path calculations.

Determination of latest completion times (LC_i). During backward pass, the following formula is used to compute latest completion times (LC_i):

$$LC_i = \min_j \ (LC_j - D_{ij})$$

Node 9: For the last node 9, $LC_9 = ES_9 = 27$

Node 8: $LC_8 = LC_9 - D_{8,9} = 27 - 7 = 20$

Node 7: $LC_7 = LC_9 - D_{7,9} = 27 - 13 = 14$

Node 6: $LC_6 = \min_{j=8,9} \ (LC_j - D_{6,j})$

$$= \min (LC_8 - D_{6,8}, LC_9 - D_{6,9})$$

$$= \min (20 - 4, 27 - 3)$$

$$= 16$$

Node 5: $LC_5 = LC_8 - D_{5,8} = 20 - 5 = 15$

Node 4: $LC_4 = \min_{j=6,7} \ (LC_j - D_{4,j})$

$$= \min (LC_6 - D_{4,6}, LC_7 - D_{4,7})$$

$$= \min (16 - 7, 14 - 2)$$

$$= 9$$

Similarly, the LC_i values for all other nodes are summarized in Figure 10.4.

An activity (i, j) is said to be *critical* if all the following conditions are satisfied.

$$ES_i = LC_i, \qquad ES_j = LC_j, \qquad ES_j - ES_i = LC_j - LC_i = D_{ij}$$

By applying the above conditions to the activities in Figure 10.4, the critical activities are identified and are shown in the same figure with thick lines on them. The corresponding critical path is 1–3–4–6–8–9 (B–D–H–K–N). The project completion time is 27 months.

(c) *Total floats:* It is the amount of time that the completion time of an activity can be delayed without affecting the project completion time. Therefore,

$$TF_{ij} = LC_j - ES_i - D_{ij} = LC_j - (ES_i + D_{ij}) = LC_j - EC_{ij}$$

where, EC_{ij}, the earliest completion of the activity (i, j), Also,

$$TF_{ij} = LS_{ij} - ES_i$$

where LS_{ij}, the latest start of the activity (i, j), is

$$LS_{ij} = LC_j - D_{ij}$$

(d) *Free floats:* It is the amount of time that the activity completion time can be delayed without affecting the earliest start time of immediate successor activities in the network.

$$FF_{ij} = ES_j - ES_i - D_{ij} = ES_j - (ES_i + D_{ij}) = ES_j - EC_{ij}$$

The calculation of total floats and free floats of the activities are summarized in Table 10.4.

Table 10.4 Summary of Total Floats and Free Floats

Activity (i, j)	Duration (D_{ij})	Total float (TF_{ij})	Free float (FF_{ij})
1–2	2	6	0
1–3	6	0	0
1–4	4	5	5
2–5	6	7	0
2–6	8	6	6
3–4	3	0	0
3–6	3	7	7
4–6	7	0	0
4–7	2	3	0
5–8	5	7	7
6–8	4	0	0
6–9	3	8	8
7–9	13	3	3
8–9	7	0	0

Any critical activity will have zero total float and zero free float. Based on this property also, one can determine the critical activities. From Table 10.4, one can check that the total floats and free floats for the activities (1, 3), (3, 4), (4, 6), (6, 8) and (8, 9) are zero. Hence, they are critical activities. The corresponding critical path is 1–3–4–6–8–9 (B–D–H–K–N).

Example 10.4 The details of a project are as shown in Table 10.5. Find the critical path and the corresponding project completion time.

Table 10.5 Data of Example 10.4

Activity	Immediate predecessor(s)	Duration (weeks)
A	–	4
B	–	3
C	–	2
D	A, B, C	5
E	A, B, C	6
F	D	7
G	D, E	6
H	D, E	9
I	F	4
J	G	6
K	H	8

Solution The network of this example is shown in Figure 10.5. The critical path calculations are also summarized in the same figure. From Figure 10.5, the critical path is A–E–H–K and the corresponding project completion time is 27 weeks.

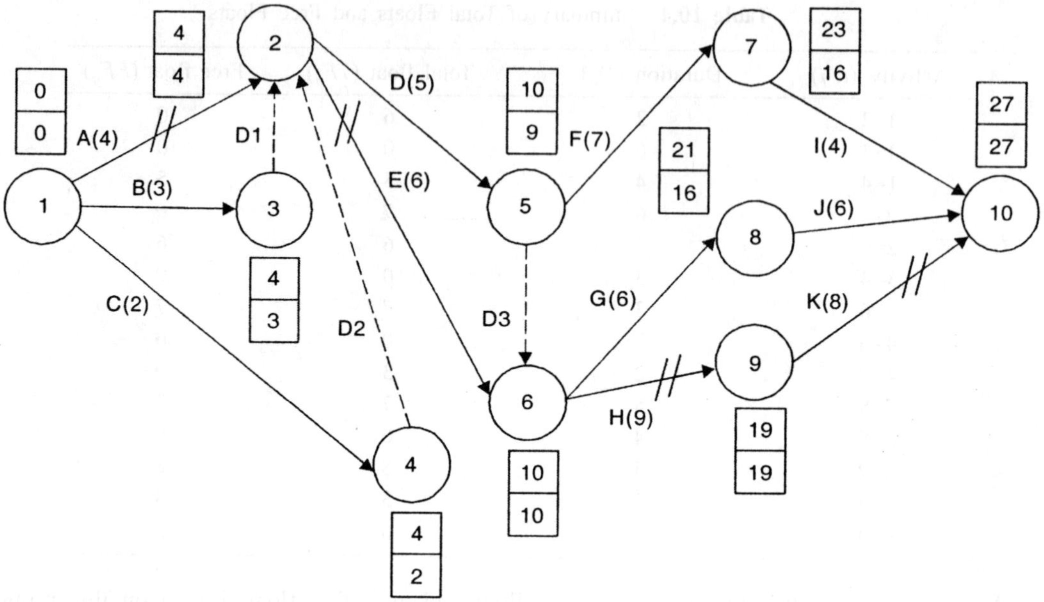

Note: D1, D2 and D3 are dummy activities.

Figure 10.5   Project network and critical path calculations of Example 10.4.

Example 10.5   A project consists of activities from A to J as shown in Table 10.6. The immediate predecessor(s) and the duration in weeks of each of the activities are given in the same table. Draw the project network and, find the critical path and the corresponding project completion time. Also, find the total float as well as free float for each of the non-critical activities.

Table 10.6   Data of Example 10.5

Activity	Immediate predecessor(s)	Duration (weeks)
A	–	4
B	–	3
C	A, B	2
D	A, B	5
E	B	6
F	C	4
G	D	3
H	F, G	7
I	F, G	4
J	E, H	2

Solution   The project network for the given problem is constructed as in Figure 10.6. The critical path calculations are shown in the same figure. The critical path is A–D–G–H–J and the corresponding project completion time is 21 weeks. The total float and the free float of each of the non-critical activities are shown in Table 10.7.

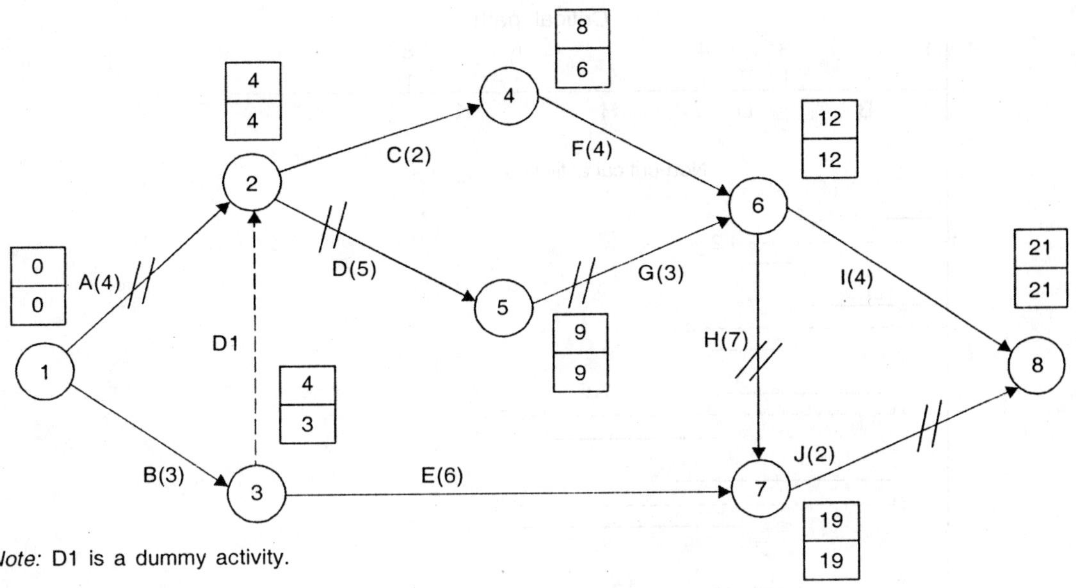

Figure 10.6 Project network and CPM calculations of Example 10.5.

Table 10.7 Details of Floats of Non-Critical Activities

Non-critical activity	Total float (weeks)	Free float (weeks)
B	1	0
C	2	0
E	10	10
F	2	2
I	5	5

10.5 GANTT CHART (TIME CHART)

The next stage after completing network calculations is to draw Gantt chart (time chart). The start time and completion time of each and every activity will be represented on this chart. This chart gives clear calendar schedule for the whole project.

This chart is also used for resource levelling purpose. When there are limitations on the available resource(s) (manpower, equipment, money, etc.), using this chart, one can adjust the schedule of non-critical activities depending upon their total floats to minimize the peak requirement of resource(s). This helps to level the resource requirements smoothly throughout the project execution. Even if there is no restriction on any resource, it is usual practice to smooth out the resource requirements by resource levelling technique.

The time chart for Example 10.3 is shown in Figure 10.7.

In Figure 10.7, the top-most horizontal line represents the schedule for various critical activities: (1, 3), (3, 4), (4, 6), (6, 8) and (8, 9). Horizontal dotted lines represent the total time span over

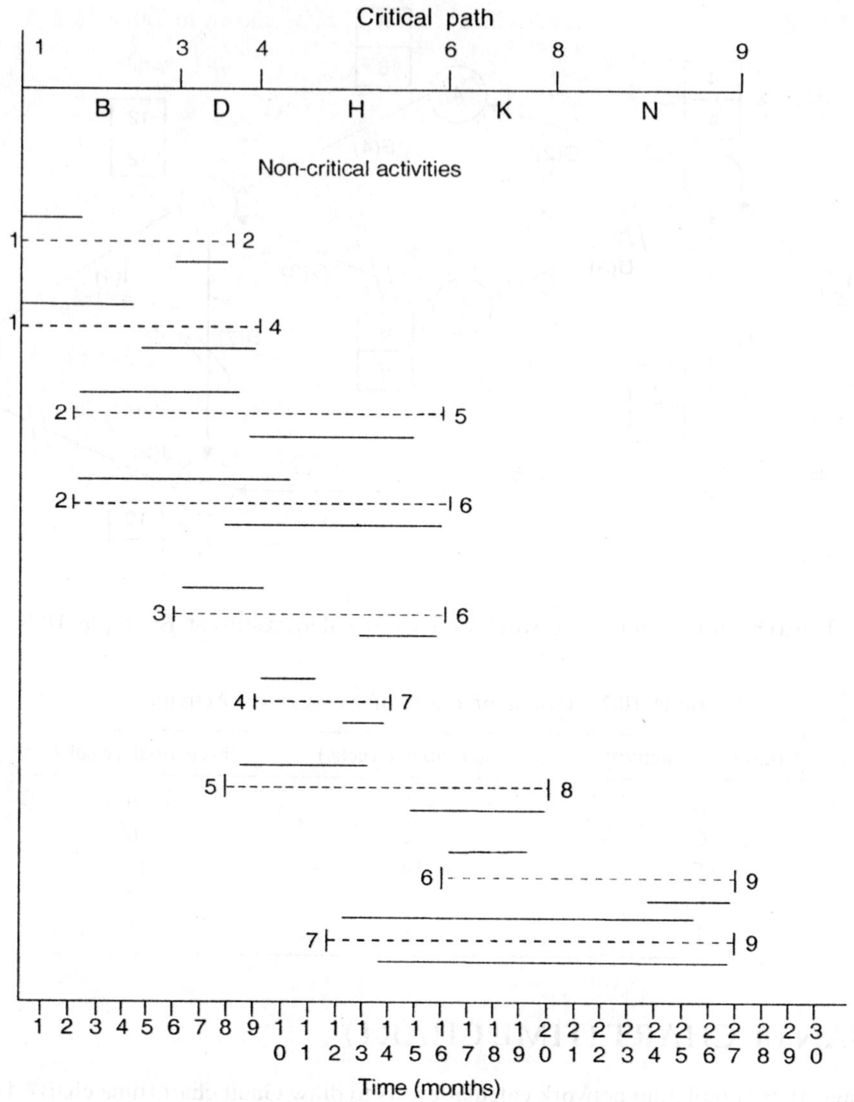

Figure 10.7 Gantt chart (time chart).

which a non-critical activity can be performed. As stated earlier, a non-critical activity will have float in excess of its time duration. So, it is possible to adjust the start and completion time of any non-critical activity over its entire range shown by the dotted line without delaying the project completion time. Infinite number of schedules are possible for the non-critical activities. But to give a finite number of schedules, we consider only two types of schedules: earliest start schedule and latest start schedule. These are shown by continuous horizontal lines over and below the dotted lines, respectively. The time range for execution of a given non-critical activity is represented by the dotted line.

Example 10.6 A project consists of activities from A to H as shown in Table 10.8. The immediate predecessor(s) and the duration in months of each of the activities are given in the same table.

(a) Draw the project network and, find the critical path and the corresponding project completion time.

(b) Also, draw a Gantt-chart/Time chart for this project.

Table 10.8 Data of Example 10.5

Activity	Immediate predecessor(s)	Duration (months)
A	–	5
B	–	2
C	A	3
D	C	4
E	C	2
F	B	4
G	D	7
H	E, F	6

Solution

(a) The project network for the given problem is as shown in Figure 10.8. The critical path calculations are shown in the same figure. In Figure 10.8, the critical path is A–C–D–G and the corresponding project completion time is 19 months.

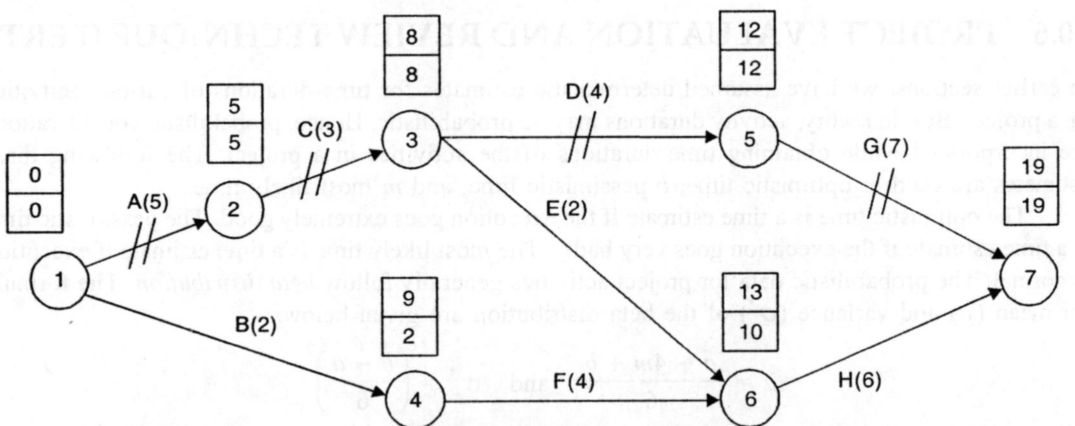

Figure 10.8 Project network with CPM calculations of Example 10.6.

(b) The Gantt-chart for the project shown in Figure 10.8 is given in Figure 10.9. In Figure 10.9, the critical activities (A, C, D and G) are represented at the top. One can notice the fact that there is no time gap between the completion of one critical activity and the start of another critical activity. The total span during which each non-critical activity can be executed is shown by a dotted line. A continuous line on the left hand side just above each dotted line shows the earliest start schedule of that non-critical activity. Another continuous line on the right hand side just below each dotted line shows the latest start schedule of that non-critical activity.

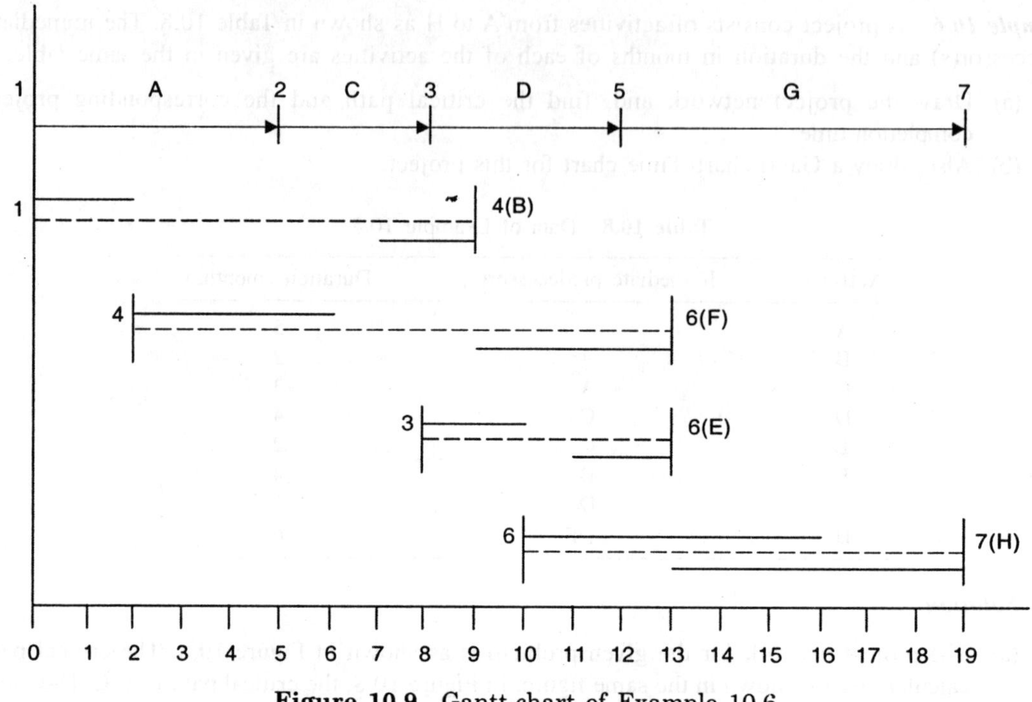

Figure 10.9 Gantt-chart of Example 10.6.

10.6 PROJECT EVALUATION AND REVIEW TECHNIQUE (PERT)

In earlier sections, we have assumed deterministic estimates for time durations of various activities in a project. But, in reality, activity durations may be probabilistic. Hence, probabilistic considerations are incorporated while obtaining time durations of the activities in a project. The following three estimates are used: a optimistic time; b pessimistic time; and m most likely time.

The optimistic time is a time estimate if the execution goes extremely good. The pessimistic time is a time estimate if the execution goes very badly. The most likely time is a time estimate if execution is normal. The probabilistic data for project activities generally follow *beta distribution*. The formula for mean (μ) and variance (σ^2) of the beta distribution are given below:

$$\mu = \frac{a + 4m + b}{6} \quad \text{and} \quad \sigma^2 = \left(\frac{b-a}{6}\right)^2$$

The range for the time estimates is from a to b. The most likely time will be anywhere in the range from a to b. The *expected project completion time* is $\Sigma \mu_i$, where μ_i is the expected duration of the ith critical activity. The *variance of the project completion time* is $\sum_i \sigma_i^2$, where σ_i^2 is the variance of the ith critical activity in the critical path.

As a part of statistical analysis, we may be interested in knowing the probability of completing the project on or before a given due date (C) or we may be interested in knowing the expected project completion time if the probability of completing the project is given. For this, the beta distribution is approximated to standard normal distribution whose statistic is given below:

$$z = \frac{x - \mu}{\sigma}$$

where x is the actual project completion time, μ is the expected project completion time (sum of the expected durations of the critical activities) and σ is the standard deviation of the expected project completion time (square root of the sum of the variances of all the critical activities). Therefore, $P\,(x \le C)$ represents the probability that the project will be completed on or before the C time units. This can be converted into the standard normal statistic z as:

$$P\left(\frac{x - \mu}{\sigma} \le \frac{C - \mu}{\sigma}\right) = P\left(z \le \frac{C - \mu}{\sigma}\right).$$

Example 10.7   Consider Table 10.9 summarizing the details of a project involving 11 activities.

Table 10.9   Details of Project with 11 Activities

Activity	Predecessor(s)	Duration (weeks)		
		a	m	b
A	–	6	7	8
B	–	1	2	9
C	–	1	4	7
D	A	1	2	3
E	A, B	1	2	9
F	C	1	5	9
G	C	2	2	8
H	E, F	4	4	4
I	E, F	4	4	10
J	D, H	2	5	14
K	I, G	2	2	8

(a) Construct the project network.
(b) Find the expected duration and variance of each activity.
(c) Find the critical path and the expected project completion time.
(d) What is the probability of completing the project on or before 25 weeks?
(e) If the probability of completing the project is 0.84, find the expected project completion time.

Solution   (a) The project network is shown in Figure 10.10.

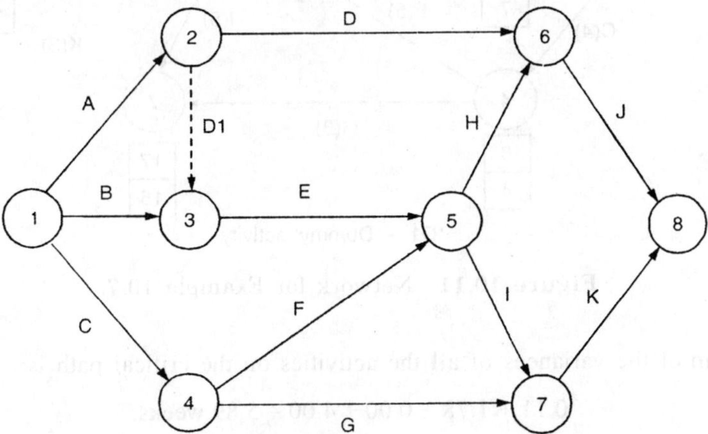

Figure 10.10   Network for Example 10.7.

(b) The expected duration and variance of each activity are shown in Table 10.10.

Table 10.10 Computations of Expected Duration and Variance

Activity	Duration (weeks)			Mean duration	Variance
	a	m	b		
A	6	7	8	7	0.11
B	1	2	9	3	1.78
C	1	4	7	4	1.00
D	1	2	3	2	0.11
E	1	2	9	3	1.78
F	1	5	9	5	1.78
G	2	2	8	3	1.00
H	4	4	4	4	0.00
I	4	4	10	5	1.00
J	2	5	14	6	4.00
K	2	2	8	3	1.00

(c) The calculations of critical path based on expected durations are summarized in Figure 10.11. The critical path is A–D1–E–H–J and the corresponding project completion time is 20 weeks.

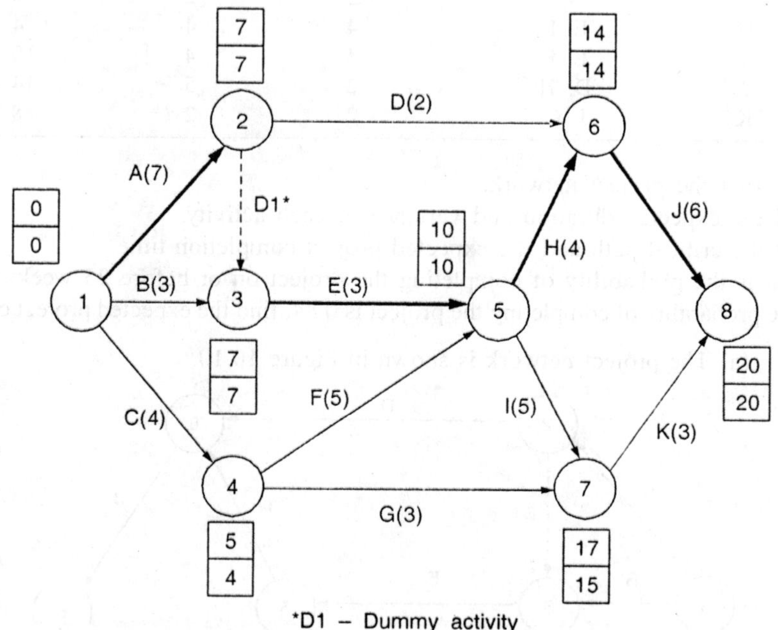

Figure 10.11 Network for Example 10.7.

(d) The sum of the variances of all the activities on the critical path is:

$$0.11 + 1.78 + 0.00 + 4.00 = 5.89 \text{ weeks.}$$

Therefore $\sigma = \sqrt{5.89} = 2.43$ weeks. Also

$$P(x \le 25) = P\left(\frac{x - \mu}{\sigma} \le \frac{25 - 20}{2.43}\right)$$

$$= P(z \le 2.06) = 0.9803$$

This value is obtained from standard normal table. Therefore, the probability of completing the project on or before 25 weeks is 0.9803.

(e) We also have $P(x \le C) = 0.84$. Therefore,

$$P\left(\frac{x - \mu}{\sigma} \le \frac{C - \mu}{\sigma}\right) = 0.84$$

$$P\left(z \le \frac{C - 20}{2.43}\right) = 0.84$$

From the standard normal table, the value of z is 0.99, when the cumulative probability is 0.84. Therefore,

$$\frac{C - 20}{2.43} = 0.99 \quad \text{or} \quad C = 22.4 \text{ weeks}$$

The project will be completed in 22.41 weeks (approximately 23 weeks) if the probability of completing the project is 0.84.

Example 10.8   Consider the data of a project summarized in Table 10.11.

Table 10.11   Data of Example 10.8

Activity	Immediate Predecessor(s)	Duration (weeks)		
		a	m	b
A	–	4	4	10
B	–	1	2	9
C	–	2	5	14
D	A	1	4	7
E	A	1	2	3
F	A	1	5	9
G	B, C	1	2	9
H	C	4	4	4
I	D	2	2	8
J	E, G	6	7	8

(a) Construct the project network.
(b) Find the expected duration and the variance of each activity.
(c) Find the critical path and the expected project completion time.
(d) What is the probability of completing the project on or before 35 weeks?

Solution

(a) The project network for the given problem is shown in Figure 10.12.

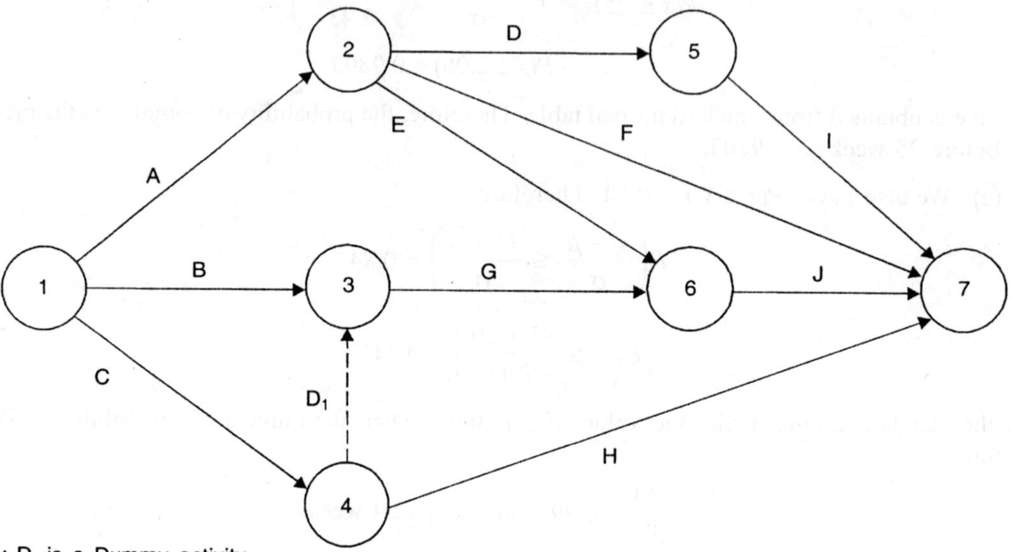

Note: D_1 is a Dummy activity

Figure 10.12 Project network of Example 10.8.

(b) The mean and variance of each activity in the project are shown in Table 10.12.

Table 10.12 Data of Example 10.8

Activity	Immediate predecessor(s)	Duration (weeks)			Mean	Variance
		a	m	b	(Weeks)	(Weeks)
A	–	4	4	10	5	1.00
B	–	1	2	9	3	1.78
C	–	2	5	14	6	4.00
D	A	1	4	7	4	1.00
E	A	1	2	3	2	0.11
F	A	1	5	9	5	1.78
G	B, C	1	2	9	3	1.78
H	C	4	4	4	4	0.00
I	D	2	2	8	3	1.00
J	E, G	6	7	8	7	0.11

(c) The critical path calculations are shown in Figure 10.13.

From Figure 10.13, the critical path is C–D_1–G–J and the corresponding expected project completion time is 16 weeks.

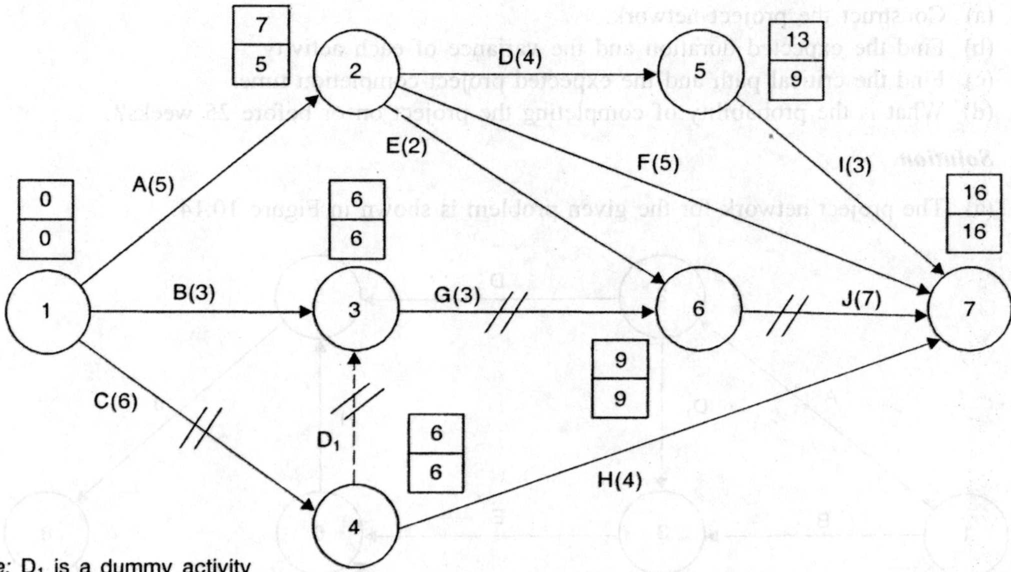

Note: D₁ is a dummy activity

Figure 10.13 Project network with critical path calculations.

(d) The probability of completing the project on or before 35 weeks is computed as shown below:

The sum of the variance of the critical activities = 4 + 1.78 + 0.11 = 5.89 weeks
Therefore, the standard deviation, $\sigma = (5.89)^{0.5} = 2.427$ weeks

$$P(X \leq 35) = P[(X - \mu)/\sigma \leq (35 - 16)/2.427]$$
$$= P(Z \leq 7.82) = 0.9999.$$

Example 10.9 The details of a project consisting of activities A to K are summarized in Table 10.13.

Table 10.13 Data of Example 10.9

Activity	Immediate Predecessor(s)	Duration (weeks)		
		a	m	b
A	–	6	7	8
B	–	1	2	9
C	–	1	4	7
D	A	1	2	3
E	A, B	1	2	9
F	C	1	5	9
G	C	2	2	8
H	E, F	4	4	4
I	E, F	4	4	10
J	D, H	2	5	14
K	I, G	2	2	8

(a) Construct the project network.
(b) Find the expected duration and the variance of each activity.
(c) Find the critical path and the expected project completion time.
(d) What is the probability of completing the project on or before 25 weeks?

Solution

(a) The project network for the given problem is shown in Figure 10.14.

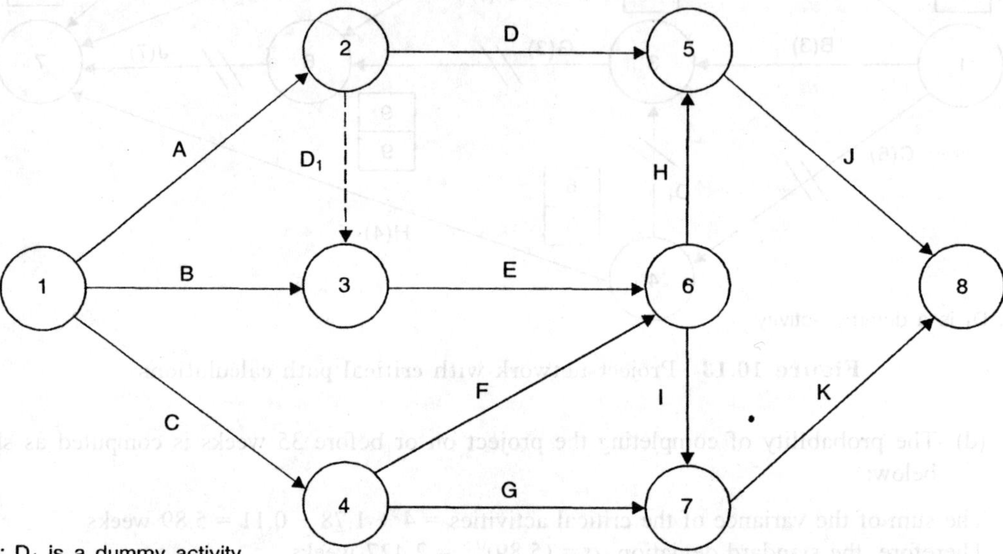

Note: D_1 is a dummy activity

Figure 10.14   Project network of Example 10.9.

(b) The mean and variance of each activity in the project are shown in Table 10.14.

Table 10.14   Data of Example 10.9

Activity	Immediate predecessor(s)	Duration (weeks)			Mean (Weeks)	Variance (Weeks)
		a	m	b		
A	–	6	7	8	7	0.11
B	–	1	2	9	3	1.78
C	–	1	4	7	4	1.00
D	A	1	2	3	2	0.11
E	A, B	1	2	9	3	1.78
F	C	1	5	9	5	1.78
G	C	2	2	8	3	1.00
H	E, F	4	4	4	4	0.00
I	E, F	4	4	10	5	1.00
J	D, H	2	5	14	6	4.00
K	I, G	2	2	8	3	1.00

(c) The critical path calculations are shown in Figure 10.15. From Figure 10.15, the critical path is A–D$_1$–E–H–J and the corresponding expected project completion time is 20 weeks.

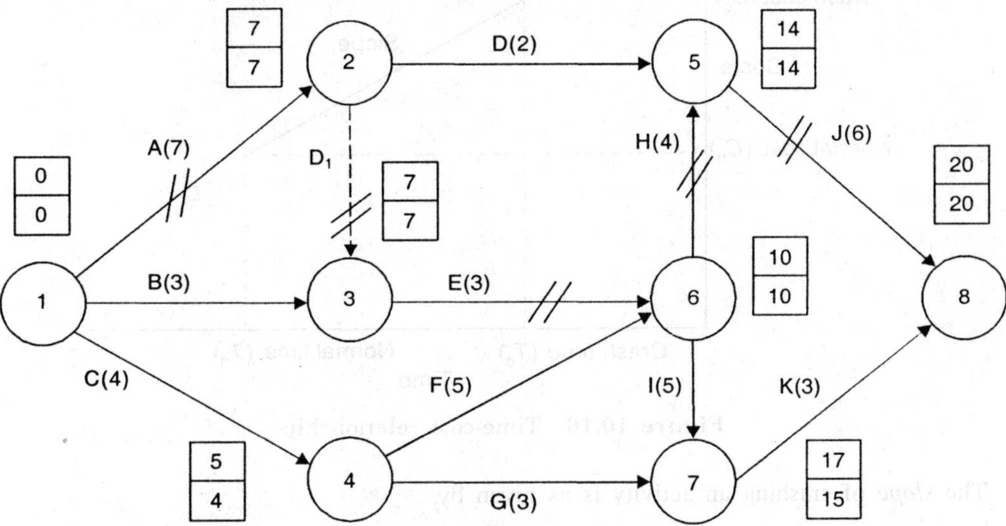

Note: D$_1$ is a dummy activity

Figure 10.15 Critical path calculations.

(d) The probability of completing the project on or before 25 weeks is computed as shown below:

The sum of the variance of the critical activities = 0.11 + 1.78 + 0 + 4 = 5.89 weeks

Therefore, the standard deviation, $\sigma = (5.89)^{0.5} = 2.427$ weeks

$$P(X \leq 25) = P[(X - \mu)/\sigma \leq (25 - 20)/2.427]$$
$$= P(Z \leq 2.06) = 0.9803.$$

10.7   CRASHING OF PROJECT NETWORK

In any project network, the first stage is to determine critical path with normal activity timings. Then, the execution of various activities can be expedited, if necessary. This is called *crashing of activity timings*. In many situations, we may be interested in finding the least possible project completion time if crashing of activities is permitted. While crashing a particular activity, there is a lower limit beyond which it is not possible to reduce its time anymore. This is called *crash limit* of that activity. So, each and every activity will have two time estimates—normal time and crash time. *Normal time* is the time taken to execute an activity under normal circumstances. *Crash time* is the minimum duration of an activity beyond which it is not possible to reduce it anymore.

The cost associated with the normal time is called *normal cost* and the cost associated with the crash time is called *crash cost*. It is obvious that the crash cost should be more than the normal cost. This concept of time-cost relationship is shown in Figure 10.16.

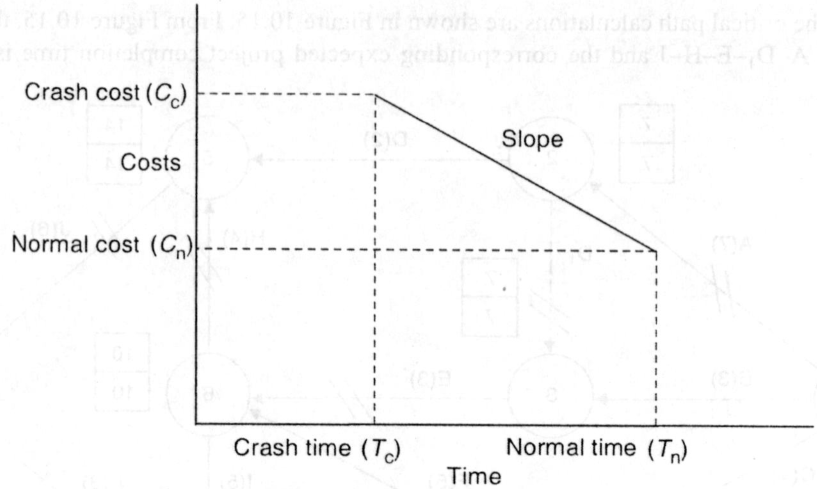

Figure 10.16 Time-cost relationship.

The *slope* of crashing an activity is as given by

$$\text{Slope} = \frac{C_c - C_n}{T_n - T_c}$$

The slope of crashing an activity means the incremental cost of expediting that activity per unit period.

The crashing of a network increases the direct cost because of expedition of activities. But it results in decreased project completion time. This in turn reduces indirect costs, like cost of supervision, security personnel salaries, etc. Hence, there is a trade-off between the direct cost and indirect cost when a project is crashed. So the objective is to crash the project until it becomes uneconomical.

10.7.1 General Guidelines for Network Crashing

Following are the guidelines for network crashing:

1. The network diagram should be drawn with normal activity durations, indicating the following on the network:

   - Duration of each activity
   - Early and late events times for each node
   - Free float for each activity.

   Next, the critical path (or paths) and the corresponding project completion time are determined.

2. If there is only one critical path, then crash the least-cost activity (least slope activity) from among the critical activities within the crash limit and free float limit of that activity.

3. For the activity which is selected based on the minimum cost slope, compression limit for the activity time is given by the following formula:

   $$\text{Compression limit} = \min (\text{Crash limit, Free float limit})$$

Crash limit is the difference between the current duration and the crash duration of the selected activity. Free float limit is determined from among the non-critical activities by reducing the duration of the non-critical activity with the least slope by one-time unit. Then find the free floats of all non-critical activities before and after reduction in time, and identify the non-critical activities for which the free floats after time reduction are reduced by just one unit. Free float limit is the smallest free float (before time reduction) among such non-critical activities.

If there are more than one critical path, then select a common critical activity if it has the least cost slope, and crash it within the compression limit.

If there is no common critical activity with the least cost slope for crashing, then choose a critical activity with the least cost slope in each path. Then they can be crashed within their compression limits.

4. Based on the decision to crash an activity or a set of activities, activity durations are accordingly revised. Revised network is then drawn and the critical path is determined. Now the project duration and the costs can be calculated. Further crashing can be done according to the procedure discussed above. Crashing of project network can be classified into the following two types.

   (a) Crashing the project network until it is not possible to crash anymore (alternatively, we can call this as crashing the project at any cost).
   (b) Crashing the project network until it is uneconomical to crash anymore.

The generalized approach presented previously can be applied without any modification to the first approach, i.e. crashing the project network at any cost. In the second approach, where trade-off between direct cost and indirect cost is considered, the amount of time compressed at a time in any critical activity is just one unit or the compression limit of that activity. In some cases, the usage of compression limit will involve excessive calculations. In such cases, the reduction using unit by unit approach will be efficient.

10.7.2   Crashing of Project Network with Costs Trade-off

In this section, the crashing of project network with costs trade-off is presented with a numerical example.

Example 10.10   Consider the data of a project as shown in Table 10.15.

Table 10.15   Example 10.10

Activity	Normal time (weeks)	Normal cost (Rs.)	Crash time (weeks)	Crash cost (Rs.)
1–2	13	700	9	900
1–3	5	400	4	460
1–4	7	600	4	810
2–5	12	800	11	865
3–2	6	900	4	1130
3–4	5	1000	3	1180
4–5	9	1500	6	1800

If the indirect cost per week is Rs. 160, find the optimal crashed project completion time.

Solution The normal and crash details along with slopes are summarized in Table 10.16.

Table 10.16 Normal Time, Crash Time and Slope Details

Activity	Normal time (weeks)	Normal cost (Rs.)	Crash time (weeks)	Crash cost (Rs.)	Slope (Rs.)
1–2	13	700	9	900	50
1–3	5	400	4	460	60
1–4	7	600	4	810	70
2–5	12	800	11	865	65
3–2	6	900	4	1130	115
3–4	5	1000	3	1180	90
4–5	9	1500	6	1800	100

Iteration 1: The CPM calculations are shown in Figure 10.17. We also have the following results:

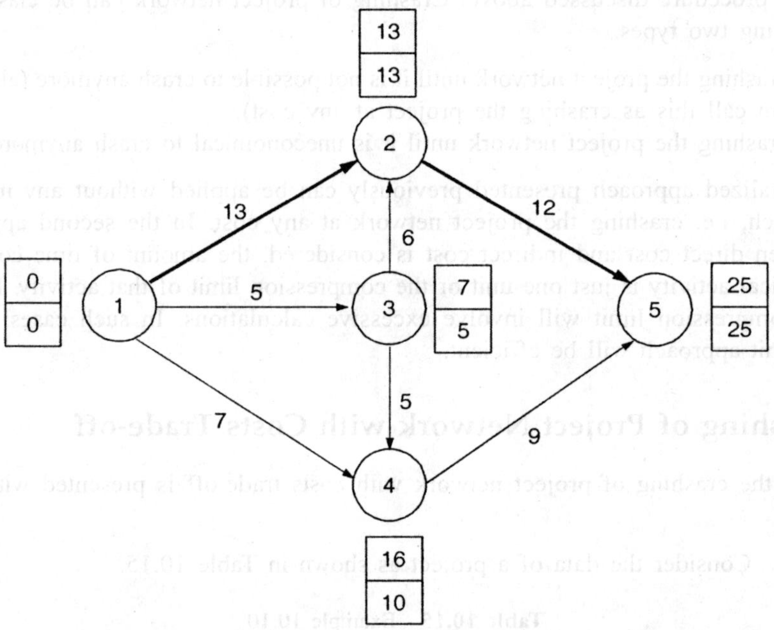

Figure 10.17 Critical path calculations of Iteration 1.

The normal project completion time = 25 weeks

Critical path = 1–2–5

Total direct cost (normal cost) = Rs. 5900

Indirect cost = Rs. 25 × Rs. 160 = Rs. 4000

Total cost = Rs. 5900 + Rs. 4000 = Rs. 9900

The details of crash limit and slope of each of the critical activities are summarized in Table 10.17.

Table 10.17 Crash Limits and Slopes of Critical Activities

Critical activity	Crash limit (weeks)	Cost slope (Rs.)
1–2	4	50*
2–5	1	65

In Table 10.17, the critical activity with minimum cost slope is 1–2. Hence, crash the duration of the activity 1–2 by one week from week 13 to week 12 as shown in Figure 10.18.

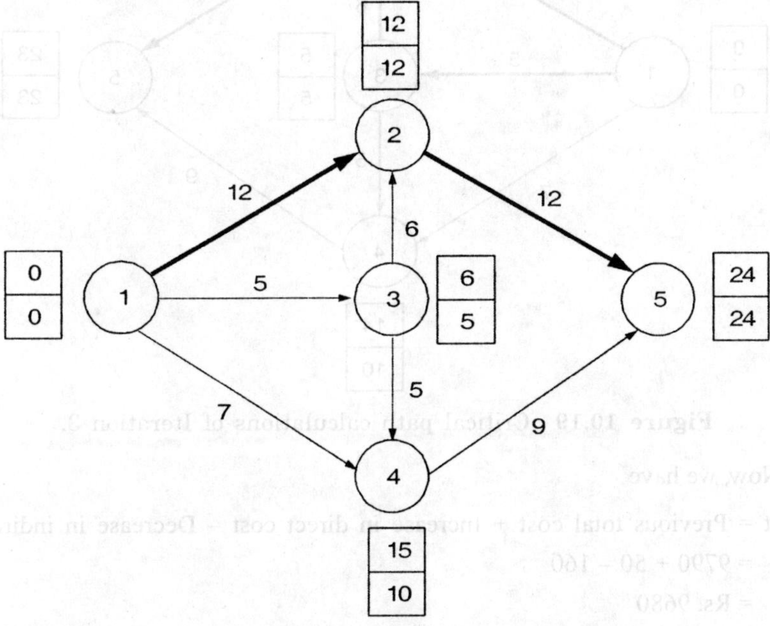

Figure 10.18 Critical path calculations of Iteration 2.

Iteration 2: We have

Total cost = Previous total cost + Increase in direct cost − Decrease in indirect cost

$$= 9900 + 50 - 160$$
$$= 9790$$

Since the cost of this iteration is less than that of the previous iteration, proceed further.

In Figure 10.18, the critical path is not altered. The details of the crash limit and cost slope of each of the critical activities on the critical path are shown in Table 10.18.

Table 10.18 Crash Limits and Slopes of Critical Activities

Critical activity	Crash limit (weeks)	Cost slope (Rs.)
1–2	3	50*
2–5	1	65

In Table 10.18, the critical activity with the minimum cost slope is 1–2. Hence, crash the duration of the activity 1–2 by 1 week from week 12 to week 11 as shown in Figure 10.19.

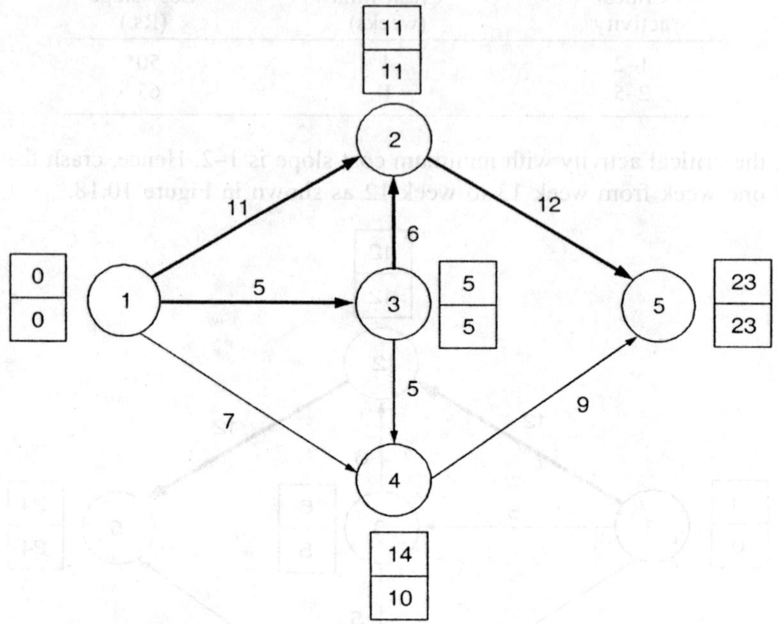

Figure 10.19 Critical path calculations of Iteration 3.

Iteration 3: Now, we have

Total cost = Previous total cost + Increase in direct cost – Decrease in indirect cost

$$= 9790 + 50 - 160$$

$$= Rs. 9680$$

Since the cost of this iteration is less than that of the previous iteration, proceed further.

In Figure 10.19, there are two critical paths: 1–2–5 and 1–3–2–5.

The details of the crash limit and cost slope of each of the critical activities on each critical path are shown in Table 10.19.

Table 10.19 Crash Limits and Slopes of Critical Activities

Critical path	Critical activity	Crash limit (weeks)	Cost slope (Rs.)
1–2–5	1–2	2	50
	2–5	1	65*
1–3–2–5	1–3	1	60
	3–2	2	115
	2–5	1	65

In Table 10.19, the activity 1–2 has the least slope of 50. If this is to be crashed, the critical activity 1–3 which has the least slope among the critical activities (parallel to the part of the critical path which contains the activity 1–2), should also be crashed simultaneously. The total cost of crashing these two critical activities by one week is Rs. 110 (Rs. 50 + Rs. 60). Contrary to this alternative, one can consider the activity 2–5 which is common to both the critical paths. Crashing of this activity by one week will cost Rs. 65.

Hence, the second alternative of crashing the activity 2–5 by one week is considered and accordingly, it is crashed as shown in Figure 10.20.

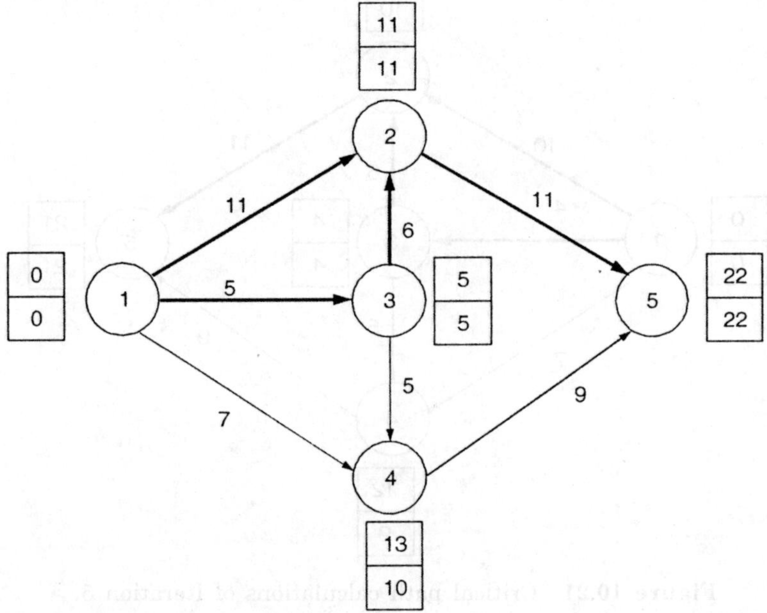

Figure 10.20 Critical path calculations of Iteration 4.

Iteration 4: Here,

Total cost = Previous total cost + Increase in direct cost − Decrease in indirect cost

$$= 9680 + 65 - 160$$

$$= \text{Rs. } 9585.$$

Since the cost of this iteration is less than that of the previous iteration, proceed further.

In Figure 10.20, there are two critical paths: 1–2–5 and 1–3–2–5. The details of the crash limit and cost slope of each of the critical activities on each critical path are shown in Table 10.20.

Table 10.20 Crash Limits and Slopes of Critical Activities

Critical path	Critical activity	Crash limit (weeks)	Cost slope (Rs.)
1–2–5	1–2	2	50*
	2–5	0	–
1–3–2–5	1–3	1	60*
	3–2	2	115
	2–5	0	–

In Table 10.20, the activity 1–2 has the least slope of 50. If this is to be crashed, the critical activity 1–3 with the least slope among the critical activities (which are parallel to the part of the critical path that contains the activity 1–2) must also be crashed simultaneously. Hence, the activities 1–2 and 1–3 are crashed simultaneously by one week as shown in Figure 10.21.

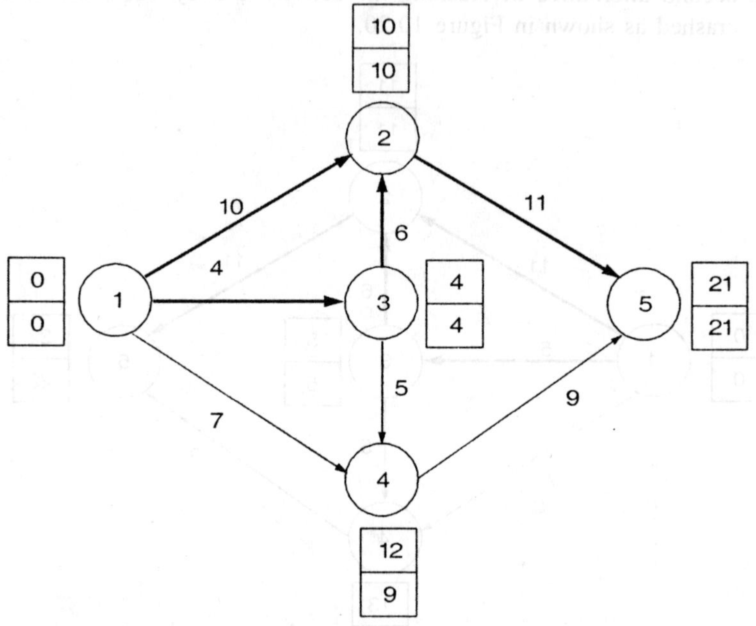

Figure 10.21   Critical path calculations of Iteration 5.

Iteration 5:   We can write

Total cost = Previous total cost + Increase in direct cost – Decrease in indirect cost

$$= 9585 + 50 + 60 - 160$$
$$= \text{Rs. } 9535.$$

Since the cost of this iteration is less than that of the previous iteration, proceed further.

The details of the crash limit and cost slope of each of the critical activities on each critical path are shown in Table 10.21.

Table 10.21   Crash Limits and Slopes of Critical Activities

Critical path	Critical activity	Crash limit (weeks)	Cost slope (Rs.)
1–2–5	1–2	1	50[*]
	2–5	0	–
1–3–2–5	1–3	0	–
	3–2	2	115[*]
	2–5	0	–

In Table 10.21, the activity 1–2 has the least slope of 50. If this is to be crashed, the critical activity 3–2 which is the only critical activity having scope for crashing and parallel to the part of the critical path (which contains the activity 1–2) must also be crashed simultaneously. Hence, the activities 1–2 and 1–3 are crashed simultaneously by one week as shown in Figure 10.22.

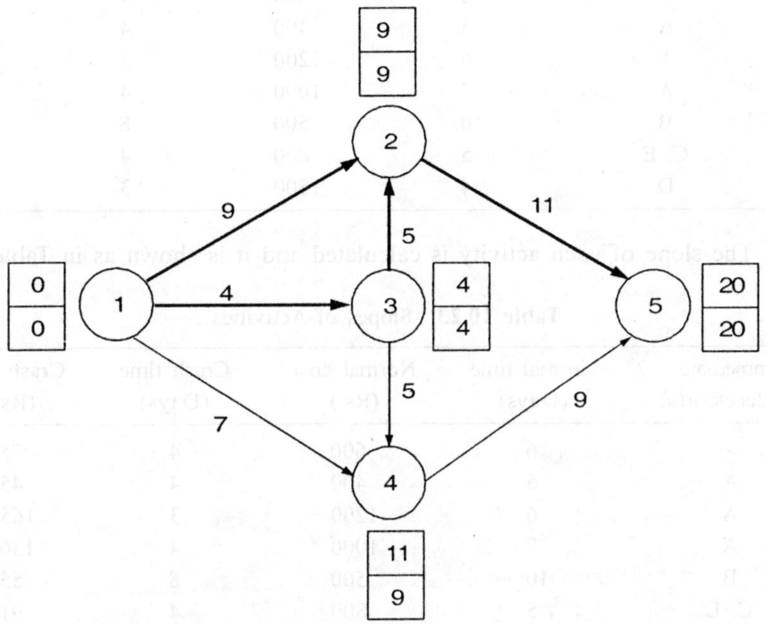

Figure 10.22 Critical path calculations of Iteration 6.

Iteration 6: For this iteration, we have

Total cost = Previous total cost + Increase in direct cost − Decrease in indirect cost

$$= 9535 + (50 + 115) - 160$$

$$= \text{Rs. } 9540.$$

Final result. Since the total cost of this iteration is more than that of the previous iteration, stop the procedure and treat the solution of the previous iteration which is shown in Figure 10.21 as the best solution for implementation. The final crashed project completion time is 21 weeks. Corresponding critical paths are: 1–2–5 and 1–3–2–5.

Example 10.11 Consider the details of a project as shown in Table 10.22. In Table 10.22, normal time in days, normal cost, crash time in days and crash cost for each of the activities in the project are given. The indirect cost of the project is Rs. 100 per day. Find the crashed duration of the project with the optimal total cost.

Table 10.22 Data of Example 10.11

Activity	Immediate predecessor(s)	Normal time (Days)	Normal cost (Rs.)	Crash time (Days)	Crash cost (Rs.)
A	–	6	600	4	750
B	A	5	400	4	450
C	A	6	1200	3	1650
D	A	7	1000	4	1360
E	B	10	500	8	550
F	C, E	5	800	4	910
G	D	4	1500	3	1660

Solution The slope of each activity is calculated and it is shown as in Table 10.23.

Table 10.23 Slopes of Activities

Activity	Immediate predecessor(s)	Normal time (Days)	Normal cost (Rs.)	Crash time (Days)	Crash cost (Rs.)	Slope (Rs.)
A	–	6	600	4	750	75
B	A	5	400	4	450	50
C	A	6	1200	3	1650	150
D	A	7	1000	4	1360	120
E	B	10	500	8	550	25
F	C, E	5	800	4	910	110
G	D	4	1500	3	1660	160

Iteration 1: The project network for the given problem is shown in Figure 10.23. In Figure 10.23, for each activity, the normal duration is taken as its execution time. As per these durations, the critical path calculations are shown in the same figure.

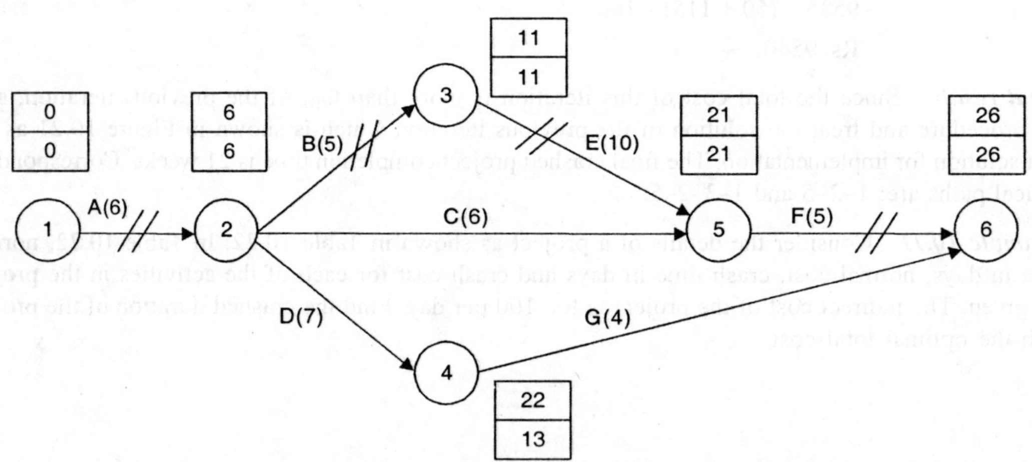

Figure 10.23 Critical path calculations with normal times.

In Figure 10.23, the critical path is A–B–E–F and the corresponding project completion time is 26 days.

$$\text{Direct cost} = \text{Total normal costs} = \text{Rs. } 6000$$
$$\text{Indirect cost} = 26 \times 100 = \text{Rs. } 2600$$
$$\text{Total cost} = \text{Rs. } 8600.$$

Iteration 2: The slope and crash limit of each of the critical activities in the critical path are summarized in Table 10.24. From Table 10.24, the least slope is Rs. 25 and the corresponding critical activity is E which has the crash limit of 2 days. Hence, crash its duration by one day as shown in Figure 10.24.

Table 10.24 Selection of Activity for Crashing

Critical activity	Slope (Rs.)	Crash limit (days)
A	75	2
B	50	1
E	25	2*
F	110	1

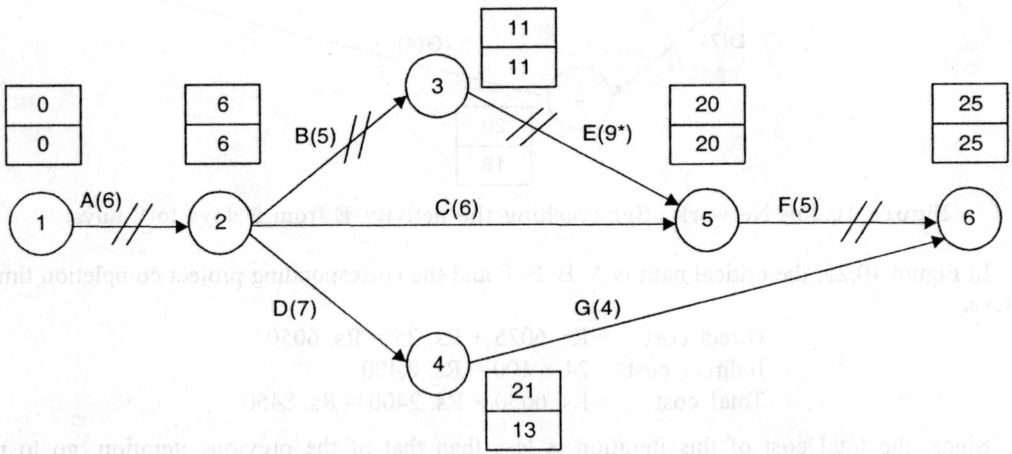

Figure 10.24 Network after crashing the activity E from 10 days to 9 days.

In Figure 10.24, the critical path is A–B–E–F and the corresponding project completion time is 25 days.

$$\text{Direct cost} = \text{Rs. } 6000 + \text{Rs. } 25 = \text{Rs. } 6025$$
$$\text{Indirect cost} = 25 \times 100 = \text{Rs. } 2500$$
$$\text{Total cost} = \text{Rs. } 6025 + \text{Rs. } 2500 = \text{Rs. } 8525.$$

Since, the total cost of this iteration is less than that of the previous iteration, go to next iteration.

Iteration 3: The slope and the crash limit of each of the activities in the critical path are shown in Table 10.25. From Table 10.25, it is identified that the next activity which is to be crashed is again E. The crashing of the activity E from 9 days to 8 days is shown in Figure 10.25.

Table 10.25 Selection of Activity for Crashing

Critical activity	Slope (Rs.)	Crash limit (days)
A	75	2
B	50	1
E	25	1*
F	110	1

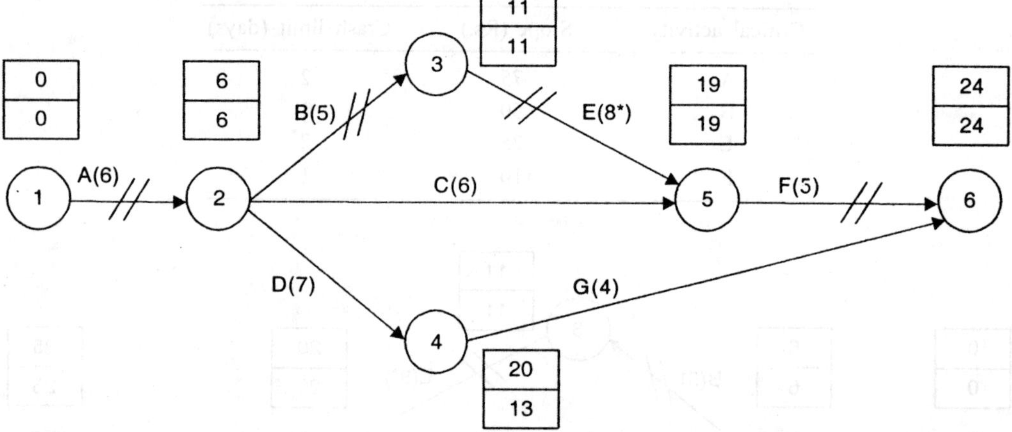

Figure 10.25 Network after crashing the activity E from 9 days to 8 days.

In Figure 10.25, the critical path is A–B–E–F and the corresponding project completion time is 24 days.

$$\text{Direct cost} \quad = \text{Rs. } 6025 + \text{Rs. } 25 = \text{Rs. } 6050$$
$$\text{Indirect cost} = 24 \times 100 = \text{Rs. } 2{,}400$$
$$\text{Total cost} \quad = \text{Rs. } 6050 + \text{Rs. } 2400 = \text{Rs. } 8450.$$

Since, the total cost of this iteration is less than that of the previous iteration, go to next iteration.

Iteration 4: The slope and the crash limit of each of the activities in the critical path are shown in Table 10.26. From Table 10.26, the next activity which is to be crashed is B. The crashing of the activity B from 5 days to 4 days is shown in Figure 10.26.

Table 10.26 Selection of Activity for Crashing

Critical activity	Slope (Rs.)	Crash limit (days)
A	75	2
B	50	1*
E	25	0
F	110	1

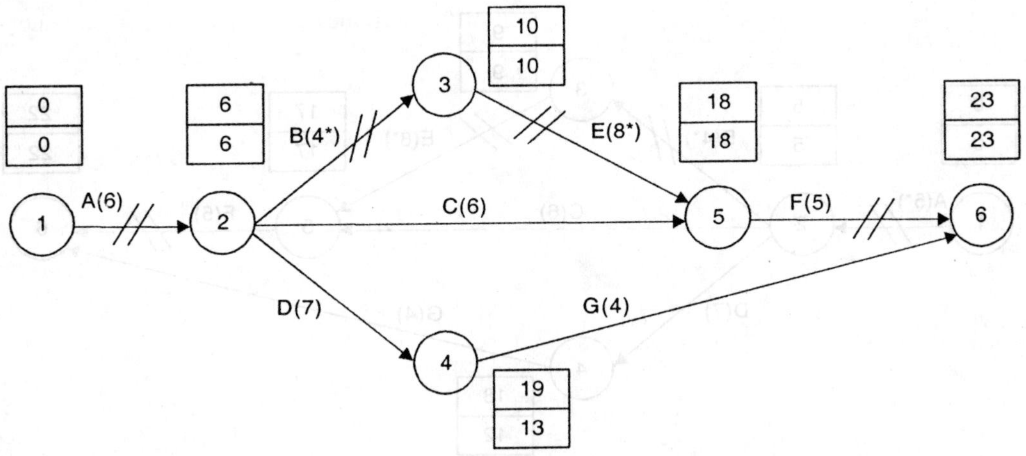

Figure 10.26 Network after crashing activity B from 5 days to 4 days.

In Figure 10.26, the critical path is A–B–E–F and the corresponding project completion time is 23 days.

$$\text{Direct cost} \quad = \text{Rs. } 6050 + \text{Rs. } 50 = \text{Rs. } 6100$$
$$\text{Indirect cost} = 23 \times 100 = \text{Rs. } 2300$$
$$\text{Total cost} \quad = \text{Rs. } 6100 + \text{Rs. } 2300 = \text{Rs. } 8400.$$

Since, the total cost of this iteration is less than that of the previous iteration, go to next iteration.

Iteration 5: The slope and the crash limit of each of the activities in the critical path are shown in Table 10.27. From Table 10.27, it is identified that the next activity which is to be crashed is A. The crashing of the activity A from 6 days to 5 days is shown in Figure 10.27.

Table 10.27 Selection of Activity for Crashing

Critical activity	Slope (Rs.)	Crash limit (days)
A	75	2*
B	50	0
E	25	0
F	110	1

In Figure 10.27, the critical path is A–B–E–F and the corresponding project completion time is 22 days.

$$\text{Direct cost} \quad = \text{Rs. } 6100 + \text{Rs. } 75 = \text{Rs. } 6175$$
$$\text{Indirect cost} = 22 \times 100 = \text{Rs. } 2200$$
$$\text{Total cost} \quad = \text{Rs. } 6175 + \text{Rs. } 2200 = \text{Rs. } 8375.$$

Since, the total cost of this iteration is less than that of the previous iteration, go to next iteration.

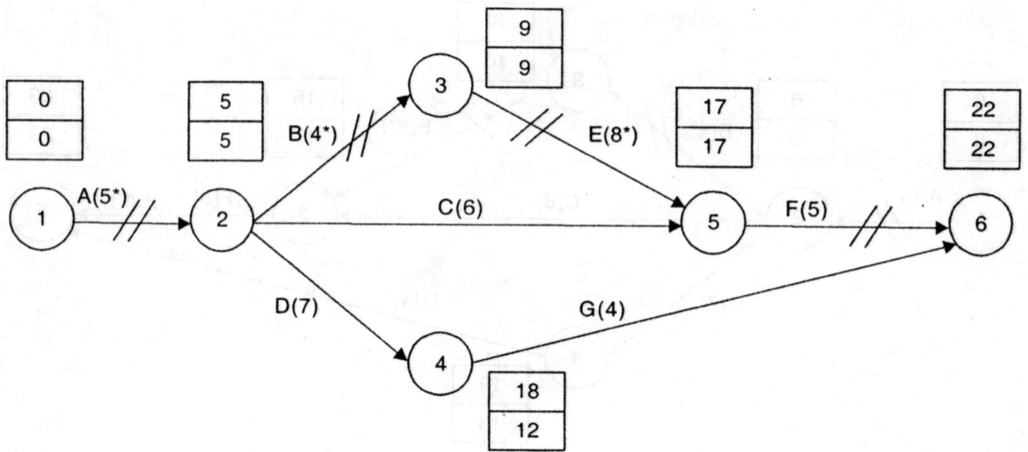

Figure 10.27 Network after crashing activity A from 6 days to 5 days.

Iteration 6: The slope and the crash limit of each of the activities in the critical path are shown in Table 10.28. From Table 10.28, it is identified that the next activity which is to be crashed is again A. The crashing of the activity A from 5 days to 4 days is shown in Figure 10.28.

Table 10.28 Selection of Activity for Crashing

Critical activity	Slope (Rs.)	Crash limit (days)
A	75	1*
B	50	0
E	25	0
F	110	1

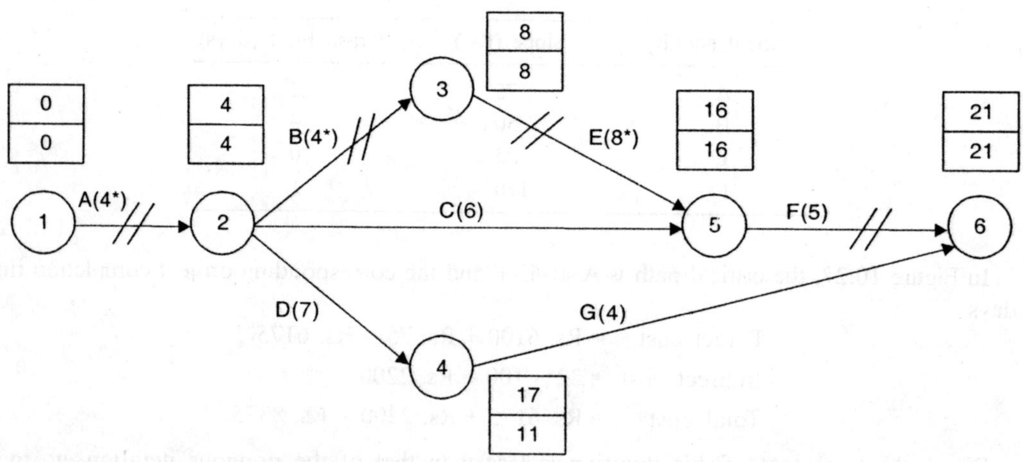

Figure 10.28 Network after crashing activity A from 5 days to 4 days.

In Figure 10.28, the critical path is A–B–E–F and the corresponding project completion time is 21 days.

$$\text{Direct cost} = \text{Rs. } 6175 + \text{Rs. } 75 = \text{Rs. } 6250$$

$$\text{Indirect cost} = 21 \times 100 = \text{Rs. } 2100$$

$$\text{Total cost} = \text{Rs. } 6250 + \text{Rs. } 2100 = \text{Rs. } 8350.$$

Since, the total cost of this iteration is less than that of the previous iteration, go to next iteration.

Iteration 7: The slope and the crash limit of each of the activities in the critical path are shown in Table 10.29. From Table 10.29, it is identified that the next activity which is to be crashed is F. The crashing of the activity F from 5 days to 4 days is shown in Figure 10.29.

Table 10.29 Selection of Activity for Crashing

Critical activity	Slope (Rs.)	Crash limit (days)
A	75	0
B	50	0
E	25	0
F	110	1*

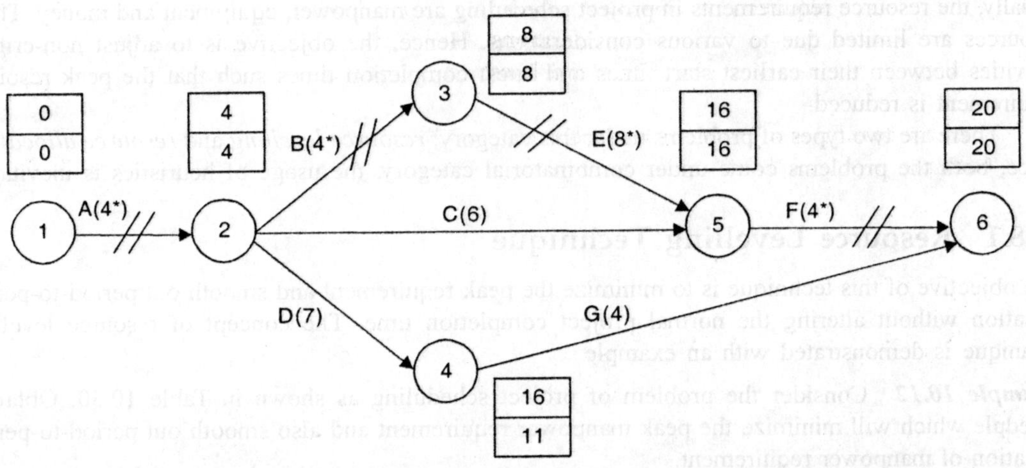

Figure 10.29 Network after crashing activity F from 5 days to 4 days.

In Figure 10.29, the critical path is A–B–E–F and the corresponding project completion time is 20 days.

$$\text{Direct cost} = \text{Rs. } 6250 + \text{Rs. } 110 = \text{Rs. } 6360$$

$$\text{Indirect cost} = 20 \times 100 = \text{Rs. } 2000$$

$$\text{Total cost} = \text{Rs. } 6360 + \text{Rs. } 2000 = \text{Rs. } 8360.$$

Since the total cost of this iteration is more than that of the previous iteration, treat the solution of the previous iteration (Iteration 6) as the best solution which is as reproduced in

Figure 10.30. The critical path in the best solution is A–B–E–F and the corresponding project completion time is 21 days.

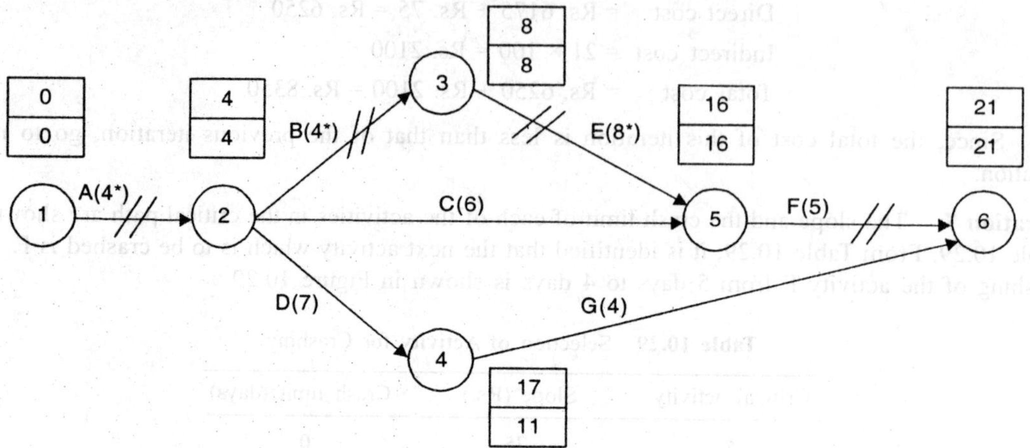

Figure 10.30 Optimally crashed project network (from Iteration 6).

10.8 PROJECT SCHEDULING WITH CONSTRAINED RESOURCES

Usually, the resource requirements in project scheduling are manpower, equipment and money. These resources are limited due to various considerations. Hence, the objective is to adjust non-critical activities between their earliest start times and latest completion times such that the peak resource requirement is reduced.

There are two types of problems under this category: *resource levelling* and *resource allocation*. Since, both the problems come under combinatorial category, the usage of heuristics is inevitable.

10.8.1 Resource Levelling Technique

The objective of this technique is to minimize the peak requirement and smooth out period-to-period variation without altering the normal project completion time. The concept of resource levelling technique is demonstrated with an example.

Example 10.12 Consider the problem of project scheduling as shown in Table 10.30. Obtain a schedule which will minimize the peak manpower requirement and also smooth out period-to-period variation of manpower requirement.

Table 10.30   Example 10.12

Activity	Duration (weeks)	Manpower requirement
1–2	8	7
1–3	6	13
1–4	8	9
2–4	12	11
2–5	4	6
3–5	4	3
4–6	10	15
5–6	10	5

Solution The project network and CPM calculations are shown in Figure 10.31.

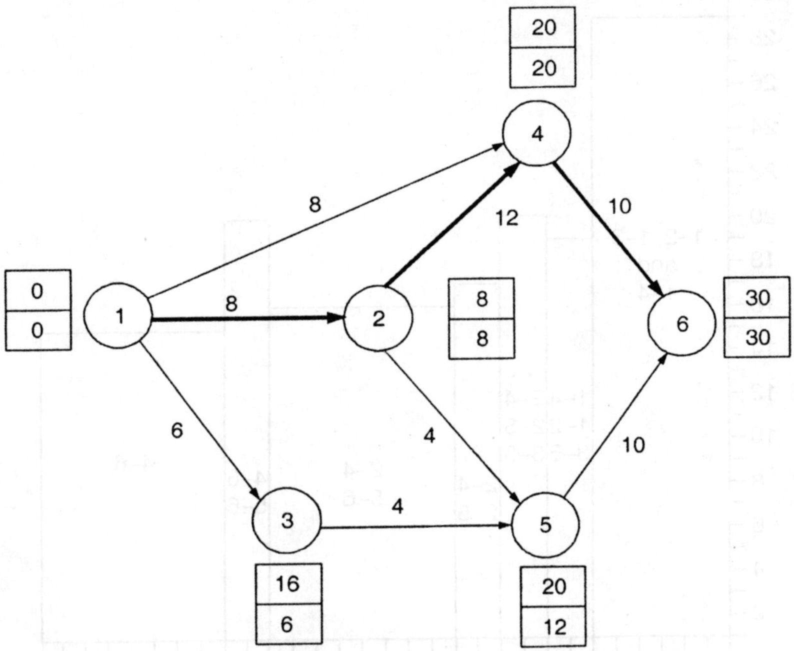

Figure 10.31 CPM calculations of Example 10.12.

Iteration 1: The earliest start schedule of the project and the corresponding manpower requirement diagram are shown in Figures 10.32(a) and 10.32(b), respectively. The peak manpower requirement is 29 and it occurs from the beginning of the project to the end of the sixth week. The activities which are scheduled during this period are: 1–2, 1–3 and 1–4. The activity 1–2 is a critical activity. So, it should not be disturbed. Between activities 1–3 and 1–4, the activity 1–4 has a slack of twelve weeks.

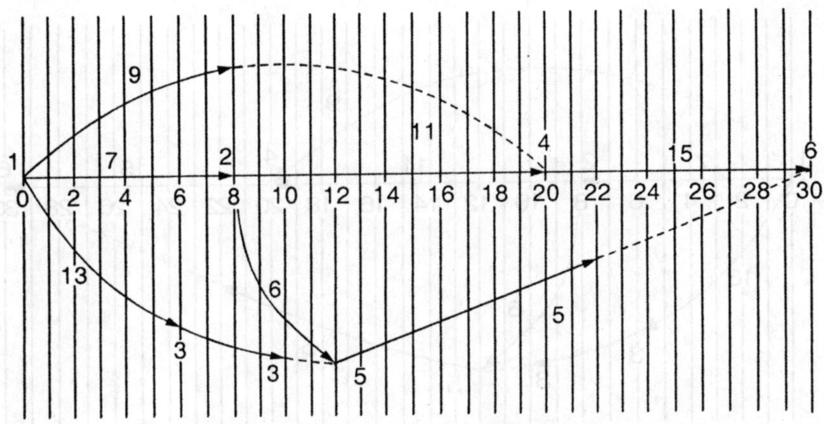

Figure 10.32(a) Network schedule of Iteration 1.

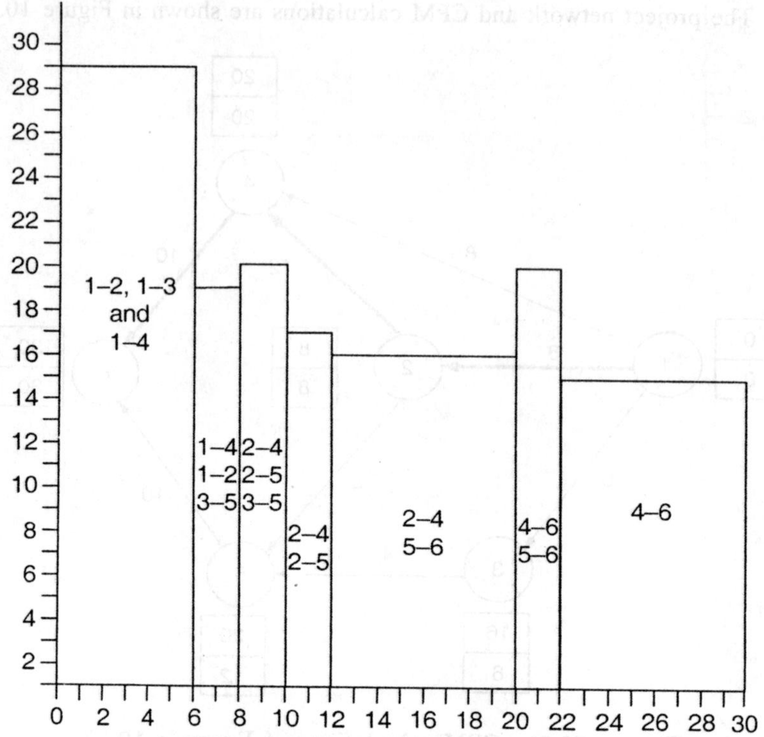

Figure 10.32(b) Resource levelling diagram (Iteration 1).

Hence, postpone it to the maximum extent (i.e. start it at the latest possible time, which means the end of the 12th week). The corresponding modifications are shown in Figures 10.33(a) and 10.33(b).

Iteration 2: In Figure 10.33(b), the peak manpower requirement is 25 and it occurs from the end of the 12th week to the end of the 20th week. The activities which are scheduled during this period are: 1–4, 2–4 and 5–6. The activity 2–4 is a critical activity. So, it should not be disturbed. Between

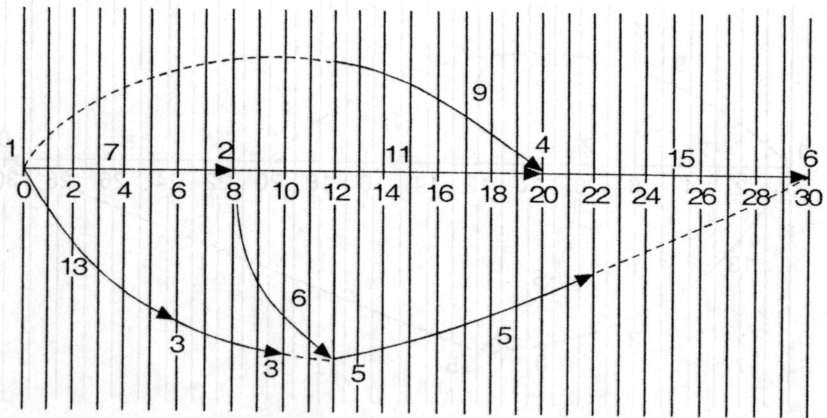

Figure 10.33(a) Network schedule of Iteration 2.

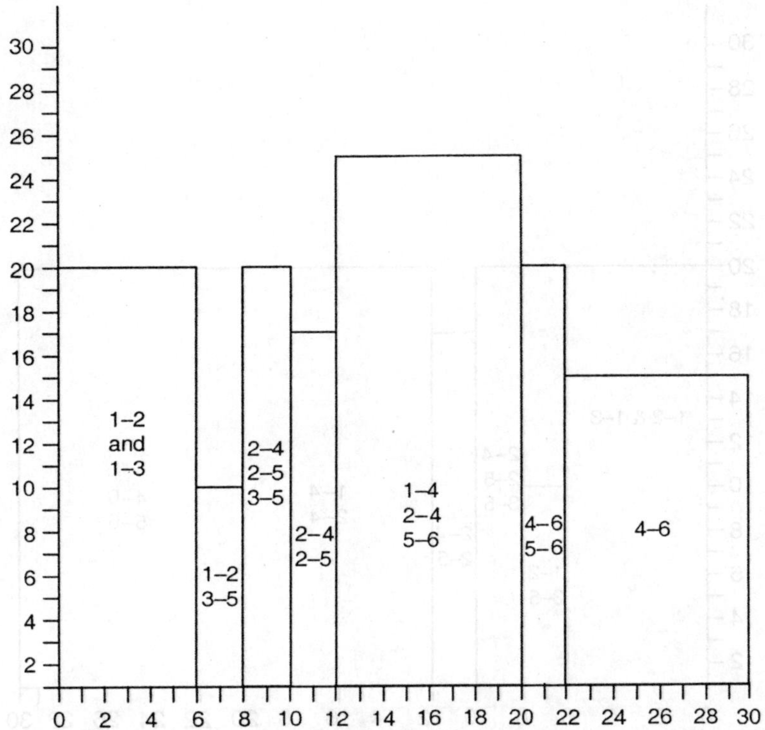

Figure 10.33(b) Resource levelling diagram (Iteration 2).

activities 1–4 and 5–6, the activity 5–6 has the maximum slack of eight weeks. Hence, postpone it to the maximum extent (start it at its latest possible time, i.e. the end of the 20th week). The corresponding modifications are shown in Figures 10.34(a) and 10.34(b).

Iteration 3: In Figure 10.34(b), the maximum manpower requirement is 20 and it is not possible to reduce it any more. So, the corresponding results, as shown in the Figure 10.34(a) is the final and it is recommended for implementation.

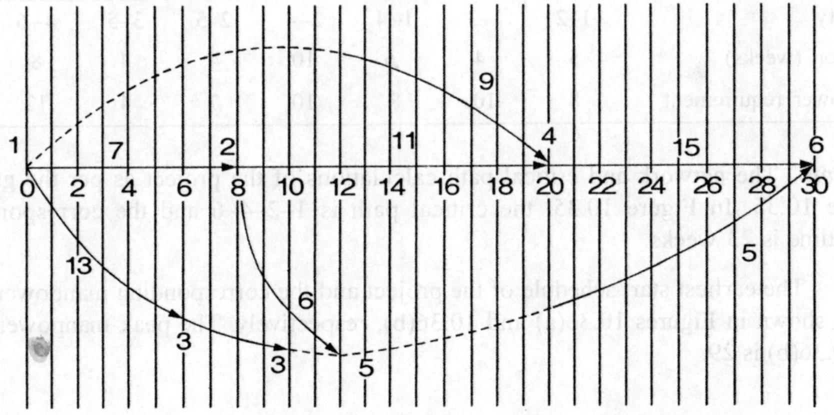

Figure 10.34(a) Network schedule of Iteration 3.

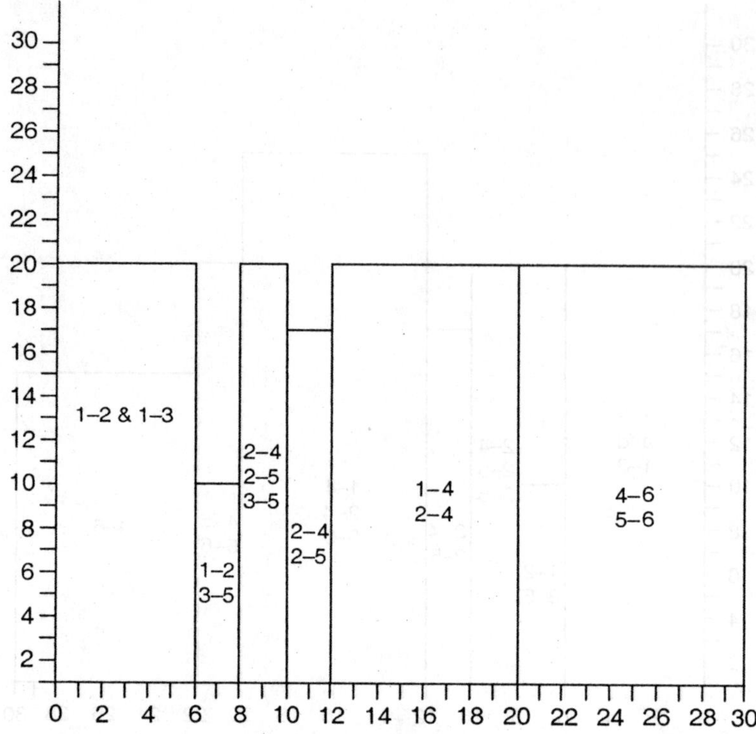

Figure 10.34(b) Resource levelling diagram (Iteration 3).

Example 10.13 A project consists of eight activities as shown in Table 10.31 The duration in weeks and the manpower requirement for each of the activities are also summarized in the same table. Find the project schedule which minimizes the peak manpower requirement and also minimizes period-to-period variation in manpower requirement.

Table 10.31 Data of Example 10.13

Activity	1–2	1–3	1–4	2–4	2–5	3–5	4–6	5–6
Duration (weeks)	5	4	6	10	4	4	8	9
Manpower requirement	8	10	8	10	7	4	12	6

Solution The network and critical path calculations of the project as per the given data are as in Figure 10.35. In Figure 10.35, the critical path is 1–2–4–6 and the corresponding project completion time is 23 weeks.

Iteration 1: The earliest start schedule of the project and the corresponding manpower requirement diagram are shown in Figures 10.36(a) and 10.36(b), respectively. The peak manpower requirement in Figure 10.36(b) is 29.

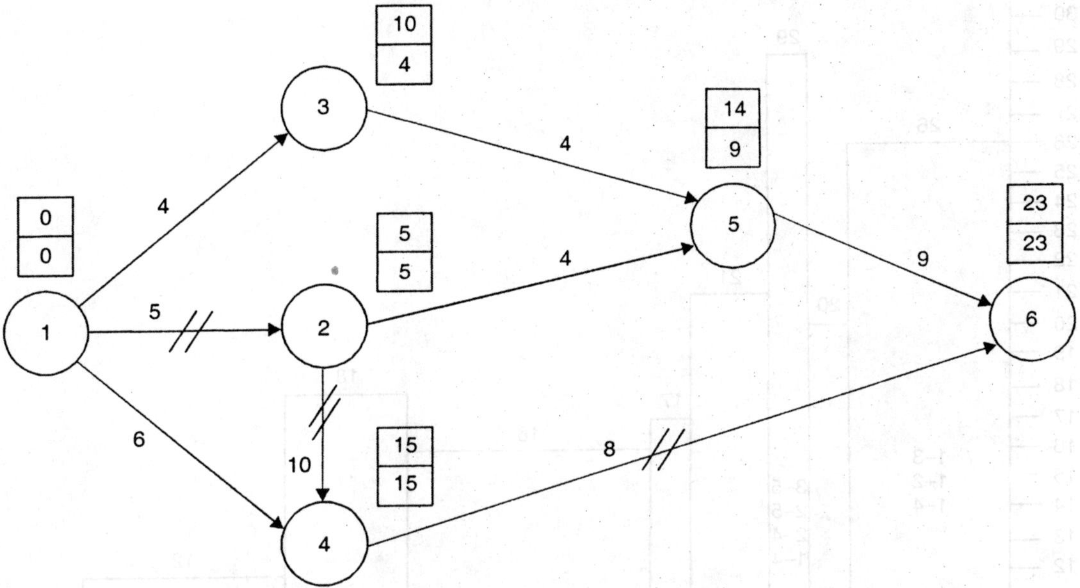

Figure 10.35 Critical path calculations of Example 10.12.

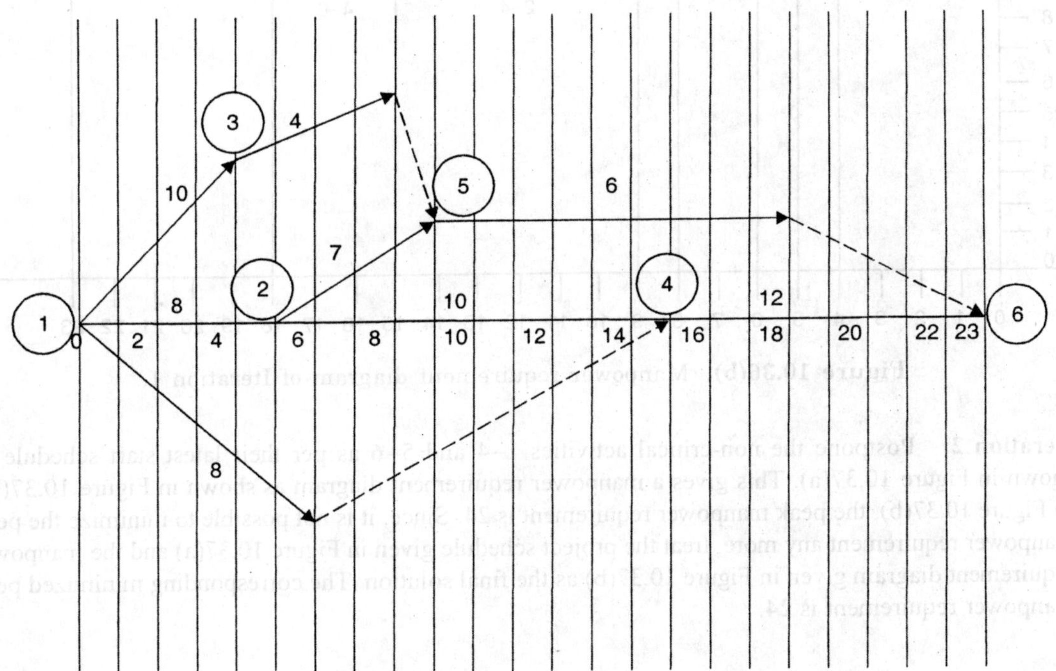

Figure 10.36(a) Network schedule of Iteration 1.

Iteration 2. Restore the non-critical activities 3–5 and 5–6 to their least-cost position schedule as shown in Figure 10.37(a). This gives a manpower requirement histogram as shown in Figure 10.37(b). In Figure 10.37(b) the peak manpower requirement is 27. Since it is not possible to minimize the peak manpower requirement any more than the project schedule given in Figure 10.37(a) and the manpower requirement diagram given in Figure 10.37(b) is the final solution. The corresponding minimized peak manpower requirement is 27.

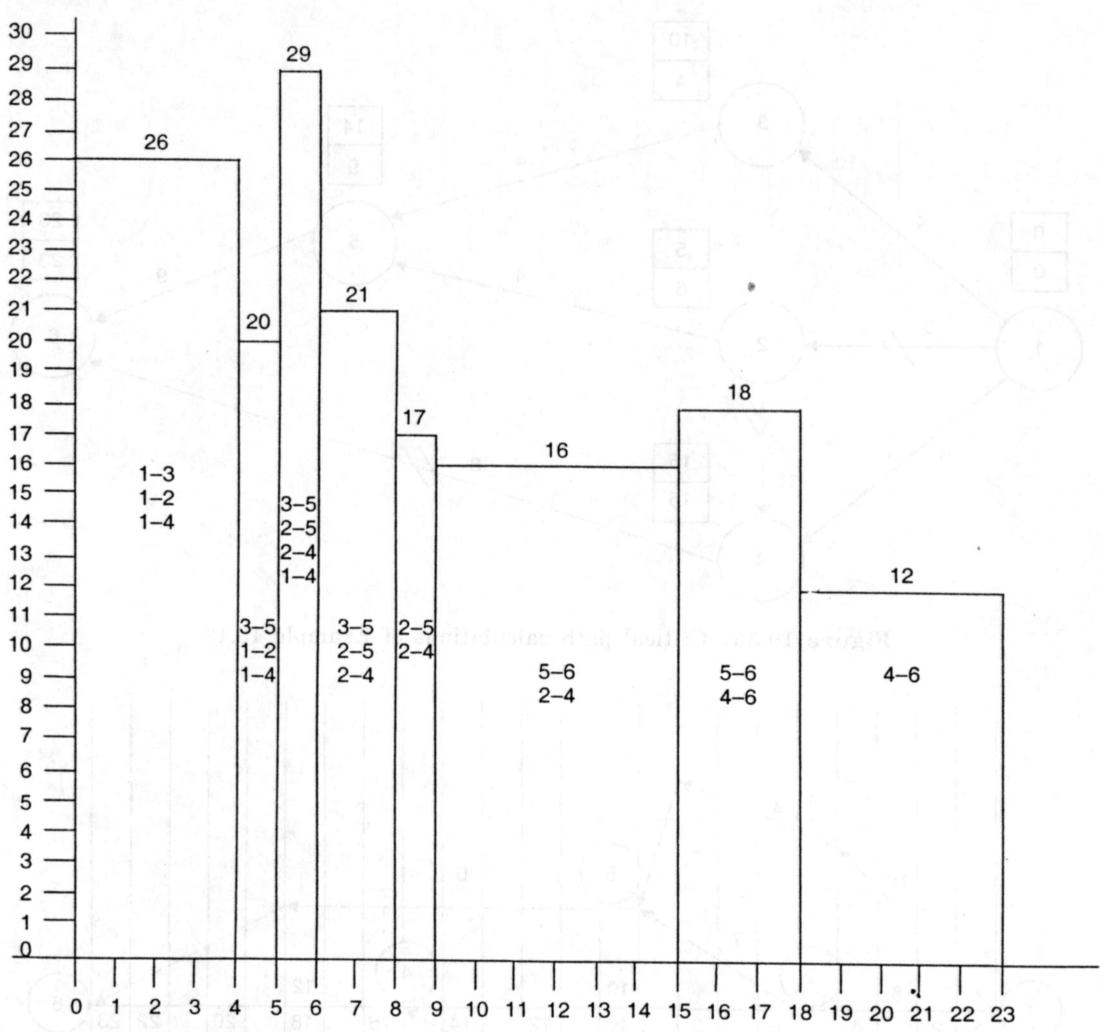

Figure 10.36(b) Manpower requirement diagram of Iteration 1.

Iteration 2: Postpone the non-critical activities 1–4 and 5–6 as per their latest start schedule as shown in Figure 10.37(a). This gives a manpower requirement diagram as shown in Figure 10.37(b). In Figure 10.37(b), the peak manpower requirement is 24. Since, it is not possible to minimize the peak manpower requirement any more, treat the project schedule given in Figure 10.37(a) and the manpower requirement diagram given in Figure 10.37(b) as the final solution. The corresponding minimized peak manpower requirement is 24.

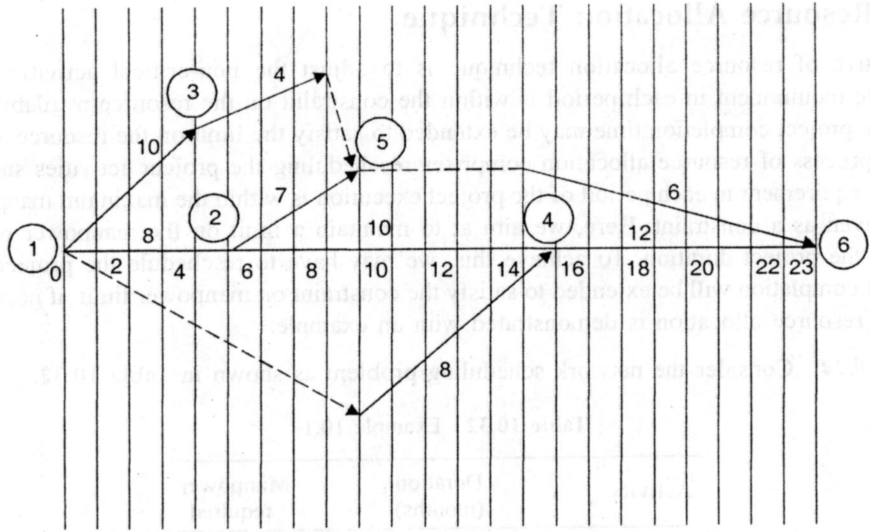

Figure 10.37(a) Network schedule of Iteration 2.

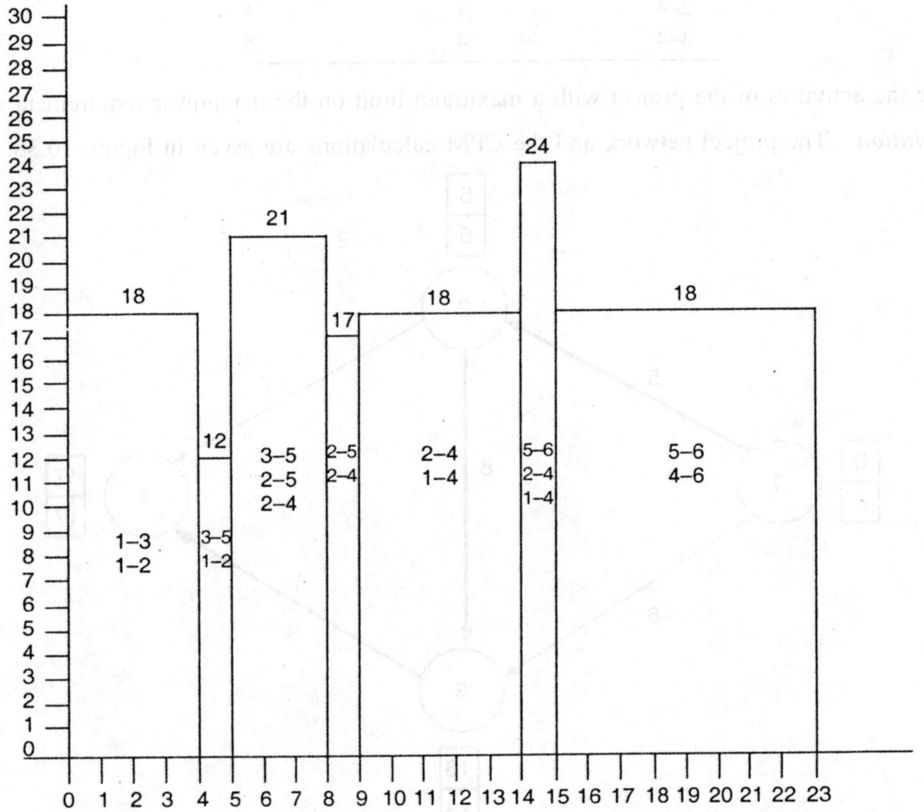

Figure 10.37(b) Manpower requirement diagram of Iteration 2.

10.8.2 Resource Allocation Technique

The objective of resource allocation technique is to adjust the non-critical activities such that the resource requirement in each period is within the constraint on the resource availability. In this process, the project completion time may be extended to satisfy the limit on the resource availability.

The process of resource allocation comprises rescheduling the project activities such that the manpower requirement in each period of the project execution is within the maximum manpower limit which is given as a constraint. Here, we aim at to maintain a limit on the manpower requirement throughout the project duration. To achieve this, we may have to reschedule the project activities. The project completion will be extended to satisfy the constraint on manpower limit, if necessary. The concept of resource allocation is demonstrated with an example.

Example 10.14 Consider the network scheduling problem as shown in Table 10.32.

Table 10.32 Example 10.14

Activity	Duration (months)	Manpower required
1–2	5	12
1–3	6	4
2–3	8	6
2–4	7	3
3–4	4	8

Schedule the activities of the project with a maximum limit on the manpower requirement as 12.

Solution The project network and the CPM calculations are given in Figure 10.38.

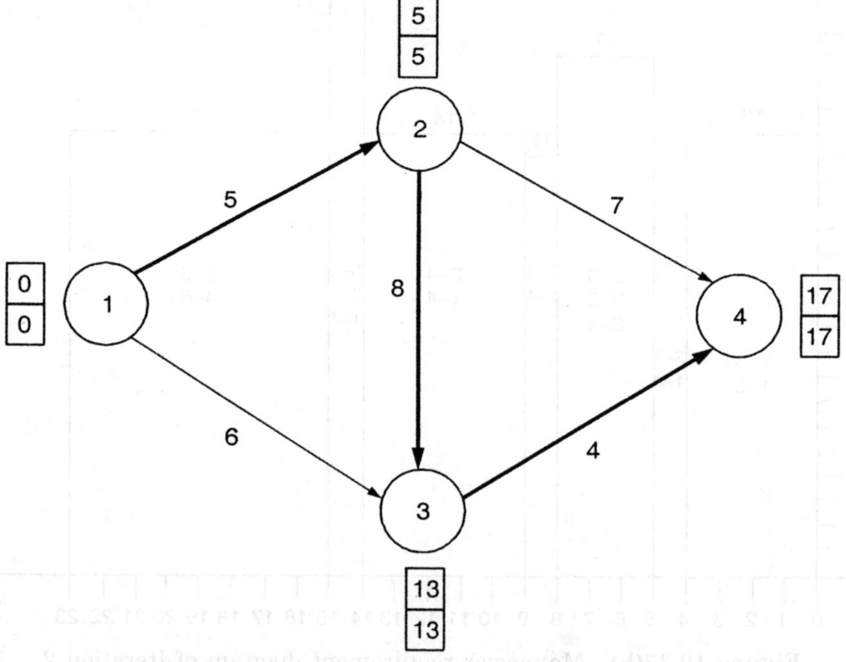

Figure 10.38 Network with CPM calculations.

Here, the critical path is 1–2–3–4 and then normal project completion time is 17 months.

Iteration 1:   The earliest start schedule of the project and the corresponding manpower requirement diagram are shown in Figures 10.39(a) and 10.39(b), respectively.

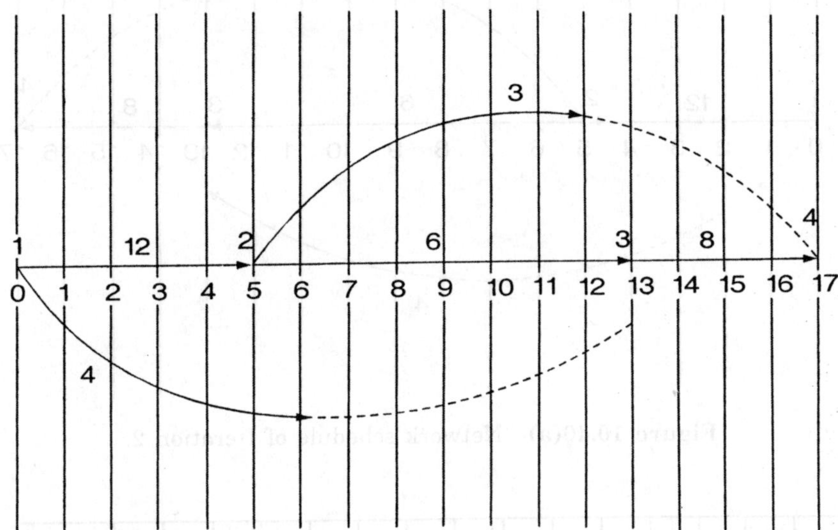

Figure 10.39(a)   Network schedule of Iteration 1.

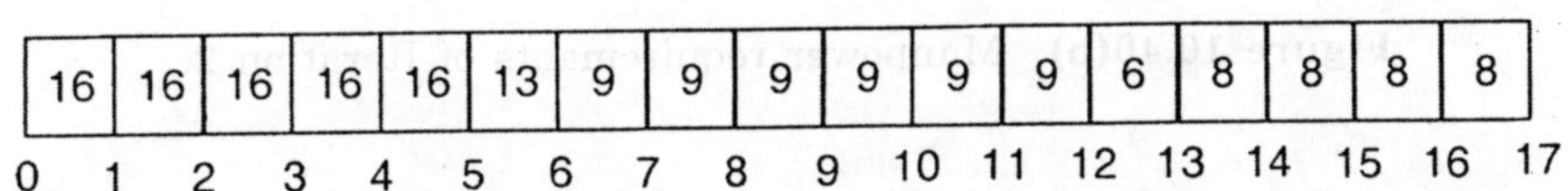

Figure 10.39(b)   Manpower requirements of Iteration 1.

The manpower requirement is represented on each activity (arc). Examine the actual manpower allocated month by month. If the actual manpower allocated as per the project schedule is more than the upper limit of 12, then postpone the non-critical activity (activities); with the most slack(s) such that the actual resource allocation on that month is less than or equal to the manpower limit. In spite of this arrangement, if the manpower allocated in any period is more than the manpower limit, postpone the critical activity (activities). Pre-emption of a scheduled activity is not permitted. If a partly scheduled activity is to be rescheduled just to meet the constraint on manpower limit, that particular activity is to be postponed fully.

Based on these guidelines, the above schedule is modified in the following manner:

Iteration 2:   In Figure 10.39(a), between the beginning of the 1st month and the end of the fifth month, activities 1–2, and 1–3 are scheduled. Total manpower requirement of these activities is more than 12. The activity 1–2 is a critical activity. So, the activity 1–3 is the only non-critical activity during this period and it has a total slack of 7 months. Hence, postpone the starting of this activity to the beginning of the 8th month. These are shown in Figure 10.40(a), and the corresponding manpower requirement diagram is shown in Figure 10.40(b).

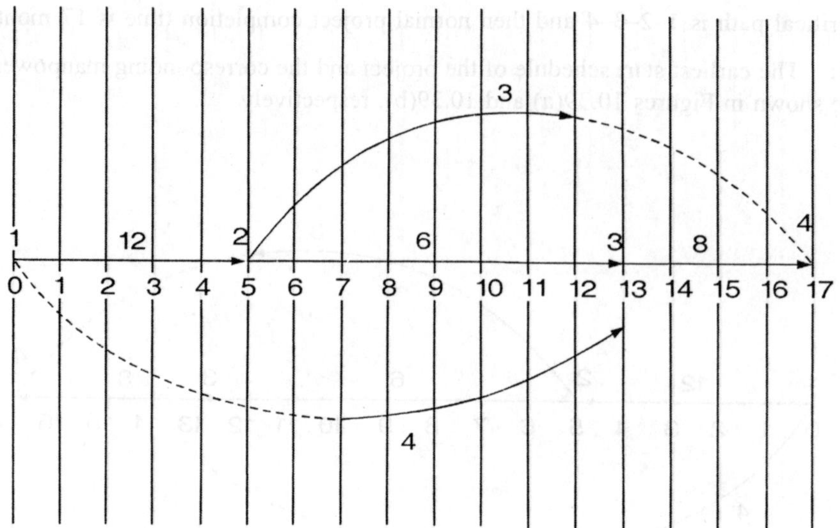

Figure 10.40(a) Network schedule of Iteration 2.

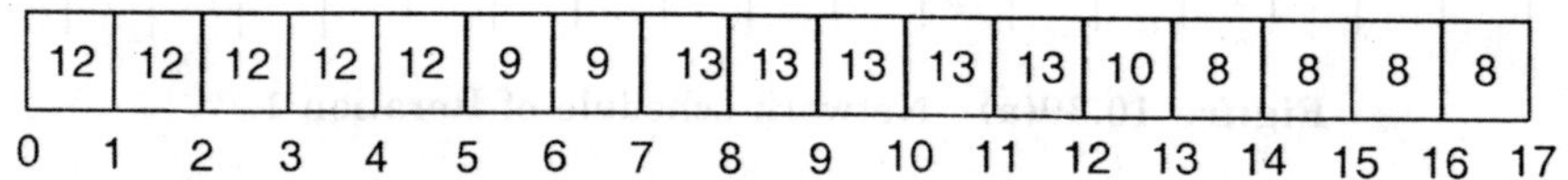

Figure 10.40(b) Manpower requirements of Iteration 2.

Iteration 3: From the beginning of the 8th month to the end of the 12th month, the activities 2–3, 1–3 and 2–4 are scheduled. The total manpower requirement of these activities is 13 which is more than the resource limit of 12. Among these activities, the activity 2–4 has the largest slack and also its latest completion time is more than that of the remaining activities which are listed above. Hence, the beginning of this activity is to be postponed to the beginning of the 13th month.

While postponing the activity 2–4 to the beginning of the 13th month, its slack is insufficient to accommodate it before the end of the 17th month. Also, the total manpower requirement for the activities 2–4, 2–3 and 1–3 will be 13 during 13th month. Hence, the critical activity 3–4 is postponed simultaneously along with the non-critical activity 2–4 by three months as shown by dotted line between the beginning of the 14th month and the end of the 16th month. In this process, the project completion time is extended to the end of the 20th month.

The corresponding manpower requirement diagram is shown in Figure 10.41(b). In Figure 10.41(b), the manpower requirements are within the maximum limit of 12 throughout the project duration. Hence, the schedule which is presented in Figure 10.41(a) is the final schedule recommended for implementation.

Example 10.15 A project consists of seven activities as shown in Table 10.33. The duration in weeks and the manpower requirement for each of the activities are also summarized in the same table. Find the project schedule which minimizes the peak manpower requirement and also minimizes period-to-period variation in manpower requirement if the maximum permitted manpower during any week is 15.

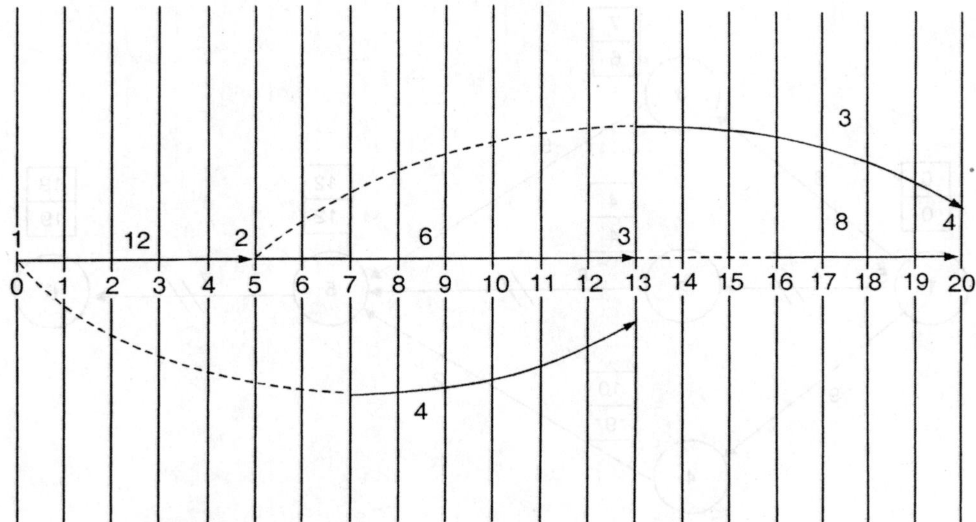

Figure 10.41(a) Network schedule of Iteration 3.

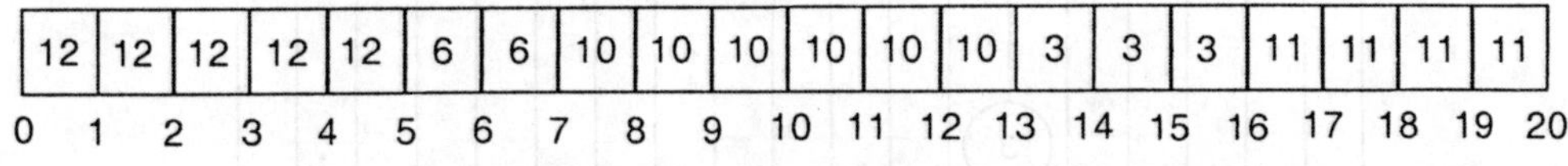

Figure 10.41(b) Manpower requirements of Iteration 3.

Table 10.33 Data of Example 10.15

Activity	1–2	1–3	1–4	2–5	3–5	4–5	5–6
Duration (weeks)	4	6	9	8	5	2	7
Manpower requirement	10	6	8	4	7	5	4

Solution The network and critical path calculations of the project as per the given data are as in Figure 10.42. In Figure 10.42, the critical path is 1–2–5–6 and the corresponding project completion time is 19 weeks.

Iteration 1: The earliest start schedule of the project and the corresponding manpower requirement diagram are shown in Figure 10.43(a) and Figure 10.43(b), respectively. The peak manpower requirement in Figure 10.43(b) is 24 which is more than the given limit of 15. Hence, go to next iteration.

Iteration 2: The starting of the non-critical activities 1–3 is postponed to the beginning of 5th week and hence, its immediate following non-critical activity 3–5 is started immediately after its completion. Similarly, the starting of the non-critical activities 1–4 is postponed to the beginning of 5th week and hence, its immediately following non-critical activity 4–5 is started immediately after its completion. The above two postponements result in the postponement of the critical activity 5–6 to the beginning of the 16th week. These are shown in Figure 10.44(a). The corresponding manpower requirement diagram is shown in Figure 10.44(b). From Figure 10.44(b), the peak manpower requirement is 19 which is more than the given maximum of 15. Hence, go to next iteration.

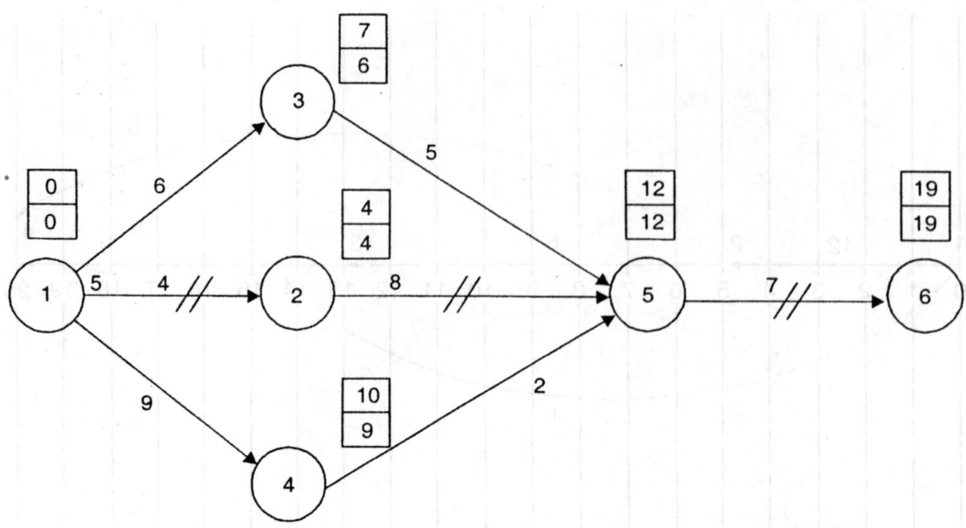

Figure 10.42   Critical path calculations of Example 10.14.

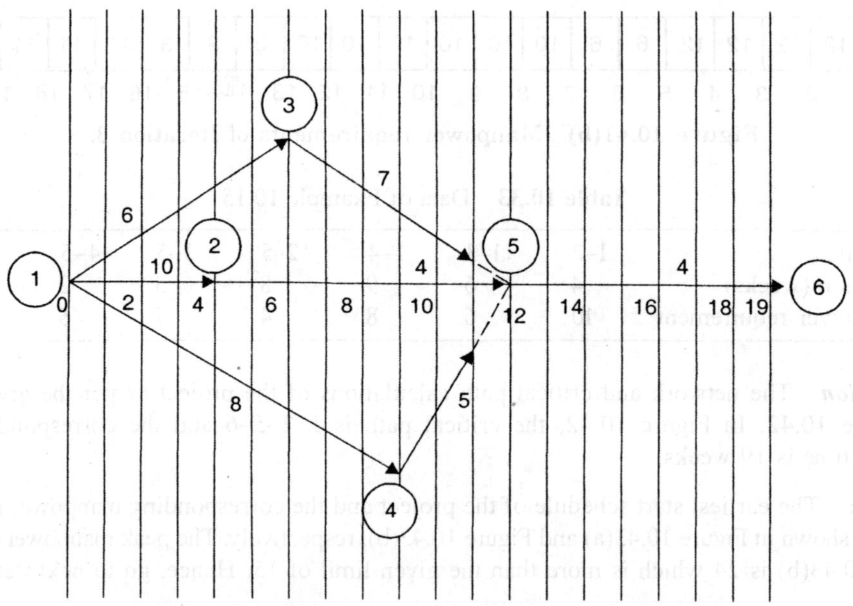

Figure 10.43(a)   Network schedule of Iteration 1.

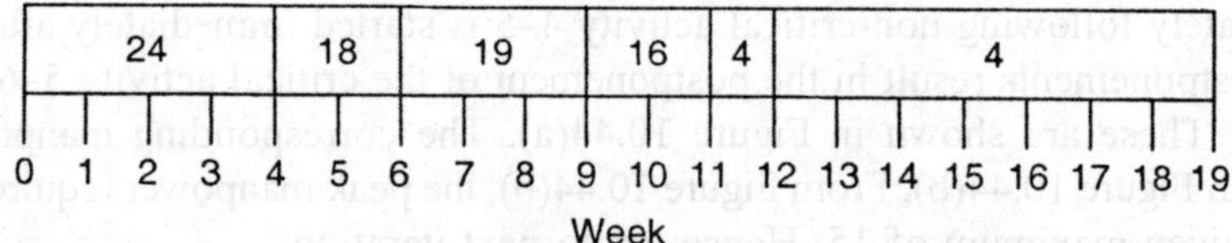

Figure 10.43(b)   Manpower requirement diagram of Iteration 1.

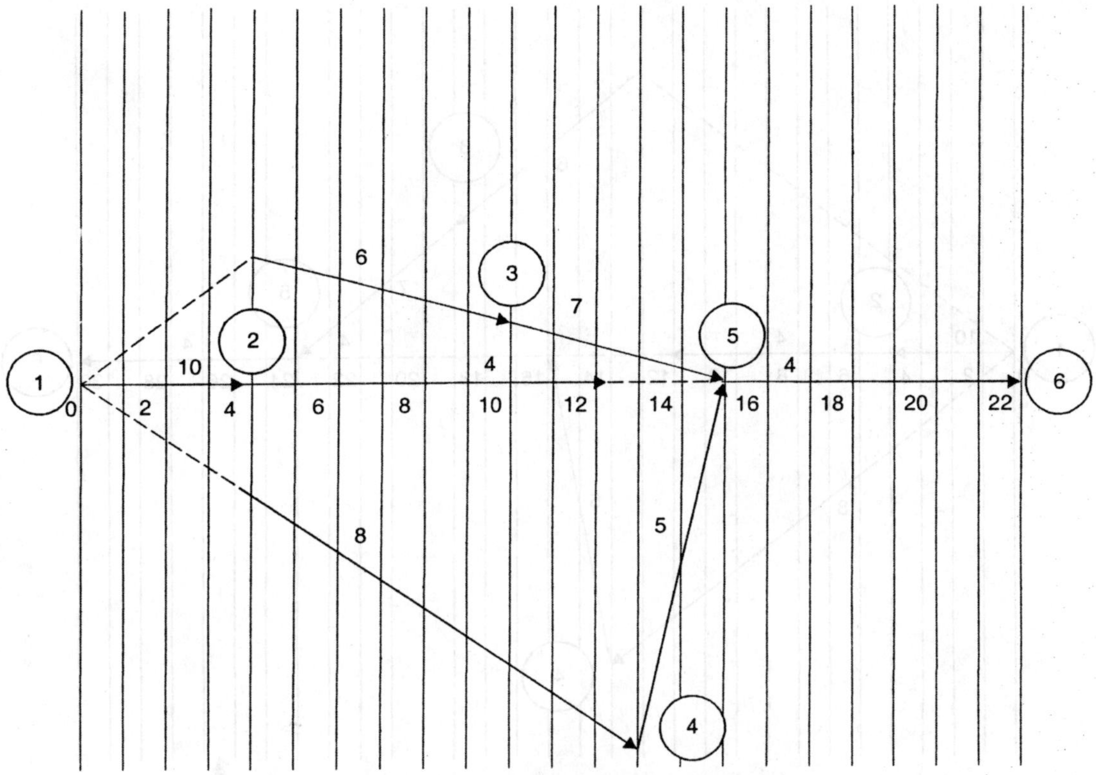

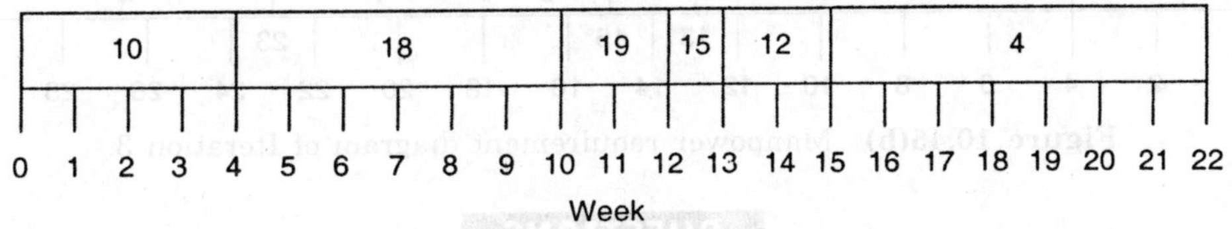

Figure 10.44(a) Network schedule of Iteration 2.

Figure 10.44(b) Manpower requirement diagram of Iteration 2.

Iteration 3: Now, the starting of the non-critical activity 1–3 is postponed to the beginning of the 13th week and its immediate following non-critical activity is scheduled immediately after this activity. This change results in the postponement of the starting of the critical activity 5–6 to the beginning of the 24th week. The revised scheduled as per these changes is shown in Figure 10.45(a) and the corresponding manpower diagram is shown in Figure 10.45(b). From Figure 10.45(b), the peak manpower requirement is 14 which is less than the given maximum limit of 15. Hence, stop the procedure. The schedule given in Figure 10.45(a) is the final and the corresponding project completion time is 30 weeks.

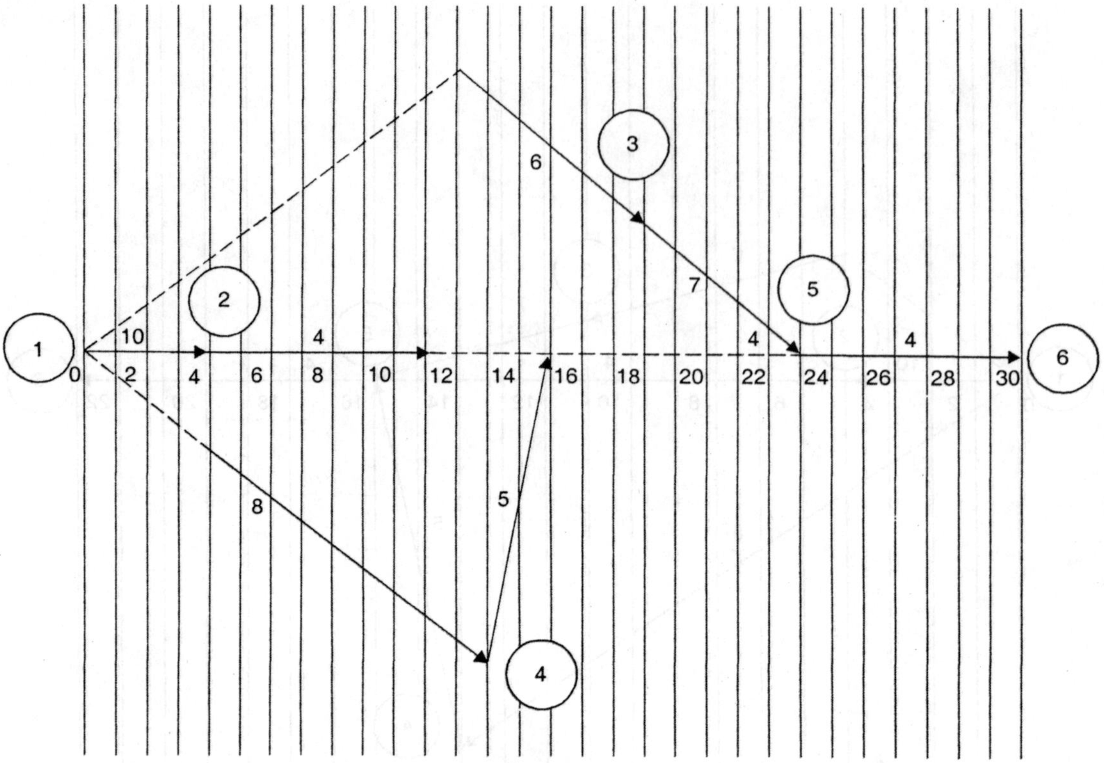

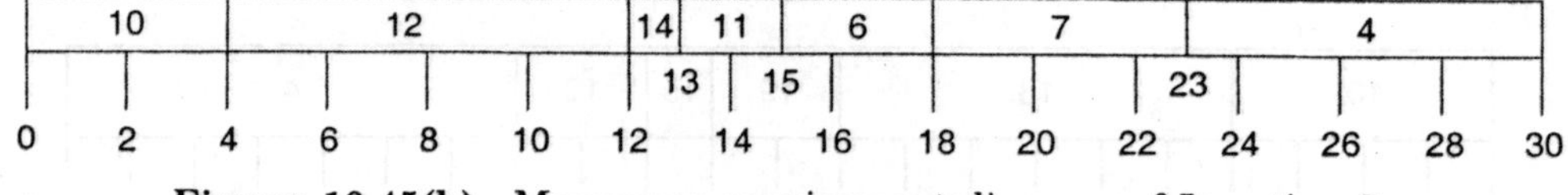

Figure 10.45(a) Network schedule of Iteration 3.

Figure 10.45(b) Manpower requirement diagram of Iteration 3.

QUESTIONS

1. Define 'project', and give some application areas of project management. Explain different phases of project management.

2. Distinguish between CPM and PERT.

3. Discuss the guidelines for constructing a project network.

4. Define the following: (a) total float, (b) free float, and (c) critical path.

5. What are the time estimates used in PERT?

6. Discuss the cost trade-off in project crashing.

7. Distingush between resource levelling and resource allocation.

8. A construction company has listed down various activities that are involved in constructing a building. These are summarized along with predecessor(s) details in the following table.

Activity	Immediate predecessor(s)
A	–
B	–
C	A
D	B
E	A, B
F	C, D
G	F, B
H	E, G
I	H, G
J	I, F
K	J, L
L	A
M	K

Draw a project network for the above project.

9. Consider the details of a project as shown in the table:

Activity	Immediate predecessor(s)	Duration (months)
A	–	4
B	–	8
C	–	5
D	A	4
E	A	5
F	B	7
G	B	4
H	C	8
I	C	3
J	D	6
K	E	5
L	F	4
M	G	12
N	H	7
O	I	10
P	J, K, L	5
Q	M, N, O	8

(a) Construct the CPM network.
(b) Determine the critical path.
(c) Compute total floats and free floats for non-critical activities.

10. The activities involved in Alpha Garment Manufacturing Company are listed with their time estimates as in the following table:

Activity	Description	Immediate predecessor(s)	Duration (days)
A	Forecast sales volume	–	10
B	Study competitive market	–	7
C	Design item and facilities	A	5
D	Prepare production plan	C	3
E	Estimate cost of production	D	2
F	Set sales price	B, E	1
G	Prepare budget	F	14

Draw the network for the given activities and carry out the critical path calculations.

11. Consider the following data of a project.

Activity	Predecessor(s)	Duration (weeks)		
		a	m	b
A	–	3	5	8
B	–	6	7	9
C	A	4	5	9
D	B	3	5	8
E	A	4	6	9
F	C, D	5	8	11
G	C, D, E	3	6	9
H	F	1	2	9

(a) Construct the project network.
(b) Find the expected duration and variance of each activity.
(c) Find the critical path and the expected project completion time.
(d) What is the probability of completing the project on or before 30 weeks?
(e) If the probability of completing the project is 0.9, find the expected project completion time.

12. Consider the following table summarizing the details of a project:

Activity	Predecessor(s)	Duration (weeks)		
		a	m	b
A	–	4	4	10
B	–	1	2	9
C	–	2	5	14
D	A	1	4	7
E	A	1	2	3
F	A	1	5	9
G	B, C	1	2	9
H	C	4	4	4
I	D	2	2	8
J	E, G	6	7	8
K	F, H	2	2	8
L	F, H	5	5	5
M	I, J, K	1	2	9
N	L	6	7	8

(a) Construct the project network.

(b) Find the expected duration and variance of each activity.

(c) Find the critical path and the expected project completion time.

(d) What is the probability of completing the project on or before 35 weeks?

(e) If the probability of completing the project is 0.85, find the expected project completion time.

13. Consider the following data of project.

Activity	Normal time (weeks)	Normal cost (Rs.)	Crash time (weeks)	Crash cost (Rs.)
1–2	7	600	4	840
1–3	11	200	9	First week: Rs. 70 Second week: Rs. 80
2–3	10	800	8	1000
2–4	6	500	4	760
2–5	16	100	9	380
3–4	6	200	4	360
3–5	9	500	4	960
4–5	8	300	5	500

If the indirect cost per week is Rs. 300, find the optimal crashed project completion time.

14. Consider the data of a project as shown in the following table:

Activity	Normal time (weeks)	Normal cost (Rs.)	Crash time (weeks)	Crash cost (Rs.)
1–2	8	800	5	950
1–3	5	500	3	700
1–4	9	600	6	1050
2–5	10	900	8	1300
3–5	5	700	3	1100
3–6	6	1200	5	1500
4–6	7	1300	5	1400
5–7	2	400	1	500
6–7	4	500	2	900

If the indirect cost per week is Rs. 300, find the optimal crashed project completion time.

15. Consider the problem of project scheduling as shown in the following table:

Activity	Duration (weeks)	Manpower requirement
1–2	4	9
1–3	8	5
2–3	10	7
2–4	6	6
3–4	9	8
3–5	5	12
4–5	7	7

Obtain a schedule which will minimize the peak manpower requirement and also smooth out period-to-period variation of manpower requirement.

16. Consider the problem of project scheduling as shown in the following table:

Activity	Duration (weeks)	Manpower requirement
1–2	5	8
1–3	4	10
1–4	6	8
2–4	10	10
2–5	4	7
3–5	4	4
4–6	8	12
5–6	9	6

Obtain a schedule which will **minimize** the peak manpower requirement and also smooth out period-to-period variation of **manpower** requirement.

17. Consider the network scheduling **problem** as shown in the following table:

Activity	Duration (months)	Manpower requirement
1–2	6	10
1–3	4	6
1–4	9	8
2–6	5	4
3–6	8	7
4–6	2	5
6–7	7	4

Schedule the activities of the project with a maximum limit on the manpower requirement as 15.

DECISION THEORY 11

11.1 INTRODUCTION

Decision making is an integral part of any business organization. The process involves selecting the best among several decisions through a proper evaluation of the parameters of each decision environment. The types of decisions can be classified into the following categories.

1. Decision under certainty
2. Decision under risk
3. Decision under uncertainty

These are explained in the following sections.

11.2 DECISION UNDER CERTAINTY (DETERMINISTIC DECISION)

If the availability of information for a decision environment is perfect, then the decision taken under such environment is called *decision under certainty* or *deterministic decision*. Some examples of this decision environment are listed below:

1. If the coefficients of the objective function and the constraints of the well-known product mix problem are constants (deterministic quantities), then the optimal solution of the product mix problem is an example of deterministic decision (decision under certainty).
2. In maintenance management, if the annual maintenance cost and the annual operating cost of an equipment are known in advance, which are not subject to any variation in future, then these quantities are called as deterministic quantities and the corresponding decision on the economic life of the equipment is an example of decision under certainty.
3. In economic evaluation of alternatives, if the annual revenues of different alternatives are known in advance which are not subject to any variation in future, then these quantities are deterministic quantities and the corresponding decision on the selection of the best alternative is an example of decision under certainty.

11.3 DECISION UNDER RISK

If the availability of information for a decision environment is partial, then the decision taken under such environment is called as *decision under risk*. In this case, the information of the decision environment can be described in the form of a probability distribution. The parameters of the probability distribution can be estimated based on the past data of the decision environment. A few examples of this decision environment are listed below.

1. In queueing theory, the arrival pattern and the service pattern can be approximated to some known distributions. Then a suitable empirical queueing model can be identified from the existing literature to find the solution of the model. This is an example of probabilistic situation or decision under risk.

2. In economic evaluation of alternatives, if the annual revenues of different alternatives are described in the form of a probability distribution (either discrete or some known distribution), then the mean and the variance of the annual revenue of each of the alternatives can be estimated from the past data which can be used for further analysis. This is a case of decision under risk.

Two approaches can be used to deal with the decision environment with risk. They are: expected value criterion and expected value combined with variance criterion.

11.3.1 Expected Value Criterion

In this approach, first the expected value of each alternative is computed. Then an alternative with the best expected value is selected for implementation. The expected value criterion is demonstrated through a numerical example as presented below.

Example 11.1   Consider the details of two competing alternatives as shown in Table 11.1. The initial outlay of each of the alternatives is Rs. 10,00,000. The life of each alternative is 10 years. Find the best alternative, when the interest rate is 0%.

Table 11.1   Annual Revenue of Alternatives

Alternative 1		Alternative 2	
Annual revenue (Rs.)	Probability	Annual revenue (Rs.)	Probability
3,00,000	0.3	4,00,000	0.1
4,00,000	0.4	5,00,000	0.5
5,00,000	0.3	6,00,000	0.4

Solution   The expected annual revenue of the Alternative i (ER_i) is computed using the following formula:

$$ER_i = \frac{\sum_{j=1}^{3} R_{ij} p_{ij}}{\sum_{j=1}^{3} p_{ij}}$$

where, R_{ij} is the annual revenue of the jth alternative during the ith year and p_{ij} is the probability of occurrence of the annual revenue of the jth alternative during the year i. Therefore, the expected annual revenues of the given alternatives are:

$$ER_1 = \text{Rs. } 4,00,000 \quad \text{and} \quad ER_2 = \text{Rs. } 5,30,000$$

Since the initial outlays and lives of the alternatives are equal and the interest rate is 0%, the Alternative 2 should be selected, because its expected annual revenue is higher than that of the Alternative 1.

11.3.2 Expected Value Combined with Variance Criterion

The expected value combined with variance criterion is demonstrated through a numerical example presented below.

Example 11.2   Consider the details of two competing alternatives as shown in Table 11.2. The initial outlay of each of the alternatives is Rs. 10,00,000. The life of each alternative is 10 years. Find the best alternative when the interest rate is 0%.

Table 11.2   Mean and Variance of Each Alternative

	Alternative j (Rs.)	
	1	2
Expected annual revenue ($\overline{X}_j$)	5,00,000	4,00,000
Variance of annual revenue (σ_j^2)	10,00,000	10,000

Solution   Since the initial outlays and lives of the alternatives are equal and the interest rate is 0%, it is sufficient to compare the coefficient of variations ($CV_j, j = 1$ and 2) of the annual revenues of the alternatives. Then

$$CV_j = \frac{\sigma j}{\overline{X}_j}$$

The coefficient of variations of Alternative 1 and Alternative 2 are computed as shown below:

$$CV_1 = \frac{(10,00,000)^{1/2}}{5,00,000} = \frac{1}{500} \quad \text{and} \quad CV_2 = \frac{(10,000)^{1/2}}{4,00,000} = \frac{1}{4000}$$

Since the coefficient of variation of Alternative 2 is lesser than that of Alternative 1, Alternative 2 ranks first in terms of selection of the best alternative.

11.4   DECISION UNDER UNCERTAINTY

If the availability of information for a decision environment is incomplete, then the decision taken under such an environment is called as *decision under uncertainty*. Under such situation, the information of the decision environment cannot be described in the form of a probability distribution. Consider a game in which, there are two players each having a set of alternatives to exercise. For each of the combinations of the alternatives of the players, there will be an associated outcome. Both the players will be deciding their options simultaneously to optimize the outcome of the game into their favour. Here, it is not possible to describe any probability distribution to define the decision environment. This is an example of the decision under uncertainty.

There are five criteria for decision making under uncertainty which are listed below.

These are explained in the following sections:

1. Laplace criterion
2. Maximin criterion
3. Minimax criterion
4. Savage minimax regret criterion
5. Hurwicz criterion

11.4.1 Laplace Criterion

Consider the profit (loss) matrix of an uncertain situation as shown in Table 11.3.

Table 11.3   Format of Profit (Loss) Matrix

Future state

	b_1	b_2	...	b_j	...	b_n
a_1	$q(a_1, b_1)$	$q(a_1, b_2)$	...	$q(a_1, b_j)$	...	$q(a_1, b_n)$
a_2	$q(a_2, b_1)$	$q(a_2, b_2)$	...	$q(a_2, b_j)$	...	$q(a_2, b_n)$
$\vdots$	$\vdots$	$\vdots$		$\vdots$		$\vdots$
Action $\quad a_i$	$q(a_i, b_1)$	$q(a_i, b_2)$	...	$q(a_i, b_j)$	...	$q(a_i, b_n)$
$\vdots$	$\vdots$	$\vdots$		$\vdots$		$\vdots$
a_n	$q(a_m, b_1)$	$q(a_m, b_2)$	...	$q(a_m, b_j)$	...	$q(a_m, b_n)$

In Table 11.3, a_i represents the ith action (for $i = 1, 2, ..., m$), b_j represents the jth future state ($j = 1, 2, ..., n$), and $q(a_i, b_j)$ represents the outcome (profit/loss) with respect to the ith action and the jth future state. Based on the principle of insufficient reasoning, the probability of occurrence of each future state is $1/n$. Therefore,

$$\text{The expected outcome of the action } i = \sum_{j=1}^{n} \frac{1}{n} q(a_i, b_j).$$

In the case of maximization (and minimizations problems) the action which has the maximum expected profit (loss or cost) is to be selected as the best action for implementation.

Example 11.3   A retail store desires to determine the optimal daily order size for a perishable item. The store buys the perishable item at the rate of Rs. 80 per kg and sells at the rate of Rs. 100 per kg. If the order size is more than the demand, the excess quantity can be sold at Rs. 70 per kg in a secondary market; otherwise, the opportunity cost for the store is Rs. 15 per kg for the unsatisfied portion of the demand. Based on the past experience, it is found that the demand varies from 50 kg to 250 kg in steps of 50 kg. The possible values of the order size are from 75 kg to 300 kg in steps of 75 kg. Determine the optimal order size which will maximize the daily profit of the store.

Solution   We have

Purchase price of the perishable item = Rs. 80/kg

Selling price of the perishable item in the primary market = Rs. 100/kg

Selling price of the perishable item in the secondary market = Rs. 70/kg

Profit in the primary market = 100 – 80 = Rs. 20/kg

Profit in the secondary market = 70 – 80 = Rs. –10/kg

Opportunity cost of not meeting the demand = Rs. 15/kg

The formula for the daily net profit (NP) is:

$$NP = D_j \times 20 - (Q_i - D_j) \times 10, \qquad \text{if } Q_i \geq D_j$$
$$= Q_i \times 20 - (D_j - Q_i) \times 15, \qquad \text{if } D_j > Q_i$$

where Q_i is the ith order size and D_j is the demand of the jth future state. The corresponding outcomes (daily net profits) are summarized in Table 11.4.

Table 11.4 Summary of Daily Net Profit

Order size (Q_i)	Demand (D_j)				
	50	100	150	200	250
75	$50 \times 20 - 25 \times 10 = 750$	$75 \times 20 - 25 \times 15 = 1125$	$75 \times 20 - 75 \times 15 = 375$	$75 \times 20 - 125 \times 15 = -375$	$75 \times 20 - 175 \times 15 = -1125$
150	$50 \times 20 - 100 \times 10 = 0$	$100 \times 20 - 50 \times 10 = 1500$	$150 \times 20 = 3000$	$150 \times 20 - 50 \times 15 = 2250$	$150 \times 20 - 100 \times 15 = 1500$
225	$50 \times 20 - 175 \times 10 = -750$	$100 \times 20 - 125 \times 10 = 750$	$150 \times 20 - 75 \times 10 = 2250$	$200 \times 20 - 25 \times 10 = 3750$	$225 \times 20 - 25 \times 15 = 4125$
300	$50 \times 20 - 250 \times 10 = -1500$	$100 \times 20 - 200 \times 10 = 0$	$150 \times 20 - 150 \times 10 = 1500$	$200 \times 20 - 100 \times 10 = 3000$	$250 \times 20 - 50 \times 10 = 4500$

The expected daily net profit with respect to each order size is computed as below:

$$E(Q_1) = \frac{1}{5}(750 + 1125 + 375 - 375 - 1125) = 150$$

$$E(Q_2) = \frac{1}{5}(0 + 1500 + 3000 + 2250 + 1500) = 1650$$

$$E(Q_3) = \frac{1}{5}(-750 + 750 + 2250 + 3750 + 4125) = 2025$$

$$E(Q_4) = \frac{1}{5}(-1500 + 0 + 1500 + 3000 + 4500) = 1500$$

The expected daily net profit is maximum when the order size is Q_3, which is equal to 225 kg. So, the store should place an order of 225 kg of the perishable item, daily.

11.4.2 Maximin Criterion

According to this criterion, the minimum guaranteed return (profit/revenue) is maximized. First, the minimum outcome of each row irrespective of the columns is determined. Then, the action with respect to the row corresponding to the maximum of these minimum outcomes is selected as the best action.

Example 11.4 Consider Example 11.3 and determine the best order size using the maximin criterion.

Solution The net profit matrix of Example 11.3 is reproduced in Table 11.5.

Table 11.5   Summary of Daily Net Profit

Demand (D_j)

		50	100	150	200	250	Minimum
	75	750	1125	375	−375	−1125	−1125
Order size	150	0	1500	3000	2250	1500	0*
(Q_i)	225	−750	750	2250	3750	4125	−750
	300	−1500	0	1500	3000	4500	−1500

*Maximin value

In Table 11.5, the maximin value is for row 2. Hence, the action corresponding to row 2 is the best action. This means that the optimal daily order size of the perishable item should be 150 kg.

11.4.3   Minimax Criterion

This criterion is based on how to make the best out of the worst possible conditions. In other words, the cost should be minimized. First, the maximum outcome of each row irrespective of the columns is determined. Then, the action with respect to the row corresponding to the minimum of these maximum outcomes is selected as the best action.

Example 11.5   Consider Table 11.4 of Example 11.3 and treat it as a cost matrix and determine the best action using the minimax criterion.

Solution   The net profit matrix of Example 11.3 is assumed as cost matrix and the corresponding table is shown in Table 11.6.

Table 11.6   Summary of Cost Matrix

Future state j

		b_1	b_2	b_3	b_4	b_5	Maximum
	a_1	750	1125	375	−375	−1125	1125*
Action i	a_2	0	1500	3000	2250	1500	3000
	a_3	−750	750	2250	3750	4125	4125
	a_4	−1500	0	1500	3000	4500	4500

*Minimax value.

In Table 11.6, the minimax value is for row 1. Hence the Action a_1 is the best action.

11.4.4   Savage Minimax Regret Criterion

Sometime, the minimax (maximin) criterion may yield a misleading result because of selecting the action with respect to the minimum of the maximum values of the rows (the maximum of the minimum values of the rows). This is avoided using *savage minimax regret criterion.*

In this criterion, for each column j, where j varies from 1 to n, the regret values $r(a_i, b_j)$ is computed for all values of i varying from 1 to m as shown below. The collection of such values is known as *regret table*:

$$r(a_i, b_j) = \max_{a_k} [q(a_k, b_j)] - q(a_i, b_j), \qquad i = 1, 2, 3,..., m,$$

(if the problem is a maximization type)

$$= q(a_i, b_j) - \min_{a_k} [q(a_k, b_j)], \qquad i = 1, 2, 3,..., m,$$

(if the problem is a minimization type)

In the case of maximization problem, $r(a_i, b_j)$ is the difference between highest outcome for a given column j and each of the values of $q(a_i, b_j)$ in the same column j for i varying from 1 to m.

In the case of minimization problem, $r(a_i, b_j)$ is the difference between each of the values of $q(a_i, b_j)$ in the column j for i varying from 1 to m and the least outcome for the same column j. Then, apply the regular minimax criterion to the regret table to obtain the best action irrespective of the type of the problem (minimization/maximization).

Example 11.6 Consider the data of Example 11.3 and obtain the best action based on the savage minimax regret criterion.

Solution The daily net profit data of the given problem is summarized in Table 11.7. This is a maximization problem. So, the following formula can be used to determine the regret values which are summarized in Table 11.8.

$$r(a_i, b_j) = \max_{a_k} [q(a_k, b_j)] - q(a_i, b_j), \qquad i = 1, 2, 3,..., m$$

Table 11.7   Summary of Daily Net Profit

		Demand (D_j)				
		50	100	150	200	250
	75	750	1125	375	−375	−1125
	150	0	1500	3000	2250	1500
Order size (Q_i)	225	−750	750	2250	3750	4125
	300	−1500	0	1500	3000	4500

Table 11.8   Regret Table

		Demand (D_j)					
		50	100	150	200	250	Maximum
	75	0	375	2625	4125	5625	5625
	150	750	0	0	1500	3000	3000
Order size (Q_i)	225	1500	750	750	0	375	1500*
	300	2250	1500	1500	750	0	2250

*Minimax value.

Application of the minimax rule to Table 11.8 yields the minimax value for row 3. Hence, the action corresponding to the third row is the best action. This means that the optimal daily order size of the perishable item should be 225 kg.

11.4.5 Hurwicz Criterion

In the case of maximization problem, maximax criterion will give the optimistic solution and maximin criterion will give the pessimistic solution. In the case of minimization problem, minimin criterion will give the optimistic solution and minimax criterion will give the pessimistic solution.

Hurwicz criterion aims to strike a balance between the extreme optimism and the extreme pessimism by assigning a weight of α to the optimism and $1 - \alpha$ for the pessimism. The range of α is from 0 to 1. Depending on the decision-making situation, the value of α will be decided. In the absence of any bias either towards optimism or pessimism, one can assume 0.5 for α.

The formula to get the expected weighted outcome for each row (WO_i) is shown as below:

$$WO_i = \alpha \max_{j=1,\,2,\dots,\,n} (a_i, b_j) + (1 - \alpha) \min_{j=1,\,2,\dots,\,n} (a_i, b_j) \qquad \text{(for maximization problem)}$$

$$= \alpha \min_{j=1,\,2,\dots,\,n} (a_i, b_j) + (1 - \alpha) \max_{j=1,\,2,\dots,\,n} (a_i, b_j) \qquad \text{(for minimization problem)}$$

In the case of maximization (minimization) problem, the row which has the maximum (minimum) weighted outcome is selected as the best action.

Example 11.7 Consider the data of Example 11.3 and obtain the best action based on the Hurwicz criterion.

Solution The daily net profit data of the given problem is summarized in Table 11.9. This is a maximization problem. So, the following formula is used to determine the weighted outcome of each row by assuming 0.5 for α and the weighted outcomes are summarized in the same table.

$$WO_i = \alpha \max_{j=1,\,2,\dots,\,n} (a_i, b_j) + (1 - \alpha) \min_{j=1,\,2,\dots,\,n} (a_i, b_j)$$

Table 11.9 Summary of Daily Net Profit and Weighted Outcome

Demand (D_j)

		50	100	150	200	250	Maximum	Minimum	Weighted outcome
	75	750	1125	375	−375	−1125	1125	−1125	0.00
Order size	150	0	1500	3000	2250	1500	3000	0	1500.00
(Q_i)	225	−750	750	2250	3750	4125	4125	−750	1687.50*
	300	−1500	0	1500	3000	4500	4500	−1500	1500.00

*Maximin weighted average outcome.

Since it is a maximization problem, the row with the maximum weighted outcome should be selected as the best action. The maximum value of the weighted outcome is for row a_3. Hence the action corresponding to row 3 is the best action. This means that the optimal daily order size of the perishable item should be 225 kg.

11.5 DECISION TREE

In reality, decision making will be in multistages. The decision at Stage i depends on the decision at the $(i-1)$th stage. For each and every stage, there will be a set of alternatives. Again for each alternative, there will be a set of chance events. A *square symbol* is used to represent a decision point in a given stage from which a set of decision alternatives will emanate. A *circle symbol* is used to represent a chance point for a given alternative from which a set of chance events will emanate.

So, the *decision tree* is a tree which consists of a set of nodes either in the form of squares or circles. The information in this tree will help the decision makers to perform an analysis before arriving at optimum decision. This concept is illustrated in Example 11.8.

Example 11.8 A computer hardware company is planning to start a multi-centre computer software training organization. The company feels that such an attempt would create an awareness of computer usage among public and employees of different organizations which in turn will improve the demand for computer hardware.

The company has decided to start training centres as per any one of the following options or in combination of them.

1.   Starting training centres only in urban areas
2.   Starting training centres in semi-urban areas
3.   Starting training centres in other countries.

There are three possible chance events for the demands of its services: *high demand, medium demand* and *low demand*.

The company has listed the following alternatives for consideration:

(a)   Starting training centres in urban areas, semi-urban areas and in other countries simultaneously and continuing the business for the next five consecutive years by investing Rs. 50 crores.
(b)   First starting training centres in urban areas only and continuing the business for the next two years by investing Rs. 10 crores.

Then, based on the demand, the company will decide to do any one of the following:

1.   On high demand: The company may start training centres in semi-urban areas and other countries simultaneously and continue the business for another three years by investing Rs. 60 crores, (or) the company may not expand the business for the next three years.
2.   On medium demand: The company may start training centres only in semi-urban areas and continue the business for another three years by investing Rs. 30 crores, or the company may not expand the business for the next three years.
3.   On low demand: The company may not expand the business for the next three years.

Again, there are three possible chance events (high, medium and low) for the demands of its services for each of the decision points in the second stage.

The entire problem with the values of probability of occurrences of the chance events in both stages and the resultant annual revenues generated (in crores of rupees) are shown in Figure 11.1, wherein revenues are shown within brackets and the investments are shown with a minus sign. The decimal values represent probabilities of occurrences of the chance events.

Determine the best investment decision using decision tree.

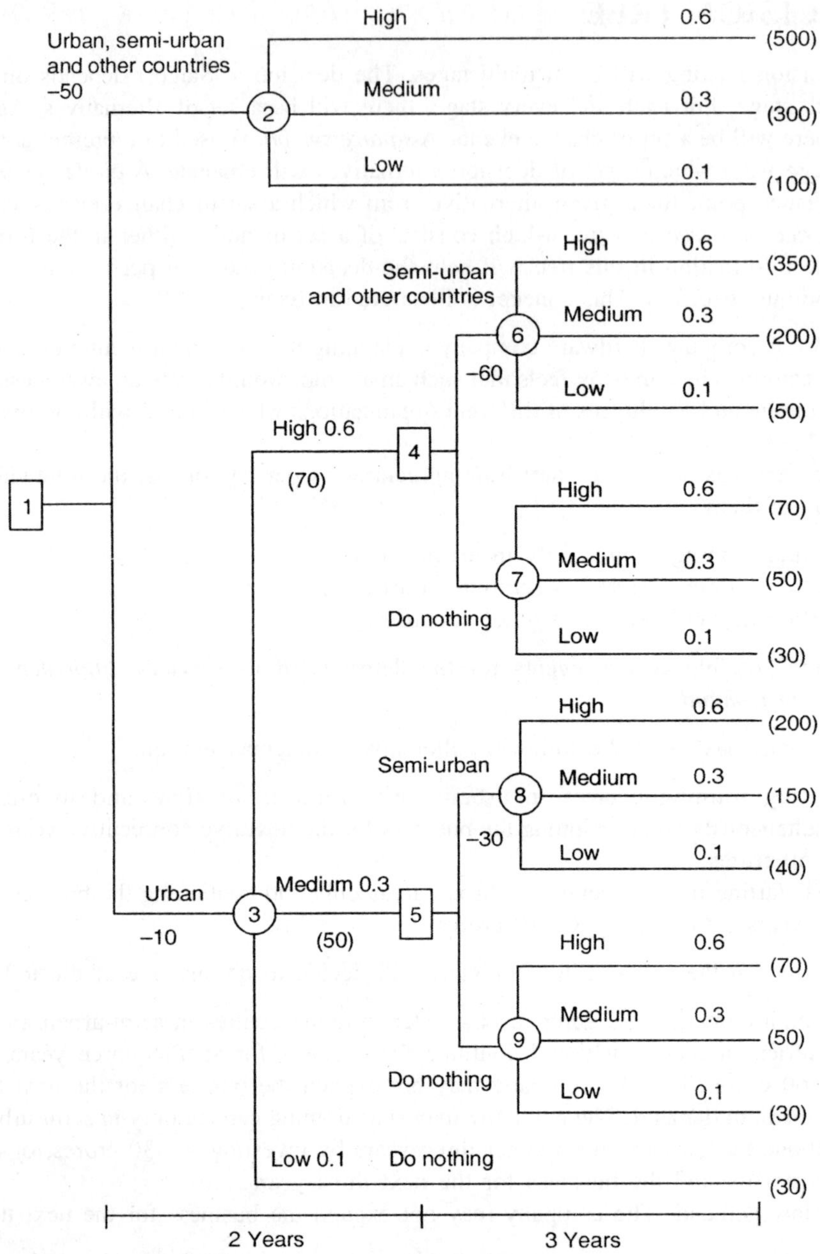

Figure 11.1   Decision tree of the given problem.

Solution   The expected net revenue for each of the alternatives is computed as shown below:

Alternative 1.   Starting training centres in urban areas, semi-urban areas and in other countries simultaneously and continuing the business for the next five consecutive years by making an initial investment of Rs. 50 crores, we get

$$\text{Expected revenue/year} = 500 \times 0.6 + 300 \times 0.3 + 100 \times 0.1 = \text{Rs. 400 crores}$$

and

$$\text{Expected net revenue at the decision node 1 for 5 years} = 400 \times 5 - \text{Investment}$$
$$= 2000 - 50$$
$$= \text{Rs. 1,950 crores.}$$

Alternative 2. Starting training centres in semi-urban and other countries and continuing the business for the next three years, if initially training centres are started only at urban areas and the demand is high during the first-two years.

Expected net-revenue at the decision point 4, for the last three years

$$= (350 \times 0.6 + 200 \times 0.3 + 50 \times 0.1)(3) - 60$$
$$= \text{Rs. 765 crores.}$$

Alternative 3. Continue the business for the next three years without starting training centres in semi-urban areas and other countries, if initially training centres are started only at urban areas and the demand is high during the first-two years. We have

Expected net revenue at the decision point 4, for the last three years

$$= (70 \times 0.6 + 50 \times 0.3 + 30 \times 0.1) \, (3)$$
$$= \text{Rs. 180 crores.}$$

At decision point 4, the expected net revenue of Alternative 2 is more than that of Alternative 3. Hence, Alternative 2 is selected at this decision point. The revenue at decision point 4 is Rs. 765 crores. So, all the branches after decision point 4 can be replaced by a single branch with the revenue of Rs. 765 crores for 3 years.

Alternative 4. Starting training centres in urban areas and continuing business for two years on high demand, selecting the successive best choice of the second-stage decision for the next 3 years.

Expected net revenue with respect to the branch emanating from the chance node 3 for 5 years

$$= (70 \times 0.6) \times 2 + 765$$
$$= \text{Rs. 849 crores.}$$

Alternative 5. Starting training centres in semi-urban areas and continuing the business for the next three years if initially training centres are started only at urban areas and the demand is medium during the first-two years, we get

Expected net revenue at the decision point 5 for the last three years

$$= (200 \times 0.6 + 150 \times 0.3 + 40 \times 0.1)(3) - 30$$
$$= \text{Rs. 477 crores.}$$

Alternative 6. Continue the business for the next three years without starting training centres in semi-urban areas and if initially training centres are started only at urban areas and the demand is medium during the first-two years, we have

Expected net revenue at the decision point 5 for the last three years

$$= (70 \times 0.6 + 50 \times 0.3 + 30 \times 0.1)(3)$$
$$= \text{Rs. 180 crores.}$$

At decision point 5, the expected net revenue of Alternative 5 is more than that of Alternative 6. Hence, Alternative 5 is selected at this decision point. The revenue at the decision point 5 is Rs. 477 crores. Hence, all the branches that are emanating from this decision point can be replaced by a single branch with the revenue of Rs. 477 crores for 3 years.

Alternative 7. Starting training centres in urban areas and continuing business for two years and on medium demand, use the successive best choice of the second-stage decision for the next 3 years. Now,

Expected net revenue with respect to the branch emanating from chance node 3 for 5 years

$$= (50 \times 0.3) \times 2 + 477$$
$$= \text{Rs. 507 crores.}$$

Alternative 8. Starting training centres in urban areas and continuing business for 5 years without any further expansion, We get

Expected net revenue with respect to branch emanating from chance node 3 for 5 years

$$= (30 \times 0.1) \times 5$$
$$= \text{Rs. 15 crores.}$$

Alternative 9. Starting training centres in urban areas and selecting the best option in the following stage, we have

Expected net revenue at the decision point 1 for five years

$$= (849 + 507 + 15) - 10$$
$$= \text{Rs. 1361 crores.}$$

Since the expected net-revenue at decision point 1 is greater for Alternative 1 than that of Alternative 9, it is suggested to start training centres in urban areas, semi-urban areas and in other countries simultaneously and continue the business for the next five consecutive years by investing Rs. 50 crores, initially.

All the net revenues (NR) as obtained above are shown in Figure 11.2 for better understanding.

Hypothetical discussion

To demonstrate the decision making process in the second stage, a hypothetical situation of this problem is explained below.

If the net-revenue of Alternative 9 at the decision point 1 is more than that of Alternative 1, then start training centres in urban areas and continue the business for 2 years. Then,

    (a) if the demand is high, start the training centres in semi-urban areas and other countries and, carry out the business for the next 3 years.

    (b) if the demand is medium, start the training centres in semi-urban areas and carry out the business for the next 3 years.

    (c) if the demand is low, carry out the business for the next 3 years without any further expansion.

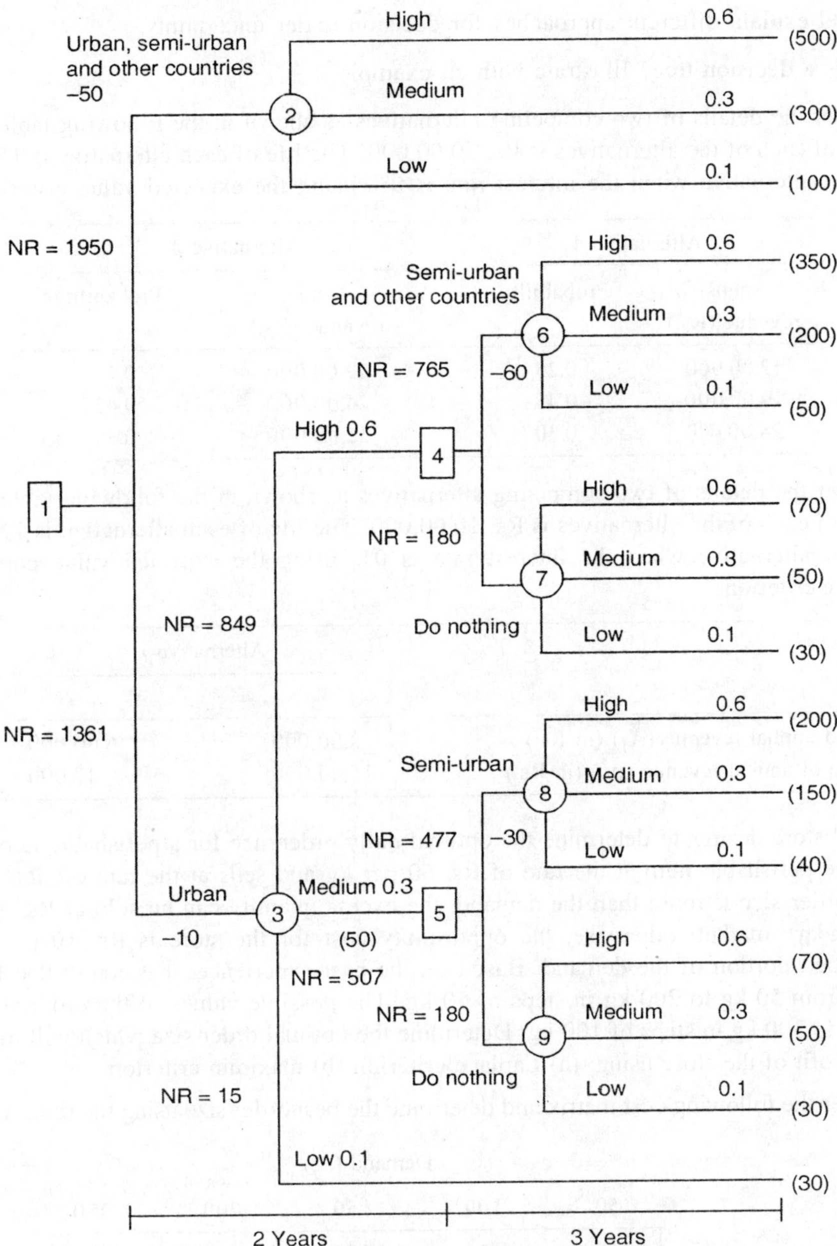

Figure 11.2 Decision tree with net revenues.

QUESTIONS

1. What are the types of decisions? Explain them in brief with suitable examples.

2. List the approaches for decision under risk and explain with examples.

3. List and explain different approaches for decision under uncertainty.

4. What is a decision tree? Illustrate with an example.

5. Consider the details of two competing alternatives as shown in the following table. The initial outlay of each of the alternatives is Rs. 50,00,000. The life of each alternative is 15 years. Find the best alternative when the interest rate is 0% using the expected value criterion.

Alternative 1		Alternative 2	
Annual revenue (Rs.)	Probability	Annual revenue (Rs.)	Probability
12,00,000	0.25	22,00,000	0.20
19,00,000	0.45	24,00,000	0.45
28,00,000	0.30	32,00,000	0.35

6. Consider the details of two competing alternatives as shown in the following table. The initial outlay of each of the alternatives is Rs. 10,00,000. The life of each alternative is 12 years. Find the best alternative when the interest rate is 0% using the expected value combined with variance criterion.

	Alternative j	
	1	2
Expected annual revenue ($\overline{X}_j$) (in Rs.)	8,00,000	6,00,000
Variance of annual revenue (σ_j^2) (in Rs.)	15,00,000	12,000

7. A retail store desires to determine the optimal daily order size for a perishable item. The store buys the perishable item at the rate of Rs. 60 per kg and sells at the rate of Rs. 90 per kg. If the order size is more than the demand, the excess quantity can be sold at Rs. 75 per kg in a secondary market; otherwise, the opportunity cost for the store is Rs. 10 per kg for the unsatisfied portion of the demand. Based on the past experience, it is found that the demand varies from 50 kg to 200 kg in steps of 50 kg. The possible values of the order size are from 100 kg to 300 kg in steps of 100 kg. Determine the optimal order size which will maximize the daily profit of the store using: (a) Laplace criterion (b) maximin criterion.

8. Consider the following cost matrix and determine the best order size using the minimax criterion.

		Demand (D_j)				
		50	100	150	200	250
Order size (Q_i)	75	50	125	375	375	125
	150	40	500	100	250	500
	225	750	550	250	750	125
	300	500	40	500	400	540

9. Consider the following data on daily net profit. Find the best order size based on the savage minimax regret criterion.

Demand (D_j)

		50	100	150	200	250
	75	950	1200	575	−675	−1425
Order size	150	50	1700	2000	2250	1600
(Q_i)	225	−850	850	2550	3550	4525
	300	−1800	600	1800	2000	5000

Also, obtain the best order size based on the Hurwicz criterion.

10. A company owns a lease on a certain property. It may sell the lease for Rs. 75,000 or may drill the said property for oil. Various possible drilling results are as under along with the probabilities of happening and rupee consequences.

Possible results	Probability	Rupee consequences
Dry well	0.10	−1,00,000
Gas well only	0.40	45,000
Oil and gas combination	0.30	98,000
Oil well	0.20	1,99,000

Draw a decision tree for the above problem and determine whether the company should drill or sell.

11. Mr. Ramesh is interested in developing and marketing a new drug. The cost of extensive research to develop the drug would be Rs. 1,00,000. The manager of the research programme said that there is 60% chance that the drug will be developed successfully. The market potential is assessed as follows with the present value of profits.

Market conditions	Probability	Present value of profits (in Rs.)
Large market potential	0.1	5,50,000
Moderate market potential	0.5	2,40,000
Low market potential	0.4	80,000

The present value figures do not include the cost of research. While Mr. Ramesh considering this proposal, another similar proposal came up which also had required the investment of Rs. 1,00,000. The present value of profits for the second proposal was Rs. 1,20,000. The return on the investment in the second proposal is almost certain.

(a) Draw a decision tree for Mr. Ramesh indicating all choices and events.
(b) What decision Mr. Ramesh should take regarding the investment of Rs. 1,00,000?

GAME THEORY ⑫

12.1 INTRODUCTION

Any business situation involves competition. Effective decision making plays a vital role in such situation mainly because decisions will have direct impact on the revenue earning potentials of business organizations.

Decision making situations may be classified into the following three types: Deterministic situation, Probabilistic situation and Uncertainty situation.

Deterministic situation. Under this situation, the problem data representing the situation are constants. They do not vary with respect to time or any other basis.

Probabilistic situation. Under this situation, the problem data representing the situation do not remain constant. They vary due to chance, and it is possible to represent the data in the form of some probability distribution. Later, these probability distributions become components of the decision making model of the situation.

Uncertainty situation. When the problem data are subjected to variation and it is not possible to represent them in the form of any probability distribution, this situation is called uncertain situation. *Game theory* is an example of the uncertainty situation.

12.1.1 Terminologies of Game Theory

The different terminologies that are associated with the game theory are now defined.

Players. There are two players in a game. The players may be any two companies (for example, Company A and Company B) competing for tenders, two countries planning for trade gains in a third country, two persons bidding in a game, etc.

Strategy. It is a course of action taken by a player, for example, giving computer furnitures free of cost, giving 20% additional hardware, giving special price, etc., while selling computer hardware.

The strategy can be further classified into pure strategy and mixed strategy. Let, m be the number of strategies of Player A. n be the number of strategies of Player B, p_i be the probability of selection of the Alternative i of Player A, $i = 1, 2, 3,..., m$. q_j be the probability of selection of the Alternative j of Player B, for $j = 1, 2, 3,..., n$. The sum of the probabilities of selection of various alternatives of each of the players is equal to 1 as shown below.

$$\sum_{i=1}^{m} p_i = 1 \quad \text{and} \quad \sum_{j=1}^{n} q_j = 1.$$

424

Pure strategy.   If a player selects a particular strategy with a probability of 1, then that strategy is known as a *pure strategy*. This means that the player is selecting that particular strategy alone ignoring his remaining strategies.

If Player A follows a pure strategy, then only one of the p_i values will be equal to 1 and the remaining p_i values will be equal to 0. A sample set of probabilities of selection of the alternatives for Player A is shown below:

$$p_1 = 0, \qquad p_2 = 1, \qquad p_3 = 0$$

The sum of these probabilities is equal to 1. That is

$$p_1 + p_2 + p_3 = 0 + 1 + 0 = 1.$$

Mixed strategy.   If a player follows more than one strategy, then the player is said to follow a *mixed strategy*. But the probability of selection of the individual strategies will be less than one and their sum will be equal to one. If Player B follows a mixed strategy, then a sample set of probabilities of selection of mixed strategy is shown below:

$$q_1 = 0.65, \qquad q_2 = 0, \qquad q_3 = 0.35$$

It is clear that the sum of the probabilities is equal to 1. That is

$$q_1 + q_2 + q_3 = 0.65 + 0 + 0.35 = 1.$$

Payoff matrix $[a_{ij}]$***.*** Each combination of the alternatives of Players A and B is associated with an outcome, a_{ij}. If a_{ij} is positive, then it represents a gain to Player A, and a loss to Player B. If it is negative, it represents a loss to A and a gain to B. A sample payoff matrix of Player A is as shown in Table 12.1.

Table 12.1   Sample Payoff Matrix of Player A

		1	2	…	j	…	n
	1	a_{11}	a_{12}	…	a_{1j}	…	a_{1n}
	2	a_{21}	a_{22}	…	a_{2j}	…	a_{2n}
	⋮	⋮	⋮		⋮		⋮
Player A	i	a_{i1}	a_{i2}	…	a_{ij}	…	a_{in}
	⋮	⋮	⋮	…	⋮	…	⋮
	m	a_{m1}	a_{m2}	…	a_{mj}	…	a_{mn}

(Columns headed "Player B": 1, 2, …, j, …, n.)

Maximin principle.   This principle maximizes the minimum guaranteed gains of Player A. The minimum gains with respect to different alternatives of A, irrespective of B's alternatives are obtained first. The maximum of these minimum gains is known as the *maximin value* and the corresponding alternative is called as *maximin strategy*.

Minimax principle.   This principle minimizes the maximum losses. The maximum losses with respect to different alternatives of Player B, irrespective of Player A's alternatives, are obtained first. The minimum of these maximum losses is known as the *minimax value* and the corresponding alternative is called as *minimax strategy*.

Saddle point. In a game, if the maximin value is equal to the minimax value, then the game is said to have a *saddle point*. The intersecting cell corresponding to these values is known as the *saddle point*. If the game has a saddle point, then each player has a *pure strategy*.

Value of the game. If the game has a saddle point, then the value of the cell at the saddle point is called the *value of the game*; otherwise, the value of the game is computed based on expected value calculations which will be explained later.

Two-person zero-sum game. In a game with two players, if the gain of one player is equal to the loss of another player, then that game is called *two-person zero-sum game*.

12.2 GAME WITH PURE STRATEGIES

As we know, if a game has a saddle point, then the game is said to have pure strategy for each of the players. Determination of such pure strategy of each of the players is illustrated in the following example.

Example 12.1 Two computer hardware manufacturing companies, Company A and Company B, competing for supplying computers to a government department. Each company has listed its strategies for selling computers.

The strategies of Company A are:

    (a) Giving special price
    (b) Giving 20% worth of additional hardwares
    (c) Supplying computer furnitures free of cost

The strategies of Company B are:

    (a) Giving special price
    (b) Giving 30% worth of additional hardwares
    (c) Giving free training to the users of the organization which is buying hardwares.

The estimated gains (+)/losses (−) of Company A for various possible combinations of the alternatives of both companies are summarized in Table 12.2. In the table, the rows represent the alternatives of Company A and the columns represent the alternatives of Company B. Each cell value of Table 12.2 represents the estimated gain (+)/loss (−) of Company A in lakhs of rupees for the corresponding alternatives of companies A and B. A positive cell entry denotes the gain to Company A (loss to Company B) and a negative cell entry represents the loss to Company A (gain to Company B). Thus, Table 12.2 is called as the payoff matrix with respect to Company A.

Determine the optimal strategy/strategies for Company A and Company B.

Table 12.2 Payoff Matrix of Company A

		Company B		
		1	2	3
	1	20	15	22
Company A	2	35	45	40
	3	18	20	25

Solution   If Company A selects strategy 1 and Company B selects strategy 2, the corresponding pay-off to Company A is Rs. 15 lakhs. This means that Company B would be losing Rs. 15 lakhs.

In Table 12.2, all the cell entries are positive. So, whatever may be the selected strategy of Company B, Company A will always have a gain. But the magnitude of the gain varies for different combinations of rows and columns. Though Company B has no chance of winning the game, it can try to minimize the gain to Company A, so that in the long run, Company B will have a competitive advantage. If there are some negative cell entries in Table 12.2, then there is a chance of winning the game for Company B, provided Company A selects any of the strategies corresponding to those negative cell values.

As per the payoff matrix shown in Table 12.2, Company A is called as *maximin player* and Company B is called as *minimax player*, which means Company A is maximizing its minimum guaranteed gain and Company B is minimizing its maximum loss. Table 12.2 is reproduced in Table 12.3 with necessary calculations, where

$$\text{Maximin value} = \text{Minimax value} = 35$$

Hence, the game has a saddle point at the cell corresponding to Row 2 and Column 1. The value of the game is Rs. 35 lakhs.

The optimal probabilities of selection of the strategies of Company A and Company B are as given below:

$$A(p_1, p_2, p_3) = A(0, 1, 0) \quad \text{and} \quad B(q_1, q_2, q_3) = B(1, 0, 0)$$

Table 12.3   Payoff Matrix of Player A with Maximin and Minimax Values

		Company B			
		1	2	3	Row minimum
	1	20	15	22	15
Company A	2	35	45	40	35 (maximin)
	3	18	20	25	18
Column maximum		35 (minimax)	45	40	

Company A should always select its second strategy with a probability of 1 and it should select all other strategies with zero probability. Company B should always select its first strategy with a probability of 1 and it should select all other strategies with zero probability.

If any one of the two companies deviates from its optimal strategic combination(s), the other player will have a gain. This means that Company A should always select its second alternative (giving 20% worth of additional hardwares). If it deviates from this alternative, Company B will gain in terms of its reduced loss, since all the entries of Figure 12.2 are positive values (loss for Company B will be less than Rs. 35 lakhs).

As per the optimal solution, Company B should always select its alternative 1 (giving special price). If it deviates from this strategy, Company A will have an additional gain over and above the value of the game. Under this situation, the gain for Company A will be more than Rs. 35 lakhs.

Note: If one of the companies follows the optimal solution as per the theory of the game and the other company does not use any such guidelines but follows a random strategy, definitely the first company will be in an advantageous position because the second company deviates from the optimal strategy, as determined using the theory of game.

12.3   GAME WITH MIXED STRATEGIES

If a game has no saddle point, then the game is said to have mixed strategies.

To demonstrate the determination of mixed strategies, consider the 2×2 payoff matrix with respect to Player A which has no saddle point (Table 12.4).

Table 12.4   A Game with Mixed Strategies

		B	
		1	2
	1	a	b
A			
	2	c	d

Algorithm to determine mixed strategies

Step 1: Find the absolute value of $a - b$ (i.e. $|a - b|$) and write it against row 2.

Step 2: Find the absolute value of $c - d$ (i.e. $|c - d|$) and write it against row 1.

Step 3: Find the absolute value of $a - c$ (i.e. $|a - c|$) and write it against column 2.

Step 4: Find the absolute value of $b - d$ (i.e. $|b - d|$) and write it against column 1.

The results of the above steps are summarized in Table 12.5. The absolute values are called as oddments.

Table 12.5   Payoff Matrix with Oddments

		B		Oddments
		1	2	
	1	a	b	$|c - d|$
A	2	c	d	$|a - b|$
Oddments		$|b - d|$	$|a - c|$	

Step 5: Compute the probabilities of selection of the alternatives of Player A (p_1 and p_2) and that of Player B (q_1 and q_2).

$$p_1 = \frac{|c-d|}{|a-b| + |c-d|}$$

$$p_2 = \frac{|a-b|}{|a-b| + |c-d|}$$

$$q_1 = \frac{|b-d|}{|a-c| + |b-d|}$$

$$q_2 = \frac{|a-c|}{|a-c| + |b-d|}$$

The value of the game can be computed using any one of the following formulae:

$$V = \frac{a|c-d| + c|a-b|}{|a-b| + |c-d|}$$

$$= \frac{b|c-d| + d|a-b|}{|a-b| + |c-d|}$$

$$= \frac{a|b-d| + b|a-c|}{|a-c| + |b-d|}$$

$$= \frac{c|b-d| + d|a-c|}{|a-c| + |b-d|}.$$

Example 12.2   Consider payoff matrix (Table 12.6) with respect to Player A and solve it optimally:

Table 12.6   Example 12.2

	B	
	1	2
A 1	6	9
2	8	4

Solution. The maximin and minimax values of the given problem are shown in Table 12.7.

Table 12.7   Maximin and Minimax Values

	B			
	1	2		
A 1	6	9	6	(maximin)
2	8	4	4	
	8	9		
	(minimax)			

In this problem, the maximin value (6) is not equal to the minimax value (8). Hence, the game has no saddle point. Under this situation, the formulae given in this section are to be used to find the mixed strategies of the players and also the value of the game. The computations of oddments of the game are summarized in Table 12.8.

Table 12.8 Payoff Matrix with Oddments

		B		
		1	2	Oddments
A	1	6	9	4
	2	8	4	3
Oddments		5	2	

Let p_1 and p_2 be the probabilities of selection of Alternative 1 and of Alternative 2, respectively of Player A. Also q_1 and q_2 be the probabilities of selection of Alternative 1 and of Alternative 2 of Player B, respectively. Then, we have

$$p_1 = \frac{|c-d|}{|a-b| + |c-d|} = \frac{4}{3+4} = \frac{4}{7}$$

$$p_2 = \frac{|a-b|}{|a-b| + |c-d|} = \frac{3}{3+4} = \frac{3}{7}$$

$$q_1 = \frac{|b-d|}{|a-c| + |b-d|} = \frac{5}{2+5} = \frac{5}{7}$$

$$q_2 = \frac{|a-c|}{|a-c| + |b-d|} = \frac{2}{2+5} = \frac{2}{7}$$

where, the value of the game is

$$V = \frac{a|c-d| + c|a-b|}{|a-b| + |c-d|} = \frac{(6)(4) + (8)(3)}{4+3} = \frac{48}{7}$$

Hence, the strategies of Player A is: A(4/7, 3/7) and of Player B: B(5/7, 2/7).

$$\text{The value of the game} = \frac{48}{7} = 6\frac{6}{7}.$$

12.4 DOMINANCE PROPERTY

In some games, it is possible to reduce the size of the payoff matrix by eliminating redundant rows (or columns). If a game has such redundant rows (or columns), those rows or columns are dominated by some other rows (or columns), respectively. Such property is known as *dominance property*.

Dominance property for rows

   (a) In the payoff matrix of Player A, if all the entries in a row (X) are greater than or equal to the corresponding entries of another row (Y), then row Y is dominated by row X. Under such situation, row Y of the payoff matrix can be deleted.

   (b) In the payoff matrix of Player A, if each of the sum of the entries of any two rows (sum of the entries of row X and row Y) is greater than or equal to the corresponding entry of a third row (Z), then row Z is dominated by row X and row Y. Under such situation, row Z of the payoff matrix can be deleted.

Dominance property for columns

   (a) In the payoff matrix of Player A, if all the entries in a column (X) are lesser than or equal to the corresponding entries of another column (Y), then column Y is dominated by column X. Under such situation, the column Y of the payoff matrix can be deleted.

   (b) In the payoff matrix of Player A, if each of the sum of the entries of any two columns (sum of the entries of column X and column Y) is lesser than or equal to the corresponding entry of a third column (Z), then column Z is dominated by columns X and Y. Under such situation, column Z of the payoff matrix can be deleted.

Example 12.3 Players A and B play a game in which each player has three coins (20p, 25p and 50p). Each of them selects a coin without the knowledge of the other person. If the sum of the values of the coins is an even number, A wins B's coin. If that sum is an odd number, B wins A's coin.

   (a) Develop a payoff matrix with respect to Player A.
   (b) Find the optimal strategies for the players.

Solution The payoff matrix with respect to Player A is shown in Table 12.9. The maximin and minimax values are also indicated in the same table.

In Table 12.9, the maximin value (-20) is not equal to the minimax value (20). Hence, the game has no saddle point. As a result, the game has mixed strategies.

Table 12.9 Payoff Matrix (Example 12.3)

		Player B			
		I 20p	II 25p	III 50p	
	I 20p	20	−20	50	−20 (maximin)
Player A	II 25p	−25	25	−25	−25
	III 50p	20	−50	50	−50
		20	25	50	
		(minimax)			

Check for dominance property. Row III is dominated by row I and hence row III is to be deleted. The resultant matrix after deleting row III is shown in Table 12.10.

Table 12.10   Payoff Matrix after Deleting Row III

Player B

	I 20p	II 25p	III 50p
Player A I 20p	20	−20	50
II 25p	−25	25	−25

In Table 12.10, the column III is dominated by the column I and hence, column III is to be deleted. The resultant matrix after deleting column III is shown in Table 12.11. The oddments of the rows and the columns are presented in the same table.

Table 12.11   Payoff Matrix after Deleting Column III

Player B

	I 20p	II 25p	oddments
Player A I 20p	20	−20	50
II 25p	−25	25	40
oddments	45	45	

Let p_i be the probability of selection of the Alternative i by Player A, where $i = 1, 2$, and q_j be the probability of selection of the Alternative j by Player B, where $j = 1, 2$. Then, we have

$$p_1 = \frac{50}{50 + 40} = \frac{5}{9}, \qquad p_2 = \frac{40}{50 + 40} = \frac{4}{9}$$

and

$$q_1 = \frac{45}{45 + 45} = \frac{1}{2}, \qquad q_2 = \frac{45}{45 + 45} = \frac{1}{2}$$

where, value of the game is

$$V = \frac{20 \times 50 - 25 \times 40}{50 + 40} = 0$$

Hence, the strategies of Player A and Player B, and the value of the game are:

$$A\left(\frac{5}{9}, \frac{4}{9}, 0\right), \qquad B\left(\frac{1}{2}, \frac{1}{2}, 0\right), \qquad V = 0.$$

Example 12.4   Consider the 4 × 4 game played by Players A and B (Table 12.12) and solve it optimally.

Table 12.12   Example 12.4

Player B

		1	2	3	4
	1	6	2	4	8
	2	2	−1	1	12
Player A	3	2	3	3	9
	4	5	2	6	10

Solution   In Table 12.13, the computations for finding the maximin and minimax values of the given game are shown. The maximin value (2) is not equal to the minimax value (3). Hence, the game has no saddle point.

Table 12.13   Payoff Matrix with Maximin and Minimax Values

Player B

		1	2	3	4	Row minimum
	1	6	2	4	8	2 (maximin)
	2	2	−1	1	12	−1
Player A	3	2	3	3	9	2 (maximin)
	4	5	2	6	10	2 (maximin)
Column maximum		6	3	6	12	
			(minimax)			

Check for dominance property.   In Table 12.13, for each row, the sum of the values in column 1 and column 2 is lesser than or equal to the corresponding value in the column 4. Hence, column 1 and column 2 dominate column 4. So, column 4 is to be deleted and the corresponding result is shown in Table 12.14.

Table 12.14   Payoff Matrix after Deleting Column 4

Player B

		1	2	3
	1	6	2	4
	2	2	−1	1
Player A	3	2	3	3
	4	5	2	6

Here, for each column, the sum of the values in row 1 and row 3 is greater than or equal to the corresponding value in row 2. Hence, row 1 and row 3 dominate row 2. Hence, row 2 is to be deleted and the corresponding result is shown in Table 12.15.

Table 12.15 Payoff Matrix after Deleting Row 2

		Player B		
		1	2	3
	1	6	2	4
Player A	3	2	3	3
	4	5	2	6

Here, the values in column 2 are less than or equal to the corresponding values in column 3. Hence, column 2 dominates column 3. So, column 3 is to be deleted and the corresponding result is shown in Table 12.16.

Table 12.16 Payoff Matrix after Deleting Column 3

		Player B	
		1	2
	1	6	2
Player A	3	2	3
	4	5	2

Now, the values in row 1 are greater than or equal to the corresponding values in row 4, in Table 12.16. Hence, row 1 dominates row 4. So, row 4 is to be deleted and the corresponding result is shown in Table 12.17.

Table 12.17 Payoff Matrix after Deleting Row 4

		Player B		
		1	2	Oddments
	1	6	2	1
Player A	3	2	3	4
Oddments		1	4	

Here, the oddments are shown. Based on these values, the probabilities of selection of the alternatives that are present in Table 12.17 are computed. Let p_i be the probability of selection of Alternative i by Player A, where $i = 1, 3$, and q_j be the probability of selection of Alternative j by Player B, where $j = 1, 2$. Then

$$p_1 = \frac{1}{1+4} = \frac{1}{5}$$

$$p_3 = \frac{4}{1+4} = \frac{4}{5}$$

$$q_1 = \frac{1}{1+4} = \frac{1}{5}$$

$$q_2 = \frac{4}{1+4} = \frac{4}{5}$$

$$V = \frac{(6)(1) + (2)(4)}{1+4} = \frac{14}{5} = 2.8$$

The optimal solution of the problem is:

$$A\left(\frac{1}{5}, 0, \frac{4}{5}, 0\right), \qquad B\left(\frac{1}{5}, \frac{4}{5}, 0, 0\right), \qquad V = 2.8.$$

Example 12.5 Solve the following 3×5 game using dominance property (Table 12.18).

Table 12.18 Example 12.5

Player B

Player A	1	2	3	4	5
1	2	5	10	7	2
2	3	3	6	6	4
3	4	4	8	12	1

Solution The computations of maximin and minimax values of the game are summarized in Table 12.19.

Table 12.19 Payoff Matrix with Maximin and Minimax Values

Player B

Player A	1	2	3	4	5	Row minimum
1	2	5	10	7	2	2
2	3	3	6	6	4	3 (maximin)
3	4	4	8	12	1	1
Column maximum	4	5	10	12	4 (minimax)	

Here, the maximin value (3) is not equal to the minimax value (4). Hence, the game has no saddle point.

Check for dominance property. In Table 12.19, column 1 dominates columns 2, 3 and 4. Hence, columns 2, 3 and 4 are to be deleted. The resulting data is given in Table 12.20.

Table 12.20   Payoff Matrix after Deleting Columns 2, 3 and 4

Player B

	1	5
1	2	2
Player A 2	3	4
3	4	1

Here, row 2 dominates row 1. Hence, row 1 is to be deleted. The resulting data is shown in Table 12.21.

Table 12.21   Payoff Matrix with Oddments

Player B

	1	5	Oddments
2	3	4	3
Player A 3	4	1	1
Oddments	3	1	

Based on these oddments values, the probabilities of selection of the alternatives that are present in Table 12.21 are computed: Let p_i be the probability of selection of Alternative i by Player A, where, $i = 2, 3$, and q_j be the probability of selection of Alternative j by Player B, $j = 1, 5$. Then

$$p_2 = \frac{3}{3+1} = \frac{3}{4}$$

$$p_3 = \frac{1}{3+1} = \frac{1}{4}$$

$$q_1 = \frac{3}{3+1} = \frac{3}{4}$$

$$q_5 = \frac{1}{3+1} = \frac{1}{4}$$

$$V = \frac{(3)(3)+(4)(1)}{3+1} = \frac{13}{4} = 3.25$$

The optimal solution of the problem is:

$$A\left(0, \frac{3}{4}, \frac{1}{4}\right), \qquad B\left(\frac{3}{4}, 0, 0, 0, \frac{1}{4}\right), \qquad V = 3.25$$

Example 12.6 Consider the 4×4 game as shown in Table 12.22 which represents the payoff matrix of the Player A. Solve it optimally.

Table 12.22 Payoff Matrix for Example 12.6

		Player B			
		1	2	3	4
	1	3	2	4	0
Player A	2	3	4	2	4
	3	4	2	4	1
	4	3	4	3	4

Solution The check for saddle point is done as shown in Table 12.23. Since, the maximin value (3) is not equal to the minimax value (4), the game has no saddle point. This means that the game has mixed strategies for both the players.

Table 12.23 Payoff Matrix with Check for Saddle Point

		Player B				
		1	2	3	4	Minimum
	1	3	2	4	0	0
	2	3	4	2	4	2
Player A	3	4	2	4	1	1
	4	3	4	3	4	3 (maximin)
maximum		4	4	4	4	

(minimax)

Check for dominance property. In Table 12.23, row 3 dominates row 1 and the corresponding reduced payoff matrix is shown in Table 12.24. In Table 12.24, column 3 dominates column 1 and the corresponding reduced payoff matrix is shown in Table 12.25. In Table 12.25, row 4 dominates row 2 and the corresponding reduced payoff matrix is shown in Table 12.26. In Table 12.26, column 4 dominates column 2 and the corresponding reduced payoff matrix (2×2 matrix) is shown in Table 12.27. The oddments of rows and columns of the 2×2 matrix are shown in the same matrix.

Table 12.24 Payoff Matrix after Deleting Row 1

		Player B			
		1	2	3	4
	2	3	4	2	4
Player A	3	4	2	4	1
	4	3	4	3	4

Table 12.25   Payoff Matrix after Deleting Column 1

Player B

		2	3	4
	2	4	2	4
Player A	3	2	4	1
	4	4	3	4

Table 12.26   Payoff Matrix after Deleting Row 2

Player B

		2	3	4
Player A	3	2	4	1
	4	4	3	4

Table 12.27   Payoff Matrix after Deleting Column 2

Player B

		3	4	Oddments
Player A	3	4	1	1
	4	3	4	3
Oddments		3	1	

$$p_3 = 1/(1 + 3) = 1/4$$
$$p_4 = 3/(1 + 3) = 3/4$$
$$q_3 = 3/(3 + 1) = 3/4$$
$$q_4 = 1/(3 + 1) = 1/4$$

$$\text{Value of the game, } V = \frac{4 \times 1 + 3 \times 3}{3 + 1} = 13/4$$

Therefore, the optimal solution of the game is:

$$A(0, 0, 1/4, 3/4)$$
$$B(0, 0, 3/4, 1/4)$$
$$V = 13/4.$$

Example 12.7   There are two players in a game, Player A and Player B. Each of them randomly shows selected fingers of his right hand. If the sum of the number of fingers shown by both the players is an even number, then the Player B has to give money in rupees equivalent to the number of fingers shown by him to the Player A; if the sum of the number of fingers shown by both the players is an odd number, then the Player A has to give money in rupees equivalent to the number of fingers shown by him to the Player B; Construct the payoff matrix with respect to the Player A and find the optimal solution for this game.

Solution   The payoff matrix with respect to the Player A as per the guidelines given in the problem is shown in Table 12.28. One verify the fact that this game has no saddle point and hence, the players have mixed strategies for selecting their alternatives.

Table 12.28   Payoff Matrix of Example 12.7

	1	2	3	4	5
1	1	-1	3	-1	5
2	-2	2	-2	4	-2
3	1	-3	3	-3	5
4	-4	2	-4	4	-4
5	1	-5	3	-5	5

Check for dominance property.   In Table 12.28, row 1 dominates row 3 and row 5, and row 2 dominates row 4. The resultant payoff matrix is shown in Table 12.29. In Table 12.29, column 1 dominates column 3 and column 5, and column 2 dominates column 4. The resultant payoff matrix is shown in Table 12.30.

Table 12.29   Payoff Matrix after Deleting Rows 3, 4, and 5

	1	2	3	4	5
1	1	-1	3	-1	5
2	-2	2	-2	4	-2

Table 12.30   Payoff Matrix after Deleting Columns 3, 4 and 5

	1	2	Oddments
1	1	-1	4
2	-2	2	2
Oddments	3	3	

$$p_1 = 4/(4 + 2) = 2/3$$
$$p_2 = 2/(4 + 2) = 1/3$$
$$q_1 = 3/(3 + 3) = 1/2$$
$$q_2 = 3/(3 + 3) = 1/2$$

Value of the game, $V = \dfrac{1 \times 4 + (-2) \times 2}{2 + 4} = 0$

Therefore, the optimal solution of the game is:

$$A(2/3,\ 1/3,\ 0,\ 0,\ 0)$$
$$B(1/2,\ 1/2,\ 0,\ 0,\ 0)$$
$$V = 0.$$

12.5 GRAPHICAL METHOD FOR $2 \times n$ OR $m \times 2$ GAMES

Some games will be of specialized nature, like $2 \times n$ or $m \times 2$ games. The payoff matrix of the $2 \times n$ game will contain 2 rows and n columns, whereas the payoff matrix of the $m \times 2$ game will contain m rows and 2 columns. If there is no saddle point for these games, one can solve them using graphical method.

Algorithm for $2 \times n$ game

Step 1: Reduce the size of the payoff matrix of Player A by applying the dominance property, if it exists.

Step 2: Let, x be the probability of selection of Alternative 1 by Player A and $1 - x$ be the probability of selection of Alternative 2 by Player A. Derive the expected gain function of Player A with respect to each of the alternatives of Player B.

Step 3: For each of the gain functions which are derived in Step 2, find the value of the gain, when x is equal to 0 as well as 1.

Step 4: Plot the gain functions on a graph by assuming a suitable scale. Keep x on X-axis and the gain on Y-axis).

Step 5: Since the Player A is a maximin player, find the *highest intersection point* in the *lower boundary* of the graph. Let it be the maximin point.

Step 6: If the number of lines passing through the maximin point is only two, form a 2×2 payoff matrix from the original problem by retaining only the columns corresponding to those two lines and go to step 8; otherwise go to step 7.

Step 7: Identify any two lines with opposite slopes passing through that point. Then form a 2×2 payoff matrix from the original problem by retaining only the columns corresponding to those two lines which are having opposite slopes.

Step 8: Solve the 2×2 game using oddments and find the strategies for Players A and B and also the value of the game.

Algorithm for $m \times 2$ game

Step 1: Reduce the size of the payoff matrix of Player A by applying the dominance property, if it exists.

Step 2: Let y be the probability of selection of Alternative 1 by Player B and $1 - y$ be the probability of selection of Alternative 2 by Player B. Derive the expected gain function of Player B with respect to each of alternatives of Player A.

Step 3: For each of the gain functions which are derived in step 2, find the value of the gain when y is equal to 0 as well as 1.

Step 4: Plot the gain functions on a graph by assuming a suitable scale. Keep y on X-axis and the gain on Y-axis.

Step 5: Since B is a minimax player, find the *lowest intersection point* in the *upper boundary* of the graph. Let it be the minimax point.

Step 6: If the number of lines passing through the minimax point is only two, form a 2×2 payoff matrix from the original problem by retaining only the rows corresponding to those two lines and go to step 8; otherwise, go to step 7.

Step 7: Identify any two lines with opposite slopes passing through that point. Then form a 2×2 payoff matrix from the original problem by retaining only the rows corresponding to those two lines which are having opposite slopes.

Step 8: Solve the 2×2 game using oddments and find the strategies for Player A and Player B and also the value of the game.

Example 12.8 Consider the payoff matrix of Player A as shown in Table 12.31 and solve it optimally using graphical method.

Table 12.31 Example 12.8

Player B

	1	2	3	4	5
Player A 1	3	0	6	−1	7
2	−1	5	−2	2	1

Solution Table 12.32 shows the payoff matrix with maximin and minimax values. Here, the maximin value (−1) is not equal to the minimax value (2). Hence, the game has no saddle point. As a result, the players will have mixed strategies. Since the game has only two rows, it can be solved using graphical method.

Table 12.32 Payoff Matrix with Maximin and Minimax Values

Player B

	1	2	3	4	5	Row minimum
Player A 1	3	0	6	−1	7	−1 (maximin)
2	−1	5	−2	2	1	−2
Column maximum	3	5	6	2	7	
				(minimax)		

Checking dominance property. In the payoff matrix as shown in Table 12.32, column 5 is dominated by column 1. Similarly, column 2 is dominated by column 4. Hence, delete column 2 and column 5. The resultant payoff matrix after deleting columns 2 and 5 is shown in Table 12.33.

Table 12.33 Payoff Matrix after Deleting Columns 2 and 5

Player B

	1	3	4
Player A 1	3	6	−1
2	−1	−2	2

Let, x be the probability of selection of Alternative 1 by Player A and $1 - x$ be the probability of selection of Alternative 2 by Player A. Therefore, the expected payoff to Player A with respect to different alternatives of Player B is summarized in Table 12.34.

Table 12.34   Expected Payoff Functions of Player A

B's alternative	A's expected payoff function
1	$3x - (1 - x) = 4x - 1$
3	$6x - 2(1 - x) = 8x - 2$
4	$-x + 2(1 - x) = -3x + 2$

The computations of the expected payoff of Player A with respect to each of the alternatives of Player B, when x is equal to 0 as well as 1, are summarized in Table 12.35.

Table 12.35   Expected Gain of Player A

B's alternative	A's expected payoff function	A's expected gain	
		$x = 0$	$x = 1$
1	$4x - 1$	-1	3
3	$8x - 2$	-2	6
4	$-3x + 2$	2	-1

Now, the expected gain functions of Player A with respect to different alternatives of Player B are plotted in Figure 12.1.

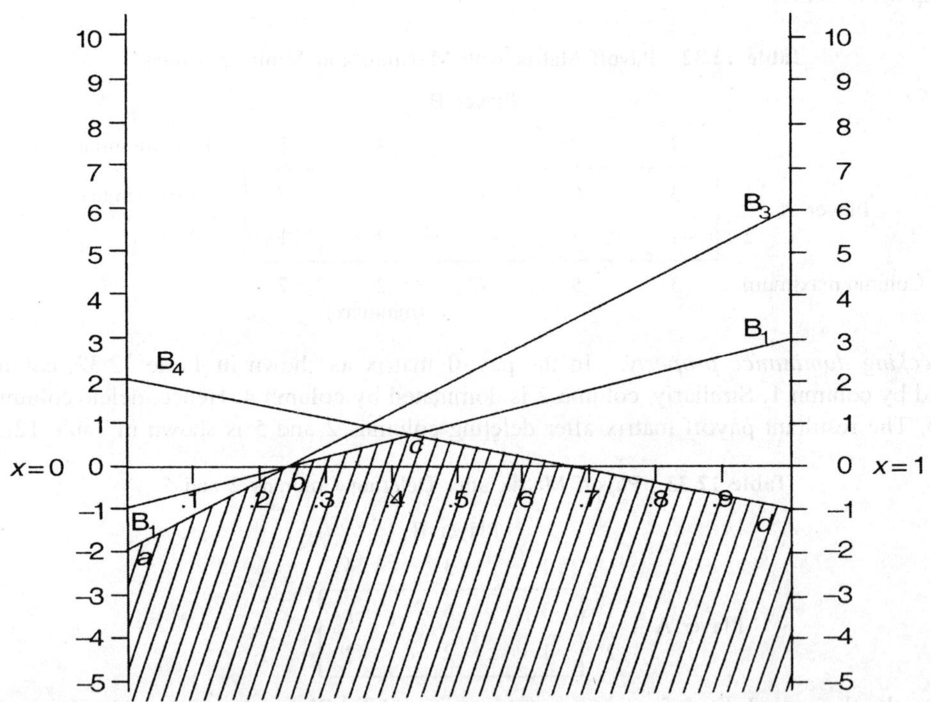

Figure 12.1   Graph with A's payoff function.

Since, A is a maximin type player, identify the highest intersection point in the lower boundary of the graph. The lower boundary consists of the intersection points a, b, c and d. Out of these points, point c is at the highest level. Hence, the corresponding solution is the optimal solution to the given problem.

In Figure 12.1, the line B_1 and line B_4 pass through the point c. Hence, form a 2×2 payoff matrix for Player A by retaining columns B_1 and B_4 as shown in Table 12.36. The oddments of the rows and columns are shown in the same figure.

Table 12.36 2×2 Payoff Matrix with Oddments

		Player B		
		1	4	Oddments
Player A	1	3	−1	3
	2	−1	2	4
Oddments		3	4	

Let p_i be the probability of selection of Alternative i by Player A, where $i = 1, 2$, and q_j be the probability of selection of Alternative j by Player B, where $j = 1, 4$. Then

$$p_1 = \frac{3}{3+4} = \frac{3}{7}$$

$$p_2 = \frac{4}{3+4} = \frac{4}{7}$$

$$q_1 = \frac{3}{3+4} = \frac{3}{7}$$

$$q_4 = \frac{4}{3+4} = \frac{4}{7}$$

$$V = \frac{(3)(3) - (1)(4)}{3+4} = \frac{5}{7}$$

The strategies of Player A and Player B, and the value of the game are:

$$A\left(\frac{3}{7}, \frac{4}{7}\right), \quad B\left(\frac{3}{7}, 0, 0, \frac{4}{7}, 0\right) \quad V = \frac{5}{7}.$$

Example 12.9 Consider the payoff matrix of Player A as shown in Table 12.37 and solve it optimally using the graphical method.

Table 12.37 Example 12.9

		Player B				
		1	2	3	4	5
Player A	1	3	6	8	4	4
	2	−7	4	2	10	2

Solution Table 12.38 shows the payoff matrix with maximin and minimax values. Here, the maximin value (3) is equal to the minimax value (3). Hence, the game has a saddle point. As a result, each player has a pure strategy and the corresponding results are shown below.

$$A(1, 0), \qquad B(1, 0, 0, 0, 0) \qquad \text{Value of the game, } V = 3$$

However, the game is solved using graphical method to have a better insight of that method with the known results.

Table 12.38 Payoff Matrix with Maximin and Minimax Values

		1	2	3	4	5	
	1	3	6	8	4	4	3 (maximin)
Player A	2	-7	4	2	10	2	-7
		3	6	8	10	4	

(minimax)

Player B (column headers); Player A (row labels)

In the payoff matrix shown in Table 12.38, column 2 is dominated by column 5. The resultant payoff matrix after deleting column 2 is shown in Table 12.39.

Table 12.39 Payoff Matrix after Deleting Column 2

		1	3	4	5
	1	3	8	4	4
Player A	2	-7	2	10	2

Player B (column headers)

Let x be the probability of selection of Alternative 1 by Player A and $1 - x$ be the probability of selection of Alternative 2 by Player A. Therefore, the expected payoff to Player A with respect to different alternatives of Player B are summarized in Table 12.40.

Table 12.40 Expected Payoff Functions of Player A

B's alternative	A's expected payoff function
1	$3x + (-7)(1 - x) = 10x - 7$
3	$8x + 2(1 - x) = 6x + 2$
4	$4x + 10(1 - x) = -6x + 10$
5	$4x + 2(1 - x) = 2x + 2$

The computations of expected payoff of Player A with respect to each of the alternatives of Player B, when x is equal to 0 and 1, are summarized in Table 12.41.

Table 12.41 Expected Gain of Player A

B's alternative	A's expected payoff function	A's expected gain $x = 0$	$x = 1$
1	$10x - 7$	-7	3
3	$6x + 2$	2	8
4	$-6x + 10$	10	4
5	$2x + 2$	2	4

The expected gain functions of Player A with respect to different alternatives of Player B are plotted in Figure 12.2.

Since, A is a maximin player, identify the highest intersection point in the lower boundary of the graph. The lower boundary consists of the intersection points a and b. Out of these two points,

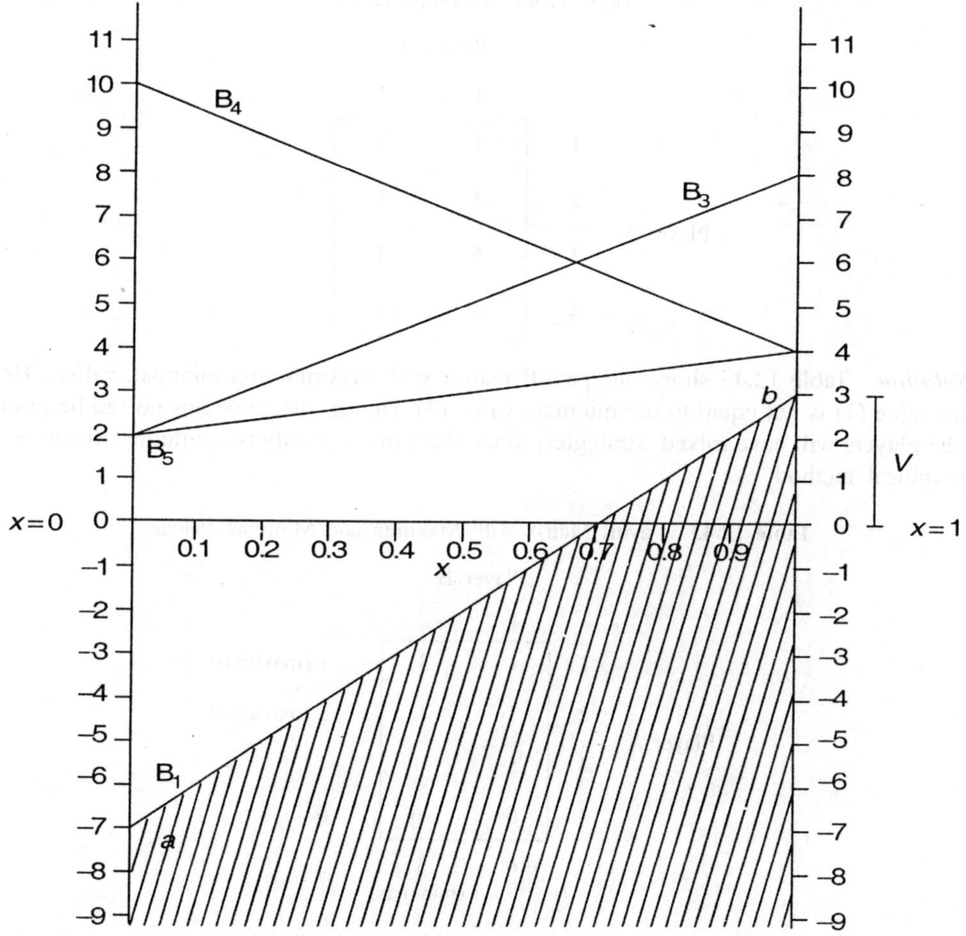

Figure 12.2 Graph with A's payoff function.

b is at the highest level. Hence, the corresponding solution is the optimal solution to the given problem. With respect to the intersection point *b*, value of the game is 3 (the vertical distance of the intersection point *b* from the horizontal line).

The probability of selection of Alternative 1 by Player A is 1 (i.e. $x = 1$). This is the horizontal distance to the point *b* from the left vertical line. Therefore, the probability of selection of Alternative 2 by Player A is 0 (i.e. $1 - x = 0$).

Since the maximin intersection point *b* is at the rightmost vertical line, the line B_1 is the only line of the Player B passing through it. So, the probability of selection of the Alternative 1 by the Player B is 1 and the probabilities of selection of the remaining alternatives by Player B are zero.

The final results of the game are:

$$A\ (1,\ 0), \qquad B\ (1,\ 0,\ 0,\ 0,\ 0), \qquad \text{Value of the game, } V = 3$$

Example 12.10 Consider the payoff matrix of Player A as shown in Table 12.42 and solve it optimally using graphical method.

Table 12.42 Example 12.10

Player B

	1	2
1	1	3
2	3	1
3	5	−1
4	6	−6

Player A

Solution Table 12.43 shows the payoff matrix with maximin and minimax values. Here, the maximin value (1) is not equal to the minimax value (3). Hence, the game has no saddle point. As a result, the players will have mixed strategies. Since the game has only two columns, it can be solved using graphical method.

Table 12.43 Payoff Matrix with Maximin and Minimax Values

Player B

	1	2	
1	1	3	1 (maximin)
2	3	1	1 (maximin)
3	5	−1	−1
4	6	−6	−6
	6	3	
		(minimax)	

Player A

None of the rows of the game can be deleted using the dominance property. Let *y* be the probability of selection of Alternative 1 by Player B and $1-y$ be the probability of selection of

Alternative 2 by Player B. Therefore, the expected payoff functions of B with respect to different alternatives of A are summarized as in Table 12.44.

Table 12.44 Expected Payoff Functions of Player B

A's alternative	B's expected loss (+)/gain (−) function
1	$y + 3(1 - y) = -2y + 3$
2	$3y + (1 - y) = 2y + 1$
3	$5y - (1 - y) = 6y - 1$
4	$6y - 6(1 - y) = 12y - 6$

Note: A positive value of the function indicates loss to B and a negative value of the function indicates gain to B.

The computations of expected loss (+)/gain (−) of B with respect to each of the alternatives of A, when y is equal to 0 as well as 1 are summarized in Table 12.45.

Table 12.45 Expected Loss (+)/Gain (−) of Player B

A's alternative	B's expected loss/ gain function	B's expected function value	
		$y = 0$	$y = 1$
1	$-2y + 3$	3	1
2	$2y + 1$	1	3
3	$6y - 1$	−1	5
4	$12y - 6$	−6	6

Now, the expected functions of Player B with respect to different alternatives of Player A are plotted in Figure 12.3.

Since B is a minimax player, identify the lowest intersection point in the upper boundary of the graph. The upper boundary consists of the intersection points *a*, *b*, *c* and *d*. Out of these points, point *b* is at the lowest level. Hence, the corresponding solution is the optimal solution to the given problem. In Figure 12.3, the lines A_1, A_2 and A_3 pass through point *b*. Among these lines, select any two lines having opposite slopes. As per this guideline, the line A_1 and the line A_3 are selected.

Hence, form a 2×2 payoff matrix for Player A by retaining only the rows A_1 and A_3 as shown in Table 12.46. The oddments of the rows and columns are also shown in the Table.

Table 12.46 2×2 Payoff Matrix with Oddments

		Player B		
		1	2	Oddments
Player A	1	1	3	6
	3	5	−1	2
	Oddments	4	4	

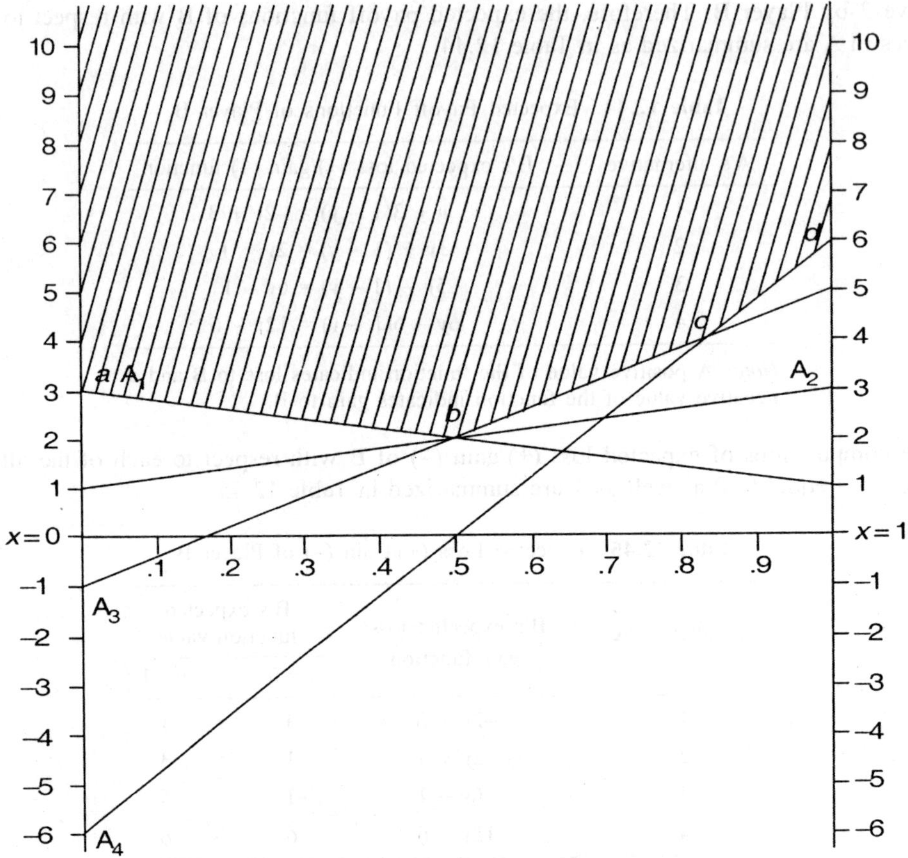

Figure 12.3 Graph with B's payoff functions.

Let, p_i be the probability of selection of Alternative i by Player A, $i = 1, 3$ and q_j be the probability of selection of Alternative j by Player B, $j = 1, 2$. Then

$$p_1 = \frac{6}{2+6} = \frac{3}{4}, \quad p_3 = \frac{2}{6+2} = \frac{1}{4}$$

$$q_1 = \frac{4}{4+4} = \frac{1}{2}, \quad q_2 = \frac{4}{4+4} = \frac{1}{2}$$

$$V = \frac{1 \times 6 + 5 \times 2}{6+2} = 2$$

Hence, the strategies of A and B, and the value of the game are:

$$A\left(\frac{3}{4}, 0, \frac{1}{4}, 0\right), \quad B\left(\frac{1}{2}, \frac{1}{2}\right), \quad V = 2$$

Example 12.11 Consider the payoff matrix with respect to the Player A as shown in Table 12.47. Solve this game optimally using graphical method.

Table 12.47 Payoff Matrix of Example 12.11

	1	2	3	4	5
1	4	2	1	7	3
2	2	7	8	1	5

Let, x be the probability of selection of the Alternative 1 by the Player A

Therefore, $(1-x)$ is the probability of selection of the Alternative 2 by the Player A.

The expected payoff to the Player A with respect to different alternatives of the Player B are summarized in Table 12.48. In the same table, the values of the expected payoff when $x = 0$ as well as when $x = 1$ are summarized.

Table 12.48 Expected Payoff Functions and their Values of Player A

B's alternative	A's expected payoff function	x	
		0	1
1	$4x + 2(1-x) = 2x + 2$	2	4
2	$2x + 7(1-x) = -5x + 7$	7	2
3	$x + 8(1-x) = -7x + 8$	8	1
4	$7x + (1-x) = 6x + 1$	1	7
5	$3x + 5(1-x) = -2x + 5$	5	3

These functions are plotted as shown in Figure 12.4. Since the Player A is maximizing his minimum guaranteed return, the highest point from the lower region of these functions is the point of intersection of the functions with respect to the Alternatives B_1 and B_3. Hence, the corresponding 2×2 payoff matrix is shown in Table 12.49 along with the oddments for rows and columns.

Table 12.49 2×2 Payoff Matrix

	1	3	Oddments
1	4	1	6
2	2	8	3
Oddments	7	2	

$$p_1 = 6/(6+3) = 2/3$$
$$p_2 = 3/(6+3) = 1/3$$
$$q_1 = 7/(7+2) = 7/9$$
$$q_3 = 2/(7+2) = 2/9$$

Value of the game, $V = \dfrac{4 \times 6 + 2 \times 3}{6 + 3} = 30/9 = 10/3$

Therefore, the optimal solution of the game is:

$$A(2/3, 1/3)$$
$$B(7/9, 0, 2/9, 0, 0)$$
$$V = 10/3.$$

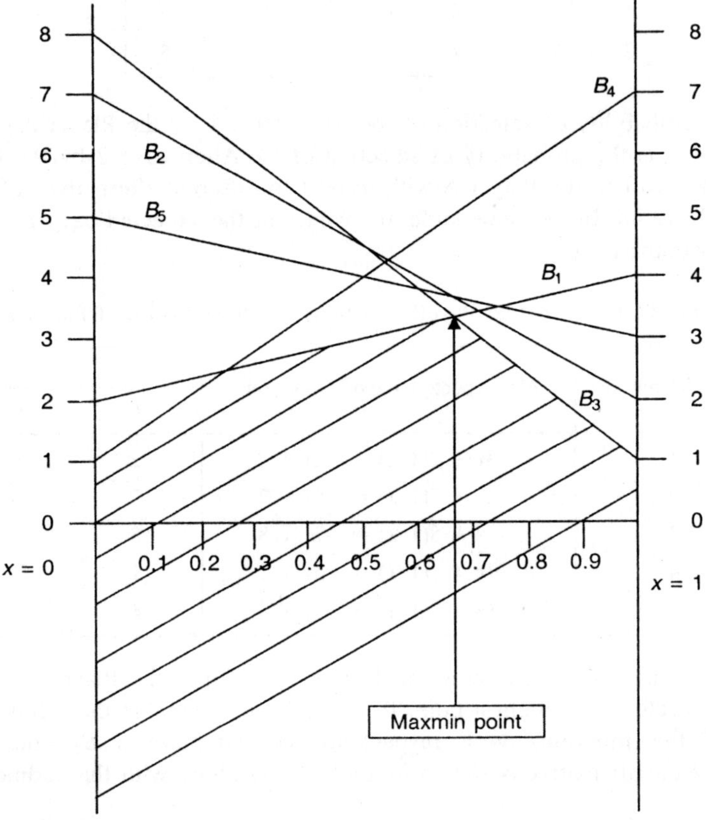

Figure 12.4   Graph with A's payoff functions of Example 12.11.

Example 12.12   Consider the payoff matrix with respect to the Player A as shown in Table 12.50. Solve this game optimally using graphical method.

Table 12.50   Payoff Matrix of Example 12.12

	1	2
1	4	6
2	3	7
3	5	−4
4	8	−5

Let, y be the probability of selection of the Alternative 1 by the Player B.

Therefore, $(1-y)$ is the probability of selection of the Alternative 2 by the Player B.

The expected payoff to the Player B with respect to different alternatives of the Player A are summarized in Table 12.51. In the same table, the values of the expected payoff when $y = 0$ as well as when $y = 1$ are summarized.

Table 12.51   Expected Payoff Functions and their Values of Player B

A's alternative	B's expected payoff function	y	
		0	1
1	$4y + 6(1 - y) = -2y + 6$	6	4
2	$3y + 7(1 - y) = -4y + 7$	7	3
3	$5y - 4(1 - y) = 9y - 4$	-4	5
4	$8y - 5(1 - y) = 13y - 5$	-5	8

These functions are plotted as shown in Figure 12.5. Since the Player B is minimizing his maximum loss, the lowest point from the upper region of these functions is the point of intersection of the functions with respect to the Alternatives A_1 and A_4. Hence, the corresponding 2×2 payoff matrix is shown in Table 12.52 along with the oddments for rows and columns.

Table 12.52 2×2 Payoff Matrix

	1	2	Oddments
1	4	6	13
4	8	-5	2
Oddments	11	4	

$$p_1 = 13/(13 + 2) = 13/15$$
$$p_4 = 2/(13 + 2) = 2/15$$
$$q_1 = 11/(11 + 4) = 11/15$$
$$q_2 = 4/(11 + 4) = 4/15$$

$$\text{Value of the game, } V = \frac{4 \times 13 + 8 \times 2}{13 + 2} = 68/15.$$

Therefore, the optimal solution of the game is:

$$A(13/15, 0, 0, 2/15)$$
$$B(11/15, 4/15)$$
$$V = 68/15$$

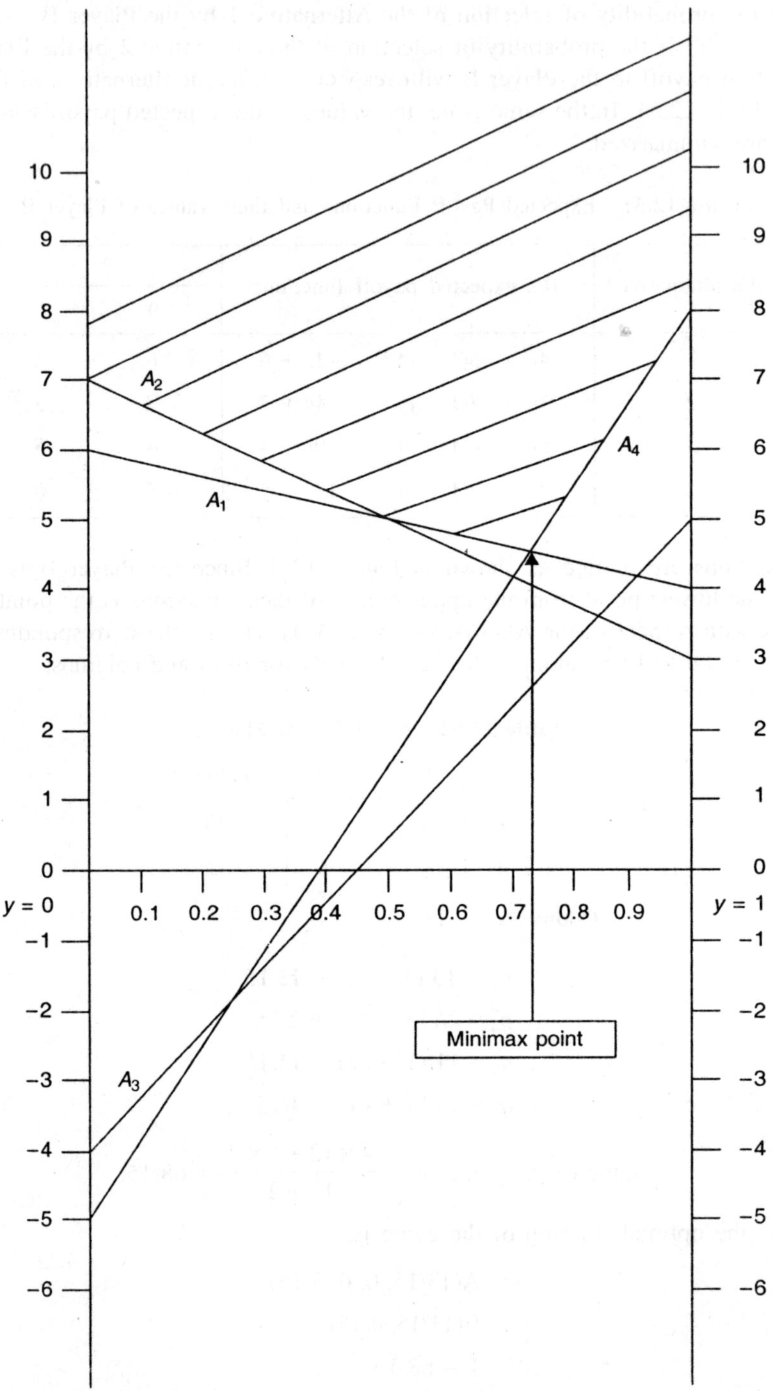

Figure 12.5 Graph with B's payoff functions of Example 12.12.

12.6 LINEAR PROGRAMMING APPROACH FOR GAME THEORY

The games can be classified into the following types:

(a) Generalized game of size $m \times n$ with a saddle point

(b) 2×2 game with or without saddle point

(c) $2 \times n$ game without saddle point

(d) $m \times 2$ game without saddle point

(e) 3×3 game and games of higher sizes without saddle point.

The first four types of games (a)–(d) can be solved using the methods discussed in earlier sections. But, the games under the case (e) (i.e. 3×3 game and games of higher sizes without saddle point) need a specialized procedure, namely *linear programming method*.

Guidelines for developing linear programming model

Let payoff matrix with respect to Player A be $[a_{ij}]$ shown as in Table 12.53.

Table 12.53 Generalized Payoff Matrix of Player A

Player B

		1	2	$\cdots$	j	$\cdots$	n
	1	a_{11}	a_{12}	$\cdots$	a_{1j}	$\cdots$	a_{1n}
	2	a_{21}	a_{22}	$\cdots$	a_{2j}	$\cdots$	a_{2n}
Player A	$\vdots$	$\vdots$	$\vdots$		$\vdots$		$\vdots$
	i	a_{i1}	a_{i2}	$\cdots$	a_{ij}	$\cdots$	a_{in}
	$\vdots$	$\vdots$	$\vdots$		$\vdots$		$\vdots$
	m	a_{m1}	a_{m2}	$\cdots$	a_{mj}	$\cdots$	a_{mn}

Also, suppose,

m = Total number of alternatives for Player A

n = Total number of alternatives for Player B

a_{ij} = Payoff to Player A if A selects his Alternative i and Player B selects his Alternative j

V = Value of the game (i.e. expected gain for A. So, A will try to maximize V and Player B will try to minimize V)

p_i = Probability of selection of Alternative i by Player A, where i = 1, 2, 3,..., m.

q_j = Probability of selection of Alternative j by Player B, where j = 1, 2, 3,..., n.

Development of linear programming model with respect to Player A. The expected gain to A with respect to the selection of each of the alternatives of B is presented in Table 12.54.

Table 12.54   Expected Gain Functions of Player A

Player B's alternative	Expected gain function to Player A
1	$a_{11}p_1 + a_{21}p_2 + \cdots + a_{i1}p_i + \cdots + a_{m1}p_m = \sum\limits_{i=1}^{m} a_{i1}p_i$
2	$a_{12}p_1 + a_{22}p_2 + \cdots + a_{i2}p_i + \cdots + a_{m2}p_m = \sum\limits_{i=1}^{m} a_{i2}p_i$
$\vdots$	$\vdots$
j	$a_{1j}p_1 + a_{2j}p_2 + \cdots + a_{ij}p_i + \cdots + a_{mj}p_m = \sum\limits_{i=1}^{m} a_{ij}p_i$
$\vdots$	$\vdots$
n	$a_{1n}p_1 + a_{2n}p_2 + \cdots + a_{in}p_i + \cdots + a_{mn}p_m = \sum\limits_{i=1}^{m} a_{in}p_i$

Descriptive model for Player A.   Since, A is a maximin player, he tries to maximize the value of the game. This is achieved by maximizing the minimum of the functions given in the last column of Table 12.54.

$$\max \left[\min \left(\sum_{i=1}^{m} a_{i1}p_i, \ \sum_{i=1}^{m} a_{i2}p_i, ..., \ \sum_{i=1}^{m} a_{ij}p_i, ..., \ \sum_{i=1}^{m} a_{in}p_i \right) \right]$$

subject to

$$p_1 + p_2 + \cdots + p_i + \cdots + p_m = 1$$

as the sum of the probabilities of selection of alternatives of A should be equal to 1. Also,

$$p_1, p_2, ..., p_i, ..., p_m \geq 0$$

Since, the objective function of this model is not in linear form, it is called a *descriptive model*.

Linear model for player A.   Since it is not possible to work with the descriptive model given in the previous section, the same is converted into a linear model by using the following transformation. Also, let,

$$V = \min \left(\sum_{i=1}^{m} a_{i1}p_i, \ \sum_{i=1}^{m} a_{i2}p_i, ..., \ \sum_{i=1}^{m} a_{ij}p_i, ..., \ \sum_{i=1}^{m} a_{in}p_i \right)$$

The linear model with respect to the above substitution is:

$$\text{Maximize } Z = V$$

subject to

$$\sum_{i=1}^{m} a_{i1}p_i \geq V$$

$$\sum_{i=1}^{m} a_{i2}p_i \geq V$$

$$\vdots$$

$$\sum_{i=1}^{m} a_{ij}p_i \geq V$$

$$\vdots$$

$$\sum_{i=1}^{m} a_{in}p_i \geq V$$

$$p_1 + p_2 + \cdots + p_i + \cdots + p_m = 1$$

$$p_i \geq 0, \qquad i = 1, 2,..., m$$

The system of constraints of the above model can be simplified by dividing it by V as follows:

$$\sum_{i=1}^{m} a_{i1}p_i/V \geq 1$$

$$\sum_{i=1}^{m} a_{i2}p_i/V \geq 1$$

$$\vdots$$

$$\sum_{i=1}^{m} a_{ij}p_i/V \geq 1$$

$$\vdots$$

$$\sum_{i=1}^{m} a_{in}p_i/V \geq 1$$

$$\frac{p_1}{V} + \frac{p_2}{V} + \cdots + \frac{p_i}{V} + \cdots + \frac{p_m}{V} = \frac{1}{V}$$

$$p_i \geq 0, \qquad i = 1, 2,..., m$$

In this modified model:

   (i) if the value of the game is less than zero, the type of each constraint will get changed;

  (ii) if the value of the game is zero, the terms of the constraints will become infinite.

Therefore, to avoid these problems, a constant K which is equal to the absolute value of the maximum of the negative values of the payoff matrix plus one, is added to each of the entries in the payoff matrix. After solving the problem, the true value of the game is obtained by subtracting K from the value of the game. If all the values of the payoff matrix are greater than zero then there is no need to add a constant K to each of the entries of the payoff matrix.

Let $p_i/V = X_i$, $i = 1, 2,..., m$. Therefore,

$$\max V = \min \frac{1}{V}$$

$$= \min \left(\frac{p_1}{V} + \frac{p_2}{V} + \cdots + \frac{p_i}{V} + \cdots + \frac{p_m}{V} \right)$$

$$= \min (X_1 + X_2 + \cdots + X_i + \cdots + X_m)$$

Then, by substituting the above function in the model, we get the revised model as:

$$\text{Minimize } Z_1 = X_1 + X_2 + \cdots + X_i + \cdots + X_m$$

subject to

$$\sum_{i=1}^{m} a_{i1} X_i \geq 1$$

$$\sum_{i=1}^{m} a_{i2} X_i \geq 1$$

$$\vdots$$

$$\sum_{i=1}^{m} a_{ij} X_i \geq 1$$

$$\vdots$$

$$\sum_{i=1}^{m} a_{in} X_i \geq 1$$

$$X_i \geq 0, \qquad i = 1, 2, \ldots, m$$

The values for p_i, where $i = 1, 2, \ldots, m$ and V are obtained using the following formulae.

$$V = \frac{1}{Z_1}$$

$$p_i = VX_i, \qquad i = 1, 2, \ldots, m$$

Development of linear programming model with respect to Player B. The expected loss (+)/gain (−) to Player B with respect to the selection of each of the alternatives of Player A is presented in Table 12.55.

Table 12.55 Expected Payoff Function of Player B

Player A's alternative	Expected loss (+)/gain (−) function to Player B
1	$a_{11}q_1 + a_{12}q_2 + \cdots + a_{1j}q_j + a_{1n}q_n = \displaystyle\sum_{j=1}^{n} a_{1j}q_j$
2	$a_{21}q_1 + a_{22}q_2 + \cdots + a_{2j}q_j + a_{2n}q_n = \displaystyle\sum_{j=1}^{n} a_{2j}q_j$
$\vdots$	$\vdots$
i	$a_{i1}q_1 + a_{i2}q_2 + \cdots + a_{ij}q_j + a_{in}q_n = \displaystyle\sum_{j=1}^{n} a_{ij}q_j$
$\vdots$	$\vdots$
m	$a_{m1}q_1 + a_{m2}q_2 + \cdots + a_{mj}q_j + a_{mn}q_n = \displaystyle\sum_{j=1}^{n} a_{mj}q_j$

Descriptive model for Player **B.**   Since B is a minimax player, he(she) tries to minimize the value of the game. This is achieved by minimizing the maximum of the functions given in the last column of the Table 12.55. Therefore,

$$\min \left[\max \left(\sum_{j=1}^{n} a_{1j}q_j, \ \sum_{j=1}^{n} a_{2j}q_j, ..., \ \sum_{j=1}^{n} a_{ij}q_j, ..., \ \sum_{j=1}^{n} a_{mj}q_j \right) \right]$$

subject to

$$q_1 + q_2 + \cdots + q_j + \cdots + q_n = 1$$

as the sum of the probabilities of selection of alternatives of B should be equal to 1.

$$q_1, q_2, ..., q_j, ..., q_n \geq 0$$

Since, the objective function of this model is not in linear form, it is called as a descriptive model.

Linear model for Player **B.**   Since it is not possible to work with the descriptive model given in the previous section, it is converted into a linear model by using the following transformation. Let

$$V = \max \left(\sum_{j=1}^{n} a_{1j}q_j, \ \sum_{j=1}^{n} a_{2j}q_j, ..., \ \sum_{j=1}^{n} a_{ij}q_j, ..., \ \sum_{j=1}^{n} a_{mj}q_j \right)$$

The linear model with respect to the above substitution is:

$$\text{Minimize } Z = V$$

subject to

$$\sum_{j=1}^{n} a_{1j}\, q_j \leq V$$

$$\sum_{j=1}^{n} a_{2j}\, q_j \leq V$$

$$\sum_{j=1}^{n} a_{ij}\, q_j \leq V$$

$$\vdots \qquad \vdots$$

$$\sum_{j=1}^{n} a_{mj}\, q_j \leq V$$

$$q_1 + q_2 + \cdots + q_j + \cdots + q_n = 1$$

$$q_j \geq 0, \qquad j = 1, 2, ..., n$$

The system of constraints of the above model can be simplified by division with V. Therefore, we get, the constraints:

$$\sum_{j=1}^{n} a_{1j}\frac{q_j}{V} \leq 1$$

$$\sum_{j=1}^{n} a_{2j}\frac{q_j}{V} \leq 1$$

$$\sum_{j=1}^{n} a_{ij}\frac{q_j}{V} \le 1$$

$$\vdots \qquad \vdots$$

$$\sum_{j=1}^{n} a_{mj}\frac{q_j}{V} \le 1$$

$$\frac{q_1}{V} + \frac{q_2}{V} + \cdots + \frac{q_j}{V} + \cdots + \frac{q_n}{V} = \frac{1}{V}$$

$$q_j \ge 0, \quad j = 1, 2,..., n$$

In this modified model:

    (i) if the value of game is less than zero, the type of each constraint will get changed;

    (ii) if the value of game is zero, the terms of the constraints will become infinite.

Therefore, to avoid these problems, each of the entries of the payoff matrix is added with a constant K which is equal to the absolute value of the maximum of the negative values of the payoff matrix plus one.

After solving the problem, the true value of the game is obtained by subtracting K from the value of the game. If all the values of the payoff matrix are greater than zero, there is no need to add a constant K to each of the entries of the payoff matrix.

Let $q_j/V = Y_j$, where $j = 1, 2, 3,..., n$. Then

$$\min V = \max \frac{1}{V}$$

$$= \max \left(\frac{q_1}{V} + \frac{q_2}{V} + \cdots + \frac{q_j}{V} + \cdots + \frac{q_n}{V} \right)$$

$$= \max (Y_1 + Y_2 + \cdots + Y_j + \cdots + Y_n)$$

By substituting the above function in the model, we get the revised model as presented below:

$$\text{Maximize } Z_2 = Y_1 + Y_2 + Y_3 + \cdots + Y_j + \cdots + Y_n$$

subject to

$$\sum_{j=1}^{n} a_{1j}Y_1 \le 1$$

$$\sum_{j=1}^{n} a_{2j}Y_2 \le 1$$

$$\vdots \qquad \vdots$$

$$\sum_{j=1}^{n} a_{ij}Y_3 \le 1$$

$$\vdots \qquad \vdots$$

$$\sum_{j=1}^{n} a_{mj} Y_j \leq 1$$

$$Y_j \geq 0, \qquad j = 1, 2,..., n$$

The values for q_j, $j = 1, 2, 3,..., n$ and V are obtained using the following formulae.

$$V = \frac{1}{Z_2} \qquad q_j = V Y_j, \qquad j = 1, 2, 3,..., n.$$

Example 12.13   Table 12.56 represents the payoff matrix with respect to Player A. Solve it optimally using linear programming method.

Table 12.56   Example 12.13

Player B

	1	2	3
Player A 1	1	−1	−1
2	−1	−1	3
3	−1	2	−1

Solution   The maximin and the minimax values are computed based on Table 12.57.

Table 12.57   Payoff Matrix with Maximin and Minimax Values

Player B

	1	2	3	Row minimum
1	1	−1	−1	−1 (maximin)
Player A 2	−1	−1	3	−1 (maximin)
3	−1	2	−1	−1 (maximin)
Column maximum	1	2	3	
	(minimax)			

Here, maximin value (−1) is not equal to the minimax value (1). Hence, the game has no saddle point and in turn, the game is said to have mixed strategies. Also, one can verify the fact that the payoff matrix cannot be reduced using dominance property. Hence, the game should be solved using the linear programming method.

Since the payoff matrix has negative values, the absolute value of the most negative value plus 1, i.e. ($K = 1 + 1$) is added to each of the entries of the payoff matrix and the corresponding revised payoff matrix is shown in Table 12.58.

Table 12.58   Payoff Matrix after Adding K to Each Entry

Player B

		1	2	3
	1	3	1	1
Player A	2	1	1	5
	3	1	4	1

Since the linear programming formulation with respect to Player B will have only "≤" constraints, it is advisable to develop a model for Player B. The solution of the problem can be obtained in the following ways:

- Solve the model and obtain the strategies of B and the value of the game.
- Obtain the strategies of A from the optimal table of B using the concept of duality.
- The true value of the game is obtained by subtracting K from the value of the modified game.

The last step is required because K (2) is added to all the cell entries of the payoff matrix at the initial stage. Let, a_{ij} be the payoff to Player A if A selects his Alternative i and Player B selects his Alternative j. V be the value of the game (i.e. expected gain for Player A. So, A will try to maximize V and B will try to minimize V). p_i be the probability of selection of Alternative i by Player A where, $i = 1, 2$ and 3. q_j be the probability of selection of Alternative j by Player B where, $j = 1, 2$ and 3.

Development of linear programming model with respect to Player B.   The expected loss (+)/gain (−) function to B with respect to the selection of each of the alternatives of A is presented below along with the necessary condition of $q_1 + q_2 + q_3 = 1$.

$$3q_1 + q_2 + q_3 \leq V$$

$$q_1 + q_2 + 5q_3 \leq V$$

$$q_1 + 4q_2 + q_3 \leq V$$

$$q_1 + q_2 + q_3 = 1$$

Dividing the above set of constraints by V, we get

$$3\frac{q_1}{V} + \frac{q_2}{V} + \frac{q_3}{V} \leq 1$$

$$\frac{q_1}{V} + \frac{q_2}{V} + 5\frac{q_3}{V} \leq 1$$

$$\frac{q_1}{V} + 4\frac{q_2}{V} + \frac{q_3}{V} \leq 1$$

$$\frac{q_1}{V} + \frac{q_2}{V} + \frac{q_3}{V} = \frac{1}{V}$$

Substituting $q_j/V = Y_j, j = 1, 2, 3$ in the above system of constraints, we have

$$3Y_1 + Y_2 + Y_3 \leq 1$$

$$Y_1 + Y_2 + 5Y_3 \leq 1$$

$$Y_1 + 4Y_2 + Y_3 \leq 1$$

$$Y_1 + Y_2 + Y_3 = \frac{1}{V}$$

As per the guidelines for constructing linear programming model for Player B, the linear programming model from the above system of constraints is presented below.

$$\text{Maximize } Z_2 = Y_1 + Y_2 + Y_3$$

subject to

$$3Y_1 + Y_2 + Y_3 \leq 1$$

$$Y_1 + Y_2 + 5Y_3 \leq 1$$

$$Y_1 + 4Y_2 + Y_3 \leq 1$$

$$Y_1, Y_2 \text{ and } Y_3 \geq 0$$

Converting the above generalized model into a standard model yields:

$$\text{Maximize } Z_2 = Y_1 + Y_2 + Y_3 + 0S_1 + 0S_2 + 0S_3$$

subject to

$$3Y_1 + Y_2 + Y_3 + S_1 = 1$$

$$Y_1 + Y_2 + 5Y_3 + S_2 = 1$$

$$Y_1 + 4Y_2 + Y_3 + S_3 = 1$$

$$Y_1, Y_2, Y_3, S_1, S_2, S_3 \geq 0$$

The starting initial table is shown in Table 12.59.

Table 12.59 Iteration 1

CB_i	Basis	1	1	1	0	0	0	Solution	Ratio
		Y_1	Y_2	Y_3	S_1	S_2	S_3		
0	S_1	3	1	1	1	0	0	1	$1/3 = 0.33$**
0	S_2	1	1	5	0	1	0	1	$1/1 = 1$
0	S_3	1	4	1	0	0	1	1	$1/1 = 1$
	Z_j	0	0	0	0	0	0	0	
	$C_j - Z_j$	1*	1	1	0	0	0		

*Key column   **Key row

In Table 12.59, Y_1 is selected as the entering variable and the leaving variable is S_1. The next iteration is shown in Table 12.60.

Table 12.60   Iteration 2

| CB_i | Basis | 1 | 1 | 1 | 0 | 0 | 0 | Solution | Ratio |
		Y_1	Y_2	Y_3	S_1	S_2	S_3		
1	Y_1	1	1/3	1/3	1/3	0	0	1/3	1
0	S_2	0	2/3	14/3	−1/3	1	0	2/3	1
0	S_3	0	11/3	2/3	−1/3	0	1	2/3	2/11
	Z_j	1	1/3	1/3	1/3	0	0	1/3	
	$C_j - Z_j$	0	2/3	2/3	−1/3	0	0		

Here, Y_2 is selected as the entering variable and the leaving variable is S_3. The next iteration is shown in Table 12.61.

Table 12.61   Iteration 3

| CB_i | Basis | 1 | 1 | 1 | 0 | 0 | 0 | Solution | Ratio |
		Y_1	Y_2	Y_3	S_1	S_2	S_3		
1	Y_1	1	0	3/11	4/11	0	−1/11	3/11	1
0	S_2	0	0	50/11	−3/11	1	−2/11	6/11	3/25
1	Y_2	0	1	2/11	−1/11	0	3/11	2/11	1
	Z_j	1	1	5/11	3/11	0	2/11	5/11	
	$C_j - Z_j$	0	0	6/11	−3/11	0	−2/11		

In Table 12.61, the entering variable is Y_3 and the leaving variable is S_2. The next iteration is shown in Table 12.62.

Table 12.62   Iteration 4

| CB_i | Basis | 1 | 1 | 1 | 0 | 0 | 0 | Solution |
		Y_1	Y_2	Y_3	S_1	S_2	S_3	
1	Y_1	1	0	0	19/50	−3/50	−2/25	6/25
1	Y_3	0	0	1	−3/50	11/50	−1/25	3/25
1	Y_2	0	1	0	−2/25	−1/25	7/25	4/25
	Z_j	1	1	1	6/25	3/25	4/25	13/25
	$C_j - Z_j$	0	0	0	−6/25	−3/25	−4/25	

In Table 12.62, since all the $C_j - Z_j$ values are less than or equal to zero, the optimality is reached and the solution of the model is:

$$Y_1 = \frac{6}{25}, \quad Y_2 = \frac{4}{25}, \quad Y_3 = \frac{3}{25} \text{ and } Z_2 = \frac{13}{25}$$

Compute the value of V and, q_1, q_2 and q_3 using the following formulae.

$$V = \frac{1}{Z_2} \quad \text{and} \quad q_j = VY_j, \quad j = 1, 2, 3$$

Therefore,

$$V = \frac{1}{Z_2} = \frac{1}{(13/25)} = \frac{25}{13}$$

where, value of the original game $= \dfrac{25}{13} - K = \dfrac{25}{13} - 2 = -\dfrac{1}{13}$

and

$$q_1 = \frac{25}{13} \frac{6}{25} = \frac{6}{13}$$

$$q_2 = \frac{25}{13} \frac{4}{25} = \frac{4}{13}$$

$$q_3 = \frac{25}{13} \frac{3}{25} = \frac{3}{13}$$

From the optimal Table 12.62, based on the concept of duality, the values of X_1, X_2 and X_3 are obtained as shown in Table 12.63.

Table 12.63 Solution of Player A

Basic variable in the initial table	S_1	S_2	S_3
Corresponding dual variable	X_1	X_2	X_3
$-(C_j - Z_j)$ from Table 12.62	6/25	3/25	4/25

Now, the solutions of Player A are:

$$X_1 = \frac{6}{25}, \quad X_2 = \frac{3}{25}, \quad X_3 = \frac{4}{25}, \quad Z_1 = \frac{13}{25}$$

Compute the value of V and, p_1, p_2 and p_3 using the following formulae.

$$V = \frac{1}{Z_1}, \quad p_i = VX_j, \quad i = 1, 2, 3$$

Therefore,

$$V = \frac{1}{Z_1} = \frac{1}{13/25} = \frac{25}{13}$$

The value of the original game $= \dfrac{25}{13} - K = \dfrac{25}{13} - 2 = -\dfrac{1}{13}$

$$p_1 = \frac{25}{13} \frac{6}{25} = \frac{6}{13}$$

$$p_2 = \frac{25}{13} \frac{3}{25} = \frac{3}{13}$$

$$p_3 = \frac{25}{13} \frac{4}{25} = \frac{4}{13}$$

The strategies of Players A and B are summarized as:

$$A\left(\frac{6}{13}, \frac{3}{13}, \frac{4}{13}\right) \quad \text{and} \quad B\left(\frac{6}{13}, \frac{4}{13}, \frac{3}{13}\right)$$

where value of the original game is $-1/13$.

Example 12.14 Consider a game in which the payoff matrix of the Player A is as shown in Table 12.64. Solve this game optimally using linear programming.

Table 12.64 Payoff Matrix of Player A for Example 12.14

		Player B		
		1	2	3
	1	6	8	2
Player A	2	8	2	10
	3	4	10	12

Solution The existence of saddle point is being verified as shown in Table 12.65. In Table 12.65, the maximin value (2) is not equal to the minimax value (8). This proves that the game has no saddle point and hence, the players have mixed strategies.

Table 12.65 Check for Saddle Point

		Player B			
		1	2	3	Minimum
	1	6	8	2	2 (maximin)
Player A	2	8	2	10	2 (maximin)
	3	4	10	12	4
Maximum		8	10	12	
		(minimax)			

Linear programming model with respect to Player B

Let, q_j be the probability of selection of the jth alternative by the Player B.

V be the value of the game

The Player B has to minimize his maximum loss.

The expected loss (+)/gain (–) function to B with respect to the selection of each of the alternatives of the Player A is presented below along with the necessary condition of $q_1 + q_2 + q_3 = 1$.

$$6q_1 + 8q_2 + 2q_3 \leq V$$
$$8q_1 + 2q_2 + 10q_3 \leq V$$
$$4q_1 + 10q_2 + 12q_3 \leq V$$
$$q_1 + q_2 + q_3 = 1$$

Dividing each of the terms in the above system of constraints by V, we get

$$6(q_1/V) + 8(q_2/V) + 2(q_3/V) \leq 1$$
$$8(q_1/V) + 2(q_2/V) + 10(q_3/V) \leq 1$$
$$4(q_1/V) + 10(q_2/V) + 12(q_3/V) \leq 1$$
$$q_1/V + q_2/V + q_3/V = 1/V$$

In the above set of constraints, by substituting $q_j/V = Y_j$ for $j = 1$, 2, and 3, we get

$$6Y_1 + 8Y_2 + 2Y_3 \leq 1$$
$$8Y_1 + 2Y_2 + 10Y_3 \leq 1$$
$$4Y_1 + 10Y_2 + 12Y_3 \leq 1$$
$$Y_1 + Y_2 + Y_3 = 1/V$$

As per the guidelines for constructing linear programming for Player B, the linear programming model from the above system of constraints is presented below. Actually, this model is the dual of the linear programming model to determine the strategies of the Player A.

$$\text{Maximize } Z_2 = Y_1 + Y_2 + Y_3$$

subject to

$$6Y_1 + 8Y_2 + 2Y_3 \leq 1$$
$$8Y_1 + 2Y_2 + 10Y_3 \leq 1$$
$$4Y_1 + 10Y_2 + 12Y_3 \leq 1$$
$$Y_1, Y_2 \text{ and } Y_3 \geq 0$$

The canonical form of this dual problem is as follows:

$$\text{Maximize } Z = Y_1 + Y_2 + Y_3 + 0S_1 + 0S_2 + 0S_3$$

subject to

$$6Y_1 + 8Y_2 + 2Y_3 + S_1 \leq 1$$
$$8Y_1 + 2Y_2 + 10Y_3 + S_2 \leq 1$$
$$4Y_1 + 10Y_2 + 12Y_3 + S_3 \leq 1$$
$$Y_1, Y_2, Y_3, S_1, S_2 \text{ and } S_3 \geq 0$$

The initial table and different iterations of simplex method applied to this problem are shown in Table 12.66.

The optimal values of Y_j, for $j = 1, 2$ and 3 are:

$$Y_1 = 1/10, \quad Y_2 = 9/190, \quad Y_3 = 1/95 \text{ and } Z_2 = 3/19.$$

Table 12.66 Initial Table and Different Iterations of Example 12.14

CB_i	C_j / Basic variable	1 Y_1	1 Y_2	1 Y_3	0 S_1	0 S_2	0 S_3	Solution	Ratio
0	S_1	6	8	2	1	0	0	1	1/6
0	S_2	8	2	10	0	1	0	1	1/8*
0	S_3	4	10	12	0	0	1	1	1/4
	Z_j	0	0	0	0	0	0	0	
	$C_j - Z_j$	1**	1	1	0	0	0		
0	S_1	0	13/2	−11/2	1	−3/4	0	1/4	1/26*
1	Y_1	1	1/4	5/4	0	1/8	0	1/8	1/2
0	S_3	0	9	7	0	−1/2	1	1/2	1/18
	Z_j	1	1/4	5/4	0	1/8	0	1/8	
	$C_j - Z_j$	0	3/4**	−1/4	0	−1/8	0		
1	Y_2	0	1	−11/13	2/13	−3/26	0	1/26	−
1	Y_1	1	0	19/13	−1/26	2/13	0	3/26	3/28
0	S_3	0	0	190/13	−18/13	7/13	1	2/13	1/95*
	Z_j	1	1	8/13	3/26	1/26	0	2/13	
	$C_j - Z_j$	0	0	5/13**	−3/26	−1/26	0		
1	Y_2	0	1	0	7/95	−8/95	11/190	9/190	
1	Y_1	1	0	0	1/10	1/10	−1/10	1/10	
1	Y_3	0	0	1	−9/95	7/190	13/190	1/95	
	Z_j	1	1	1	3/38	1/19	1/38	3/19	
	$C_j - Z_j$	0	0	0	−3/38	−1/19	−1/38		

Note: * Key row   ** key column

$$V = 1/Z_2 = 1/(3/19) = 19/3$$
$$q_j = VY_j, \text{ for } j = 1, 2 \text{ and } 3$$
$$q_1 = VY_1 = (19/3) \times (1/10) = 19/30$$
$$q_2 = VY_2 = (19/3) \times (9/190) = 3/10$$
$$q_3 = VY_3 = (19/3) \times (1/95) = 1/15$$

From the final iteration of (dual problem) Table 12.66, the solution of the primal problem (solution for the Player A) is obtained as shown in Table 12.67.

Table 12.67 Determination of Solution of Primal (Solution of Player A)

Basic variable in the initial table of the dual problem	S_1	S_2	S_3
$-(C_j - Z_j)$ of the final table of the dual problem	3/38	1/19	1/38
Corresponding primal variable	X_1	X_2	X_3

From the Table 12.67, the optimal values of X_1, X_2 and X_3 are as shown below:

$$X_1 = 3/38, \; X_2 = 1/19 \text{ and } X_3 = 1/38$$
$$p_i = VX_i, \text{ for } i = 1, 2 \text{ and } 3$$
$$p_1 = (19/3) \times (3/38) = 1/2$$
$$p_2 = (19/3) \times (1/19) = 1/3$$
$$p_3 = (19/3) \times (1/38) = 1/6$$

Therefore, the solution of the game is:

$$A \; (1/2, \; 1/3, \; 1/6)$$
$$B \; (19/30, \; 3/10, \; 1/15)$$
$$\text{Value of the game, } V = 19/3$$

QUESTIONS

1. Explain the following terminologies of game theory:

    (a) Players
    (b) Strategy
    (c) Maximin principle
    (d) Minimax principle
    (e) Saddle point
    (f) Value of the game
    (g) Two-person zero-sum game
    (h) Dominance property.

2. Two machine-tool companies, A and B, competing for supplying a CNC lathe to a new factory. Each company has listed its alternatives/strategies for selling machine tools. The strategies of Company A are as listed below:

    (a) Giving special price
    (b) Giving 15% worth of additional tools
    (c) Supplying some work holding device free of cost.

   The strategies/alternatives of Company B are as follows:

    (a) Giving special price
    (b) Giving 20% worth of additional tools
    (c) Giving free training to the users of the organization which is buying the machine.

The estimated gains (+)/losses (–) in lakhs of rupees of Company A for various possible combinations of the alternatives of both the companies are summarized in the following table. Find the optimal strategies for each of the companies.

Company B

		1	2	3
	1	40	45	50
Company A	2	20	45	60
	3	25	30	30

3. Find the optimum strategies of the players in the following games.

(a)

B

		1	2	3
	1	30	20	40
A	2	55	50	60
	3	60	30	40

(b)

B

		1	2	3
	1	10	5	20
A	2	65	50	40
	3	55	25	30

4. Players A and B play a game in which each player has three coins [25p, 50p and 100p (one rupee)]. Each of them selects a coin without the knowledge of the other person. If the sum of the values of the coins is an even number, A wins B's coin. If that sum is an odd number, B wins A's coin.

(a) Develop a payoff matrix with respect to Player A.
(b) Find the optimal strategies for the players.

5. Consider the following payoff matrix with respect to Player A and solve it optimally.

B

		1	2
	1	10	8
A	2	6	12

6. Consider the following payoff matrix with respect to Player A and solve it optimally.

B

		1	2
	1	8	18
A	2	16	9

7. Solve the following games:

B

		1	2	3
	1	–3	4	2
A	2	7	8	5
	3	6	2	9

8. Consider the following 4 × 4 game played by Players A and B and solve it optimally.

<table>
<tr><td></td><td></td><td colspan="4" align="center">Player B</td></tr>
<tr><td></td><td></td><td>1</td><td>2</td><td>3</td><td>4</td></tr>
<tr><td rowspan="4">Player A</td><td>1</td><td>12</td><td>4</td><td>8</td><td>16</td></tr>
<tr><td>2</td><td>4</td><td>−2</td><td>2</td><td>24</td></tr>
<tr><td>3</td><td>4</td><td>6</td><td>6</td><td>18</td></tr>
<tr><td>4</td><td>10</td><td>4</td><td>12</td><td>20</td></tr>
</table>

9. Solve the following 3 × 5 game using dominance property.

<table>
<tr><td></td><td></td><td colspan="5" align="center">Player B</td></tr>
<tr><td></td><td></td><td>1</td><td>2</td><td>3</td><td>4</td><td>5</td></tr>
<tr><td rowspan="3">Player A</td><td>1</td><td>6</td><td>15</td><td>30</td><td>21</td><td>6</td></tr>
<tr><td>2</td><td>3</td><td>3</td><td>6</td><td>6</td><td>4</td></tr>
<tr><td>3</td><td>12</td><td>12</td><td>24</td><td>36</td><td>3</td></tr>
</table>

10. Consider the payoff matrix of Player A as shown below and solve it optimally using graphical method.

<table>
<tr><td></td><td></td><td colspan="5" align="center">Player B</td></tr>
<tr><td></td><td></td><td>1</td><td>2</td><td>3</td><td>4</td><td>5</td></tr>
<tr><td rowspan="2">Player A</td><td>1</td><td>4</td><td>2</td><td>5</td><td>−6</td><td>6</td></tr>
<tr><td>2</td><td>7</td><td>−9</td><td>7</td><td>4</td><td>8</td></tr>
</table>

11. Consider the payoff matrix of Player A and solve it optimally using graphical method.

<table>
<tr><td></td><td></td><td colspan="5" align="center">Player B</td></tr>
<tr><td></td><td></td><td>1</td><td>2</td><td>3</td><td>4</td><td>5</td></tr>
<tr><td rowspan="2">Player A</td><td>1</td><td>7</td><td>8</td><td>4</td><td>6</td><td>8</td></tr>
<tr><td>2</td><td>−8</td><td>6</td><td>1</td><td>9</td><td>6</td></tr>
</table>

12. Consider the payoff matrix of Player A as given below and solve it optimally using graphical method.

<table>
<tr><td></td><td></td><td colspan="2" align="center">Player B</td></tr>
<tr><td></td><td></td><td>1</td><td>2</td></tr>
<tr><td rowspan="4">Player A</td><td>1</td><td>5</td><td>3</td></tr>
<tr><td>2</td><td>6</td><td>4</td></tr>
<tr><td>3</td><td>2</td><td>−7</td></tr>
<tr><td>4</td><td>9</td><td>−8</td></tr>
</table>

13. Consider the following payoff matrix with respect to Player A.

    (a) Develop a linear programming model with respect to Player A to find his optimal strategies.

    (b) Develop a linear programming model with respect to Player B to find his optimal strategies.

Player B

		1	2	$\cdots$	j	$\cdots$	n
	1	a_{11}	a_{12}	$\cdots$	a_{1j}	$\cdots$	a_{1n}
	2	a_{21}	a_{22}	$\cdots$	a_{2j}	$\cdots$	a_{2n}
Player A	$\vdots$	$\vdots$	$\vdots$		$\vdots$		$\vdots$
	i	a_{i1}	a_{i2}	$\cdots$	a_{ij}	$\cdots$	a_{in}
	$\vdots$	$\vdots$	$\vdots$		$\vdots$		$\vdots$
	m	a_{m1}	a_{m2}	$\cdots$	a_{mj}	$\cdots$	a_{mn}

14. The following table represents the payoff matrix with respect to Player A. Solve it optimally using linear programming method.

Player B

		1	2	3
	1	2	3	-4
Player A	2	5	-2	6
	3	2	6	3

15. The following table represents the payoff matrix with respect to Player A. Solve it optimally using linear programming method.

Player B

		1	2	3
	1	40	50	-70
Player A	2	10	25	-10
	3	100	30	60

REPLACEMENT AND MAINTENANCE ANALYSIS 13

13.1 INTRODUCTION

Organizations providing goods and services use several facilities like equipment and machineries to carry out their operations. In addition to these facilities, there are several other items which are essential to facilitate the functioning of the organizations. All such facilities are to be continuously monitored for their efficient functioning; otherwise the quality of service of those facilities will be poor. In addition to the quality of service of the facilities, the cost of their operation and maintenance would increase with the passage of time. Hence, it is an absolute necessity to maintain the equipments in good operating conditions with economical cost. Hence, we need an integrated approach to minimize the cost of maintenance. In certain cases, the equipment will become obsolete over a period of time.

If a firm wants to survive the competition, it has to decide on whether to replace the outdated equipment or to retain it, by taking the cost of maintenance and operation into account.

There are two basic reasons for considering the replacement of an equipment. They are: *physical impairment* or malfunctioning of various parts and *obsolescence* of the equipment. Physical impairment refers only to changes in the physical condition of the equipment itself. This would lead to a decline in the value of the service rendered by the equipment, increased operating cost of the equipment, increased maintenance cost of the equipment or a combination of these costs. Obsolescence is caused due to improvement in the existing tools and machinery mainly when the technology becomes advanced.

Therefore, it becomes uneconomical to continue production with the same equipment under any of the above situations. Hence, the equipments are to be periodically replaced.

Sometimes, the capacity of existing facilities may be inadequate to meet the current demand. Under such situations, the following two alternatives will be considered.

1. Replacement of the existing equipment with a new one.
2. Augment the existing one with an additional equipment.

13.2 TYPES OF MAINTENANCE

Maintenance activity can be classified into two types, viz. *preventive maintenance* and *breakdown maintenance*. Preventive maintenance (PM) is the periodical inspection and service activities which are aimed to detect potential failures and perform minor adjustments or repairs which will prevent major operating problems in future. Breakdown maintenance is the repair which is generally done after the equipment breaks down. It is often an emergency which will have an associated penalty in terms of increasing the cost of maintenance and downtime cost of equipment. Preventive maintenance will reduce such costs up to a certain extent. Beyond that, the cost of the preventive maintenance will be more when compared to the cost of the breakdown maintenance. The total cost, which is the sum of the preventive maintenance cost and the breakdown maintenance cost, will go on decreasing with an increase in the level of maintenance up to a point, beyond which the total cost will start increasing. The level of maintenance corresponding to the minimum total cost is the optimal level of maintenance. These concepts are demonstrated in Figure 13.1.

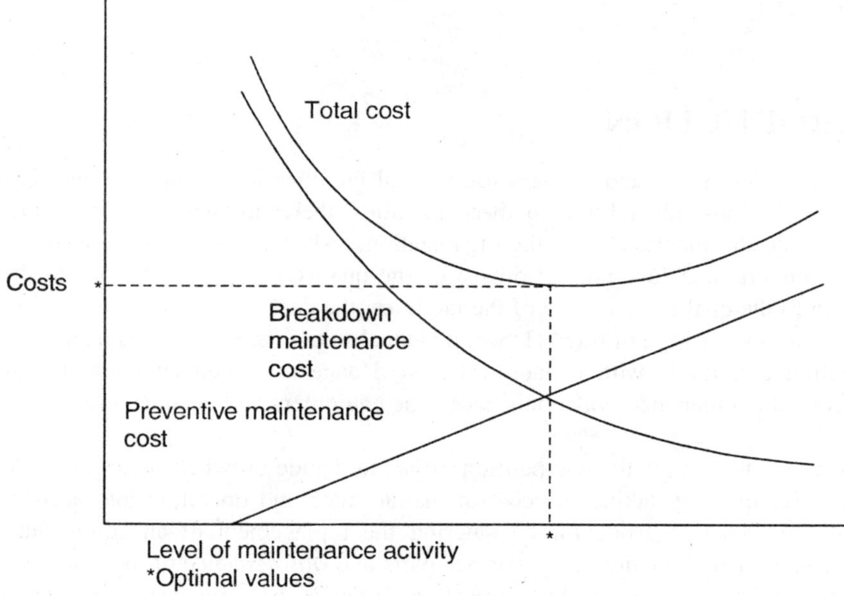

Figure 13.1   Trade-off between maintenance costs with respect to level of maintenance activity.

13.3   TYPES OF REPLACEMENT PROBLEM

The replacement study can be classified into two categories:

    (a) Replacement of assets that deteriorate with time (replacement due to gradual failure, due to wear and tear of the components of the machines). This can be further classified into the following types.

       (i) Determination of economic life of an asset

      (ii) Replacement of an existing asset with a new asset.

    (b) Simple probabilistic model for assets which fail completely (replacement due to sudden failure).

13.4   DETERMINATION OF ECONOMIC LIFE OF AN ASSET

Any asset will have the following cost components.

    (a) Capital recovery cost (average first cost), computed from the first cost (purchase price) of the asset.

(b) Average operating and maintenance cost

(c) Total cost which is the sum of capital recovery cost (average first cost), and average operating and maintenance cost.

A typical shape of each of the above costs with respect to life of the asset is shown in Figure 13.2.

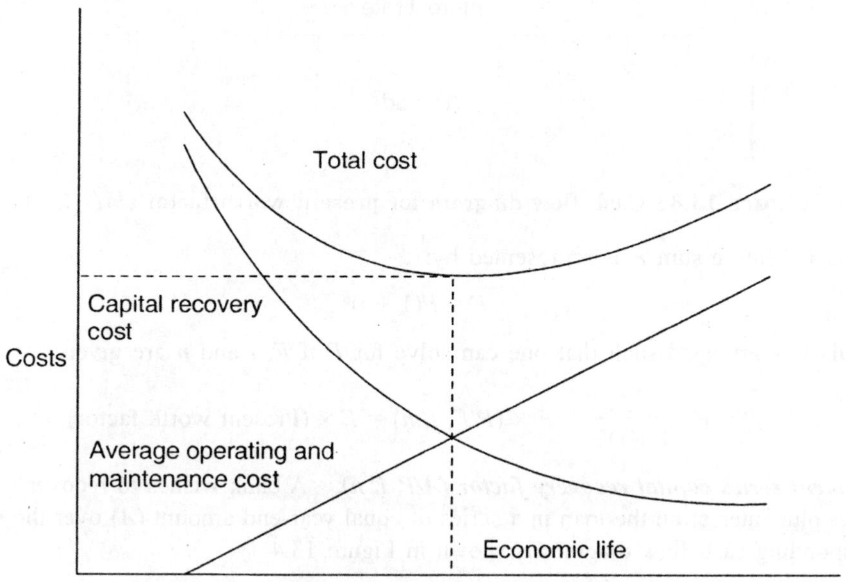

Figure 13.2 Chart showing economic life.

In Figure 13.2, when the life of the machine increases, it is clear that the capital recovery cost (average first cost) goes on decreasing, and the average operating and maintenance cost goes on increasing. From the beginning, the total cost goes on decreasing up to a particular life of the asset and then it starts increasing. The point where the total cost is the minimum is called the *economic life* of the asset.

13.4.1 Basics of Interest Formulae

If the interest rate is more than zero per cent, then interest formulae are used to determine the economic life. The replacement alternatives are generally evaluated based on annual equivalent criterion.

Present worth factor (P/F, i, n).   If an amount P is invested now with the amount earning interest at the rate i per year, a cash flow diagram showing the future sum (F) accumulated after n years is presented in Figure 13.3. In this diagram,

$$P = \text{Principal sum at year 0}$$
$$F = \text{Future sum of } P \text{ at the end of the } n\text{th year}$$
$$i = \text{Annual interest rate}$$
$$n = \text{Number of interest periods}$$

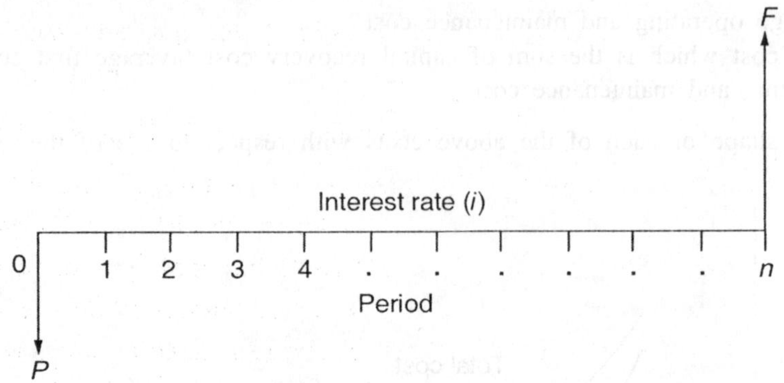

Figure 13.3 Cash flow diagram for present worth factor (*P/F, i, n*).

The formula for future sum F is represented by:

$$F = P(1 + i)^n$$

This formula is rearranged such that one can solve for P if F, i and n are given:

$$P = F \ \frac{1}{(1 + i)^n} = F \times (P/F, \ i, \ n) = F \times (\text{Present worth factor})$$

Equal payment series capital recovery factor (A/P, i, n). A bank wishes to recover a loan paid to its customer plus interest on the loan in a series of equal year-end amount (A) over the next n years. The corresponding cash flow diagram is shown in Figure 13.4.

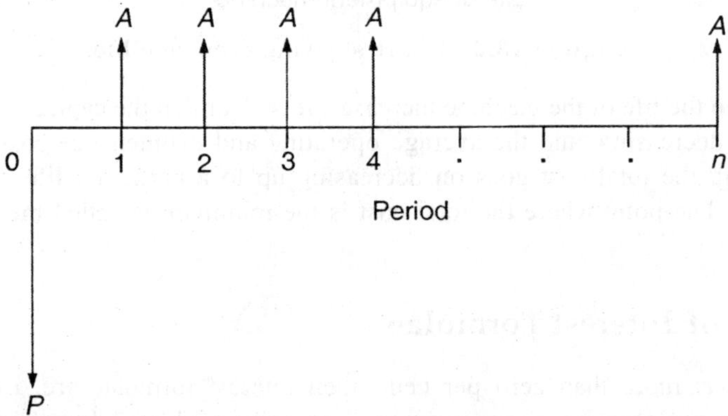

Figure 13.4 Cash flow diagram for equal payment series capital recovery factor (*A/P, i, n*).

Here, A is the annual equivalent amount which occurs at the end of every year from year 1 through year n. The formula to compute A is given by:

$$A = P \ \frac{i(1 + i)^n}{(1 + i)^n - 1}$$
$$= P \ [A/P, \ i, \ n]$$
$$= P \ (\text{equal payment series capital recovery factor})$$

The present worth factor, $(P/A, i, n)$ and the annual equivalent factor/capital recovery factor, $(A/P, i, n)$ are used in the examples which are presented in the next section, when the interest rate is more than 0%.

13.4.2   Examples of Determination of Economic Life of an Asset

In this section, based on the fundamentals of the interest formulae which are presented in the previous section, some examples of determination of economic life of an asset are presented.

Example 13.1   A firm is considering replacement of an equipment whose first cost is Rs. 4000 and the scrap value is negligible at the end of any year. Based on experience, it has been found that the maintenance cost is zero during the first year and it is Rs. 1000 for the second year. It increases by Rs. 300 every year thereafter.

  (a)   When should the equipment be replaced if $i = 0\%$?
  (b)   When should the equipment be replaced if $i = 12\%$?

Solution   (a) Here, first cost = Rs. 4000 and maintenance cost is Re. 0 during the first year and it is Rs. 1000 during the second year. Then, it increases by Rs. 300 every year thereafter. These are summarized in Column B of Table 13.1.

Table 13.1   Calculations to Determine Economic Life

First cost = Rs. 4000,		Interest rate = 0%			
End of year (n)	Maintenance cost at end of year	Summation of maintenance costs	Average cost of maintenance through the given year	Average first cost if replaced at the given year end	Average total cost through the given year
		(Σ B)	(C/A)	(4000/A)	(D + E)
A	B (Rs.)	C (Rs.)	D (Rs.)	E (Rs.)	F (Rs.)
1	0	0	0	4000	4000
2	1000	1000	500	2000	2500
3	1300	2300	767	1333	2100
4	1600	3900	975	1000	1975
5	1900	5800	1160	800	1960*
6	2200	8000	1333	667	2000
7	2500	10500	1500	571	2071

*Economic life of the machine = 5 years

Column C summarizes the summation of maintenance costs for each replacement period. The value corresponding to any end-of-year in this column represents the total maintenance cost for using the equipment till the end of that particular year. Therefore,

$$\text{Average total cost} = \frac{\text{First cost (FC)} + \text{Summation of maintenance cost}}{\text{Replacement period}}$$

$$= \frac{\text{FC}}{n} + \frac{\text{Column C}}{n}$$

$$= \left(\begin{array}{c}\text{Average first cost}\\\text{for the given period}\end{array}\right) + \left(\begin{array}{c}\text{Average maintenance}\\\text{cost for the given period}\end{array}\right)$$

$$\text{Column F} = \text{Column E} + \text{Column D}$$

The value corresponding to any end-of-year (n) in Column F represents the average total cost of using the equipment till the end of that particular year.

For this problem, the average total cost decreases till the end of the year 5 and then it increases. Hence, the optimal replacement period is 5 years, i.e. the economic life of the equipment is 5 years.

(b) When interest rate, $i = 12\%$: When the interest rate is more than 0%, the steps to get the economic life are summarized with reference to Table 13.2.

Table 13.2 Calculations to Determine Economic Life

	First cost = Rs. 4000,			Interest rate = 12%			
End of year (n)	Maintenance cost at end of year	(P/F, 12%, n)	Present worth as of beginning of year 1 of maintenance costs	Summation of present worth of maintenance costs through the given year	Present worth of cumulative maintenance cost and first cost	(A/P, 12%, n)	Annual equivalent total cost through the given year
			(B × C)	Σ D	(E + Rs. 4000)		(F × G)
A	B (Rs.)	C	D (Rs.)	E (Rs.)	F (Rs.)	G	H (Rs.)
1	0	0.8929	0	0	4000	1.1200	4480
2	1000	0.7972	797	797	4797	0.5917	2838
3	1300	0.7118	925	1722	5722	0.4163	2382
4	1600	0.6355	1017	2739	6739	0.3292	2219
5	1900	0.5674	1078	3817	7817	0.2774	2168*
6	2200	0.5066	1115	4932	8932	0.2432	2172
7	2500	0.4524	1131	6063	10,063	0.2191	2205
8	2800	0.4039	1131	7194	11,194	0.2013	2253

*Economic life of the machine = 5 years.

The steps are summarized as follows:

1. Discount the maintenance costs to the beginning of the year 1.

$$\text{Column D} = \text{Column B} \times \frac{1}{(1 + i)^n}$$

$$= \text{Column B} \times (P/F,\ i,\ n)$$

$$= \text{Column B} \times \text{Column C}$$

2. Find the summation of present worth of maintenance costs through given year, i.e.

$$\text{Column E} = \Sigma \text{ Column D}$$

3. Find Column F by adding the first cost of Rs. 4000 to Column E.
4. Find the annual equivalent total cost through the given years.

$$\text{Column H} = \text{Column F} \times \frac{i(1+i)^n}{(1+i)^n - 1}$$

$$= \text{Column F} \times (A/P, \ 12\%, \ n)$$

$$= \text{Column F} \times \text{Column G}$$

5. Identify the end of year for which the annual equivalent total cost is minimum in Column H.

For this problem, the annual equivalent total cost is minimum at the end of year 5. Hence, the economic life of the equipment is 5 years.

Example 13.2   Table 13.3 gives the operation cost, maintenance cost and salvage value at the end of every year of a machine whose purchase value is Rs. 12,000. Find the economic life of the machine assuming:

(a) the interest rate as 0%.
(b) the interest rate as 15%.

Table 13.3   Example 13.2

End of year (*n*)	Operation cost at the end of year (Rs.)	Maintenance cost at the end of year (Rs.)	Salvage value at the end of year (Rs.)
1	2000	2500	8000
2	3000	3000	7000
3	4000	3500	6000
4	5000	4000	5000
5	6000	4500	4000
6	7000	5000	3000
7	8000	5500	2000
8	9000	6000	1000

Solution   (a) When i = 0%: The calculations to determine the economic life of the machine for 0% interest rate are summarized in Table 13.4. Since, the operation cost and maintenance cost occur at the end of every year, they can be added to get total cost of operation and maintenance in each year as shown in Column D of Table 13.4.

$$\text{Average annual cost through year} = \frac{\text{Cumulative operation and maintenance cost} + \text{First cost} - \text{Salvage value}}{\text{End of year}}$$

Therefore,

$$\text{Column H} = \frac{\text{Column E} + 12{,}000 - \text{Column F}}{\text{Column A}}$$

Table 13.4 Summary of Calculations when $i = 0\%$

First Cost = Rs. 12,000,   Interest rate = 0%

End of year (n)	Operation cost at the end of year	Maintenance cost at the end of year	Sum of operation and maintenance cost at the end of year	Cumulative sum of operation and maintenance cost through the given year	Salvage value at end of year	Total cost through the given year	Average annual cost
			(B + C)	Σ D		(E + 12000 – F)	G/A
A	B (Rs.)	C (Rs.)	D (Rs.)	E (Rs.)	F (Rs.)	G (Rs.)	H (Rs.)
1	2000	2500	4500	4500	8000	8500	8500
2	3000	3000	6000	10,500	7000	15,500	7750*
3	4000	3500	7500	18,000	6000	24,000	8000
4	5000	4000	9000	27,000	5000	34,000	8500
5	6000	4500	10,500	37,500	4000	45,500	9100
6	7000	5000	12,000	49,500	3000	58,500	9750
7	8000	5500	13,500	63,000	2000	73,000	10,429
8	9000	6000	15,000	78,000	1000	89,000	11,125

*Economic life of the machine = 2 years

In Column H, the average annual cost is minimum when n is equal to 2. Hence, the economic life of the machine is 2 years.

(b) When $i = 15\%$: First cost = Rs. 12,000, Interest rate = 15%. The other details are summarized in Table 13.5 along with regular calculations for determining the economic life of the machine.

$$\text{Total annual equivalent cost} = \left[\left(\begin{array}{c} \text{Cumulative sum of present} \\ \text{worth as of beginning of} \\ \text{year 1 of operation and} \\ \text{maintenance cost} \end{array} \right) + \left(\begin{array}{c} \text{First} \\ \text{cost} \end{array} \right) - \left(\begin{array}{c} \text{Present worth} \\ \text{as of beginning} \\ \text{of year 1 of} \\ \text{salvage value} \end{array} \right) \right] \times (A/P, 15\%, n)$$

Therefore,

$$\text{Column L} = (\text{Column G} + 12000 - \text{Column I}) \times \text{Column K}$$
$$= \text{Column J} \times \text{Column K}$$

In column L, the annual equivalent total cost is minimum when n is equal to 2. Hence, the economic life of the machine is 2 years.

Example 13.3 A specific requirement of a company can be met either by Machine A or Machine B. Machine A costs Rs. 10,000. The maintenance cost during the first year of its operation is Rs. 2000 and it increases by Rs. 800 every year thereafter up to its fourth year of operation. The maintenance cost during the fifth year is Rs. 6000 and it increases by Rs. 1000 every year thereafter.

Machine B, which has the same capacity as that of Machine A, is priced at Rs. 15,000. The maintenance costs of the Machine B are estimated at Rs. 2000 for the first year and an equal yearly increment of Rs. 750 thereafter.

If the money is worth 15% per year, which machine should be purchased? (Assume that the scrap value of each of the machines is negligible at any year.)

Table 13.5 Calculations to Determine Economic Life

First Cost = Rs. 12,000,   Interest Rate = 15%

End of year (n)	Operation cost at the end of year	Maintenance cost at the end of year	Sum of operation and maintenance costs at the end of year n	$(P/F, 15\%, n)$	Present worth as of beginning of year 1 of sum of operation and maintenance costs	Cumulative sum of Column F through the given year	Salvage value at the end of year	Present worth as of beginning of year 1 of salvage value	Total present worth	$A/P, 15\%, n$	Average annual total cost through the given year
			(B + C)		(D × E)	Σ F		(H × E)	(G + 12000 − I)		(J × K)
A	B (Rs.)	C (Rs.)	D (Rs.)	E	F (Rs.)	G (Rs.)	H (Rs.)	I (Rs.)	J (Rs.)	K	L (Rs.)
1	2000	2500	4500	0.8696	3913	3913	8000	6957	8956	1.1500	10,299.40
2	3000	3000	6000	0.7561	4537	8450	7000	5293	15,157	0.6151	9323.07*
3	4000	3500	7500	0.6575	4931	13,381	6000	3945	21,436	0.4380	9388.97
4	5000	4000	9000	0.5718	5146	18,527	5000	2859	27,668	0.3503	9692.10
5	6000	4500	10,500	0.4972	5221	23,748	4000	1989	33,759	0.2983	10,070.31

*Economic life of the machine = 2 years.

Solution

Step 1: *Determination of economic life and corresponding annual equivalent total cost of Machine A*

The details of Machine A are summarized in Table 13.6 along with usual calculations to determine its economic life.

Table 13.6 Calculations to Determine Economic Life of Machine A

First cost = Rs. 10,000,   Interest rate = 15%

End of year (n)	Maintenance cost at the end of year	$(P/F, 15\%, n)$	Present worth as of beginning of year 1 of maintenance costs	Summation of present worth of maintenance costs through the given year	Column E + Rs. 10,000	$(A/P, 15\%, n)$	Annual equivalent total cost through the given year
			(B × C)	ΣD			(F × G)
A	B (Rs.)	C	D (Rs.)	E (Rs.)	F (Rs.)	G	H (Rs.)
1	2000	0.8696	1739	1739	11739	1.1500	13500
2	2800	0.7561	2117	3856	13856	0.6151	8523
3	3600	0.6575	2367	6223	16223	0.4380	7106
4	4400	0.5718	2516	8739	18739	0.3503	6564
5	6000	0.4972	2983	11722	21722	0.2983	6480*
6	7000	0.4323	3026	14748	24748	0.2642	6538
7	8000	0.3759	3007	17755	27755	0.2404	6672
8	9000	0.3269	2942	20697	30697	0.2229	6842

*Economic life of Machine A = 5 years.

Column B of Table 13.6 summarizes the yearly maintenance costs of Machine A. First cost of Machine A is equal to Rs. 10,000. Therefore,

Annual equivalent total cost of Machine A (Column H)

$$= \left[\left(\begin{matrix} \text{Summation of present worth of maintenance} \\ \text{cost through the given year} \end{matrix} \right) + \left(\begin{matrix} \text{First} \\ \text{cost} \end{matrix} \right) \right] \times (A/P,\ 15\%,\ n)$$

$$= (\text{Column E} + \text{Rs. 10,000}) \times \text{Column G}$$

$$= \text{Column F} \times \text{Column G}$$

In Column H, the minimum annual equivalent total cost occurs when n is equal to 5. Hence, the economic life of Machine A is 5 years and the corresponding annual equivalent total cost is Rs. 6480.

Step 2: *Determination of economic life and corresponding annual equivalent total cost of Machine B.*

The details of Machine B are summarized in Table 13.7 along with the usual calculations to determine its economic life. The Column B of Table 13.7 summarizes the yearly maintenance costs of Machine B. First cost of Machine B is equal to Rs. 15,000.

Table 13.7 Calculations to Determine Economic Life of Machine B

First cost = Rs. 15,000, Interest rate = 15%

End of year (n)	Maintenance cost at the end of year	(P/F, 15%, n)	Present worth as of beginning of year 1 of maintenance costs	Summation of present worth of maintenance costs through the given year	Column E + Rs. 15000	(A/P, 15%, n)	Annual equivalent total cost through the given year
			(B × C)	ΣD			(F × G)
A	B (Rs)	C	D (Rs.)	E (Rs.)	F (Rs.)	G	H (Rs.)
1	2000	0.8696	1739	1739	16,739	1.1500	19,250
2	2750	0.7561	2079	3818	18,818	0.6151	11,575
3	3500	0.6575	2301	6119	21,119	0.4380	9250
4	4250	0.5718	2430	8549	23,549	0.3503	8249
5	5000	0.4972	2486	11,035	26,035	0.2983	7766
6	5750	0.4323	2486	13,521	28,521	0.2642	7535
7	6500	0.3759	2443	15,964	30,964	0.2404	7444
8	7250	0.3269	2370	18,334	33,334	0.2229	7430*
9	8000	0.2843	2274	20,608	35,608	0.2096	7463
10	8750	0.2472	2163	22,771	37,771	0.1993	7528

*Economic life of the machine = 8 years.

Annual equivalent total cost of Machine B (Column H)

$$= \left[\left(\begin{array}{c} \text{Summation of present worth of maintenance} \\ \text{cost through the given year} \end{array} \right) + \left(\begin{array}{c} \text{First} \\ \text{cost} \end{array} \right) \right] \times (A/P, 15\%, n)$$

$$= (\text{Column E} + \text{Rs. 15,000}) \times \text{Column G}$$

$$= \text{Column F} \times \text{Column G}$$

In Column H, the minimum annual equivalent total cost occurs when *n* is equal to 8. Hence, the economic life of Machine B is 8 years and the corresponding annual equivalent total cost is Rs. 7430.

Result: Minimum annual equivalent total cost for Machine A = Rs. 6480 and minimum annual equivalent total cost for Machine B = Rs. 7430. Since the minimum annual equivalent total cost of A is lesser than that of B, A is selected as the best machine which has the economic life of 5 years.

(*Note:* Selection of the best machine is based on the minimum annual equivalent total cost. The comparison is made over the minimum common multiple of lives of Machine A and Machine B, i.e. 40 years.)

Example 13.4 Alpha Castings Private Limited, a small scale industry purchases a generator for Rs. 20,000. The operation cost is Rs. 2000 during the first year of its operation and it increases by Rs. 1000 every year thereafter. The maintenance cost is Rs. 200 during the first year of its operation and it increases by Rs. 100 every year thereafter. The purchase of this generator is through an interest free loan sanctioned for this company by Small Scale Industrial Development Corporation. Find the economic life of the generator.

Solution The data of this problem are:

Purchase price [First Cost (FC)] of the generator = Rs. 20,000
Operation cost during its first year of operation = Rs. 2000
Annual increase in the operation cost from second year onwards = Rs. 1000
Maintenance cost during its first year of operation = Rs. 200
Annual increase in the maintenance cost from second year onwards = Rs. 100

The calculations to determine the economic life of the generator are summarized in Table 13.8. From Table 13.8, the minimum annual equivalent cost of the generator is Rs. 8283.33 which corresponds to the generator's life of 6 years. Hence, the economic life of the generator is 6 years.

Table 13.8 Calculations to Determine Economic Life of Generator

End of year (n)	Operation cost	Maintenance cost	Operation + maintenance cost ($B + C$)	Cumulative operation and maintenance cost ΣD	FC (Rs. 20,000) $+E$	Annual equivalent cost F/A
(A)	(B) (Rs.)	(C) (Rs.)	(D) (Rs.)	(E) (Rs.)	(F) (Rs.)	(G) (Rs.)
1	2000	200	2200	2200	22,200	22,200.00
2	3000	300	3300	5500	25,500	12,750.00
3	4000	400	4400	9900	29,900	9966.67
4	5000	500	5500	15,400	35,400	8850.00
5	6000	600	6600	22,000	42,000	8400.00
6	*7000*	*700*	*7700*	*29,700*	*49,700*	*8283.33**
7	8000	800	8800	38,500	58,500	8357.14
8	9000	900	9900	48,400	68,400	8550.00
9	10,000	1000	11,000	59,400	79,400	8822.22
10	11,000	1100	12,100	71,500	91,500	9150.00

Example 13.5 A company purchases a machine for Rs. 10,000. The operation cost of the machine is expected to be more or less the same during its life. The maintenance cost of the machine is Rs. 2000 during its first year of operation. It increases by Rs. 800 from second year to fourth year of its operation. During its fifth year of operation, it is Rs. 6000 and then onwards, it increases by Rs. 1,000 every year. The interest rate is 12%. Determine the economic life of the machine.

Solution The data of the machine are as given below.

First cost (FC) = Rs. 10,000
Maintenance cost during the first year of operation of the machine = Rs. 2000
Annual increase in the maintenance cost from second year to fourth year = Rs. 800
Maintenance cost during the fifth year of its operation = Rs. 6000
Annual increase in the maintenance cost from sixth year onwards = Rs. 1000
Interest = 12%

The calculations for determining the economic life of the machine are summarized in Table 13.9. From Table 13.9, the minimum average annual cost is Rs. 6319.46 which occurs for the 5th year. Hence, the economic life of the machine is 5 years.

Table 13.9 Economic Life Determination of Machine

End of year (n)	Maintenance cost	P/F, 12%, n	Present worth of maintenance cost (B × C)	Cumulative present worth of maintenance cost (ΣD)	FC (Rs. 10,000) +E	A/P, 12%, n	Annual equivalent cost (F × G)
(A)	(B) (Rs.)	(C)	(D) (Rs.)	(E) (Rs.)	(F) (Rs.)	(G)	H (Rs.)
1	2000	0.8929	1785.80	1785.80	11,785.80	1.1200	13,200.10
2	2800	0.7972	2232.16	4017.96	14,017.96	0.5917	8294.43
3	3600	0.7118	2562.48	6580.44	16,580.44	0.4163	6902.44
4	4400	0.6355	2796.20	9376.64	19,376.64	0.3292	6378.79
5	*6000*	*0.5674*	*3404.40*	*12,781.04*	*22,781.04*	*0.2774*	*6319.46**
6	7000	0.5066	3546.20	16,327.24	26,327.24	0.2432	6402.79

Example 13.6   The cost of a grinding machine purchased by an industry is Rs. 30,000. The operation cost of the grinding machine during its first year of operation is Rs. 3000 and it increases by Rs. 1000 every year thereafter. The maintenance cost of the grinding machine is Rs. 300 during its first year of operation and it increases by Rs. 100 every year thereafter. The salvage value of the grinding machine is Rs. 10,000 at the end of the first year of its operation and it decreases by Rs. 1000 every year thereafter. The interest rate is 15%. Find the economic life of the grinding machine.

Solution   The data of the grinding machine are:

Purchase price [First Cost (FC)] of the grinding machine = Rs. 30,000
Operation cost during the first year of operation of the grinding machine = Rs. 3000
Annual increase in the operation cost from second year onwards = Rs. 1000
Maintenance cost during the first year of operation of the grinding machine = Rs. 300
Annual increase in the maintenance cost from second year onwards = Rs. 100
Salvage value at the end of the first year life of the grinding machine = Rs. 10,000
Annual decrease in salvage value from the end of second year onwards = Rs. 1000
Interest rate = 15%

The workings to determine the economic life for the grinding machine are shown in Table 13.10. From Table 13.10, the minimum annual equivalent cost is Rs. 12,829.22 which occurs for the 8th year. Hence, the economic life of the grinding machine is 8 years.

Example 13.7   The cost of an office equipment in an industry is Rs. 1,00,000. The maintenance cost of the equipment is Rs. 10,000 during its first year of operation and it increases by Rs. 2000 every year thereafter. The salvage value of the equipment is Rs. 65,000 at the end of the first year of its operation and it decreases by Rs. 10,000 every year thereafter till it reduces to 0. The interest rate is 12%. Find the economic life of the office equipment.

Solution   The data of the grinding machine are:

Purchase price of the office equipment = Rs. 1,00,000
Maintenance cost during the first year of operation of office equipment = Rs. 10,000
Annual increase in the maintenance cost from second year onwards = Rs. 2000
Salvage value at the end of the first year life of the office equipment = Rs. 65,000
Annual decrease in salvage value from the end of second year onwards till it reduces to 0
= Rs. 10,000

Interest rate = 12%

Table 13.10 Workings for the Determination of Economic Life of Grinding Machine

End of year (n)	Operation cost (Rs.)	Maintenance cost (Rs.)	Operation + maintenance cost (Rs.) (B + C)	P/F, 15%, n	Present worth of (D) (D × E)	Cumulative of column F	G + 30,000	Salvage value	Present worth of column I (I × E)	Column H − Column J	A/P, 15%, n	Annual equivalent cost (K × L)
(A)	(B) (Rs.)	(C) (Rs.)	(D) (Rs.)	(E)	(F) (Rs.)	(G) (Rs.)	(H) (Rs.)	(I) (Rs.)	(J) (Rs.)	(K) (Rs.)	(L)	(M) (Rs.)
1	3000	300	3300	0.8696	2869.68	2869.68	32,869.68	10,000	8696.0	24,173.68	1.1500	27,799.73
2	4000	400	4400	0.7561	3326.84	6196.52	36,196.52	9000	6804.9	29,391.62	0.6151	18,078.79
3	5000	500	5500	0.6575	3616.25	9812.77	39,812.77	8000	5260.0	34,552.77	0.4380	15,134.11
4	6000	600	6600	0.5718	3773.88	13,586.65	43,586.65	7000	4002.6	39,584.05	0.3503	13,866.29
5	7000	700	7700	0.4972	3828.44	17,415.09	47,415.09	6000	2983.2	44,431.89	0.2983	13,254.03
6	8000	800	8800	0.4323	3804.24	21,219.33	51,219.33	5000	2161.5	49,057.83	0.2642	12,961.08
7	9000	900	9900	0.3759	3721.41	24,940.74	54,940.74	4000	1503.6	53,437.14	0.2404	12,846.29
8	10,000	1000	11,000	0.3269	3595.90	28,536.64	58,536.64	3000	980.7	57,555.94	0.2229	12,829.22*
9	11,000	1100	12,100	0.2843	3440.03	31,976.67	61,976.67	2000	568.6	61,408.07	0.2096	12,871.13
10	12,000	1200	13,200	0.2472	3263.04	35,239.71	65,239.71	1000	247.2	64,992.51	0.1993	12,953.01

The workings to determine the economic life for the office equipment are shown in Table 13.11. From Table 13.11, the minimum annual equivalent cost is Rs. 34,509.63, which occurs for the 13th year. Hence, the economic life of the office equipment is 13 years.

Table 13.11 Determination of Economic Life of Office Equipment

End of year (n)	Maintenance cost	P/F, 12%, n	Present worth of maintenance cost (B × C)	Cumulative present worth of maintenance cost (ΣD)	Salvage value	Present worth of salvege value (F × C)	E + 1,00,000 – G	A/P, 12%, n	Annual equivalent cost (H × I)
(A)	(B) (Rs.)	(C)	(D) (Rs.)	(E) (Rs.)	(F) (Rs.)	(G) (Rs.)	H (Rs.)	I	J (Rs.)
1	10,000	0.8929	8929.0	8929.0	65,000	58,038.5	50,890.5	1.1200	56,997.36
2	12,000	0.7972	9566.4	18,495.4	55,000	43,846.0	74,649.4	0.5917	44,170.05
3	14,000	0.7118	9965.2	28,460.6	45,000	32,031.0	96,429.6	0.4163	40,143.64
4	16,000	0.6355	10168.0	38,628.6	35,000	22,242.5	1,16,386.1	0.3292	38,314.30
5	18,000	0.5674	10213.2	48,841.8	25,000	14,185.0	1,34,656.8	0.2774	37,353.80
6	20,000	0.5066	10132.0	58,973.8	15,000	7599.0	1,51,374.8	0.2432	36,814.35
7	22,000	0.4523	9950.6	68,924.4	5000	2261.5	1,66,662.9	0.2191	36,515.84
8	24,000	0.4039	9693.6	78,618.0	0	0.0	1,78,618.0	0.2013	35,955.80
9	26,000	0.3606	9375.6	87,993.6	0	0.0	1,87,993.6	0.1877	35,286.40
10	28,000	0.322	9016.0	97,009.6	0	0.0	1,97,009.6	0.1770	34,870.70
11	30,000	0.2875	8625.0	1,05,634.6	0	0.0	2,05,634.6	0.1684	34,628.87
12	32,000	0.2567	8214.4	1,13,849.0	0	0.0	2,13,849.0	0.1614	34,515.23
13	*34,000*	*0.2292*	*7792.8*	*1,21,641.8*	*0*	*0.0*	*2,21,641.8*	*0.1557*	*34,509.63**
14	36,000	0.2046	7365.6	1,29,007.4	0	0.0	2,29,007.4	0.1509	34,557.22
15	38,000	0.1827	6942.6	1,35,950.0	0	0.0	2,35,950.0	0.1468	34,637.46

13.5 SIMPLE PROBABILISTIC MODEL FOR ITEMS WHICH COMPLETELY FAIL

Electronic items like bulbs, resistors, tube lights, etc., generally fail all of a sudden, instead of gradual deterioration. The sudden failure of the item results in complete breakdown of the system. The system may contain a collection of such items or just an item like a single tube light. Hence, we use some replacement policy for such items which would minimize the possibility of complete breakdown.

The following are the replacement policies which are applicable for this situation.

Individual replacement policy.   Under this policy, an item is replaced immediately after its failure.

Group replacement policy.   Under group replacement policy, a decision is made in this regard: At what equal intervals, all the items are to be replaced simultaneously with a provision to replace the items individually which fail during the fixed group replacement period.

There is a trade-off between the individual replacement policy and the group replacement policy. Hence, for a given problem, each of the replacement policies is evaluated and the most economical policy is selected for implementation. This is explained with two numerical problems.

Example 13.8 The failure rates of 1000 street bulbs in a colony are summarized in Table 13.12.

Table 13.12 Failure Rates of Street Bulbs

End of month	1	2	3	4	5	6
Probability of failure to date	0.05	0.20	0.40	0.65	0.85	1.00

The cost of replacing an individual bulb is Rs. 60. If all the bulbs are replaced simultaneously it would cost Rs. 25 per bulb. Any one of the following two options can be followed to replace the bulbs.

   (i) Replace the bulbs individually when they fail (individual replacement policy).
   (ii) Replace all the bulbs simultaneously at fixed intervals and replace the individual bulbs as and when they fail in service during the fixed interval (group replacement policy).

Find out the optimal replacement policy, i.e. individual replacement policy or group replacement policy? If group replacement policy is optimal, then find at what equal intervals should all the bulbs be replaced?

Solution Number of bulbs in the colony, $N_0 = 1000$. Let p_i be the probability that a bulb which is new when placed in position for use, fails during the ith month of its life. Hence,

$$p_1 = 0.05 \qquad p_2 = 0.15 \qquad p_3 = 0.20$$
$$p_4 = 0.25 \qquad p_5 = 0.20 \qquad p_6 = 0.15$$

Since the sum of p_i values is equal to 1 at the end of the 6th month, the bulbs are sure to fail during the 6th month.

Let us assume that,

   (i) bulbs that fail during a month are replaced just before the end of the month and
   (ii) the actual percentage of failures during a month for a sub-group of bulbs with the same age is same as the expected percentage of failures during the month for that sub-group of bulbs.

Let N_i be the number of bulbs replaced at the end of the ith month. Then

N_0 = Number of bulbs replaced at the end of the month-0 (or at the beginning of the 1st month)

$\quad = 1000$

N_1 = Number of bulbs replaced at the end of the 1st month

$\quad = N_0 p_1$

$\quad = 1000 \times 0.05$

$\quad = 50$

N_2 = Number of bulbs replaced at the end of the 2nd month

$\quad = N_0 p_2 + N_1 p_1$

$\quad = 1000 \times 0.15 + 50 \times 0.05$

$\quad = 153$

$$N_3 = N_0 p_3 + N_1 p_2 + N_2 p_1$$
$$= 1000 \times 0.20 + 50 \times 0.15 + 153 \times 0.05$$
$$= 215$$

$$N_4 = N_0 p_4 + N_1 p_3 + N_2 p_2 + N_3 p_1$$
$$= 1000 \times 0.25 + 50 \times 0.20 + 153 \times 0.15 + 215 \times 0.05$$
$$= 294$$

$$N_5 = N_0 p_5 + N_1 p_4 + N_2 p_3 + N_3 p_2 + N_4 p_1$$
$$= 1000 \times 0.20 + 50 \times 0.25 + 153 \times 0.20 + 215 \times 0.15 + 294 \times 0.05$$
$$= 290$$

$$N_6 = N_0 p_6 + N_1 p_5 + N_2 p_4 + N_3 p_3 + N_4 p_2 + N_5 p_1$$
$$= 1000 \times 0.15 + 50 \times 0.20 + 153 \times 0.25 + 215 \times 0.20 + 294 \times 0.15 + 290 \times 0.05$$
$$= 300$$

Calculation of individual replacement cost

$$\text{Expected life of each bulb} = \sum_{i=1}^{6} i p_i$$
$$= 1 \times 0.05 + 2 \times 0.15 + 3 \times 0.20 + 4 \times 0.25 + 5 \times 0.20 + 6 \times 0.15$$
$$= 3.85 \text{ months}$$

and

$$\text{Average number of failures/month} = \frac{1000}{3.85} = 260 \text{ (approx.)}$$

Therefore,

The cost of individual replacement

= No. of failures/month × Individual replacement cost/bulb

= 260 × 60

= Rs. 15,600.

Determination of group replacement cost

Cost per bulb when replaced simultaneously = Rs. 25

Cost per bulb when replaced individually = Rs. 60

Therefore, the costs of group replacement policy for several replacement periods are summarized in Table 13.13.

From Table 13.13, it is clear that the average cost/month is minimum for the third month. Hence,

The group replacement period = 3 months

Individual replacement cost per month = Rs. 15,600

Minimum group replacement cost per month = Rs. 16,693

Table 13.13 Calculations of Cost for Preventive Maintenance

End of month (n) A	Cost of replacing 1000 bulbs at a time B (Rs.)	Cost of replacing bulbs individually during given replacement period C (Rs.)	Total Cost (B + C) D (Rs.)	Average cost/month (D/A) E (Rs.)
1	25,000	$50 \times 60 = 3000$	28,000	28,000
2	25,000	$(50 + 153) \times 60 = 12,180$	37,180	18,590
3	25,000	$(50 + 153 + 215) \times 60 = 25,080$	50,080	16,693*
4	25,000	$(50 + 153 + 215 + 294) \times 60 = 42,720$	67,720	16,930
5	25,000	$(50 + 153 + 215 + 294 + 290) \times 60 = 60,120$	85,120	17,024
6	25,000	$(50 + 153 + 215 + 294 + 290 + 300) \times 60 = 78,120$	1,03,120	17,187

*Economic life of the bulb = 3 months

Since the individual replacement cost per month is lesser than the minimum group replacement cost/month, the individual replacement policy is the best and hence all the bulbs are to be replaced whenever they fail.

Example 13.9 An electronic equipment contains 500 resistors. When any resistor fails, it is replaced. The cost of replacing a resistor individually is Rs. 20. If all the resistors are replaced at the same time, the cost per resistor is Rs. 5. The percentage surviving, $S(i)$ at the end of month i is given in Table 13.14.

Table 13.14 Percent Survival Rate

Month i	0	1	2	3	4	5
$S(i)$	100	90	75	55	30	0

What is the optimum replacement plan?

Solution Let p_i be the probability of failure during the month i. Then,

$$p_1 = \frac{100 - 90}{100} = 0.10 \qquad p_2 = \frac{90 - 75}{100} = 0.15 \qquad p_3 = \frac{75 - 55}{100} = 0.20$$

$$p_4 = \frac{55 - 30}{100} = 0.25 \qquad p_5 = \frac{30 - 0}{100} = 0.30$$

It is clear that no resistor can survive beyond 5 months. Hence, a resistor which has survived for four months would certainly fail during the fifth month. We assume that the resistors failing during a month are accounted at the end of the month.

Let N_i be the number of resistors replaced at the end of the ith month. Then

$$N_0 = 500$$

$$N_1 = N_0 p_1$$
$$= 500 \times 0.1$$
$$= 50$$

$$N_2 = N_0 p_2 + N_1 p_1$$
$$= 500 \times 0.15 + 50 \times 0.1$$
$$= 80$$

$$N_3 = N_0 p_3 + N_1 p_2 + N_2 p_1$$
$$= 500 \times 0.20 + 50 \times 0.15 + 80 \times 0.10$$
$$= 116$$

$$N_4 = N_0 p_4 + N_1 p_3 + N_2 p_2 + N_3 p_1$$
$$= 500 \times 0.25 + 50 \times 0.20 + 80 \times 0.15 + 116 \times 0.10$$
$$= 159$$

$$N_5 = N_0 p_5 + N_1 p_4 + N_2 p_3 + N_3 p_2 + N_4 p_1$$
$$= 500 \times 0.30 + 50 \times 0.25 + 80 \times 0.20 + 116 \times 0.15 + 159 \times 0.10$$
$$= 212$$

Determination of individual replacement cost

$$\text{Expected life of each resistor} = \sum_{i=1}^{5} i p_i$$
$$= 1 \times 0.1 + 2 \times 0.15 + 3 \times 0.20 + 4 \times 0.25 + 5 \times 0.30$$
$$= 3.5 \text{ months}$$

and

$$\text{Average number of failures/month} = \frac{500}{3.5}$$
$$= 143 \text{ (approximately)}$$

Therefore,

The cost of individual replacement

$$= \text{No. of failures/month} \times \text{Individual replacement cost/resistor}$$
$$= 143 \times 20$$
$$= \text{Rs. } 2860.$$

Determination of group replacement cost

Cost/resistor when replaced simultaneously = Rs. 5
Cost/resistor when replaced individually = Rs. 20

The costs of group replacement policy for several replacement periods are summarized in Table 13.15.

Table 13.15 Calculations of Costs for Preventive Maintenance

End of month (n) A	Cost of replacing 500 resistors at a time B (Rs.)	Cost of replacing resistors individually during given replacement period C (Rs.)	Total cost (B + C) D (Rs.)	Average cost/month (D/A) E (Rs.)
1	2500	$50 \times 20 = 1000$	3500	3500
2	2500	$(50 + 80) \times 20 = 2600$	5100	2550
3	2500	$(50 + 80 + 116) \times 20 = 4920$	7420	2473*
4	2500	$(50 + 80 + 116 + 159) \times 20 = 8100$	10,600	2650
5	2500	$(50 + 80 + 116 + 159 + 212) \times 20 = 12,340$	14,840	2968

*Economic life of the resistor = 3 months.

From Table 13.15, it is clear that the average cost/month is minimum for the third month. Hence, the group replacement period is 3 months.

Summary: Individual replacement cost/month = Rs. 2860 and minimum group replacement cost/month = Rs. 2473. Since the minimum group replacement cost/month is lesser than the individual replacement cost/month, the group replacement policy is the best and hence all the resistors are to be replaced once in three months and the resistors which fail during this three-month period are to be replaced individually.

Example 13.10 An electronic device which is used to control the parameters of a chemical process consists of 1000 resistors. The cost of replacing a resistor individually is Rs. 25. The cost of replacing all the resistors at the same time is Rs. 6 per resistor. The probability of failure of a resistor for each of its life spans in terms of months is shown in Table 13.16. Determine the best replacement policy.

Table 13.16 Probability Distribution of Failures of Resistor

Month i	1	2	3	4	5
p_i	0.10	0.23	0.40	0.18	0.09

Solution The probabilities of failure of a resistor in different months are:

$$p_1 = 0.10, \ p_2 = 0.23, \ p_3 = 0.40, \ p_4 = 0.18 \text{ and } p_5 = 0.09$$

Let, N_i be the number of resistors replaced at the end of the ith month. Then

$N_0 = 1000$

$N_1 = N_0 p_1$

$\quad = 1000 \times 0.10$

$\quad = 100$

$N_2 = N_0 p_2 + N_1 p_1$

$\quad = 1000 \times 0.23 + 100 \times 0.10$

$\quad = 240$

$N_3 = N_0 p_3 + N_1 p_2 + N_2 p_1$

$\quad = 1000 \times 0.40 + 100 \times 0.23 + 240 \times 0.10$

$\quad = 447$

$N_4 = N_0 p_4 + N_1 p_3 + N_2 p_2 + N_3 p_1$

$\quad = 1000 \times 0.18 + 100 \times 0.40 + 240 \times 0.23 + 447 \times 0.10$

$\quad = 320$

$N_5 = N_0 p_5 + N_1 p_4 + N_2 p_3 + N_3 p_2 + N_4 p_1$

$\quad = 1000 \times 0.09 + 100 \times 0.18 + 240 \times 0.40 + 447 \times 0.23 + 320 \times 0.10$

$\quad = 339$

Determination of individual replacement cost

Expected life of each resistor $= \sum_{i=1}^{5} ip_i$

$$= 1 \times 0.10 + 2 \times 0.23 + 3 \times 0.40 + 4 \times 0.18 + 5 \times 0.09$$
$$= 2.93 \text{ months}$$

Average number of failures/month $= 1000/2.93 = 341.3 = 342$ (Approximately)

Therefore,

The cost of individual replacement/month $=$ No. of failures/month $\times$ Individual replacement cost/resistor

$$= 342 \times 25 = \text{Rs. } 8,550$$

Determination of group replacement cost

Cost/resistor when it is replaced individually $\quad =$ Rs. 25
Cost/resistor when it is replaced simultaneously $=$ Rs. 6

The cost of group replacement policy for several replacement periods are summarized in Table 13.17.

Table 13.17 Calculations of Costs for Preventive Maintenance

End of month A	Cost of replacing 1000 resistors at a time B (Rs.)	Cost of replacing resistors individually during given replacement period C (Rs.)	Total cost (B + C) D (Rs.)	Average cost per month (D/A) E (Rs.)
1	6000	$100 \times 25 = 2500$	8500	8500.00
2	6000	$(100 + 240) \times 25 = 8500$	14,500	7250.00
3	6000	$(100 + 240 + 447) \times 25 = 19{,}675$	25,675	8558.33
4	*6000*	*$(100 + 240 + 447 + 320) \times 25 = 27{,}675$*	*33,675*	*8418.75*
5	6000	$(100 + 240 + 447 + 320 + 339) \times 25 = 36{,}150$	42,150	8430.00

From Table 13.17, the minimum average cost/month is Rs. 8418.75 and the corresponding replacement period is 4 months. That is, if the group replacement policy is the best when compared to the individual replacement policy, then the group replacement period is 4 months.

Average cost per month for the individual replacement policy $\quad =$ Rs. 8550.00
Minimum average cost per month for the group replacement policy $=$ Rs. 8418.75

Since, the minimum average cost per month for the group replacement policy is lesser than the average cost per month for the individual replacement policy, the group replacement policy is the best policy.

Summary: Replace all the resistors once in 4 months and if there is any failures during this 4 month period, then replace those resistors individually during the 4 month period.

QUESTIONS

1. List and explain different types of maintenance. Discuss the reasons for replacement.

2. Define economic life of an equipment.

3. Distinguish between breakdown maintenance and preventive maintenance.

4. A firm is considering replacement of an equipment by a new equipment whose first cost is Rs. 1750 and the scrap value is negligible at any year. Based on the experience, it is found that the maintenance cost is zero during the first year and it increases by Rs. 100 every year thereafter. (a) When should the equipment be replaced if (a) $i = 0\%$? (b) $i = 12\%$?

5. A company is planning to replace an equipment by a new equipment whose first cost is Rs. 1,00,000. The operating and maintenance cost of the equipment during its first year of operation is Rs. 10,000 and it increases by Rs. 2000 every year thereafter. The resale value of the equipment at the end of the first year of its operation is Rs. 65,000 and it decreases by Rs. 10,000 every year thereafter. Find the economic life of the equipment by assuming the interest rate as 12%.

6. The table gives the operation cost, maintenance cost and salvage value at the end of every year of a machine whose purchase value is Rs. 20,000. Find the economic life of the machine assuming interest rate (a) $i = 0\%$ (b) $i = 15\%$.

End of year (n)	Operation cost at the end of the year (Rs.)	Maintenance cost at the end of the year (Rs.)	Salvage value at the end of the year (Rs.)
1	2000	200	10,000
2	3000	300	9000
3	4000	400	8000
4	5000	500	7000
5	6000	600	6000
6	7000	700	5000
7	8000	800	4000
8	9000	900	3000
9	10,000	1000	2000
10	11,000	1100	1000

7. A manufacturer is offered two machines, A and B. A is priced at Rs. 8000 and maintenance costs are estimated at Rs. 500 for the first year and an equal increment of Rs. 100 from second to fifth year and Rs. 1500 for the sixth year and an equal increment of Rs. 500 from seventh year onwards. Machine B which has the same capacity is priced at Rs. 6000. The maintenance costs of the machine B are estimated at Rs. 1000 for the first year and an equal yearly increment of Rs. 200 thereafter. If the money is worth 15% per year, which machine should be purchased? (Assume that the scrap value of each of the machines is negligible at any year.)

8. The following failure rates have been observed for a certain type of transistors in a computer.

End of week	1	2	3	4	5	6	7
Probability of failure to date	0.07	0.18	0.30	0.48	0.69	0.89	1.00

The cost of replacing any failed transistor is Rs. 10. If all the transistors are replaced simultaneously it would cost Rs. 2.50 per transistor. Any one of the following two options can be followed to replace the transistors.

(a) Replace the transistors individually when they fail (individual replacement policy).
(b) Replace all the transistors simultaneously at fixed intervals and replace the individual transistors as they fail in service during the fixed interval (group replacement policy).

Find out the optimal replacement policy, i.e. individual replacement policy or group replacement policy? If group replacement policy is optimal, then find at what equal intervals should all the transistors be replaced?

9. An electronic equipment contains 1000 resistors. When any resistor fails, it is replaced. The cost of replacing a resistor individually is Rs. 8. If all the resistors are replaced at the same time, the cost per resistor is Rs. 2. The percentage surviving, $S(i)$ at the end of month i is given below.

Month i	0	1	2	3	4	5	6
$S(i)$	100	96	89	68	37	13	0

What is the optimum replacement plan?

PRODUCTION SCHEDULING $\quad$ (14)

14.1 INTRODUCTION

Production scheduling is the allocation of resources pertaining to start and finish times for tasks. It is nothing but scheduling various jobs on a set of resources (machines) such that certain performance measures are optimized. Production scheduling can be classified into the following three categories:

1. Single-machine scheduling
2. Flow shop scheduling
3. Job shop scheduling

The single-machine scheduling problem consists of n independent jobs, each with single operation. The flow shop scheduling problem consists of n independent jobs, each with m operations. In this problem, all the jobs will have the same process sequence. The job shop scheduling problem consists of n independent jobs, each with m operations. In this problem, the jobs will have different process sequences. The objective of each of these problems is to schedule the jobs such that some performance measure(s) is (are) optimized. In this chapter, these three types of scheduling problem are presented.

14.2 SINGLE-MACHINE SCHEDULING

The single-machine scheduling problem is characterized by the following conditions:

1. A set of n independent jobs, each with single operation is available for processing at time zero.
2. Set-up time of each of the jobs is independent of its position in the sequence of jobs. So, the set-up time of each job can be included in its processing time.
3. Job descriptors are known in advance.
4. One machine is continuously available and is never kept idle when work is waiting.
5. Each job is processed till its completion without pre-emption.

Under these conditions, one can see one-to-one correspondence between a sequence of the n jobs and a permutation of the job indices 1, 2,..., n. The total number of sequences in the basic single-machine scheduling problem is $n!$ which is the number of permutation of n elements. [3] = 5 means that the job 5 is assigned to the third position in the sequence. Similarly, $d_{[4]}$ refers to the due date of the job assigned to the fourth position in the sequence.

The following three basic data are necessary to describe jobs in a deterministic single machine scheduling problem.

Processing time (t_j). It is the time required to process job j. The processing time, t_j will normally include both actual processing time and set-up time.

Ready time (r_j). It is the time at which job j is available for processing. The ready time of job j is the difference between the arrival time of that job and the time at which that job is taken for processing. In the basic model, as per condition 1, $r_j = 0$ for all jobs.

Due date (d_j). It is the time at which the processing of the job j is to be completed.

Completion time (C_j). It is the time at which the job j is actually completed in a sequence. Each and every performance measure for evaluating schedules is usually a function of completion times of jobs. Some sample performance measures are Flow time, Lateness, Tardiness, etc.

Flow time (F_j). It is the amount of time that job j spends in the system. It is the difference between the completion time and the ready time of the job j.

$$F_j = C_j - r_j$$
$$= C_j, \qquad \text{if } r_j = 0$$

Total flow time is a measure which indicates the waiting times of jobs in a system. This in turn gives some idea about in-process inventory due to a schedule.

Lateness (L_j). It is the amount of time by which the completion time of job j differs from its due date ($L_j = C_j - d_j$). Lateness can be either *positive lateness* or *negative lateness*. Positive lateness of a job means that the job is completed after its due date. Negative lateness of a job means that the job is completed before its due date. The positive lateness is a measure of poor service, while negative lateness is a measure of better service. In many situations, distinct penalties and other costs are associated with positive lateness, but generally, no benefits are associated with negative lateness. Therefore, it is often desirable to minimize only positive lateness.

Tardiness (T_j). Tardiness is the lateness of job j if it fails to meet its due date; otherwise, it is zero. It is defined as:

$$T_j = \max \{0, C_j - d_j\} = \max \{0, L_j\}$$

which means

$$T_j = C_j - d_j, \qquad \text{if } C_j > d_j;$$
$$= 0, \qquad \text{otherwise.}$$

14.2.1 Measures of Performance

The different measures of performance which are used in the single-machine scheduling problem are listed below with their formulae.

$$\text{Mean flow time, } \bar{F} = \frac{1}{n} \sum_{j=1}^{n} F_j$$

$$\text{Mean tardiness, } \bar{T} = \frac{1}{n} \sum_{j=1}^{n} T_j$$

$$\text{Maximum flow time, } F_{\max} = \max_{1 \le j \le n} \{F_j\}$$

$$\text{Maximum tardiness, } T_{\max} = \max_{1 \le j \le n} \{T_j\}$$

$$\text{Number of tardy jobs, } N_T = \sum_{j=1}^{n} f(T_j)$$

where $f(T_j) = 1$, if $T_j > 0$, and $f(T_j) = 0$, otherwise.

14.2.2 Shortest Processing Time (SPT) Rule to Minimize Mean Flow Time

In single-machine scheduling problem, sequencing the jobs in increasing order of processing time is known as *shortest processing time* (SPT) sequencing. In single-machine scheduling, sometimes we may be interested in minimizing the time spent by jobs in the system. This in turn will minimize in-process inventory. Also, we may be interested in rapid turnaround/throughput times of the jobs. The time spent by job j in the system is nothing but its flow time (F_j) and the 'rapid turnaround time' is the mean flow time $(\overline{F})$ of the jobs in the system. Shortest processing time (SPT) rule minimizes the mean flow time.

Example 14.1 The data about a set of single operation jobs that is to be processed in a CNC lathe and the processing times of the jobs in the given set of jobs are as given in Table 14.1.

Table 14.1 Example 14.1

Job, j	Processing time, t_j
1	7
2	18
3	6
4	8
5	12

Find the optimal sequence which will minimize the mean flow time and also obtain the corresponding minimum mean flow time.

Solution For the given problem, the number of jobs is equal to 5. The jobs are arranged as per the SPT ordering as shown in Table 14.2.

Table 14.2 Data as per SPT Ordering

Job, j	Processing time, t_j
3	6
1	7
4	8
5	12
2	18

Therefore, the job sequence which will minimize the mean flow time is 3–1–4–5–2.

Computation of optimal $\overline{F}$: The computations of flow time for the optimal sequence (3–1–4–5–2) are presented in Table 14.3.

Table 14.3 Computations of Flow Time

Job, j	Processing time, t_j	Completion time, $C_j(F_j)$
3	6	6
1	7	13
4	8	21
5	12	33
2	18	51

Since the ready time r_j is zero for all the values of j, the flow time $(F_j) = C_j$ for all j. Therefore,

$$\overline{F} = \frac{1}{5} \sum_{j=1}^{5} F_j = \frac{1}{5}(6 + 13 + 21 + 33 + 51)$$

$$= \frac{1}{5}(124)$$

$$= 24.8 \text{ hours}$$

Therefore, the optimal mean flow time = 24.8 hours.

14.2.3 Weighted Shortest Processing Time (WSPT) Rule to Minimize Weighted Mean Flow Time

Sometimes, the jobs in a single-machine scheduling problem will not have equal importance. Under such situation, each job is assigned with a weight, w_j. The mean flow time which is computed after incorporating w_j is called *weighted mean flow time* and its formula is:

$$\overline{F}_w = \frac{\displaystyle\sum_{j=1}^{n} w_j F_j}{\displaystyle\sum_{j=1}^{n} w_j}$$

WSPT Rule. In single-machine scheduling problem, sequencing the jobs in increasing order of weighted processing time is known as *weighted shortest processing time* (WSPT) sequencing. The weighted processing time of a job is obtained by dividing its processing time by its weight.

Example 14.2 Consider Example 14.1 with additional data on weights assigned to different jobs as shown in Table 14.4.

Table 14.4 Data

Job, j	Processing time, t_j	Weight, w_j
1	7	1
2	18	2
3	6	1
4	8	2
5	12	3

Determine the sequence which will minimize the weighted mean flow time of this problem. Also find the corresponding weighted mean flow time.

Solution   The weighted processing times of various jobs are summarized in Table 14.5 by using the formula t_j/w_j, $j = 1, 2, 3, 4$ and 5.

Table 14.5   Details of Weighted Processing Times

Job, j	Processing time, t_j	Weight, w_j	t_j/w_j
1	7	1	7
2	18	2	9
3	6	1	6
4	8	2	4
5	12	3	4

Next, the jobs are arranged in the increasing order of t_j/w_j (i.e. WSPT ordering). From the above table one can verify the following relation.

$$\frac{t_4}{w_4} \le \frac{t_5}{w_5} \le \frac{t_3}{w_3} \le \frac{t_1}{w_1} \le \frac{t_2}{w_2}$$

Therefore, the optimal sequence which will minimize the weighted mean flow time is 4–5–3–1–2.

$\overline{F}_w$ *Calculation:*   The details of weighted flow time calculations of various jobs with respect to the optimal sequence are summarized in Table 14.6.

Table 14.6   Calculations of Weighted Flow Times

Job, j	Processing time, t_j	C_j (F_j)	Weight, w_j	$F_j w_j$
4	8	8	2	16
5	12	20	3	60
3	6	26	1	26
1	7	33	1	33
2	18	51	2	102

Therefore,

$$\overline{F}_w = \frac{\sum_{j=1}^{5} w_j F_j}{\sum_{j=1}^{5} w_j} = \frac{16 + 60 + 26 + 33 + 102}{2 + 3 + 1 + 1 + 2} = 26.33 \text{ hours.}$$

14.2.4   Earliest Due Date (EDD) Rule to Minimize Maximum Lateness

The *lateness* (L_j) of a job is defined as the difference between the completion time and the due date of that job.

$$L_j = C_j - d_j$$

L_j can be either positive or negative values. The *maximum job lateness* (L_{max}) which is known as the maximum job tardiness (T_{max}) is minimized by earliest due date (EDD) sequencing. In single machine scheduling problem, sequencing of jobs in the increasing order of due date is known as *earliest due date* rule.

Example 14.3   Consider the single-machine scheduling problem as shown in Table 14.7.

Table 14.7   Example 14.3

Job, j	Processing time, t_j	Due date, d_j
1	12	17
2	16	14
3	10	12
4	9	20
5	14	18
6	17	27

Determine the sequence which will minimize the maximum lateness (L_{max}). Also, determine the corresponding L_{max} with respect to the optimal sequence.

Solution   The EDD sequence (i.e. jobs in the increasing order of due dates) of the given problem is 3–2–1–5–4–6. This sequence gives the minimum value for L_{max}.

Computation of L_{max}:   The details of computations of lateness values of different jobs as per the EDD sequence are summarized in Table 14.8.

Table 14.8   Lateness Values of Jobs

Job, j (EDD sequence)	Processing time, t_j	Completion time, C_j	Due date, d_j	Lateness, L_j
3	10	10	12	0
2	10	20	14	6
1	12	32	17	15
5	14	46	18	28
4	9	55	20	35
6	17	72	27	45

In the above table, the maximum lateness (L_{max}) is 45. This is the optimized (minimized) value of L_{max}. The L_{max} value of any other non-EDD sequence will not be less than 45.

14.2.5   Model to Minimize Total Tardiness

In this section a mathematical model to minimize the total tardiness in the single-machine scheduling is presented.* Consider a problem involving n independent jobs which can be processed on a single

---

*Panneerselvam, 1991.

machine. One of the measures of this problem is minimizing the total tardiness of the jobs. The basic data for the single machine scheduling problem is presented in Table 14.9.

Table 14.9 Generalized Data of Single-Machine Scheduling Problem

Job, i	Processing time, t_i	Due date, d_i
1	t_1	d_1
2	t_2	d_2
⋮	⋮	⋮
i	t_i	d_i
⋮	⋮	⋮
n	t_n	d_n

Let C_i be the completion time of the job i in a given schedule. T_j, tardiness of the job assigned to the jth position in a given schedule. Then

$$Y_{ij} = 1, \quad \text{if job } i \text{ is assigned to the } j\text{th position;}$$
$$= 0, \quad \text{otherwise.}$$

The objective of the problem is to sequence the jobs such that the total tardiness is minimized. Assume that all the jobs are available at time zero, i.e. ready time, $r_i = 0$ for all $i = 1, 2, 3,..., n$. Under this situation,

$$C_i = (\text{Completion time of the job in the sequence immediately preceding to the job } i) + t_i$$
$$= C_{\text{immediately preceding job to the job } i} + t_i$$

and

$$T_j = \max [0, (\text{Completion time of the job at the position } j - \text{Due date of the job at the position } j)]$$

A descriptive model to minimize the total tardiness is presented below:

$$\text{Minimize } Z = \sum_{j=1}^{n} \max [0, (\text{Completion time of the job at the position } j$$
$$- \text{Due date of the job at the position } j)]$$

$$= \sum_{j=1}^{n} \max \left[0, \sum_{i=1}^{n}\sum_{q=1}^{j} t_i Y_{iq} - \sum_{i=1}^{n} d_i Y_{ij} \right]$$

subject to

$$\sum_{j=1}^{n} Y_{ij} = 1, \quad i = 1, 2, 3,..., n$$

$$\sum_{i=1}^{n} Y_{ij} = 1, \quad j = 1, 2, 3,..., n$$

where

$$Y_{ij} = 0 \text{ or } 1, \quad i = 1, 2, 3,..., n \quad \text{and} \quad j = 1, 2, 3,..., n$$

For easy understanding of the above problem, the different possible variables of Y_{ij} are presented in Table 14.10.

Table 14.10 Combinations of 0–1 Variables

Position in the sequence

		1	2	...	j	...	n
	1	Y_{11}	Y_{12}	...	Y_{1j}	...	Y_{1n}
	2	Y_{21}	Y_{22}	...	Y_{2j}	...	Y_{2n}
Job	$\vdots$	$\vdots$	$\vdots$		$\vdots$		$\vdots$
	i	Y_{i1}	Y_{i2}	...	Y_{ij}	...	Y_{in}
	$\vdots$	$\vdots$	$\vdots$		$\vdots$		$\vdots$
	n	Y_{n1}	Y_{n2}	...	Y_{nj}	...	Y_{nn}

Linear integer and 0–1 programming model. The nonlinear objective function of the descriptive model is converted into a linear function by replacing the objective term by T_j (tardiness of the job assigned to the jth position in a given schedule) as shown below along with additional constraints Set 3 and Set 5.

$$\text{Minimize } Z = \sum_{j=1}^{n} T_j$$

subject to

$$\sum_{j=1}^{n} Y_{ij} = 1, \qquad i = 1, 2, 3, ..., n \tag{Set 1}$$

$$\sum_{i=1}^{n} Y_{ij} = 1, \qquad j = 1, 2, 3, ..., n \tag{Set 2}$$

$$T_j \geq \sum_{i=1}^{n} \sum_{q=1}^{j} t_i Y_{iq} - \sum_{i=1}^{n} d_i Y_{ij}, \qquad j = 1, 2, 3, ..., n \tag{Set 3}$$

where,

$$Y_{ij} = 0 \text{ or } 1, \quad i = 1, 2, 3, ..., n \quad \text{and} \quad j = 1, 2, 3, ..., n \tag{Set 4}$$

$$T_j \geq 0, \quad \text{for } j = 1, 2, 3, ..., n \tag{Set 5}$$

In this model, the following points may be noted:

1. The objective function minimizes the total tardiness.
2. The constraints in Set 1 guarantee that each job is assigned to exactly one position in the schedule.
3. The constraints in Set 2 guarantee that each position in the schedule is assigned exactly one job.
4. The constraints in Set 3 and Set 5, jointly guarantee that T_j is more than zero if the job in the position j is late; otherwise it is zero.
5. So, the constraint Set 3 and Set 5 are 'either-or constraint sets'.
6. In this model, for a general problem of size n, the number of variables is $(n^2 + n)$ and the number of constraints is $3n$.

Example 14.4 Consider the data of a single-machine scheduling problem with 3 jobs as shown in Table 14.11.

Table 14.11 Example 14.4

Job, i	Processing time, t_i	Due date, d_i
1	5	10
2	3	5
3	8	15

Find the optimal sequence of jobs and the corresponding total tardiness using the linear integer 0–1 programming model.

Solution The development of the model for the given problem is left as an exercise. Application of the model to this problem gives the following results: $Y_{12} = Y_{21} = Y_{33} = 1$, $L_3 = 1$ and other variables are zero. From this result, one can infer the best sequence as 2–1–3. The corresponding total tardiness is 1.

14.2.6 Introduction to Branch-and-Bound Technique to Minimize Mean Tardiness

The problem of minimizing mean tardiness in a single-machine scheduling comes under combinatorial category. For this type of problem, if we want to obtain the optimal solution, we will have to use either complete enumeration technique or branch-and-bound technique. The other name for the branch-and-bound technique is curtailed enumeration technique.

In this section, the application of the branch-and-bound technique to obtain an optimal sequence which will minimize the mean tardiness in the single-machine scheduling problem is presented. Branching is the process of partitioning a large problem into two or more subproblems and bounding is the process of calculating a lower bound on the optimal solution of a given subproblem.

The branching procedure replaces an original problem by a set of new problems that are:

1. Mutually exclusive and exhaustive subproblems of the original problem.
2. Partially solved versions of the original problem.
3. Smaller problems than the original problem.

Furthermore, the subproblems can themselves be partitioned in a similar fashion. The generalized representation of this scheme is shown in Figure 14.1.

In Figure 14.1 at root node (Level 0), P_ϕ^0 denotes a single-machine scheduling problem containing n jobs. At Level 1 the problem P_ϕ^0 is partitioned into n different subproblems $(P_1^1, P_2^1, P_3^1 \ldots P_n^1)$ by assigning each job to the last position in the sequence. Thus, P_1^1 is same as P_ϕ^0 problem, but with job 1 fixed in the last position, P_2^1 is similar, but with job 2 fixed in the last position; and so on. Clearly, these subproblems are smaller than P_ϕ^0 and each P_i^1 is a partially solved version of P_ϕ^0. In addition, the set of subproblems at the Level 1 is a mutually exclusive and exhaustive partition of P_ϕ^0. Therefore, the P_i^1 satisfies conditions (1), (2) and (3) above. Next, each of the subproblems can be partitioned further (see the Figure 14.1). For instance, P_2^1 can be partitioned into $P_{12}^2, P_{32}^2, P_{42}^2 \ldots, P_{n2}^2$.

In P_{32}^2, jobs 3 and 2 occupy the last two positions in the sequence in that order. Therefore the second-level partition P_{i2}^2 bears the same relation to P_2^1 as the first level partition P_i^1 bears to P_ϕ^0. That is, the partitions at each level satisfy conditions (1), (2) and (3). At level k, then each subproblem

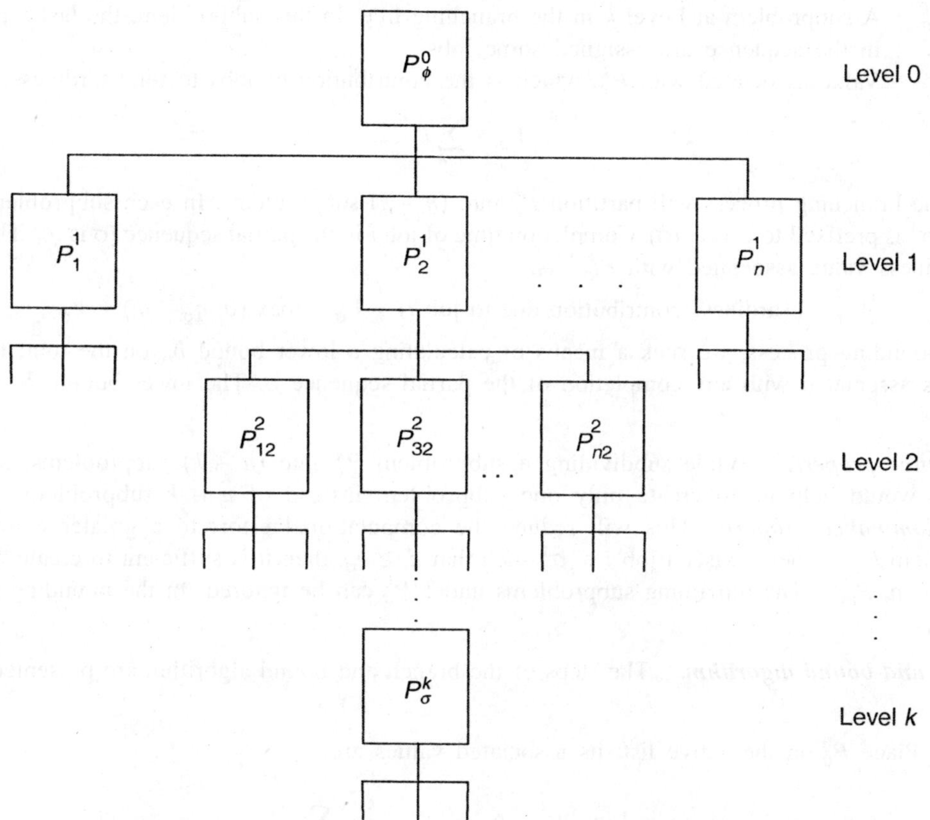

Figure 14.1 Generalized structure of branch-and-bound tree.

contains k fixed positions and can be further partitioned into $(n - k)$ subproblems, which form part of Level $(k + 1)$. If this branching procedure were to be carried out completely, there would be $n!$ subproblems at level n, each corresponding to a distinct feasible solution to the original problem. The bounding procedure computes a lower bound on the tardiness to each subproblem generated in the branching procedure. At any point in time, we compare the lower bounds of all the terminal nodes and select the node with the minimum lower bound for further branching. If there is a tie on the minimum lower bound, then select the node at the lower level (below) for further branching. If the node with the minimum lower bound lies at $(n - 1)$th level, then the optimality is reached. A complete optimal sequence can be obtained by filling the missing number in the first position of the partial sequence which is leading to the node at the $(n - 1)$th level. This process of branching and bounding is called *breadth-first search*. Before explaining these concepts applied to a problem of minimizing total tardiness penalty (mean tardiness), the terminologies are explained:

σ = Partial sequence of jobs from among the n jobs originally in the problem. In single-machine scheduling problem, a partial sequence is formed from the end of the full sequence

$i\sigma$ = Partial sequence in which σ is immediately preceded by job i

σ^1 = Complement of σ, where

$$q_\sigma = \sum_{j \in \sigma^1} t_j$$

P_σ^k = A subproblem at Level k in the branching tree. In this subproblem, the last k positions in the sequence are assigned some jobs.

V_σ = Value associated with P_σ^k, which is the contribution of jobs to total tardiness, where

$$V_\sigma = \sum_{j \in \sigma} T_j$$

The branching process will partition P_σ^k into $(n - k)$ subproblems. In each subproblem, some job $i \in \sigma^1$ is prefixed to σ (i.e. $i\sigma$). Completion time of job i in the partial sequence $i\sigma$ is q_σ. Therefore, the tardiness value associated with $P_{i\sigma}^{k+1}$ is:

$$V_{i\sigma} = \text{(Tardiness contribution due to job } i) + V_\sigma = \max(0, q_\sigma - d_i) + V_\sigma$$

In the bounding process, we seek a means of calculating a lower bound b_σ on the total tardiness penalties associated with any completion of the partial sequence σ. The lower bound, b_σ is equal to V_σ.

Dominance property. While subdividing a subproblem P_σ^k into $(n - k)$ subproblems, a careful analysis would help us to create only one subproblem instead of $n - k$ subproblems. This is called *dominance property*. This will reduce the computational effort to a greater extent. In a subproblem P_σ^k, if there exists a job $i \in \sigma^1$ such that $d_i \geq q_\sigma$, then it is sufficient to create only one subproblem, $P_{i\sigma}^{k+1}$. The remaining subproblems under P_σ^k can be ignored. In the bounding process, $V_{i\sigma} = V_\sigma$.

Branch-and-bound algorithm. The steps of the branch and bound algorithm are presented in this section.

Step 1: Place P_ϕ^0 on the active list; its associated values are:

$$V_\phi = 0 \quad \text{and} \quad q_\phi = \sum_{j=1}^{n} t_j$$

At a given stage of the algorithm, the active list consists of all the terminal nodes of the partial tree created up to that stage.

Step 2: Remove the first subproblem P_σ^k from the active list. If k is equal to $n - 1$, stop. Prefix the missing job with σ and treat it as the optimal sequence. Otherwise, check the dominance property for P_σ^k. If the property holds, go to step 3; otherwise go to step 4.

Step 3: Let the job j be the job with the largest due date in σ^1. Create the subproblem $P_{j\sigma}^{k+1}$ with

$$q_{j\sigma} = q_\sigma - t_j, \qquad V_{j\sigma} = V_\sigma, \qquad b_{j\sigma} = V_\sigma$$

Place $P_{j\sigma}^{k+1}$ on the active list, ranked by its lower bound. Return to step 2.

Step 4: Create $(n - k)$ subproblems, one for each $i \in \sigma^1$. For $P_{i\sigma}^{k+1}$, let,

$$q_{i\sigma} = q_\sigma - t_i, \quad V_{i\sigma} = V_\sigma + \max(0, q_\sigma - d_i), \qquad b_{i\sigma} = V_{i\sigma}$$

Now place each $P_{i\sigma}^{k+1}$ on the active list, ranked by its lower bound. Return to step 2.

Example 14.5 Consider the following single-machine scheduling problem with five jobs as shown in Table 14.12.

Table 14.12   Example 14.5

Job, j	Processing time, t_j	Due date, d_j
1	4	5
2	7	11
3	7	18
4	16	25
5	15	20

Determine the optimal sequence which will minimize the total tardiness using the branch-and-bound algorithm.

Solution

Step 1: Active list at Level 0 = $\{P_\phi^0\}$, $\sigma = (\phi)$, $\sigma^1 = \{1, 2, 3, 4, 5\}$, $V_\phi = 0$ and $q_\phi = \Sigma_{j=1}^{5} t_j = 49$.

Step 2: Since the current level $k(0)$ is not equal to $n - 1(4)$, check the dominance property. Also,

$$\max_{i \in \sigma^1} d_i = 25$$

Since, this maximum is not greater than q_ϕ, go to step 4.

Step 4: In this step, five subproblems are created under the root node P_ϕ^0 as shown in Figure 14.2. The details of computations of the lower bound for each of the nodes in the Figure 14.2 are summarized in Table 14.13.

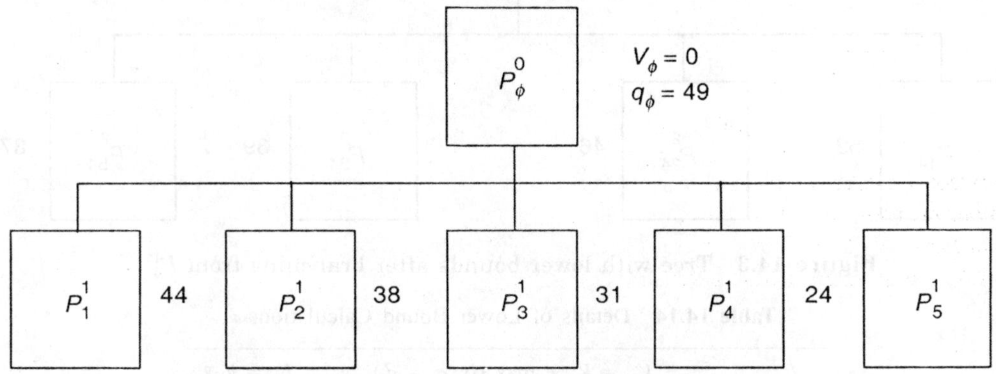

Figure 14.2   Tree with lower bounds after branching from P_ϕ^0.

Table 14.13   Details of Lower Bound Calculations

$P_{i\sigma}^1$	$V_{i\sigma} = V_\sigma + \max(0, q_\sigma - d_i)$	$b_{i\sigma} = V_{i\sigma}$
P_1^1	$0 + \max(0, 49 - 5) = 44$	44
P_2^1	$0 + \max(0, 49 - 11) = 38$	38
P_3^1	$0 + \max(0, 49 - 18) = 31$	31
P_4^1	$0 + \max(0, 49 - 25) = 24$	24
P_5^1	$0 + \max(0, 49 - 20) = 29$	29

Active list = $\{P_4^1\,(24), P_5^1\,(29), P_3^1\,(31), P_2^1\,(38), P_1^1\,(44)\}$

The value within brackets in each of the subproblems shown in the active list represents the lower bound for that subproblem. The value by the side of each node in the tree also represents the lower bound for the subproblem shown within that node. Go to step 2.

Step 2: The first subproblem in the active list of the preceding step is P_4^1. Hence, it is selected for further branching. At this node, the value of k is equal to 1 which is not equal to $n - 1(4)$. Hence, check the dominance property.

$$\sigma = \{4\},$$
$$\sigma^1 = \{1, 2, 3, 5\}$$
$$q_\sigma = 49 - 16 = 33 \qquad \text{(using, } q_{j\sigma} = q_\sigma - t_j\text{)}$$
$$\max_{i \in \sigma^1} d_i = 20$$

Since this maximum value is not greater than q_σ (33), go to step 4.

Step 4: In this step, four subproblems are created under the node P_4^1, as shown in Figure 14.3. The details of computations of the lower bound for each of the nodes in Figure 14.3 are summarized in Table 14.14.

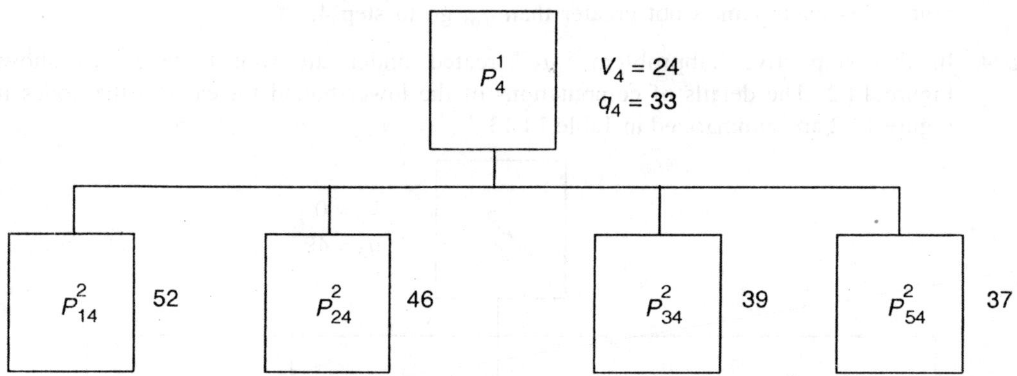

Figure 14.3 Tree with lower bounds after branching from P_4^1.

Table 14.14 Details of Lower Bound Calculations

$P_{i\sigma}^2$	$V_{i\sigma} = V_\sigma + \max (0, q_\sigma - d_i)$	$b_{i\sigma} = V_{i\sigma}$
P_{14}^2	$24 + \max (0, 33 - 5) = 52$	52
P_{24}^2	$24 + \max (0, 33 - 11) = 46$	46
P_{34}^2	$24 + \max (0, 33 - 18) = 39$	39
P_{54}^2	$24 + \max (0, 33 - 20) = 37$	37

Active list = $[P_5^1 (29), P_3^1 (31), P_{54}^2 (37), P_2^1 (38), P_{34}^2 (39), P_1^1 (44), P_{24}^2 (46), P_{14}^2 (52)]$
Go to step 2.

Step 2: Since the first subproblem in the active list is P_5^1, it is considered for further branching. The node P_5^1 occurs at Level 1 which is not equal to $n - 1(4)$. Hence, check the dominance property.

$$\sigma = 5$$

$$\sigma^1 = [1, 2, 3, 4]$$

$$q_\sigma = 49 - 15 = 34 \qquad (\text{using } q_{j\sigma} = q_\sigma - t_j)$$

$$\max_{i \in \sigma^1} d_i = 25.$$

This maximum value is not greater than q_σ (34). Therefore, go to step 4.

Step 4: In this step, four subproblems are created under the node P_5^1, as shown in Figure 14.4. The details of computations of the lower bound for each of the nodes in Figure 14.4 are summarized in Table 14.15.

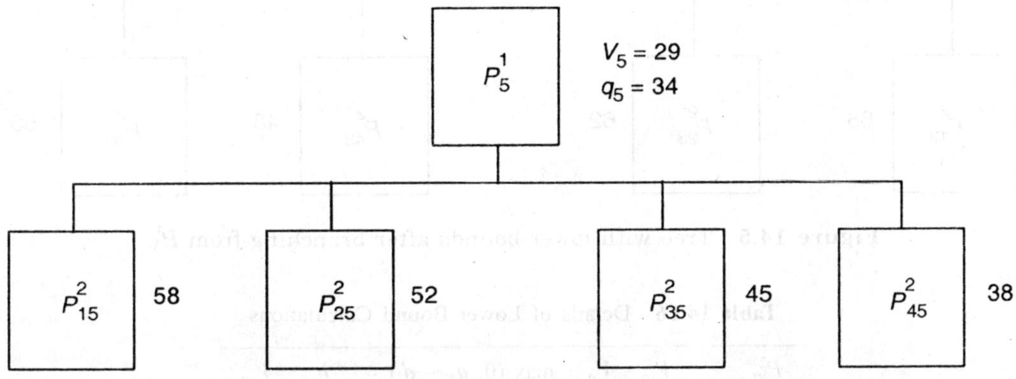

Figure 14.4 Tree with lower bounds after branching from P_5^1.

Table 14.15 Details of Lower Bound Calculations

$P_{i\sigma}^2$	$V_{i\sigma} = V_\sigma + \max(0, q_\sigma - d_i)$	$b_{i\sigma} = V_{i\sigma}$
P_{15}^2	$29 + \max(0, 34 - 5) = 58$	58
P_{25}^2	$29 + \max(0, 34 - 11) = 52$	52
P_{35}^2	$29 + \max(0, 34 - 18) = 45$	45
P_{45}^2	$29 + \max(0, 34 - 25) = 38$	38

Active list $= [P_3^1 (31), P_{54}^2 (37), P_2^1 (38), P_{45}^2 (38), P_{34}^2 (39), P_1^1 (44), P_{35}^2 (45), P_{24}^2 (46), P_{25}^2 (52),$
$P_{14}^2 (52), P_{15}^2 (58)]$

Step 2: Since the first subproblem in the active list is P_3^1, it is considered for further branching. The node P_3^1 occurs at Level 1 which is not equal to $n - 1(4)$. Hence, check the dominance property.

$$\sigma = 3, \sigma^1 = (1, 2, 4, 5)$$

$$q_\sigma = 49 - 7 = 42 \qquad (\text{using } q_{j\sigma} = q_\sigma - t_j)$$

$$\max_{i \in \sigma^1} d_i = 25$$

Since this maximum value is not greater than q_σ (42), go to step 4.

Step 4: In this step, four subproblems are created under the node P_3^1, as shown in Figure 14.5. The details of computations of the lower bound for each of the nodes in Figure 14.5 are summarized in Table 14.16.

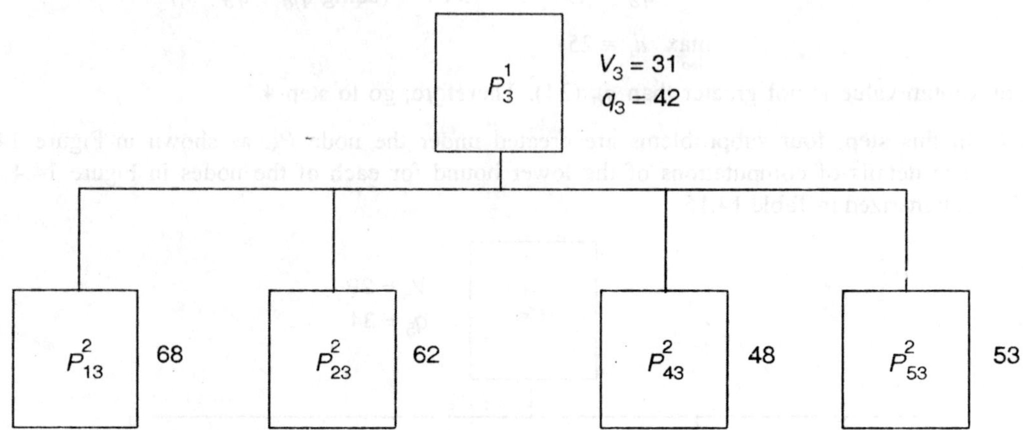

Figure 14.5 Tree with lower bounds after branching from P_3^1.

Table 14.16 Details of Lower Bound Calculations

$P_{i\sigma}^2$	$V_{i\sigma} = V_\sigma + \max\,(0,\ q_\sigma - d_i)$	$b_{i\sigma} = V_{i\sigma}$
P_{13}^2	$31 + \max\,(0,\ 42 - \ \ 5) = 68$	68
P_{23}^2	$31 + \max\,(0,\ 42 - 11) = 62$	62
P_{43}^2	$31 + \max\,(0,\ 42 - 25) = 48$	48
P_{53}^2	$31 + \max\,(0,\ 42 - 20) = 53$	53

Active list $= [P_{54}^2\,(37),\ P_2^1\,(38),\ P_{45}^2\,(38),\ P_{34}^2\,(39),\ P_1^1\,(44),\ P_{35}^2\,(45),\ P_{24}^2\,(46),\ P_{43}^2\,(48),$
$P_{14}^2\,(52),\ P_{25}^2\,(52),\ P_{53}^2\,(53),\ P_{15}^2\,(58),\ P_{23}^2\,(62),\ P_{13}^2\,(68)]$

Step 2: The first subproblem in the active list of the preceding step is P_{54}^2. Hence, it is considered for further branching. This subproblem occurs at Level 2 which is not equal to 4. Hence, check the dominance property.

$$\sigma = (5, 4)$$
$$\sigma^1 = (1, 2, 3)$$
$$q_\sigma = 33 - 15 = 18 \qquad (q_{54} = q_4 - t_5)$$
$$\max_{i \in \sigma^1} d_i = 18$$

This maximum value is equal to $q_\sigma = 18$. Hence, go to step 3.

Step 3: Job 3 has an element in σ^1, which has the highest due date. Hence, based on the dominance property, the subproblem P_{54}^2 is further partitioned with a single branch P_{354}^3 as shown in Figure 14.6.

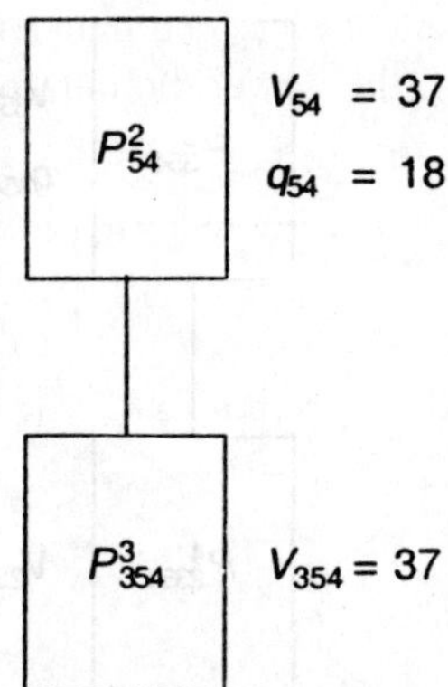

Figure 14.6 Tree with lower bounds after branching from P^2_{54}.

For $\sigma = \{5, 4\}$ and $j = 3$, the following are computed:

$$q_\sigma = 18 \quad \text{(computed in the previous step)}$$

$$q_{j\sigma} = q_\sigma - t_j = q_{54} - t_3 = 18 - 7 = 11$$

$$V_{j\sigma} = V_\sigma + \max\,[(0, (q_\sigma - d_j)]$$

$$= 37 + \max\,[0, (18 - 18)]$$

$$= 37 + 0 = 37$$

$$b_{j\sigma} = V_{j\sigma} = 37$$

Active list = $[P^3_{354}\,(37), P^1_2\,(38), P^2_{45}\,(38), P^2_{34}\,(39), P^1_1\,(44), P^2_{35}\,(45), P^2_{24}\,(46), P^2_{43}\,(48),$

$P^2_{14}\,(52), P^2_{25}\,(52), P^2_{53}\,(53), P^2_{15}\,(58), P^2_{23}\,(62), P^2_{13}\,(68)]$

Step 2: Since the first subproblem in the active list of the preceding step is P^3_{354}, it is considered for further branching. This subproblem occurs at level 3 which is not equal to 4. So, check the dominance property.

$$\sigma = (3, 5, 4)$$

$$\sigma^1 = (1, 2)$$

$$q_{354} = q_{54} - t_3 = 18 - 7 = 11$$

$$\max_{i \in \sigma^1}\; d_i = 11$$

Since the maximum value is equal to $q_\sigma(11)$, go to step 3.

Step 3: Job 2 is an element in σ^1 which has the highest due date. Hence, based on the dominance property, the subproblem P^3_{354} is further partitioned with a single branch P^4_{2354} as shown in Figure 14.7.

For $\sigma = \{3, 5, 4\}$ and $j = 2$, the following are computed:

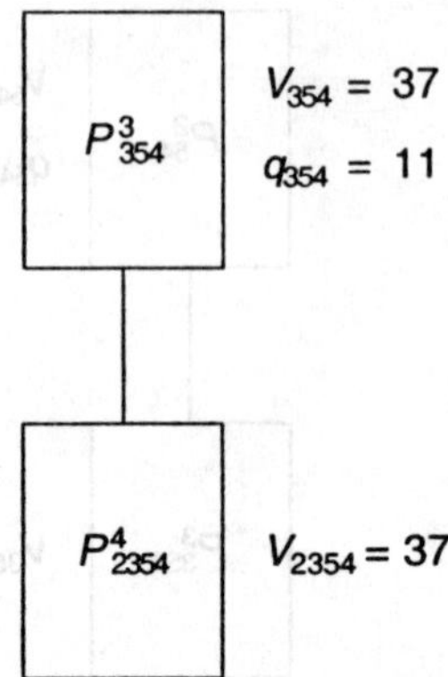

Figure 14.7 Tree with lower bounds after branching from P^3_{354}.

$$q_\sigma = 11 \quad \text{(computed in the previous step)}$$
$$q_{j\sigma} = q_\sigma - t_j = 11 - 7 = 4$$
$$V_{j\sigma} = V_\sigma + \max [0, (q_\sigma - d_j)]$$
$$= 37 + \max [0, (11 - 11)]$$
$$= 37 + 0 = 37$$
$$b_{j\sigma} = V_{j\sigma} = 37$$

Active list $= [P^4_{2354}(37), P^1_2(38), P^2_{45}(38), P^2_{34}(39), P^1_1(44), P^2_{35}(45), P^2_{24}(46), P^2_{43}(48),$

$P^2_{14}(52), P^2_{25}(52), P^2_{53}(53), P^2_{15}(58), P^2_{23}(62), P^2_{13}(68)]$

Step 2: The first subproblem in the active list of the preceding step is P^4_{2354}. This subproblem occurs at Level 4 which is equal to $n - 1(4)$. Hence, the optimality is reached. The optimal sequence can be obtained by prefixing the missing number to the partial sequence at the node P^4_{2354} as shown below. The total tardiness of this optimal sequence is 37. Thus, optimal sequence is 1–2–3–5–4. The total tardiness of the optimal sequence is verified in Table 14.17.

Table 14.17 Computation of Total Tardiness of Optimal Sequence

Job, j	Processing time, t_j	Completion time, C_j	Due date, d_j	T_j
1	4	4	5	0
2	7	11	11	0
3	7	18	18	0
5	15	33	20	13
4	16	49	25	24

Therefore,

$$\sum_{j=1}^{5} T_j = 0 + 0 + 0 + 13 + 24 = 37$$

where $T_j = \max [0, (C_j - d_j)]$. Thus, the result is verified.

The steps are summarized in the form of a complete tree as shown in Figure 14.8. The entry within each box represents a subproblem. The value by the side of each box represents the lower bound for that particular subproblem.

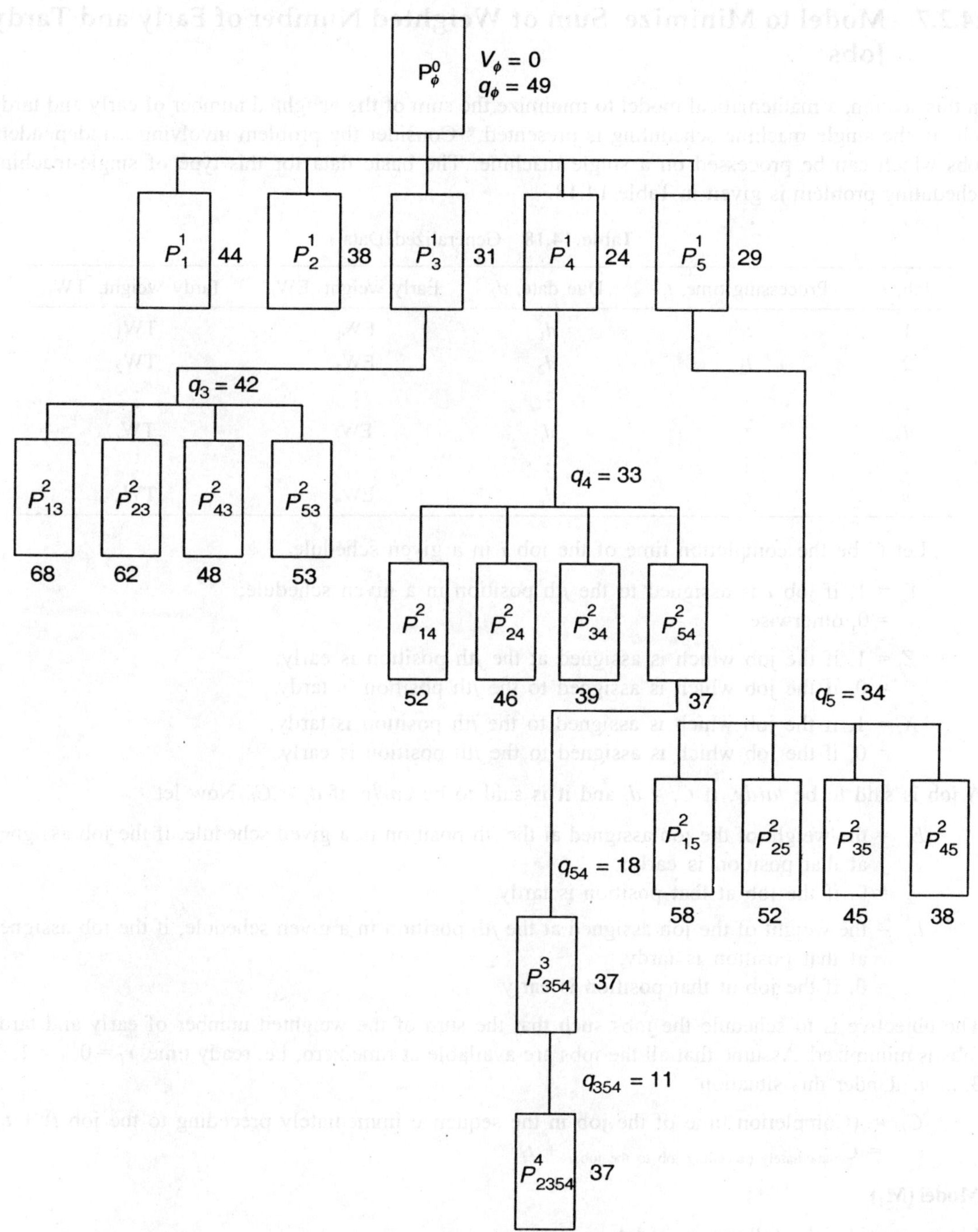

Figure 14.8 Complete tree with lower bounds.

14.2.7 Model to Minimize Sum of Weighted Number of Early and Tardy Jobs

In this section, a mathematical model to minimize the sum of the weighted number of early and tardy jobs in the single machine scheduling is presented.* Consider the problem involving n independent jobs which can be processed on a single machine. The basic data for this type of single-machine scheduling problem is given in Table 14.18.

Table 14.18 Generalized Data

Job, i	Processing time, t_i	Due date, d_i	Early weight, EW_i	Tardy weight, TW_i
1	t_1	d_1	EW_1	TW_1
2	t_2	d_2	EW_2	TW_2
$\vdots$	$\vdots$	$\vdots$	$\vdots$	$\vdots$
i	t_i	d_i	EW_i	TW_i
$\vdots$	$\vdots$	$\vdots$	$\vdots$	$\vdots$
n	t_n	d_n	EW_n	TW_n

Let C_i be the completion time of the job i in a given schedule.

Y_{ij} = 1, if job i is assigned to the jth position in a given schedule;
= 0, otherwise.

Z_j = 1, if the job which is assigned at the jth position is early;
= 0, if the job which is assigned to the jth position is tardy.

X_j = 1, if the job which is assigned to the jth position is tardy;
= 0, if the job which is assigned to the jth position is early.

A job is said to be *tardy*, if $C_i > d_i$ and it is said to be *early*, if $d_i > C_i$. Now let

E_j = the weight of the job assigned at the jth position in a given schedule, if the job assigned at that position is early;
= 0, if the job at that position is tardy.

L_j = the weight of the job assigned at the jth position in a given schedule, if the job assigned at that position is tardy;
= 0, if the job at that position is early.

The objective is to schedule the jobs such that the sum of the weighted number of early and tardy jobs is minimized. Assume that all the jobs are available at time zero, i.e. ready time, $r_i = 0$, $i = 1, 2, 3,..., n$. Under this situation.

C_i = (Completion time of the job in the sequence immediately preceding to the job i) + t_i
= $C_{\text{immediately preceding job to the job } i}$ + t_i

Model (M_1)

Let us consider the following model:

$$\text{Minimize } Q = \sum_{j=1}^{n} \left[Z_j \sum_{i=1}^{n} EW_i\, Y_{ij} + X_j \sum_{i=1}^{n} TW_i\, Y_{ij} \right]$$

*Panneerselvam and Balasubramanian, 1991.

subject to

$$\sum_{j=1}^{n} Y_{ij} = 1, \qquad i = 1, 2, 3,..., n \qquad\qquad \text{(Set 1)}$$

$$\sum_{i=1}^{n} Y_{ij} = 1, \qquad j = 1, 2, 3,..., n \qquad\qquad \text{(Set 2)}$$

$$\sum_{i=1}^{n} \sum_{q=1}^{j} t_i Y_{iq} < \sum_{i=1}^{n} d_i Y_{ij} + M(1 - Z_j), \qquad j = 1, 2, 3,..., n \qquad \text{(Set 3)}$$

$$\sum_{i=1}^{n} d_i Y_{ij} \leq \sum_{i=1}^{n} \sum_{q=1}^{j} t_i Y_{iq} + M(Z_j), \qquad j = 1, 2, 3,..., n \qquad \text{(Set 4)}$$

$$M(1 - X_j) + \sum_{i=1}^{n} \sum_{q=1}^{j} t_i Y_{iq} > \sum_{i=1}^{n} d_i Y_{ij}, \qquad j = 1, 2, 3,..., n \qquad \text{(Set 5)}$$

$$M X_j + \sum_{i=1}^{n} d_i Y_{ij} \geq \sum_{i=1}^{n} \sum_{q=1}^{j} t_i Y_{iq}, \qquad j = 1, 2, 3,..., n \qquad \text{(Set 6)}$$

where

$$Y_{ij} = 0 \text{ or } 1, \qquad i = 1, 2, 3,..., n \quad \text{and} \quad j = 1, 2, 3,..., n$$

$$Z_j = 0 \text{ or } 1, \qquad j = 1, 2, 3,..., n$$

$$X_j = 0 \text{ or } 1, \qquad j = 1, 2, 3,..., n$$

M is a very large value.

In this model, the following points should be noted:

1. The objective function minimizes the sum of the weighted number of early and tardy jobs.
2. The constraints in Set 1 guarantee that each job is assigned to exactly one position in the schedule.
3. The constraints in Set 2 guarantee that each position in a schedule is assigned to exactly one job.
4. The constraints in Set 3 and Set 4 are *either-or* constraints to decide early jobs.

   (a) If $Z_j = 1$, then the respective constraint in Set 3 will be active and the respective constraint in Set 4 will be inactive.

   (b) If $Z_j = 0$, then the respective constraint in Set 4 will be active and the respective constraint in Set 3 will be inactive.

5. The constraints Set 5 and Set 6 are also *either-or* constraints to decide tardy jobs.

   (a) If $X_j = 1$, then the respective constraint in Set 5 will be active and the respective constraint in Set 6 will be inactive.

   (b) If $X_j = 0$, then the respective constraint in Set 6 will be active and the respective constraint in Set 5 will be inactive.

Final linear model. The nonlinear objective function of the model M_1 is converted into a linear model with the help of additional constraint sets 7, 8, 9, 10, 11 and 12 as shown in the following model M_2.

Model (M_2)

Let us consider the following model:

$$\text{Minimize } Q = \sum_{j=1}^{n} [E_j + L_j)$$

subject to

Constraint Set 1 to Set 6 are same as in Model M_1.

$$M(1 - Z_j) + \sum_{i=1}^{n} \text{EW}_j \ Y_{ij} \geq E_j, j = 1, 2, 3,..., n \qquad \text{(Set 7)}$$

$$\sum_{i=1}^{n} \text{EW}_i \ Y_{ij} \leq E_j + M(1 - Z_j), j = 1, 2,..., n \qquad \text{(Set 8)}$$

$$M \ Z_j \geq E_j, j = 1, 2, 3,..., n \qquad \text{(Set 9)}$$

$$M(1 - X_j) + \sum_{i=1}^{n} \text{TW}_j \ Y_{ij} \geq L_j, j = 1, 2, 3,..., n \qquad \text{(Set 10)}$$

$$\sum_{i=1}^{n} \text{TW}_i \ Y_{ij} \leq L_j + M(1 - X_j), j = 1, 2,..., n \qquad \text{(Set 11)}$$

$$M \ X_j \geq L_j, j = 1, 2, 3,..., n \qquad \text{(Set 12)}$$

where Y_{ij}, Z_j, X_j and M are as defined in Model M_1 and

$$E_j \geq 0, \quad j = 1, 2,..., n$$
$$L_j \geq 0, \quad j = 1, 2,..., n$$

Note: Identify the smallest possible unit or unit decimal places which will occur in the left-hand side of the constraints of Set 3 and Set 5 as a result of addition/subtraction. It will be either 1 or 0.01 or 0.001 or 0.0001, etc. Let it be s. To practically implement the '<' type constraints of Set 3 into '≤' type constraints, add s to the left-hand side of each of the constraints in Set 3.

Similarly, subtract s from the left-hand side of each of the constraints in Set 5 to convert them into '≥' constraints.

In this model, the following points should be noted:

1. The constraints in Set 7 and Set 8 are either-or type constraints which guarantee the following:

   (a) If $Z_j = 1$, then the constraints in Set 7 and Set 8 are made active, i.e. E_j is assigned the early weight of the job at the jth position in a given schedule.

   (b) If $Z_j = 0$, then the constraint Set 9 is made active. Moreover, since $E_j > 0$ (non-negative constraint), E_j is assigned a zero value.

2. The constraints in Set 10, Set 11 and Set 12 are either-or type constraints which guarantee the following:

   (a) If $X_j = 1$, the constraints in Set 10 and Set 11 are made active, i.e. L_j is assigned the tardy, weight of the job at the jth position in a given schedule.

(b) If $X_j = 0$, then the constraint Set 12 is made active. Moreover, and since $L_j > 0$ (non-negative constraint), L_j is assigned a zero value.

Example 14.6 Solve the problem which is in Table 14.19 using the Model 2.

Table 14.19 Example 14.6

Job, i	Processing time, t_i	Due date, d_i	Early weight, EW_i	Tardy weight, TW_i
1	5	10	1	2
2	3	5	4	6
3	8	15	6	10

Solution The development of the model for the given problem is left as an exercise. The results of the above problem using the model are presented below:

$$Y_{13} = Y_{21} = Y_{32} = X_3 = 1, \qquad E_1 = 4, \quad E_2 = 6, \quad L_3 = 2$$

All other variables are zero.

From the above results, one can infer the best schedule as 2–3–1. The corresponding objective function value (i.e. sum of the weighted number of early and tardy jobs) is 12. The details of computations of this value are shown in Table 14.20.

Table 14.20 Details of Computations of Performance Measure

Job, j	Processing time, t_i	Completion time, C_j	Due date, d_i	Early/tardy	Corresponding weight
2	3	3	5	Early	4
3	8	11	15	Early	6
1	5	16	10	Tardy	2

Total weighted number of jobs = 12

14.3   FLOW SHOP SCHEDULING

In flow shop scheduling problem, there are n jobs, each of which requires processing on m different machines. The order in which the machines are required to process a job is called *process sequence* of that job. In flow shop scheduling problem, all the jobs will have the same process sequence. But the processing time of each operation in a job will be different from that of other jobs. If an operation is absent in a job, then the processing time of that operation in that job is assumed as zero.

The flow shop scheduling problem can be characterized as follows:

1. A set of multiple-operation jobs is available for processing at time zero (each job requires m operations and each operation requires a different machine).
2. Set-up times for the operations are sequence independent and are included in processing times.
3. Job descriptors are known in advance.
4. m different machines are continuously available.
5. Each individual operation of jobs is processed till its completion without pre-emption.

The main difference of the flow shop scheduling problem from the single-machine scheduling problem is that the inserted idle time may be advantageous in flow shop scheduling problem. Though the current machine is free, if the job from the previous machine is not released to the current machine, we cannot start processing on that job. So, the current machine has to be idle for some time. Hence, inserted idle time on some machine would lead to optimality. In flow shop scheduling problem, minimizing makespan is a dominant measure. The makespan of a schedule in the flow shop scheduling problem is the total time taken to complete processing of all the jobs in a given batch of jobs. For example, consider the flow shop problem shown in Table 14.21.

Table 14.21 Sample Flow Shop Problem

Job	Machine 1	Machine 2
1	4	2
2	9	6
3	5	3
4	8	7

If the sequence of the jobs is 4–1–3–2, then the corresponding makespan (total time) is computed, as shown in Figure 14.9.

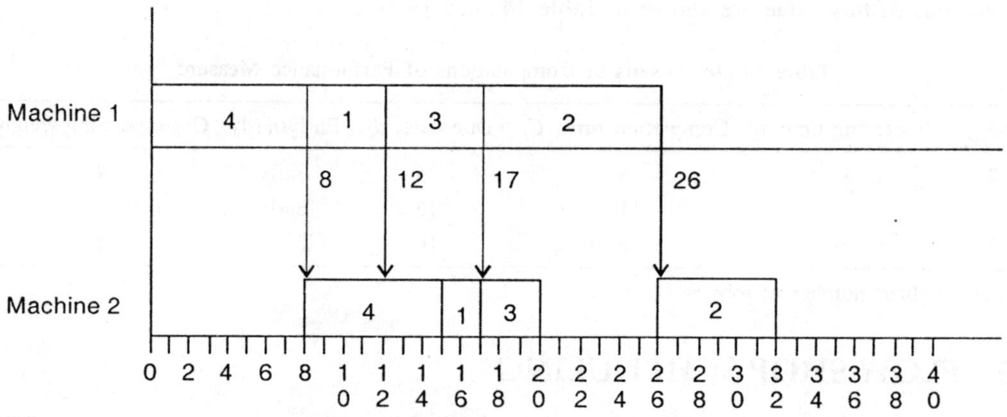

Figure 14.9 Gantt chart for 4–1–3–2 sequence.

In Figure 14.9, the makespan is 32. The inserted idle times on Machine 2 are from 0 to 8 and from 20 to 26.

Consider another sequence say 3–1–2–4. The Gantt chart for this sequence is shown in Figure 14.10. The makespan for the schedule in Figure 14.10 is 33. The Machine 2 has idle times from 0 to 5, 8 to 9, 11 to 18 and 24 to 26.

This problem has 4 jobs. Hence, 4! sequences are possible. Unlike in single machine scheduling, in flow shop scheduling, inserted idle time would minimize the makespan.

In the above two sequences, 4–1–3–2 and 3–1–2–4, the first sequence has lesser makespan. Like this, one can enumerate all 4! sequences, then select the sequence which has the minimum makespan as the optimal sequence. Since the value of $n!$ grows exponentially with n, one needs some efficient procedure to solve the problem. For large size of n, it would be difficult to solve the problem optimally. Under such situation we can use some efficient heuristic.

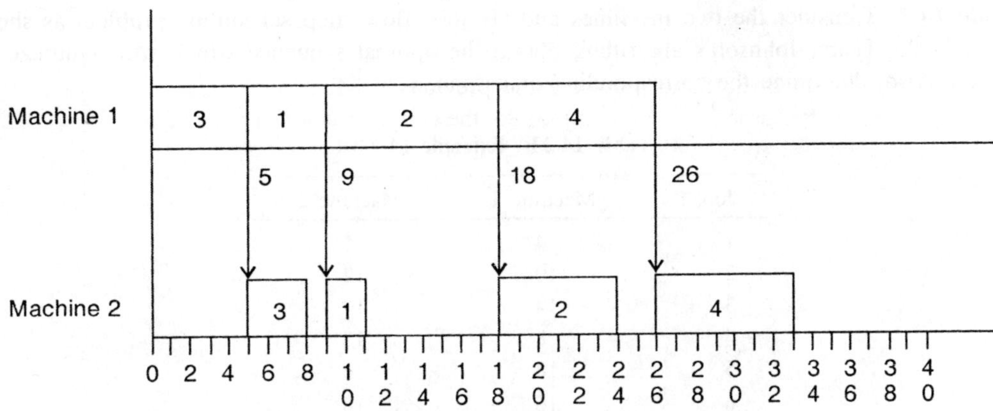

Figure 14.10   Gantt chart for 3–1–2–4 sequence.

14.3.1   Johnson's Algorithm for *n* Jobs and Two Machines Problem

As mentioned in the earlier section, the time complexity function for a general flow shop problem is exponential in nature. That means, the function grows exponentially with an increase in the problem size. But, for a problem with 2 machines and *n* jobs, Johnson had developed a polynomial algorithm to get optimal solution, i.e. in a definite time, one can get the optimal solution.

Consider the generalized format of the flow shop problem as shown in Table 14.22.

Table 14.22   Generalized Format of Flow Shop Problem

Job, i	Machine 1	Machine 2
1	t_{11}	t_{12}
2	t_{21}	t_{22}
$\vdots$	$\vdots$	$\vdots$
i	t_{i1}	t_{i2}
$\vdots$	$\vdots$	$\vdots$
n	t_{n1}	t_{n2}

In the above table, t_{ij} represents the processing time of the job i on Machine j.

Algorithm

Step 1:   Find the minimum among various t_{i1} and t_{i2}, where i varies from 1 to n.

Step 2(a):   If the minimum processing time requires Machine 1, place the associated job in the first available position in sequence. Go to step 3.

Step 2(b):   If the minimum processing time requires Machine 2, place the associated job in the last available position in sequence. Go to step 3.

Step 3:   Delete the assigned job from the table.

Step 4:   Find the minimum among the undeleted processing times in the given table and return to step 2(a) until all the positions in the sequence are filled. (Ties may be broken randomly.)

The above algorithm is illustrated using the following example.

Example 14.7 Consider the two machines and six jobs flow shop scheduling problem as shown in Table 14.23. Using Johnson's algorithm, obtain the optimal sequence which will minimize the makespan. Also, determine the corresponding makespan.

Table 14.23 Example 14.7

Job, i	Machine 1	Machine 2
1	4	6
2	10	12
3	14	10
4	8	12
5	18	6
6	16	8

Solution The workings of the algorithm are summarized in Table 14.24.

Table 14.24 Summary of Working of Johnson's Algorithm

Stage	Unscheduled jobs	Minimum, t_{ik}	Assignment	Partial sequence
1	1, 2, 3, 4, 5, 6	t_{11}	1 = [1]*	1 X X X X X
2	2, 3, 4, 5, 6	t_{52}	5 = [6]	1 X X X X 5
3	2, 3, 4, 6	t_{41}	4 = [2]	1 4 X X X 5
4	2, 3, 6	t_{62}	6 = [5]	1 4 X X 6 5
5	2, 3	t_{32}	3 = [4]	1 4 X 3 6 5
6	2	–	–	1 4 2 3 6 5

*1 = [1] means that job 1 is assigned to the position 1 in the sequence.

The optimal sequence is 1–4–2–3–6–5. The makespan is determined as shown in Table 14.25.

Table 14.25 Makespan Determination of Example 14.7

Job	Processing time				Idle time on Machine 2
	Machine 1		Machine 2		
	Time-in	Time-out	Time-in	Time-out	
1	0	4	4	10	4
4	4	12	12	24	2
2	12	22	24	36	0
3	22	36	36	46	0
6	36	52	52	60	6
5	52	70	70	76	10

From Table 14.25,

Time-in on Machine 2 = max (Machine 1 Time-out of the current job,

Machine 2 Time-out of the previous job)

The makespan for this schedule is 76. The makespan can also be obtained using the Gantt chart as shown in Figure 14.11.

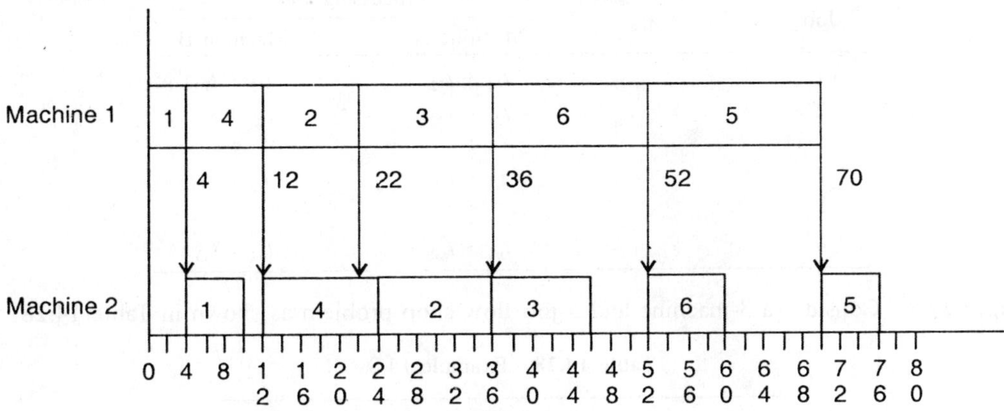

Figure 14.11   Gantt chart for Example 14.7.

14.3.2   Extension of Johnson's Algorithm for *n* Jobs and Three Machines Problem

Consider a 'three machines and *n* jobs' flow shop scheduling problem as shown in Table 14.26.

Table 14.26   Three Machines and *n* Jobs Problem

Job, i	Processing time		
	Machine 1	Machine 2	Machine 3
1	t_{11}	t_{12}	t_{13}
2	t_{21}	t_{22}	t_{23}
⋮	⋮	⋮	⋮
i	t_{i1}	t_{i2}	t_{i3}
⋮	⋮	⋮	⋮
n	t_{n1}	t_{n2}	t_{n3}

Now, if any one of the following two conditions:

$$\min_i t_{i1} \geq \max_i t_{i2} \quad \text{or} \quad \min_i t_{i3} \geq \max_i t_{i2}$$

is satisfied, then the Johnson's algorithm can be extended in the following way.

Create a hypothetical problem with two machines and *n* jobs as shown in Table 14.27. The objective is to obtain optimal sequence for the data given in the Table 14.27. Later, the makespan is to be determined for the optimal sequence by using the data of the original problem as shown in Table 14.26.

This concept of extending Johnson's algorithm to this type of problem is demonstrated using the following example.

Table 14.27 Hypothetical Problem for the Generalized Problem

Job	Processing time	
	Machine A	Machine B
1	$t_{11} + t_{12}$	$t_{12} + t_{13}$
2	$t_{21} + t_{22}$	$t_{22} + t_{23}$
$\vdots$	$\vdots$	$\vdots$
i	$t_{i1} + t_{i2}$	$t_{i2} + t_{i3}$
$\vdots$	$\vdots$	$\vdots$
n	$t_{n1} + t_{n2}$	$t_{n2} + t_{n3}$

Example 14.8 Consider a 3-machine and 5-job flow shop problem as shown in Table 14.28.

Table 14.28 Example 14.8

Job	Processing time		
	Machine 1	Machine 2	Machine 3
1	10	10	12
2	12	8	20
3	16	6	14
4	12	4	20
5	20	8	8

(a) Check whether the Johnson's algorithm can be extended to this problem.

(b) If yes, find the optimal sequence and the corresponding makespan.

Solution (a) For the given problem, we have

$$\min (t_{i1}) = 10, \qquad \max (t_{i2}) = 10$$

Since the first condition $\min (t_{i1}) \geq \max (t_{i2})$ is satisfied, without checking the second condition, one can extend the Johnson's algorithm to this problem.

(b) The hypothetical problem of the data given in Table 14.28 is presented in Table 14.29.

Table 14.29 Data for Hypothetical Problem

Job i	Processing time	
	Machine A	Machine B
1	20	22
2	20	28
3	22	20
4	16	24
5	28	16

One of the sequences giving the optimal makespan for the above problem is 4–2–1–3–5. The makespan of this sequence is determined as shown in Table 14.30.

Table 14.30 Determination of Makespan of Example 14.8

Job i	Machine 1		Machine 2		Machine 3		Idle time on Machine 2	Idle time on Machine 3
	In	Out	In	Out	In	Out		
4	0	12	12	16	16	36	12	16
2	12	24	24	32	36	56	8	0
1	24	34	34	44	56	68	2	0
3	34	50	50	56	68	82	6	0
5	50	70	70	78	82	90	14	0

The makespan for this problem is 90 units of time. The same can also be determined using the Gantt chart as shown in Figure 14.12.

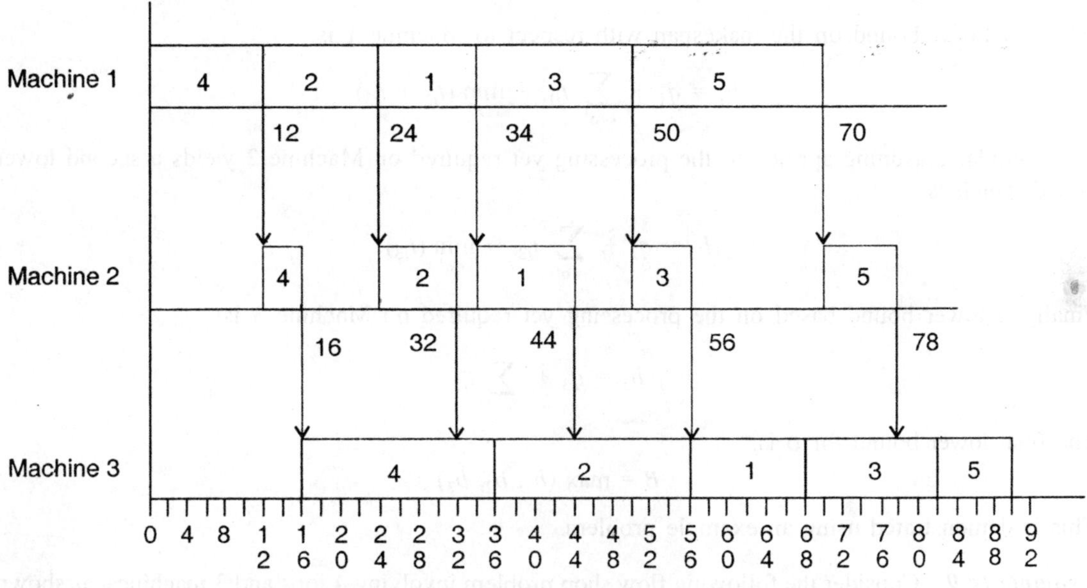

Figure 14.12 Gantt chart for Example 14.8.

14.3.3 Branch-and-Bound Method for n Jobs and m Machines

A branch-and-bound method for a generalized n jobs and m machines flow shop problem was developed by Ignall and Schrage.* The underlying branching tree has the same structure as the tree shown in the single machine scheduling problem, except that σ represents a partial permutation occurring at the beginning of the sequence instead of at the end. In other words, the job sequence is constructed in a forward direction (instead of a backward direction) while proceeding down the branching tree. For each node on the tree, a lower bound on the makespan associated with the completion of the corresponding partial sequence σ is obtained by considering the work remaining on each machine. To illustrate the procedure for $m = 3$ (m is number of machines), the required terminologies are presented.

*Baker, 1972.

Let σ be a partial permutation occurring at the beginning of the sequence, and σ^1 denotes the set of jobs that are not contained in the partial permutation σ. For a given partial sequence σ, again let q_1, q_2 and q_3 be the completion time of the last job on Machine 1, Machine 2 and Machine 3, respectively, among jobs in σ. The amount of processing yet required on Machine 1 is

$$\sum_{i \in \sigma^1} t_{i1}$$

Moreover, suppose that a particular job K is the last job in the sequence. Under this situation, after the job K is processed on Machine 1, it will take at least $(t_{k2} + t_{k3})$ time units on Machine 2 and Machine 3 before the whole schedule is completed. In the most favourable situation, the last job:

(a) encounters no delay between the completion of an operation and the start of its direct successor, and

(b) has the minimum sum $(t_{i2} + t_{i3})$ among jobs $i \in \sigma^1$.

Hence, a lower bound on the makespan with respect to Machine 1 is:

$$b_1 = q_1 + \sum_{i \in \sigma^1} t_{i1} + \min_{i \in \sigma^1} (t_{i2} + t_{i3})$$

Similar reasoning applied to the processing yet required on Machine 2 yields a second lower bound which is:

$$b_2 = q_2 + \sum_{i \in \sigma^1} t_{i2} + \min_{i \in \sigma^1} (t_{i3})$$

Finally, a lower bound based on the processing yet required on Machine 3 is:

$$b_3 = q_3 + \sum_{i \in \sigma^1} t_{i3}$$

The final lower bound for σ is:

$$B = \max (b_1, b_2, b_3)$$

This is demonstrated using an example problem.

Example 14.9 Consider the following flow shop problem involving 4 jobs and 3 machines, as shown in Table 14.31.

Table 14.31 Example 14.9

Job	Machine 1	Machine 2	Machine 3
1	4	5	11
2	12	2	6
3	8	10	14
4	11	13	3

Find the optimal makespan schedule of the above problem using branch-and-bound technique.

Solution At the root node, $\sigma = \{\phi\}$ and $\sigma^1 = \{1, 2, 3, 4\}$. The lower bound (B) for the root node = 0. The set of nodes created under the root node, P_ϕ^0 is shown in Figure 14.13.

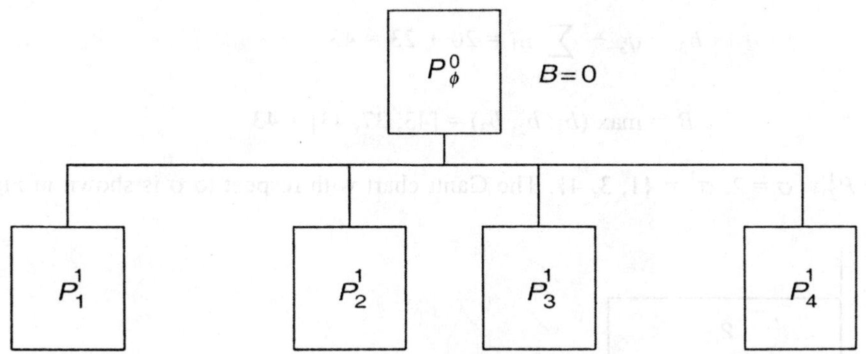

Figure 14.13   Nodes created from P_ϕ^0.

The lower bound calculations pertaining to the nodes created in Figure 14.13 are presented below:

Node P_1^1. $\sigma = 1$, $\sigma^1 = \{2, 3, 4\}$. Initially, $(q_1, q_2, q_3) = (0, 0, 0)$. The Gantt chart with respect to σ is shown in Figure 14.14.

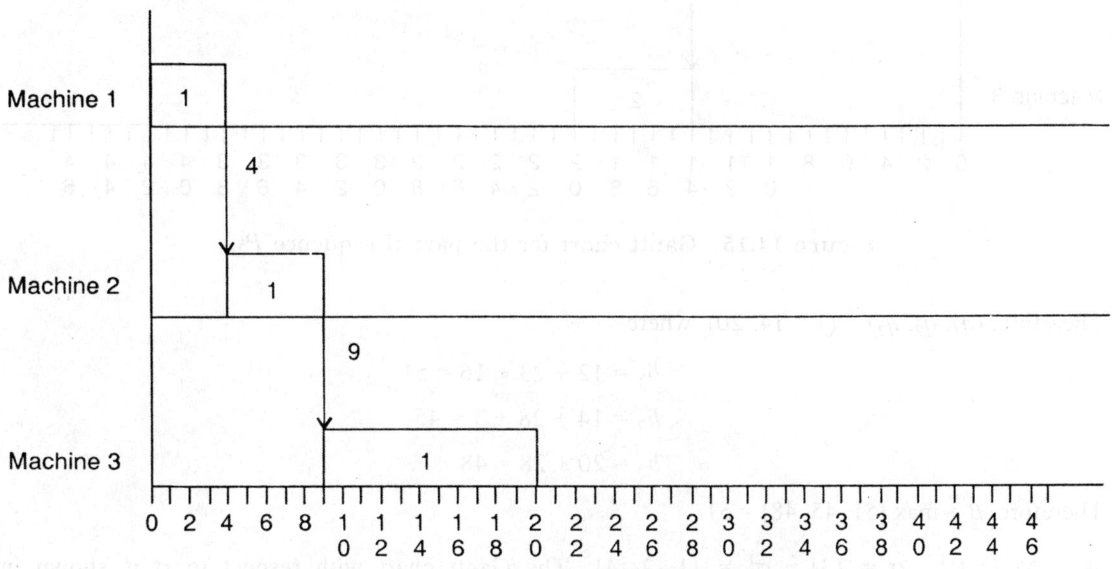

Figure 14.14   Gantt chart for the partial sequence P_1^1.

Therefore, $(q_1, q_2, q_3) = (4, 9, 20)$, where

$$b_1 = q_1 + \sum_{i \in \sigma^1} t_{i1} + \min_{i \in \sigma^1} \{t_{i2} + t_{i3}\} = 4 + 31 + 8 = 43$$

$$b_2 = q_2 + \sum_{i \in \sigma^1} t_{i2} + \min_{i \in \sigma^1} \{t_{i3}\} = 9 + 25 + 3 = 37$$

$$b_3 = q_3 + \sum_{i \in \sigma^1} t_{i3} = 20 + 23 = 43$$

$$B = \max (b_1, b_2, b_3) = [43, 37, 43] = 43$$

Node P_2^1. $\sigma = 2$, $\sigma^1 = \{1, 3, 4\}$. The Gantt chart with respect to σ is shown in Figure 14.15.

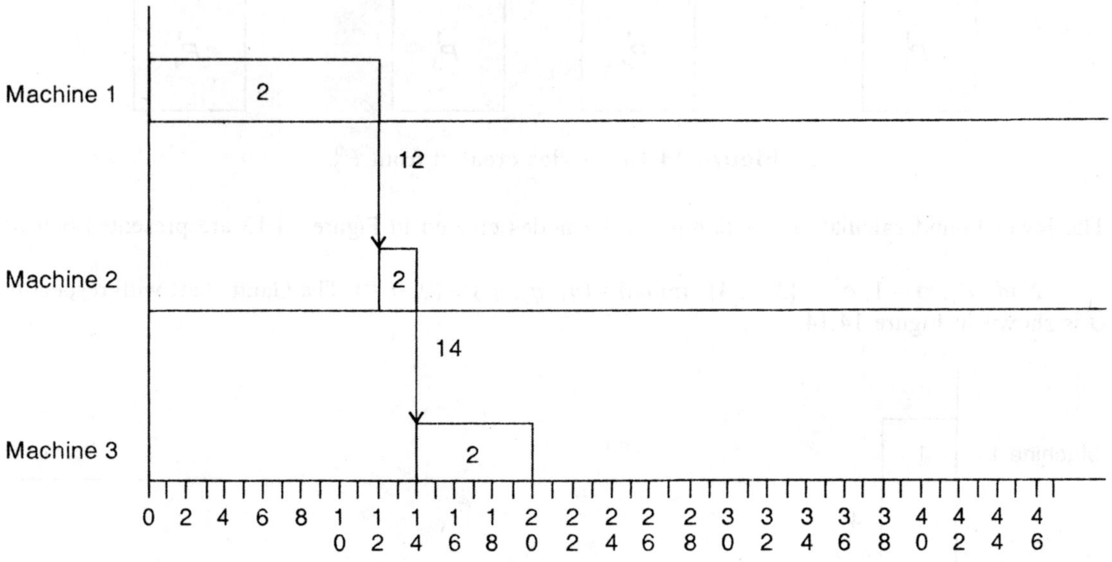

Figure 14.15   Gantt chart for the partial sequence P_2^1.

Therefore, $(q_1, q_2, q_3) = (12, 14, 20)$, where

$$b_1 = 12 + 23 + 16 = 51$$
$$b_2 = 14 + 28 + 3 = 45$$
$$b_3 = 20 + 28 = 48$$

Therefore, $B = \max [51, 45, 48] = 51$.

Node P_3^1. $\sigma = \{3\}$, $\sigma^1 = \{1, 2, 4\}$. The Gantt chart with respect to σ is shown in Figure 14.16.

Therefore, $(q_1, q_2, q_3) = (8, 18, 32)$, where

$$b_1 = 8 + 27 + 8 = 43$$
$$b_2 = 18 + 20 + 3 = 41$$
$$b_3 = 32 + 20 = 52$$

and

$$B = \max [43, 41, 52] = 52$$

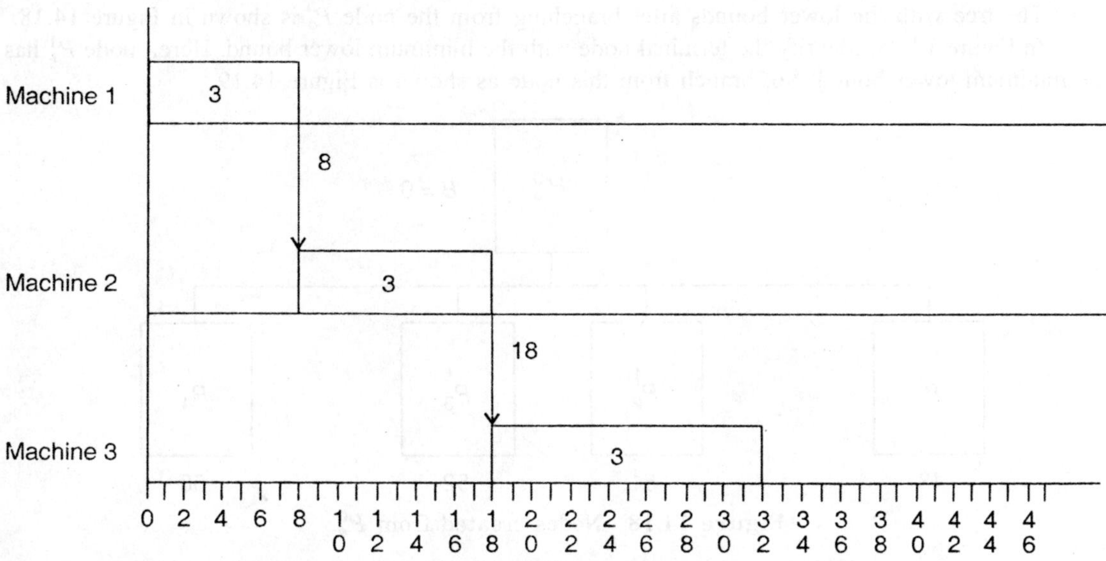

Figure 14.16   Gantt chart for the partial sequence P_3^1.

Node P_4^1. $\sigma = \{4\}$, $\sigma^1 = \{1, 2, 3\}$. The Gantt chart with respect to σ is shown in Figure 14.17. Therefore, $(q_1, q_2, q_3) = (11, 24, 27)$, where

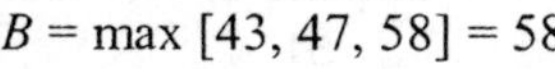

$$b_1 = 11 + 24 + 8 = 43$$
$$b_2 = 24 + 17 + 6 = 47$$
$$b_3 = 27 + 31 = 58$$

and

$$B = \max\,[43, 47, 58] = 58$$

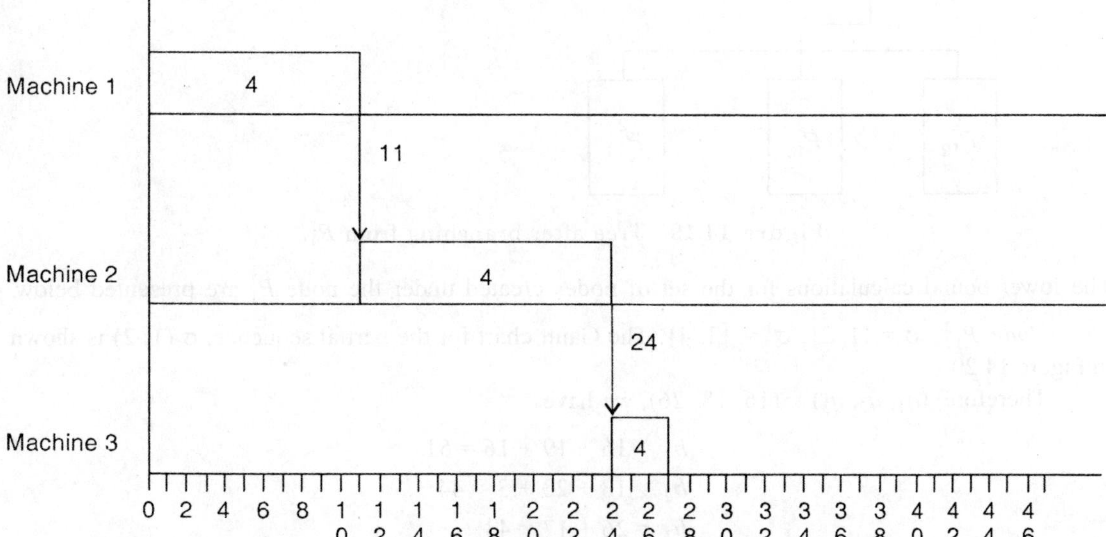

Figure 14.17   Gantt chart for the partial sequence P_4^1.

The tree with the lower bounds after branching from the node P_ϕ^0 is shown in Figure 14.18.

In Figure 14.18, identify the terminal node with the minimum lower bound. Here, node P_1^1 has the minimum lower bound. So, branch from this node as shown is Figure 14.19.

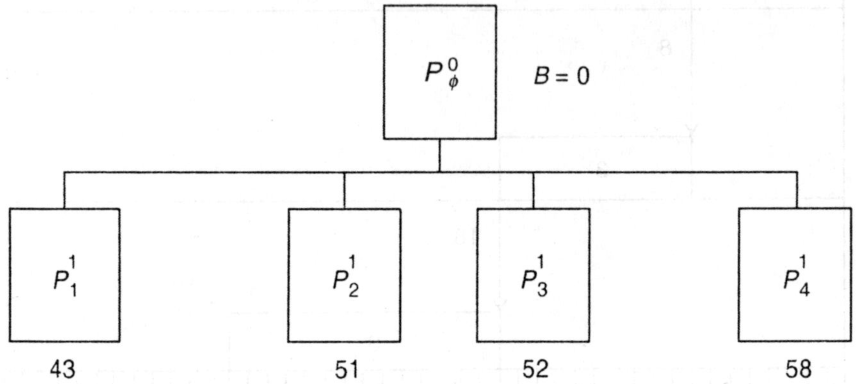

Figure 14.18   Nodes created from P_ϕ^0.

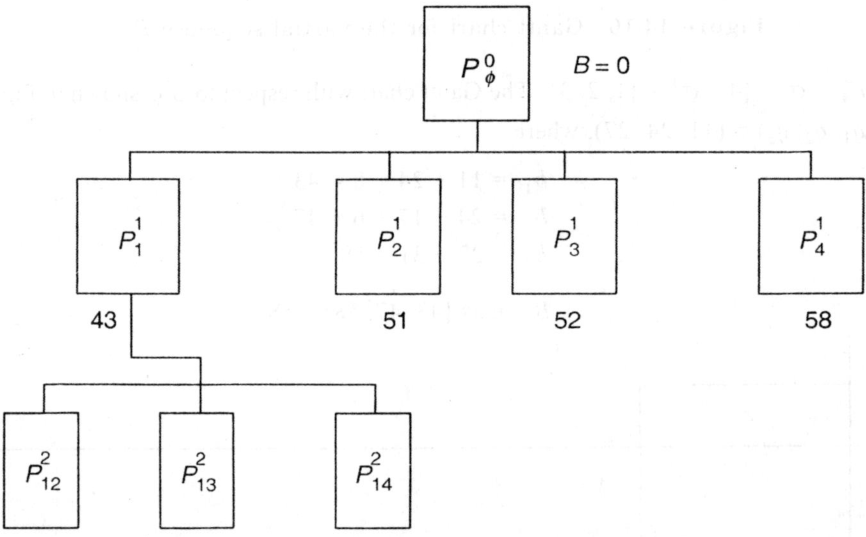

Figure 14.19   Tree after branching from P_1^1.

The lower bound calculations for the set of nodes created under the node P_1^1 are presented below.

Node P_{12}^2. $\sigma = \{1, 2\}$, $\sigma^1 = \{3, 4\}$. The Gantt chart for the partial sequence, $\sigma\,(1, 2)$ is shown in Figure 14.20.

Therefore, $(q_1, q_2, q_3) = (16, 18, 26)$, we have

$$b_1 = 16 + 19 + 16 = 51$$
$$b_2 = 18 + 23 + 3 = 44$$
$$b_3 = 26 + 17 = 43$$

and

$$B = \max\,\{51, 44, 43\} = 51$$

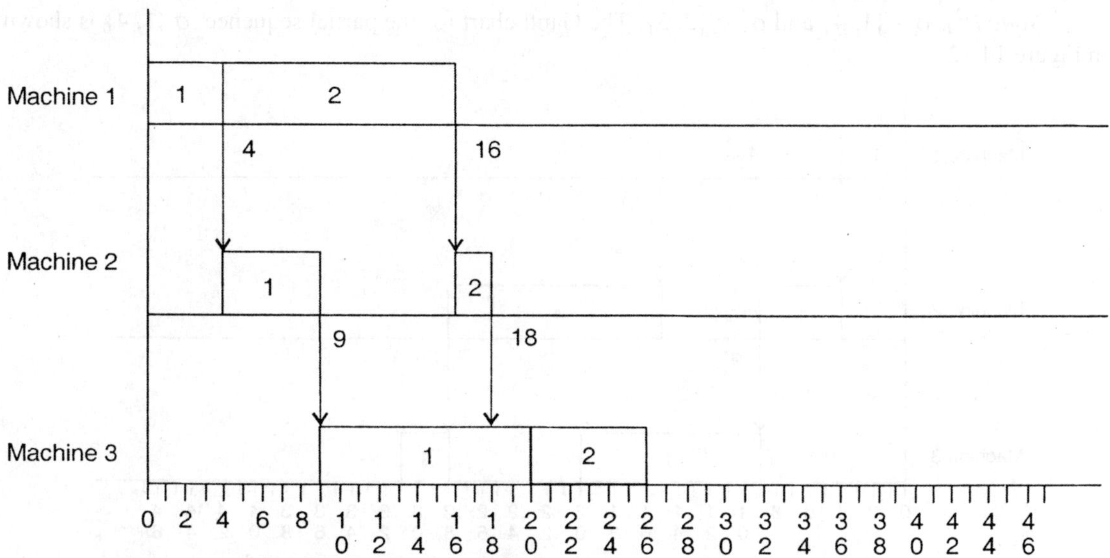

Figure 14.20 Gantt chart for the partial sequence P_{12}^2.

Node P_{13}^2. $\sigma = \{1, 3\}$, $\sigma^1 = (2, 4)$. The Gantt chart for the partial sequence, $\sigma \{1, 3\}$ is shown in Figure 14.21.

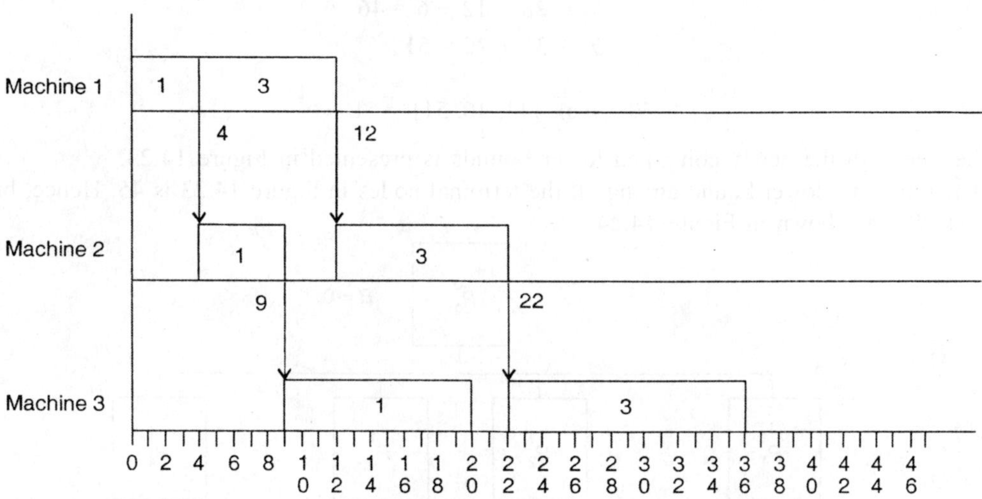

Figure 14.21 Gantt chart for the partial sequence P_{13}^2.

Therefore, $(q_1, q_2, q_3) = (12, 22, 36)$, where

$$b_1 = 12 + 23 + 8 = 43$$
$$b_2 = 22 + 15 + 3 = 40$$
$$b_3 = 36 + 9 = 45$$

and

$$B = \max \{43, 40, 45\} = 45$$

Node P_{14}^2. $\sigma = \{1, 4\}$ and $\sigma' = \{2, 3\}$. The Gantt chart for the partial sequence, $\sigma \{1, 4\}$ is shown in Figure 14.22.

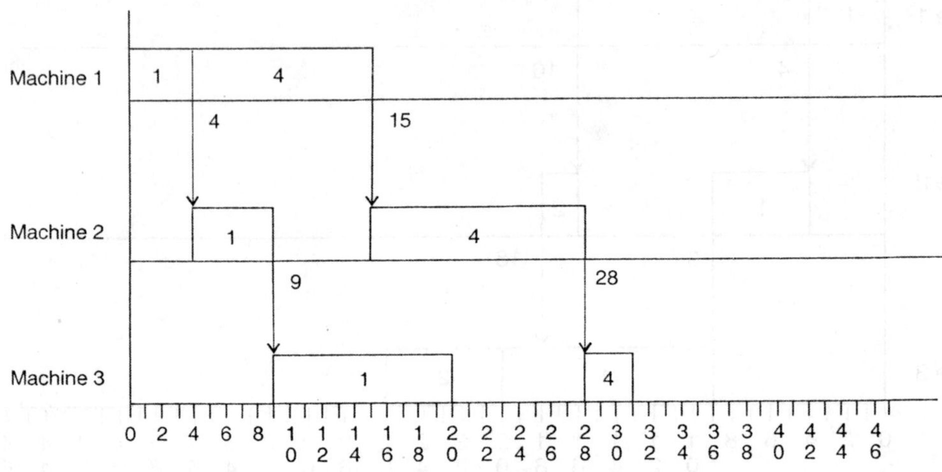

Figure 14.22 Gantt chart for the partial sequence P_{14}^2.

Therefore, $(q_1, q_2, q_3) = (15, 28, 31)$, where

$$b_1 = 15 + 20 + 8 = 43$$
$$b_2 = 28 + 12 + 6 = 46$$
$$b_3 = 31 + 20 = 51$$

and

$$B = \max \{43, 46, 51\} = 51$$

Now the tree with the newly computed lower bounds is presented in Figure 14.23.

The minimum lower bound among all the terminal nodes in Figure 14.23 is 45. Hence, branch from node P_{13}^2 as shown in Figure 14.24.

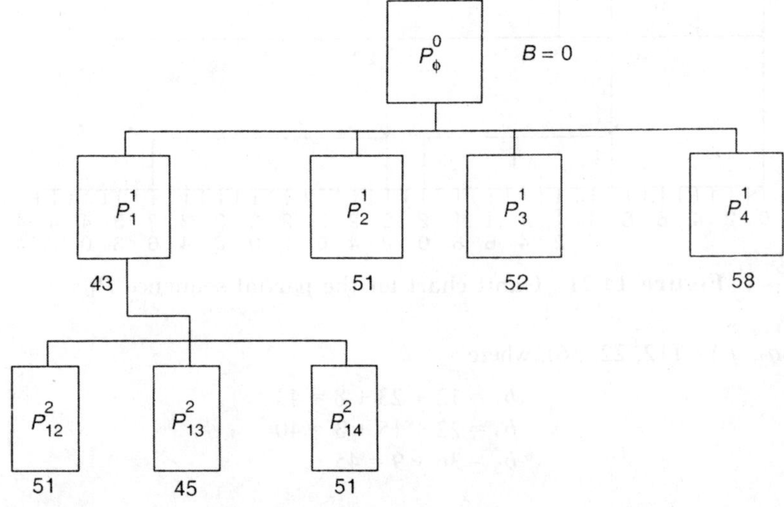

Figure 14.23 Tree with lower bounds after branching from P_1^1.

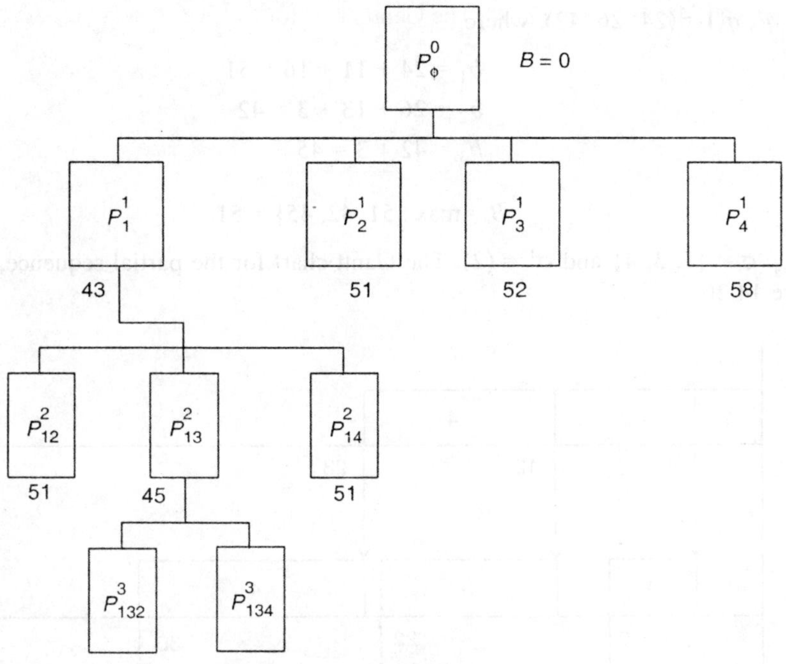

Figure 14.24   Tree after branching from P^2_{13}.

The calculations pertaining to the lower bounds of the nodes created from P^2_{13} are presented below:

Node P^3_{132}. $\sigma = \{1, 3, 2\}$ and $\sigma^1 = (4)$. The Gantt chart for the partial sequence, $\sigma \{1, 3, 2\}$ is shown in Figure 14.25.

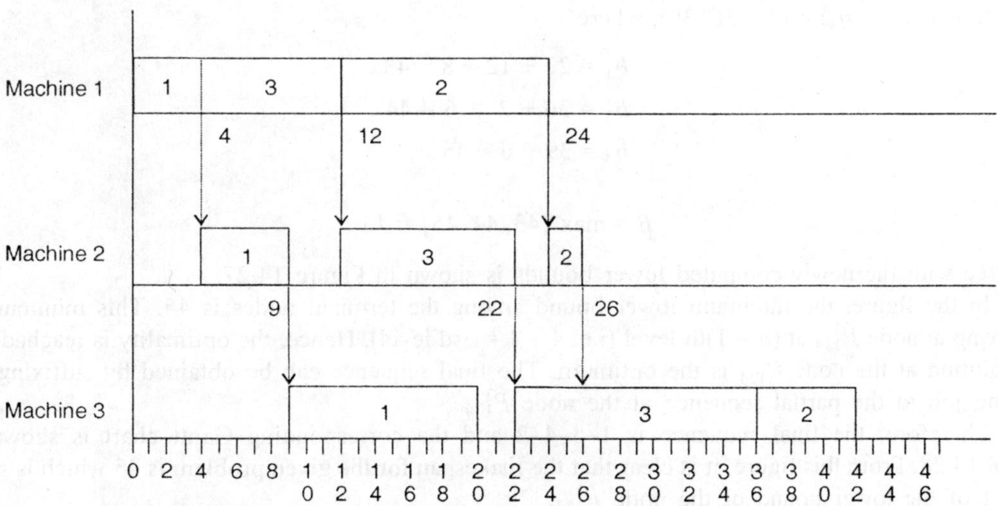

Figure 14.25   Gantt chart for the partial sequence P^3_{132}.

Therefore, $(q_1, q_2, q_3) = (24, 26, 42)$, where

$$b_1 = 24 + 11 + 16 = 51$$
$$b_2 = 26 + 13 + 3 = 42$$
$$b_3 = 42 + 3 = 45$$

and

$$B = \max \{51, 42, 45\} = 51$$

Node P^3_{134}. $\sigma = \{1, 3, 4\}$ and $\sigma^1 = (2)$. The Gantt chart for the partial sequence, $\sigma \{1, 3, 4\}$ is shown in Figure 14.26.

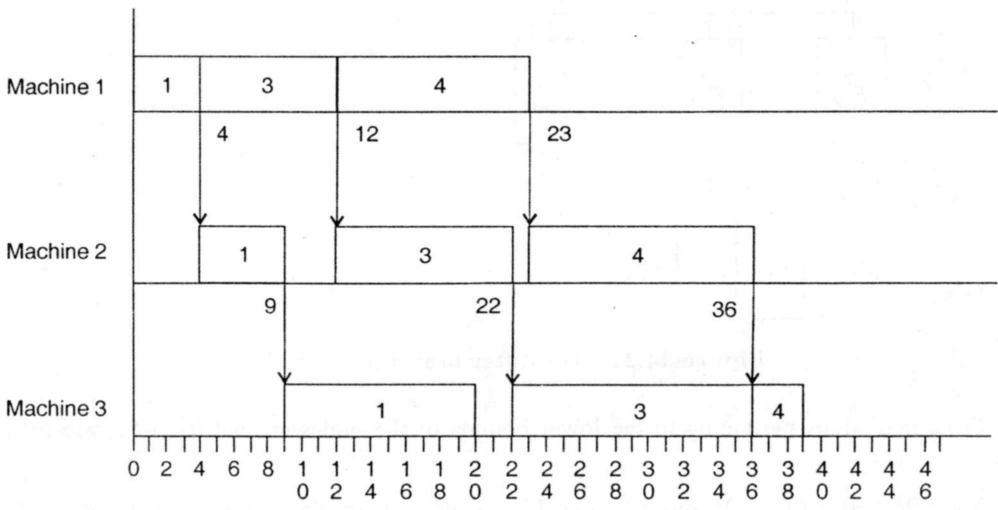

Figure 14.26 Gantt chart for the partial sequence P^3_{134}.

Therefore, $(q_1, q_2, q_3) = (23, 36, 39)$, where

$$b_1 = 23 + 12 + 8 = 43$$
$$b_2 = 36 + 2 + 6 = 44$$
$$b_3 = 39 + 6 = 45$$

and

$$B = \max \{43, 44, 45\} = 45$$

The tree with the newly computed lower bounds is shown in Figure 14.27.

In the figure, the minimum lower bound among the terminal nodes is 45. This minimum is occurring at node P^3_{134} at $(n - 1)$th level (i.e. $4 - 1 = $ 3rd level). Hence, the optimality is reached and the solution at the node P^3_{134} is the optimum. The final sequence can be obtained by suffixing the missing job to the partial sequence at the node P^3_{134}.

Therefore, the final sequence is 1–3–4–2 and the corresponding Gantt chart is shown in Figure 14.28. From this figure, it is clear that the makespan for the given problem is 45 which is same as that of the lower bound of the node P^3_{134}.

The computations of the whole procedure is summarized in Table 14.32.

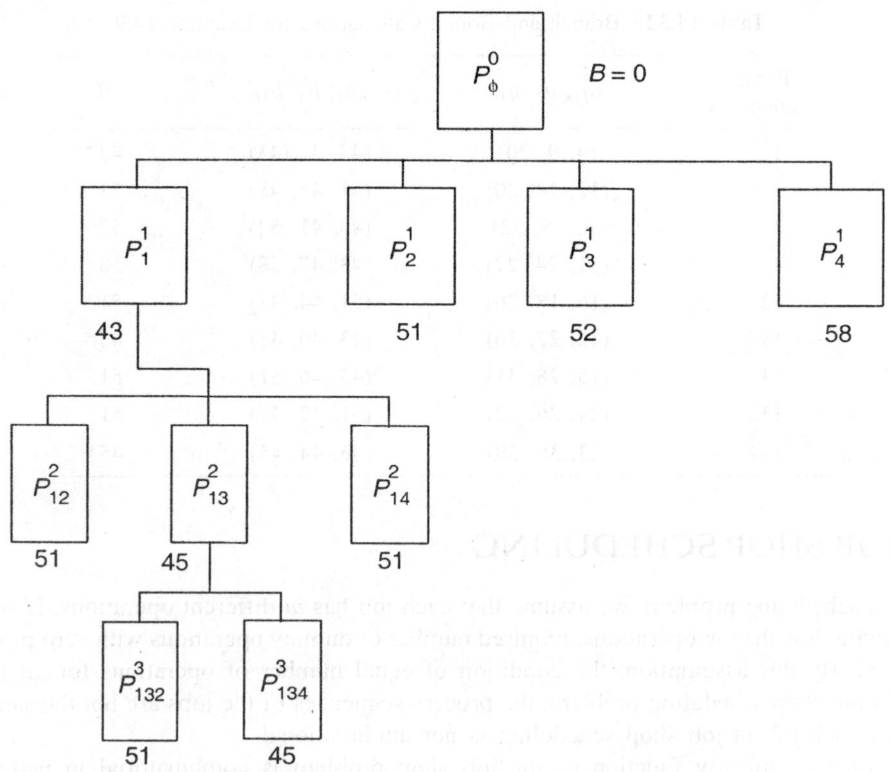

Figure 14.27 Tree with lower bounds after branching from P^2_{13}.

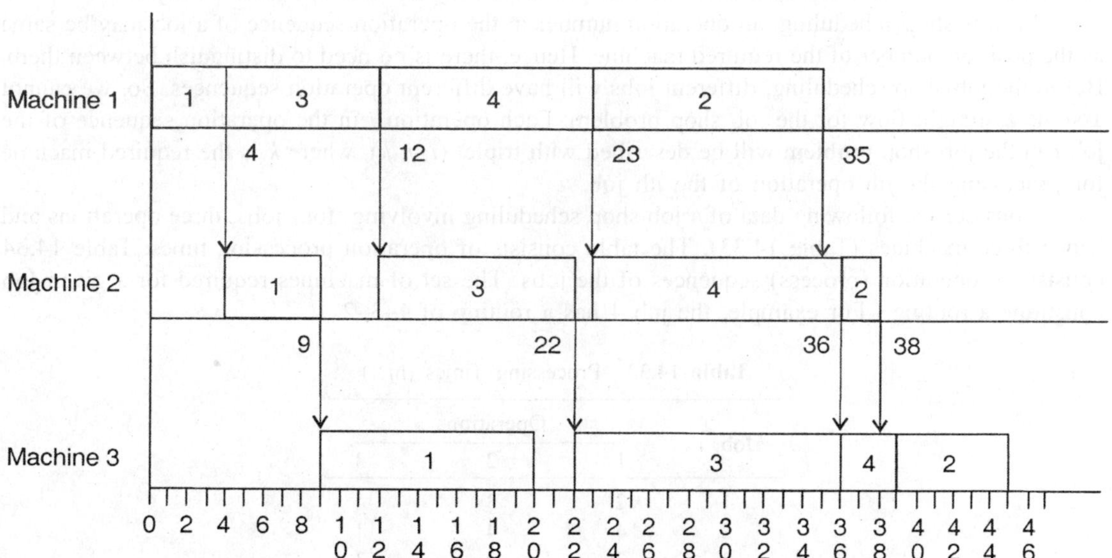

Figure 14.28 Gantt chart for Example 14.9.

Table 14.32 Branch-and-Bound Calculations for Example 14.9

Partial sequence, σ	(q_1, q_2, q_3)	(b_1, b_2, b_3)	B
1	(4, 9, 20)	(43, 37, 43)	43*
2	(12, 14, 20)	(51, 45, 48)	51
3	(8, 18, 32)	(43, 41, 52)	52
4	(11, 24, 27)	(43, 47, 58)	58
12	(16, 18, 26)	(51, 44, 43)	51
13	(12, 22, 36)	(43, 40, 45)	45*
14	(15, 28, 31)	(43, 46, 51)	51
132	(24, 26, 42)	(51, 42, 45)	51
134	(23, 36, 39)	(43, 44, 45)	45*

14.4 JOB SHOP SCHEDULING

In job shop scheduling problem, we assume that each job has m different operations. If some of the jobs are having less than m operations, required number of dummy operations with zero process times are assumed. By this assumption, the condition of equal number of operations for all the jobs is ensured. In job shop scheduling problem, the process sequences of the jobs are not the same. Hence, the flow of each job in job shop scheduling is not unidirectional.

The time complexity function of the job shop problem is combinatorial in nature. Hence, heuristic approaches are popular in this area.

Unlike the flow shop model, there is no initial machine that performs only the first operation of a job, nor is there a terminal machine that performs only the last operation of a job.

In flow shop scheduling, an operation number in the operation sequence of a job may be same as the position number of the required machine. Hence, there is no need to distinguish between them. But in the job shop scheduling, different jobs will have different operation sequences. So, we cannot assume a straight flow for the job shop problem. Each operation j in the operation sequence of the job i in the job shop problem will be described with triplet (i, j, k), where k is the required machine for processing the jth operation of the ith job.

Consider the following data of a job shop scheduling involving four jobs, three operations and hence three machines (Table 14.33). The table consists of operation processing times. Table 14.34 consists of operation (process) sequences of the jobs. The set of machines required for a given job constitute a routing. For example, the job 4 has a routing of 1–3–2.

Table 14.33 Processing Times (hrs.)

Job, i	Operations		
	1	2	3
1	2	3	4
2	4	4	1
3	2	2	3
4	3	3	1

Table 14.34 Operation Sequences

Job, i	Operations		
	1	2	3
1	1	2	3
2	3	2	1
3	2	3	1
4	1	3	2

This type of problem comes under combinatorial category. Hence, it will take longer time to solve large size problems of this type. This topic is beyond the scope of this book. But a simple version of this problem (two jobs and m machines problem) is presented in the next section.

14.4.1 Two Jobs and m Machines Job Shop Scheduling

Two jobs and m machines scheduling problem is a special problem under job shop scheduling. The problem consists of 2 jobs which require processing on m machines. The processing sequences of the jobs are not the same. Since this is a special kind under the job shop scheduling, a graphical procedure can be used to get optimum schedule.

The graphical procedure consists of the following steps:

Step 1: Construct a two dimensional graph in which x-axis represents the job 1, its sequence of operations and their processing times, and y-axis represents the job 2, its sequence of operations and their processing times (use same scale for both x-axis and y-axis).

Step 2: Shade each region where a machine would be occupied by the two jobs simultaneously.

Step 3: The processing of both jobs can be shown by a continuous line consisting of horizontal, vertical and 45 degree diagonal lines. The line is drawn from the origin and continued to the upper-right corner by avoiding the shaded regions. A diagonal line means that both jobs can be performed simultaneously. In the continuous line, a vertical line means that the job 1 keeps idle and a horizontal line means that the job 2 keeps idle. So, while drawing the line from the origin to the top-right corner, care should be taken to maximize the length of diagonal travel (sum of the lengths of 45 degree lines) which will minimize the makespan of the problem.

Using trial-and-error method, one can draw the final line which has the maximum diagonal portion. This concept is demonstrated using a numerical problem.

Example 14.10 Consider a 'two jobs and m machines job shop scheduling' problem which is presented in Table 14.35.

Use graphical method to minimize the time needed to process the following jobs on the machines shown (i.e. for each machine find the job which should be scheduled first). Also, calculate the total time elapsed to complete both jobs.

Table 14.35   Example 14.10

	Job 1		Job 2	
Sequence	**Time (hrs)**		**Sequence**	**Time (hrs)**
A	5		B	7
B	6		C	6
C	4		A	5
D	8		D	4
E	4		E	8

Solution   As per the procedure stated, the above data is presented in the form of a graph as shown in Figure 14.29. The line from the origin to the top-right corner shows the processing details and makespan. The start and completion times for both jobs are given in Table 14.36.

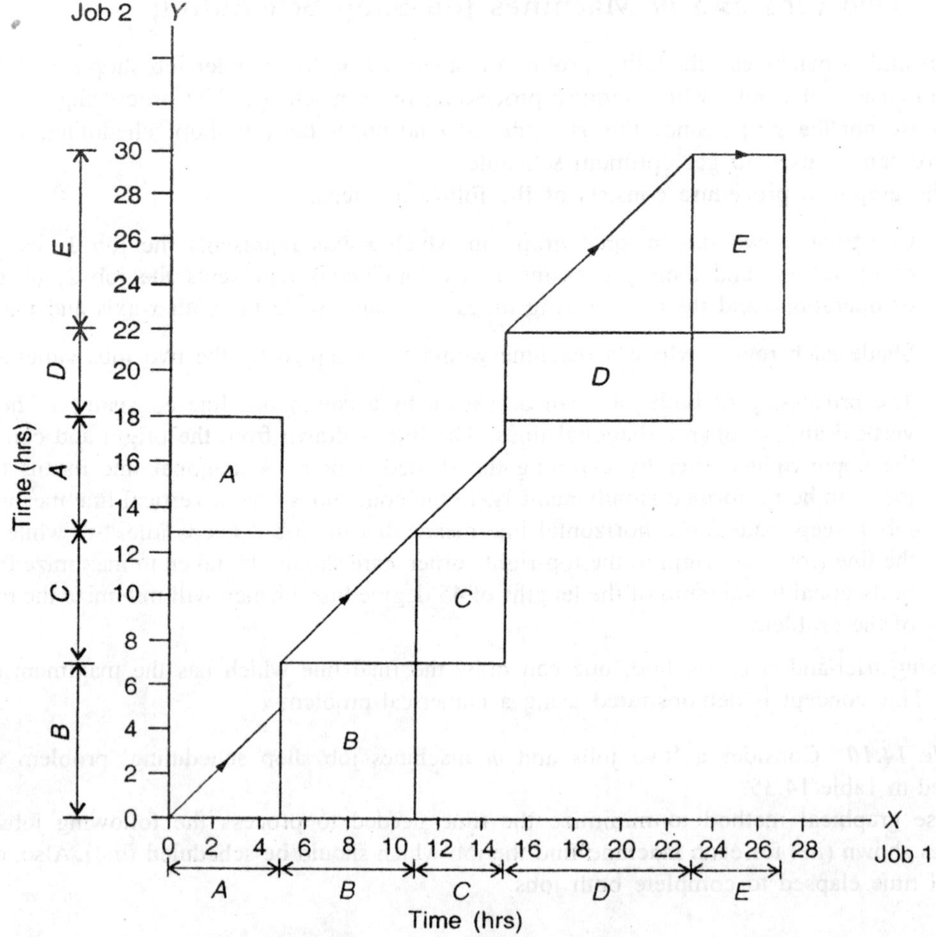

Figure 14.29   Gantt chart for Example 14.10.

Based on Figure 14.29 and Table 14.36, one can easily observe that the total idle time for job 1 is 7 hours (2 + 5). Hence, the total time for completing the job 1 is its sum of the processing times plus its idle time, i.e. 27 + 7 hrs = 34 hrs.

For job 2, the idle time is 4 hrs. Hence, the total time taken for completing the job 2 is its sum of the processing times plus its idle time, i.e. 30 + 4 hrs = 34 hrs.

The makespan is the maximum of these two quantities. That is,

$$\max (34, 34) = 34 \text{ hours}$$

Therefore, the makespan (time taken to complete both the jobs) of the given problem is 34 hours.

Table 14.36   Start and Completion Times of Jobs

Machine	Time for Job 1		Machine	Time for Job 2	
	Start	End		Start	End
A	0	5	B	0	7
B	7	13	C	7	13
C	13	17	A	13	18
D	22	30	D	18	22
E	30	34	E	26	34

QUESTIONS

1. What are the assumptions in single-machine scheduling problem?

2. Briefly discuss different measures of performance in single-machine scheduling with independent jobs.

3. Develop a mathematical model for single-machine scheduling problem to minimize mean tardiness.

4. Distinguish between single-machine scheduling and flow shop scheduling.

5. What are the assumptions in flow shop scheduling?

6. What are the assumptions in job shop scheduling problem?

7. Distinguish between flow shop problem and job shop problem.

8. Explain the procedure to solve a two jobs and m machines job shop scheduling problem.

9. Consider the following problem in single-machine scheduling with independent jobs:

Job, j	t_j	d_j	w_j
1	5	10	2
2	12	16	1
3	8	11	1
4	10	16	2
5	3	6	3
6	15	25	4
7	8	12	2
8	6	14	3

Obtain the optimal schedule for each of the following performance measures:

(a) To minimize mean flow time

(b) To minimize the maximum lateness

(c) To minimize weighted mean flow time

10. Consider the following single-machine scheduling problem.

Job, j	t_j	d_j
1	15	20
2	8	15
3	17	30
4	9	17
5	12	25

Find the optimal sequence of jobs to minimize mean tardiness using branch-and-bound technique. Also, find the optimal sequence which will minimize the maximum lateness.

11. Consider the following two machines and six jobs flow shop problem.

Job	Machine 1	Machine 2
1	5	7
2	10	8
3	8	13
4	9	7
5	6	11
6	12	10

Obtain the optimal schedule and the corresponding makespan for the above problem.

12. Consider the following 3 machines and 5 jobs flow shop problem. Check whether Johnson's rule can be extended to this problem. If so, what is the optimal schedule and the corresponding makespan?

Job	Machine 1	Machine 2	Machine 3
1	11	10	12
2	13	8	20
3	15	6	15
4	12	7	19
5	20	9	7

13. Consider the following flow shop problem.

Job	Machine 1	Machine 2	Machine 3
1	10	15	23
2	8	10	7
3	12	7	10
4	15	20	6

Find the optimal schedule for the above flow shop problem using branch-and-bound technique. Also, determine the corresponding makespan.

14. Use graphical method to minimize the time needed to process the following jobs on the machines shown (i.e. for each machine find the job which should be scheduled first). Also, calculate the total time elapsed to complete both jobs.

	Job 1		Job 2	
Sequence	Time (hrs)		Sequence	Time (hrs)
A	2		C	6
B	6		B	5
C	5		D	7
D	4		A	4
E	7		E	8

Goal Programming 15

15.1 INTRODUCTION

In many important 'real-world' decision-making situations, it may not be feasible, or desirable to reduce all the goals of an organization into a single objective. For example, instead of focussing only on optimizing profits, the organization may simultaneously be interested in maintaining a stable work force, increasing its share of market and limiting price increases.

Goal programming is an extension of linear or nonlinear programming involving an objective function with multiple objectives. While developing a goal programming model, the decision variables of the model are to be defined first. Then the managerial goals related to the problems are to be listed down and ranked in order of priority. Since it may be very difficult to rank these goals on a cardinal scale, an ordinal ranking is usually applied to each of the goals. It may not always be possible to fully achieve every goal specified by the decision-maker. Thus, goal programming is often referred to as a lexicographic procedure in which the various goals are satisfied in order of their relative importance.

Example 15.1 An office equipment manufacturer produces two kinds of products: computer covers and floppy boxes. Production of either a computer cover or a floppy box requires 1 hour of production capacity in the plant. The plant has a maximum production capacity of 10 hours per day. Because of the limited sales capacity, the maximum number of computer covers and floppy boxes that can be sold are 6 and 8 per day, respectively. The gross margin from the sale of a computer cover is Rs. 80, and Rs. 40 for a floppy box. The overtime hour should not exceed 2 hours/day. The plant manager has set the following goals arranged in order of importance.

1. To avoid any underutilization of production capacity.
2. To limit the overtime hours to 2 hours.
3. To sell as many computer covers and floppy boxes as possible. Since the gross margin from the sale of a computer cover is set at twice the amount of profit from a floppy box, he has twice as much desire to achieve the sales goal for computer covers as for the floppy boxes.
4. To minimize the overtime operation of the plant as much as possible.

Develop a goal programming model for this problem.

Solution The company manufactures two types of products, viz., computer covers and floppy boxes. Following are given:

$$\text{Production time per computer cover} = 1 \text{ hour}$$
$$\text{Production time per floppy box} = 1 \text{ hour}$$
$$\text{The maximum production time available per day} = 10 \text{ hours}$$
$$\text{Maximum demand for computer cover per day} = 6$$
$$\text{Maximum demand for floppy box per day} = 8$$
$$\text{Gross margin from sale of a computer cover} = \text{Rs. } 80$$
$$\text{Gross margin from sale of a floppy box} = \text{Rs. } 40.$$

Let X_1 and X_2 be the number of units of computer cover and floppy box, respectively, to be produced per day.

Production capacity constraint

The maximum production capacity per day is 10 hours. Number of hours of production capacity required for each computer cover is 1 hour. Number of hours of production capacity required per floppy box = 1 hour. Overtime is allowed, if necessary. The underutilization capacity is to be minimized. These are incorporated in the following constraint:

$$X_1 + X_2 + d_1^- - d_1^+ = 10$$

where

d_1^- = Underutilization of production capacity as against 10 hours of operation per day

d_1^+ = Overutilization of normal production capacity beyond 10 hours per day.

Sales constraint

The maximum sales for computer cover is 6 units per day, whereas, the maximum sales for floppy box is 8 units per day. Hence, overachievement of sales beyond the maximum sales is impossible. The corresponding sales constraints are:

$$X_1 + d_2^- = 6 \quad \text{and} \quad X_2 + d_3^- = 8$$

where

d_2^- = Underachievement of sales goal of computer covers per day

d_3^- = Underachievement of sales goal of floppy boxes per day.

Overtime constraint

The overtime hour is restricted to 2 hours per day. From the production capacity constraint, it is clear that d_1^+ is the overtime variable. Therefore,

$$d_1^+ + d_4^- - d_4^+ = 2$$

where,

d_4^- = Underachievement of overtime hours per day

d_4^+ = Overachievement of overtime hours per day.

One unit of computer cover gives a gross margin of Rs. 80 and one unit of floppy box gives a gross margin of Rs. 40. Since the production rate is same for both computer cover and floppy box (one per hour), the hourly margins of computer cover and floppy box are in the ratio of 80 : 40 or 2 : 1. Hence, it is appropriate to assign these as differential weights in Goal 3. The differential weights imply that management is relatively more concerned with the achievement of the sales goal of computer covers than that of floppy boxes.

Goal programming model

Based on the above guidelines, a goal programming model for the given problem is presented below:

$$\text{Minimize } Z = P_1 d_1^- + P_2 d_4^+ + (2P_3 d_2^- + P_3 d_3^-) + P_4 d_1^+$$

subject to

$$X_1 + X_2 + d_1^- - d_1^+ = 10$$
$$X_1 + d_2^- = 6$$
$$X_2 + d_3^- = 8$$
$$d_1^+ + d_4^- - d_4^+ = 2$$

where,

$$X_1,\ X_2,\ d_1^-,\ d_1^+,\ d_2^-,\ d_3^-,\ d_4^-,\ d_4^+ \geq 0$$

Here, P_1, P_2, P_3 and P_4 are the priority levels starting from high to low; $P_1 d_1^-$ is the objective function term for Goal 1; $P_2 d_4^+$ is the objective function term for Goal 2. $(2P_3 d_2^- + P_3 d_3^-)$ is the objective function term for Goal 3; and $P_4 d_1^+$ is the objective function term for Goal 4.

Example 15.2 A production manager is faced with the problem of job allocation to his two production teams. The production rate of Team 1 is 8 units per hour, while the production rate of Team 2 is 5 units per hour. The normal working hours for each of the teams is 40 hours per week. The production manager has prioritized the following goals for the coming week:

$P_1 =$ Avoid underachievement of the desired production level of 550 units

$P_2 =$ Overtime operation of Team 1 is limited to 5 hours

$P_3 =$ The total overtime for both teams should be minimized

$P_4 =$ Any underutilization of regular working hours of the teams should be avoided; assign differential weights according to the relative productivity of the two teams.

Formulate this problem as a goal programming model.

Solution Let X_1 be the number of hours of working of Team 1 per week, and X_2 be the number of hours of working of Team 2 per week. We also have

$$\text{Production rate of Team 1 per hour} = 8 \text{ units}$$
$$\text{Production rate of Team 2 per hour} = 5 \text{ units}$$
$$\text{Normal working hours per week of each team} = 40 \text{ hours}$$

Production volume constraint

Let

$$8X_1 + 5X_2 + d_1^- - d_1^+ = 550$$

where, d_1^- is the underachievement of production target and d_1^+ is the overachievement of production target.

Constraints for hours of operation of the teams

The normal hours of operation of each of the teams is 40 hours per week. Based on this information, the overtime constraint for the two teams are as presented below:

$$X_1 + d_2^- - d_2^+ = 40$$

where, d_2^- and d_2^+ are the underachievement and overachievement variables of the weekly normal working hours by Team 1. Also, we have the constraint

$$X_2 + d_3^- - d_3^+ = 40$$

where d_3^- and d_3^+ are the underachievement and overachievement variables of the weekly normal working hours by Team 2.

Overtime constraint for the Team 1

The overtime per week of Team 1 is restricted to 5 hours. The corresponding constraint is:

$$d_2^+ + d_4^- - d_4^+ = 5$$

Here, d_4^- and d_4^+ are the underachievement and overachievement variables of the weekly overtime by Team 1.

Minimizing underutilization of regular working hours. The underutilization of regular working hours of each team should be minimized in the ratio of the relative productivity of Teams 1 and 2 (i.e. in the ratio of 8 : 5).

Goal programming model

Based on these guidelines, a goal programming model for the given problem is presented as:

$$\text{Minimize } Z = P_1 d_1^- + P_2 d_4^+ + (P_3 d_2^+ + P_3 d_3^+) + (8P_4 d_2^- + 5P_4 d_3^-)$$

subject to

$$8X_1 + 5X_2 + d_1^- - d_1^+ = 550$$
$$X_1 + d_2^- - d_2^+ = 40$$
$$X_2 + d_3^- - d_3^+ = 40$$
$$d_2^+ + d_4^- - d_4^+ = 5$$

where,

$$X_1, X_2, d_1^-, d_1^+, d_2^-, d_2^+, d_3^-, d_3^+, d_4^-, d_4^+ \geq 0$$

Here, $P_1 d_1^-$, $P_2 d_4^+$, $(P_3 d_2^+ + P_3 d_3^+)$, and $(8P_4 d_2^- + 5P_4 d_3^-)$ is the objective function term for Goal 1, Goal 2, Goal 3 and Goal 4, respectively.

15.2 SIMPLEX METHOD FOR SOLVING GOAL PROGRAMMING

In this section, a procedure to solve goal programming problem is presented. The simplex method is suitably modified for this purpose. The modifications in the simplex method are presented as follows:

1. The purpose is to minimize the unattained portion of goals as much as possible. This is achieved by minimizing the deviational variables through the use of certain pre-emptive priority factors and deferential weights. These are represented by C_j values.

2. The pre-emptive priority factors are ordinal weights and they are not commensurable. Hence, Z_j or $C_j - Z_j$ can not be expressed by a single row. The simplex criterion becomes a matrix of $(m \times n)$ size, where m is the number of pre-emptive factors and n is the number of variables (number of decision variable + number of deviational variables).

3. In the criterion matrix, the pre-emptive factors are arranged from low priority to high priority, i.e. from top to bottom. The selection of key column must be based on the least value in the case of minimization problem/largest value in the case of maximization problem, starting from the criterion row with respect to the high priority to the low priority.

Example 15.3 Solve Example 15.1 using the simplex method.

Solution The goal programming model for Example 15.1 is reproduced as:

$$\text{Minimize } Z = P_1 d_1^- + P_2 d_4^+ + (2P_3 d_2^- + P_3 d_3^-) + P_4 d_1^+$$

subject to

$$X_1 + X_2 + d_1^- - d_1^+ = 10$$
$$X_1 + d_2^- = 6$$
$$X_2 + d_3^- = 8$$
$$d_1^+ + d_4^- - d_4^+ = 2$$

where

$$X_1, X_2, d_1^-, d_1^+, d_2^-, d_3^-, d_4^-, d_4^+ \geq 0$$

In the above model, X_1 and X_2 are decision variables and the remaining variables are deviational variables. The initial table for the problem is presented in Table 15.1.

Table 15.1 Initial Table

CB_i	Basic variable	C_j 0 X_1	0 X_2	P_1 d_1^-	$2P_3$ d_2^-	P_3 d_3^-	0 d_4^-	P_4 d_1^+	P_2 d_4^+	Solution	Ratio
P_1	d_1^-	1	1	1	0	0	0	−1	0	10	10
$2P_3$	d_2^-	1	0	0	1	0	0	0	0	6	6*
P_3	d_3^-	0	1	0	0	1	0	0	0	8	−
0	d_4^-	0	0	0	0	0	1	1	−1	2	−
$C_j - Z_j$	P_4	0	0	0	0	0	0	1	0	0	
	P_3	−2	−1	0	0	0	0	0	0	20	
	P_2	0	0	0	0	0	0	0	1	0	
	P_1	−1**	−1	0	0	0	0	1	0	10	

*Key row. **Key column.

The computation of the values for the criterion matrix is:

$$C_1 - Z_1 = 0 - (P_1 + 2P_3) = -P_1 - 2P_3$$
$$C_2 - Z_2 = 0 - (P_1 + P_3) = -P_1 - P_3$$
$$C_3 - Z_3 = P_1 - P_1 = 0$$
$$C_4 - Z_4 = 2P_3 - 2P_3 = 0$$
$$C_5 - Z_5 = P_3 - P_3 = 0$$

$$C_6 - Z_6 = 0 - 0 = 0$$
$$C_7 - Z_7 = P_4 - (-P_1) = P_4 + P_1$$
$$C_8 - Z_8 = P_2 - 0 = P_2$$

The expression for the solution column is: $10P_1 + 20P_3$.

Next, enter the coefficients of $C_j - Z_j$ and solution column in the respective cells of $C_j - Z_j$ matrix as shown in Table 15.1.

Selection of key column

First, examine the values in the $C_j - Z_j$ matrix under the solution column. Each positive entry represents the unattained portion of that goal. For P_1, the unattained level is 10 and for P_3, the unattained level is 20.

Starting from the highest unattained row (from bottom), find the most negative $C_j - Z_j$ (for minimization problem). For P_1 in $C_j - Z_j$ matrix, the most negative value is -1 under column X_1 and column X_2. To break the tie, it is necessary to examine $C_j - Z_j$ values in the next highest unattained priority row, i.e. in row P_3 of $C_j - Z_j$ matrix. The most negative value is -2 which occurs in column X_1. Therefore, column X_1 is selected as the key column.

To determine key row, divide the values under solution column by the respective values in the key column (column X_1) and then select the row with minimum positive ratio. The second row (Row d_2^-) is selected as the key row.

Obtain the next table by following the usual procedure of simplex method as shown in Table 15.2.

Table 15.2 Iteration 1

CB_i	Basic variable	C_j → 0 X_1	0 X_2	P_1 d_1^-	$2P_3$ d_2^-	P_3 d_3^-	0 d_4^-	P_4 d_1^+	P_2 d_4^+	Solution	Ratio
P_1	d_1^-	0	1	1	-1	0	0	-1	0	4	4*
0	X_1	1	0	0	1	0	0	0	0	6	–
P_3	d_3^-	0	1	0	0	1	0	0	0	8	8
0	d_4^-	0	0	0	0	0	1	1	-1	2	–
$C_j - Z_j$	P_4	0	0	0	0	0	0	1	0	0	
	P_3	0	-1	0	2	0	0	0	0	8	
	P_2	0	0	0	0	0	0	0	1	0	
	P_1	0	-1*	0	1	0	0	1	0	4	

The expressions for different $C_j - Z_j$ are derived as follows:

$$C_1 - Z_1 = 0$$
$$C_2 - Z_2 = 0 - (P_1 + P_3) = -P_1 - P_3$$
$$C_3 - Z_3 = P_1 - (P_1) = 0$$
$$C_4 - Z_4 = 2P_3 - (-P_1) = 2P_3 + P_1$$
$$C_5 - Z_5 = P_3 - P_3 = 0$$
$$C_6 - Z_6 = 0 - 0 = 0$$
$$C_7 - Z_7 = P_4 - (-P_1) = P_4 + P_1$$
$$C_8 - Z_8 = P_2$$

The expression for the solution column is: $4P_1 + 8P_3$.

In Table 15.2, one can see that the priority rows P_1 and P_3 are still with unattained goals of 4 and 8, respectively. Hence, check for the most negative value in row P_1 of the criterion matrix. The column X_2 has the only negative value of -1 in row P_1. Hence, it is selected as the key column. Thus, the ratios are obtained as shown in Table 15.2 for the selection of key row. The minimum ratio is 4. Hence, select the first row (row d_1^-) as the key row. Then, the next iteration is presented in Table 15.3.

Table 15.3 Iteration 2

CB_i	Basic variable	C_j								Solution	Ratio
		0	0	P_1	$2P_3$	P_3	0	P_4	P_2		
		X_1	X_2	d_1^-	d_2^-	d_3^-	d_4^-	d_1^+	d_4^+		
0	X_2	0	1	1	-1	0	0	-1	0	4	–
0	X_1	1	0	0	1	0	0	0	0	6	–
P_3	d_3^-	0	0	-1	1	1	0	1	0	4	4
0	d_4^-	0	0	0	0	0	1	$\boxed{1}$	-1	2	2*
$C_j - Z_j$	P_4	0	0	0	0	0	0	1	0	0	
	P_3	0	0	1	1	0	0	-1	0	4	
	P_2	0	0	0	0	0	0	0	1	0	
	P_1	0	0	1	0	0	0	0*	0	0	

The expressions for different $C_j - Z_j$ are derived as shown below:

$$C_1 - Z_1 = 0 - 0 = 0$$
$$C_2 - Z_2 = 0$$
$$C_3 - Z_3 = P_1 - (-P_3) = P_1 + P_3$$
$$C_4 - Z_4 = 2P_3 - (P_3) = P_3$$
$$C_5 - Z_5 = P_3 - P_3 = 0$$
$$C_6 - Z_6 = 0 - 0 = 0$$
$$C_7 - Z_7 = P_4 - P_3$$
$$C_8 - Z_8 = P_2 - 0 = P_2$$

The expression for the solution column is $4P_3$.

Since row P_3 in the criterion matrix has unattained goal, check its $C_j - Z_j$ coefficients. It has -1 in Column d_1^+. Hence, select it as the key column. Accordingly, the selection of key row is shown in the same table (i.e. in fourth row). The next iteration is shown in Table 15.4.

Now, $C_j - Z_j$ values are computed as under:

$$C_1 - Z_1 = 0$$
$$C_2 - Z_2 = 0$$
$$C_3 - Z_3 = P_1 - (-P_3) = P_1 + P_3$$
$$C_4 - Z_4 = 2P_3 - P_3 = P_3$$
$$C_5 - Z_5 = P_3 - P_3 = 0$$
$$C_6 - Z_6 = 0 - (-P_3 + P_4) = P_3 - P_4$$
$$C_7 - Z_7 = P_4 - P_4 = 0$$
$$C_8 - Z_8 = P_2 - (P_3 - P_4) = P_2 - P_3 + P_4$$

The expression for the solution column is: $2P_3 + 2P_4$.

Table 15.4 Iteration 3

CB_i	C_j / Basic variable	0 X_1	0 X_2	P_1 d_1^-	$2P_3$ d_2^-	P_3 d_3^-	0 d_4^-	P_4 d_1^+	P_2 d_4^+	Solution
0	X_2	0	1	1	-1	0	1	0	-1	6
0	X_1	1	0	0	1	0	0	0	0	6
P_3	d_3^-	0	0	-1	1	1	-1	0	1	2
P_4	d_1^+	0	0	0	0	0	1	1	-1	2
$C_j - Z_j$	P_4	0	0	0	0	0	-1	0	1	2
	P_3	0	0	1	1	0	1	0	-1	2
	P_2	0	0	0	0	0	0	0	1	0
	P_1	0	0	1	0	0	0	0	0	0

In criterion matrix, P_3 row has -1 for d_4^+ column. But at a higher level (P_2 level) in the same column, 1 exists. If we want to try to achieve Goal 3 (P_3 row), we will have to sacrifice Goal 2 (P_2 row). Hence, there is no need to introduce the variable d_4^+ into the basis. Similarly, the P_4 row has -1 under d_4^- column. But, at P_3 row in the same column, 1 exists. If we try to achieve Goal 4 (P_4 row), we will have to sacrifice Goal 3 (P_3 row). Hence, it is not necessary to introduce the variable d_4^- into the basis.

Since all other values in the criterion matrix are 0 or more than zero, and the problem is to attain minimization, the optimality is reached. Hence, the optimal results are:

Number of units of computer cover to be produced per day, $X_1 = 6$
Number of units of floppy box to be produced per day, $X_2 = 6$
Underutilization of sales demand of floppy box per day, $d_3^- = 2$ units
Overutilization of production capacity per day, $d_1^+ = 2$ hours.

Example 15.4   A company produces two kinds of products, A and B. Production of either A or B requires 3 hours per unit of production capacity in the plant. The plant has a maximum production capacity of 30 hours per week to manufacture these two products. The overtime hour should not exceed 5 hours/week. The plant manager has set the following goals arranged in the order of importance.

(a) To avoid any underutilization of production capacity
(b) To limit the overtime hours to 5 hours
(c) To minimize the overtime operation of the plant as much as possible.

Formulate this as a goal programming problem and then solve it by simplex method.

Solution   Let the production volume of the product A be X_1 and that of the product B be X_2.

Number of hours required/unit to produce product A = 3 hours
Number of hours required/unit to produce product B = 3 hours
Maximum normal production capacity/week = 30 hours
Upper limit for overtime hours/week = 5 hours.

Production volume constraint: The production volume constraint as per the given data is

$$3X_1 + 3X_2 + d_1^- - d_1^+ = 30$$

where, d_1^- is the underachievement of the production target and d_1^+ is the overachievement of the production target.

Overtime constraint: The overtime constraint is

$$d_1^+ + d_2^- - d_2^+ = 5$$

where, d_2^- is the underachievement of the overtime target and d_2^+ is the overachievement of the overtime target.

Goal programming model

The goal programming model for the given problem is as presented below:

$$\text{Minimize } Z = P_1 d_1^- + P_2 d_2^+ + P_3 d_1^+$$

subject to

$$3X_1 + 3X_2 + d_1^- - d_1^+ = 30$$
$$d_1^+ + d_2^- - d_2^+ = 5$$
$$X_1, X_2, d_1^-, d_2^-, d_1^+, d_2^+ \geq 0$$

The initial table of this goal programming model is shown in Table 15.5. In Table 15.5, the values of the criterion matrix are computed using $C_j - Z_j$ and the coefficients are entered as shown in the same table. In Table 15.5, the bottom most row w.r.t. P_1 is the only unattained row and the level of unattainment is 30. In this row, the highest negative coefficient is -3 which occurs for column X_1 as well as for column X_2. Since, there is no other row above this row with unattainment, the tie is broken randomly and the column X_1 is selected as the key column. The corresponding key row is d_1^-. The next iteration after performing pivot operation is shown in Table 15.6. In Table 15.6, there is no unattained row. Hence, the optimality is reached and the corresponding solution is:

$X_1 = 10$, $d_2^- = 5$ and the value for each of the other variables $(X_2, d_1^-, d_1^+, d_2^+)$ is zero.

Inference: The production volume of the product A is 10 and that of the product B is 0. Since, d_2^- is 5, the under achievement of the overtime is 5 hours. This means that the overtime is not at all used.

Table 15.5   Initial Table

CB_i	C_j Basic variable	0 X_1	0 X_2	P_1 d_1^-	0 d_2^-	P_3 d_1^+	P_2 d_2^+	Solution	Ratio
P_1	d_1^-	[3]	3	1	0	-1	0	30	10*
0	d_2^-	0	0	0	1	1	-1	5	–
	P_3	0	0	0	0	1	0	0	
$C_j - Z_j$	P_2	0	0	0	0	0	1	0	
	P_1	-3**	-3	0	0	1	0	30	

*Key row;   **Key column

Table 15.6 Iteration 1

	C_j	0	0	P_1	0	P_3	P_2	
CB_i	Basic variable	X_1	X_2	d_1^-	d_2^-	d_1^+	d_2^+	Solution
0	X_1	1	1	1/3	0	−1/3	0	10
0	d_2^-	0	0	0	1	1	−1	5
	P_3	0	0	0	0	1	0	0
$C_j - Z_j$	P_2	0	0	0	0	0	1	0
	P_1	0	0	1	0	0	0	0

QUESTIONS

1. What is goal programming? Distinguish it from linear programming.

2. List and explain different applications of goal programming.

3. Discuss the procedure for selecting a key column in goal programming.

4. A production manager is faced with the problem of job allocation between his two production teams. The production rate of Team X is 6 units per hour, while the production rate of Team Y is 10 units per hour. The normal working hours for each of the teams are 50 hours per week. The production manager has prioritized the following goals for the coming week:

   P_1: Avoid underachievement of the desired production level of 825 units.

   P_2: Any overtime operation of Team X beyond 10 hours should be avoided.

   P_3: Any overtime operation of Team Y beyond 15 hours should be avoided.

   P_4: Any underutilization of regular working hours should be avoided. Again, assign differential weights according to the relative productivity of the two teams.

   Formulate this problem as a goal programming model and solve it by simplex method.

5. A company produces two kinds of products, X and Y. Production of either X or Y requires 2 hours of production capacity in the plant. The plant has a maximum production capacity of 20 hours per week. The overtime hour should not exceed 4 hours/week. The plant manager has set the following goals arranged in the order of importance.

   (a) To avoid any underutilization of production capacity

   (b) To limit the overtime hours to 4 hours

   (c) To minimize the overtime operation of the plant as much as possible.

   Formulate this as a goal programming problem and then solve it by simplex method.

6. A company manufactures two types of products, A and B. The products are produced in three distinct steps: turning, milling and grinding. The schedule for machine hours and availability of each type are as follows:

	A (hour)	B (hour)	Availability (hour/month)
Turning	3	2	1700
Milling	2	1	1400
Grinding	1	3	1600

Profit contribution of each type is Rs. 600 and Rs. 800, respectively. The company has three equally desirable goals: minimizing the idle time in turning as well as grinding and making a monthly profit of Rs. 1,00,000. Set up the problem as a goal programming model and find the optimal solution.

7. A company manufactures three types of lamps, A, B and C. The lamps are produced in two distinct steps: turning and grinding. The schedule for machine hours and availability of each type are as follows:

	A (hour)	B (hour)	C (hour)	Availability (hour/month)
Turning	1	3	4	1800
Grinding	2	2	3	1500

Profit contribution of each type is Rs. 450, Rs. 550 and Rs. 675, respectively. The company has two equally desirable goals: minimizing the idle time in grinding and making a monthly profit of Rs. 1,20,000. Set up the problem as a goal programming model and find the optimal solution.

PARAMETRIC LINEAR PROGRAMMING

16

16.1 INTRODUCTION

In linear programming problems, generally the following quantities are assumed as constants.

(a) Objective function coefficients
(b) Right-hand side constants of constraints
(c) Technological coefficients (resource requirement vectors).

But in reality, each of these components of the linear programming problem is parameterized.

Let us consider the following linear programming formulation of a product-mix problem, where X_j is the production volume of the product j.

$$\text{Maximize } Z = 10X_1 + 5X_2$$

subject to

$$8X_1 + 2X_2 \leq 48$$
$$2X_1 + 4X_2 \leq 24$$
$$X_1 \text{ and } X_2 \geq 0$$

Here, the objective function may be modified as:

$$\text{Maximize } Z = (10 - 2t)X_1 + (5 - 3t)X_2$$

where t is a variable representing certain parameter like time period, rate of inflation, and employment level which have direct impact on the selling price per unit (C_j) of each of the products.

The right hand side constants of the constraints may be modified as:

$$[b] = \begin{bmatrix} 40 - 2t \\ 24 + 4t \end{bmatrix}$$

where t is a variable representing a parameter, viz. the productivity index of supplier firms which affect the raw material availability, number of shifts of the company which affects the machine-hour availability, employees' productivity index which affects the man-hour availability, etc.

The resource requirement vector like, P_1 for the variable X_1 may be modified as:

$$P_1 = \begin{bmatrix} 8 + 2t \\ 2 - t \end{bmatrix}$$

where t is variable representing a parameter, e.g. labour efficiency index which affects the material requirement per unit, or labour-hour requirement per hour, or machine-hour requirement per hour.

In each of the above situations, the objective is to find the ranges of the parameter values such that the optimality within each range is unaffected. This type of analysis is called as *parametric analysis*. Generally, the parametric expression for each of the situations is assumed as linear to simplify the discussion.

16.2 CHANGES IN OBJECTIVE FUNCTION COEFFICIENTS (C_j VALUES)

In this section, the parametric analysis applied to the objective function coefficients is presented.

Consider the generalized version of a linear programming problem applied to the product-mix problem as presented below.

$$\text{Maximize } Z = C_1X_1 + C_2X_2 + \cdots + C_jX_j + \cdots + C_nX_n$$

subject to

$$a_{11}X_1 + a_{12}X_2 + \cdots + a_{1j}X_j + \cdots + a_{1n}X_n \leq b_1$$
$$a_{21}X_1 + a_{22}X_2 + \cdots + a_{2j}X_j + \cdots + a_{2n}X_n \leq b_2$$
$$\vdots \qquad \vdots \qquad \vdots \qquad \vdots \qquad \vdots$$
$$a_{i1}X_1 + a_{i2}X_2 + \cdots + a_{ij}X_j + \cdots + a_{in}X_n \leq b_i$$
$$\vdots \qquad \vdots \qquad \vdots \qquad \vdots \qquad \vdots$$
$$a_{m1}X_1 + a_{m2}X_2 + \cdots + a_{mj}X_j + \cdots + a_{mn}X_n \leq b_m$$
$$X_j \geq 0, \qquad j = 1, 2, 3,\dots, n$$

In this model, C_j corresponds to the selling price per unit of Product j which may be affected by some parameter like income level of the public. This means that the selling prices of the products will be affected because of its sensitiveness towards income level of the public. Such a parameter may be called as t and its impact on the objective function coefficients is as shown below.

$$\text{Maximize } Z = (C_1 + C_1't)X_1 + (C_2 + C_2't)X_2 + \cdot + (C_j + C_j't)X_j + \cdot + (C_n + C_n't)X_n$$

subject to

$$a_{11}X_1 + a_{12}X_2 + \cdots + a_{1j}X_j + \cdots + a_{1n}X_n \leq b_1$$
$$a_{21}X_1 + a_{22}X_2 + \cdots + a_{2j}X_j + \cdots + a_{2n}X_n \leq b_2$$
$$\vdots \qquad \vdots \qquad \vdots \qquad \vdots \qquad \vdots$$
$$a_{i1}X_1 + a_{i2}X_2 + \cdots + a_{ij}X_j + \cdots + a_{in}X_n \leq b_i$$
$$\vdots \qquad \vdots \qquad \vdots \qquad \vdots \qquad \vdots$$
$$a_{m1}X_1 + a_{m2}X_2 + \cdots + a_{mj}X_j + \cdots + a_{mn}X_n \leq b_m$$
$$X_j \geq 0, \qquad j = 1, 2, 3,\dots, n$$

where C_j' constants are for j which vary from 1 to n, and t is a parameter representing the income level of the public.

The analysis of the above parametric linear programming problem is demonstrated using Example 16.1.

Example 16.1 Consider the following parametric linear programming problem.

$$\text{Maximize } Z = (10 - 2t)X_1 + (5 - 3t)X_2$$

subject to

$$8X_1 + 2X_2 \leq 48$$
$$2X_1 + 4X_2 \leq 24$$
$$X_1, X_2 \geq 0 \text{ and } t \text{ is a non-negative parameter.}$$

Perform parametric analysis with respect to the objective function coefficients and identify the ranges of t over which the optimal solution in each range is unaffected.

  Solution The canonical form of the given problem is shown below in which S_1 and S_2 are slack variables.

$$\text{Maximize } Z = (10 - 2t)X_1 + (5 - 3t)X_2$$

subject to

$$8X_1 + 2X_2 + S_1 = 48$$
$$2X_1 + 4X_2 + S_2 = 24$$
$$X_1, X_2, S_1 \text{ and } S_2 \geq 0$$

The canonical form of the above problem when t is equal to 0 is:

$$\text{Maximize } Z = 10X_1 + 5X_2$$

subject to

$$8X_1 + 2X_2 + S_1 = 48$$
$$2X_1 + 4X_2 + S_2 = 24$$
$$X_1, X_2, S_1 \text{ and } S_2 \geq 0$$

The optimal table of the above linear programming problem is shown in Table 16.1.

Table 16.1 Optimal Table of Non-parametric Problem

CB_i	C_j	10	5	0	0	
	Basic variable	X_1	X_2	S_1	S_2	Solution
10	X_1	1	0	1/7	−1/14	36/7
5	X_2	0	1	−1/14	2/7	24/7
	Z_j	10	5	15/14	5/7	480/7
	$C_j - Z_j$	0	0	−15/14	−5/7	

Modified form of the Table 16.1 when t is greater than 0, is shown in Table 16.2.

Table 16.2 Modified Form of Table 16.1 when $t \geq 0$

CB_i'	CB_i	C_j'	$-2t$	$-3t$	0	0	
		C_j	10	5	0	0	
		Basic variable	X_1	X_2	S_1	S_2	Solution
$-2t$	10	X_1	1	0	1/7	−1/14	36/7
$-3t$	5	X_2	0	1	−1/14	2/7	24/7
		Z_j	10	5	15/14	5/7	480/7
		Z_j'	$-2t$	$-3t$	$-(1/14)t$	$-(5/7)t$	$-(144/7)t$
		$(C_j + C_j') - (Z_j + Z_j')$	0	0	$-\dfrac{15}{14}+\dfrac{1}{14}t$	$-\dfrac{5}{7}+\dfrac{5}{7}t$	

The problem has the maximization-type objective function. So, to maintain the optimality of the solution in Table 16.2, the criterion row values of the non-basic variables must be less than or equal to 0. Thus,

$$(C_j + C_j') - (Z_j + Z_j') \leq 0, \, j = 3, 4$$

Then, we get

$$-\frac{15}{14} + \frac{1}{14}t \leq 0 \qquad \text{or } t \leq 15$$

and

$$-\frac{5}{7} + \frac{5}{7}t \leq 0 \qquad \text{or } t \leq 1$$

with respect to non-basic variables S_1 and S_2, respectively.

Therefore, the first effective interval of the parameter t over which the optimality is not affected is given by $0 \leq t \leq 1$. The corresponding optimal solution is:

$$Z(\text{optimum}) = \frac{480}{7} - \frac{144}{7}t$$

with

$$X_1^* = \frac{36}{7}, \qquad X_2^* = \frac{24}{7}, \qquad S_1^* = S_2^* = 0$$

When t is greater than 1, the optimality of the solution in Table 16.2 will be affected because the value of the criterion row of the non-basic variable S_2 will be positive. So, the first critical value of t is 1 (i.e. $t_1 = 1$). Therefore, the variable S_2 enters the basis. The corresponding leaving variable and the pivot element are X_2 and 2/7, respectively. These are indicated in Table 16.3 which shows the necessary portion of the Table 16.2 for further iteration.

Table 16.3   Necessary Portion of Table 16.2 when $t > 1$

CB_i'	CB_i	Basic variable	X_1	X_2	S_1	S_2	Solution
$-2t$	10	X_1	1	0	1/7	$-1/14$	36/7
$-3t$	5	X_2	0	1	$-1/14$	$\boxed{2/7}$	24/7*

*Key row   **Key column   **

The results after performing pivot operations on Table 16.3 are shown in Table 16.4.

Table 16.4   Results after Pivot Operations on Table 16.3

| | | C_j' | $-2t$ | $-3t$ | 0 | 0 | |
| | | C_j | 10 | 5 | 0 | 0 | |
CB_i'	CB_i	Basic variable	X_1	X_2	S_1	S_2	Solution
$-2t$	10	X_1	1	1/4	1/8	0	6
0	0	S_2	0	7/2	$-1/4$	1	12
		Z_j	10	5/2	5/4	0	60
		Z_j'	$-2t$	$-(1/2)t$	$-(1/4)t$	0	$-12t$
	$(C_j + C_j') - (Z_j + Z_j')$		0	$\frac{5}{2} - \frac{5}{2}t$	$-\frac{5}{4} + \frac{1}{4}t$	0	

To maintain the optimality of the solution in Table 16.4, the criterion row values of the non-basic variables must be less than or equal to 0. Thus we have,

$$(C_j + C_j') - (Z_j + Z_j') \leq 0, \quad j = 2, 3$$

$$5/2 - (5/2)t \leq 0$$

or

$$t \geq 1 \text{ w.r.t. the non-basic variable } X_2.$$

$$-5/4 + (1/4)t \leq 0$$

i.e. $t \leq 5$ w.r.t. the non-basic variable S_1.

Therefore, the second effective interval of the parameter t over which the optimality is not affected is given below:

$$1 \leq t \leq 5$$

The corresponding optimal solution is presented below:

$$Z(\text{optimum}) = 60 - 12t$$

$$X_1^* = 6, \quad S_2^* = 12, \quad X_2^* = S_1^* = 0$$

When t is greater than 5, the optimality of the solution in Table 16.4 will be affected because the value of the criterion row value of the nonbasic variable S_1 will be positive. So, the second critical value of t is 5 (i.e. $t_2 = 5$).

Therefore, in Table 16.2, the variable S_1 enters the basis. Now, the necessary portion of Table 16.2 for further iteration is shown in Table 16.5.

Table 16.5 Necessary Portion of Table 16.2 When $t > 5$

CB_i'	CB_i	Basic variable	X_1	X_2	S_1	S_2	Solution
$-2t$	10	X_1	1	0	$\boxed{1/7}$	$-1/14$	36/7*
$-3t$	5	X_2	0	1	$-1/14$	2/7	24/7

*Key row **key column **

The corresponding leaving variable and the pivot element are X_1 and 1/7, respectively as indicated in Table 16.5.

The results after performing pivot operations on Table 16.5 are shown in Table 16.6.

Table 16.6 Results after Pivot Operations on Table 16.5

CB_i'	CB_i	Basic variable	X_1	X_2	S_1	S_2	Solution
		C_j'	$-2t$	$-3t$	0	0	
		C_j	10	5	0	0	
0	0	S_1	7	0	1	$-1/2$	36
$-3t$	5	X_2	1/2	1	0	1/4	6
		Z_j	5/2	5	0	5/4	30
		Z_j'	$-(3/2)t$	$-3t$	0	$-(3/4)t$	$-18t$
		$(C_j + C_j') - (Z_j + Z_j')$	$\dfrac{15}{2} - \dfrac{1}{2}t$	0	0	$-\dfrac{5}{4} + \dfrac{3}{4}t$	

To maintain the optimality of the solution in Table 16.6, the criterion row values of the non-basic variables must be less than or equal to 0. Therefore,

$$(C_j + C_j') - (Z_j + Z_j') \leq 0, \qquad j = 1, 4$$

Then, we get

$$\frac{15}{2} - \frac{1}{2}t \leq 0 \quad \text{or} \quad t \geq 15$$

and

$$-\frac{5}{4} + \frac{3}{4}t \leq 0 \quad \text{or} \quad t \leq \frac{5}{3}$$

with respect to non-basic variables X_1 and S_2, respectively.

Therefore, the third effective interval of the parameter t over which the optimality is not affected is given by

$$t \leq \frac{5}{3} \quad \text{and} \quad t \geq 15$$

The corresponding optimal solution is:

$$Z(\text{optimum}) = 30 - 18t$$

$$S_1^* = 36, \qquad X_2^* = 6, \qquad X_1^* = 0, \qquad S_2^* = 0$$

When t is greater than 5/3, the optimality of the solution in Table 16.6 will be affected because the value of the criterion row of the non-basic variable S_2 will be positive. So, the third critical value of t is 5/3(i.e. $t_3 = 5/3$). Therefore, in Table 16.2, the variable S_2 enters the basis and the corresponding leaving variable is X_2. This combination of the entering variable (S_2) and the leaving variable X_2 has been already tried out.

When $t \leq 15$, $(C_j + C_j') - (Z_j + Z_j')$ of the X_1 column will become positive till t becomes 0. Since X_1 is a basic variable, it cannot be treated as an entering variable. So, the procedure is terminated here and the final results are summarized in Table 16.7.

Table 16.7 Summary of Results

Interval of t	Critical value, t	Optimal solution				
		Z	X_1	X_2	S_1	S_2
$t \leq 1$	1	$(480/7) - (144/7)t$	36/7	24/7	0	0
$1 \leq t \leq 5$	5	$60 - 12t$	6	0	0	12
$t \leq 5/3$	5/3	$30 - 18t$	0	6	36	0
$t \geq 15$	15	$30 - 18t$	0	6	36	0

16.3 CHANGES IN RIGHT-HAND SIDE CONSTANTS (B_i VALUES) OF CONSTRAINTS

In this section, the parametric analysis applied to the right-hand side constants is presented.

Consider the generalized version of a linear programming problem applied to the product-mix problem as presented in Section 16.2.

In that model, if all the constraints correspond to different raw materials, then the right-hand side constants may be affected by some parameter such as, inflation rate. This means that the supply of the raw materials will be affected because of the sensitivity of raw material prices towards inflation rate. Such parameter may be called as t and its impact on the right hand side constants is as shown below:

$$\text{Maximize } Z = C_1X_1 + C_2X_2 + \cdots + C_jX_j + \cdots + C_nX_n$$

subject to

$$a_{11}X_1 + a_{12}X_2 + \cdots + a_{1j}X_j + \cdots + a_{1n}X_n \leq b_1 + b_1't$$

$$a_{21}X_1 + a_{22}X_2 + \cdots + a_{2j}X_j + \cdots + a_{2n}X_n \leq b_2 + b_2't$$

$$\vdots \qquad \vdots \qquad \qquad \vdots \qquad \qquad \vdots \qquad \vdots$$

$$a_{i1}X_1 + a_{i2}X_2 + \cdots + a_{ij}X_j + \cdots + a_{in}X_n \leq b_i + b_i't$$

$$\vdots \qquad \vdots \qquad \qquad \vdots \qquad \qquad \vdots \qquad \vdots$$

$$a_{m1}X_1 + a_{m2}X_2 + \cdots + a_{mj}X_j + \cdots + a_{mn}X_n \leq b_m + b_m't$$

$$X_j \geq 0, \qquad j = 1, 2, 3, ..., n$$

where, b_i' is a constant for i varies from 1 to m and t is a parameter representing the inflation rate. The analysis of the above parametric linear programming problem is demonstrated through Example 16.2.

Example 16.2 Consider the following parametric linear programming problem:

$$\text{Maximize } Z = 6X_1 + 8X_2$$

subject to

$$5X_1 + 10X_2 \leq 60 + 10t$$

$$4X_1 + 4X_2 \leq 40 - 10t$$

$X_1, X_2 \geq 0$ and t is a parameter with unrestricted sign.

Perform parametric analysis with respect to the right-hand side constants of the constraints and identify the ranges of t over which the optimal solution in each range is unaffected.

Solution The canonical form of the given problem is shown below in which S_1 and S_2 are slack variables.

$$\text{Maximize } Z = 6X_1 + 8X_2$$

subject to

$$5X_1 + 10X_2 + S_1 = 60 + 10t$$

$$4X_1 + 4X_2 + S_2 = 40 - 10t$$

$$X_1, X_2, S_1 \text{ and } S_2 \geq 0$$

The canonical form of the above problem, when t is equal to 0, is:

$$\text{Maximize } Z = 6X_1 + 8X_2$$

subject to

$$5X_1 + 10X_2 + S_1 = 60$$

$$4X_1 + 4X_2 + S_2 = 40$$

$$X_1, X_2, S_1 \text{ and } S_2 \geq 0$$

The optimal table of the above linear programming problem is shown in Table 16.8.

Table 16.8 Optimal Table of Non-parametric Problem

CB_i	C_j Basic variable	6 X_1	8 X_2	0 S_1	0 S_2	Solution
8	X_2	0	1	1/5	−1/4	2
6	X_1	1	0	−1/5	1/2	8
	Z_j	6	8	2/5	1	64
	$C_j - Z_j$	0	0	−2/5	−1	

Any change in the right-hand side constants will affect the feasibility of the solution, but not the optimality of the solution.

The revised values of the right-hand side constants can be obtained for the parametric problem using the following formula.

$$[b_p] = [B^{-1}][b]$$

where

$[b_p]$ = Right-hand side constants matrix of the modified table with respect to the parametric problem

$[B^{-1}]$ = Technological coefficient matrix in the optimal table of the non-parametric problem with respect to the basic variables of the initial table

$[b]$ = Right-hand side constants matrix of the original parametric problem.

For the given problem, we have

$$[b_p] = \begin{bmatrix} \dfrac{1}{5} & -\dfrac{1}{4} \\ -\dfrac{1}{5} & \dfrac{1}{2} \end{bmatrix} \begin{bmatrix} 60 + 10t \\ 40 - 10t \end{bmatrix} = \begin{bmatrix} 2 + \left(\dfrac{9}{2}\right)t \\ 8 - 7t \end{bmatrix}$$

The corresponding modified table of the parametric problem is shown in Table 16.9.

Table 16.9 Modified Table of the Parametric Problem

CB_i	C_j Basic variable	6 X_1	8 X_2	0 S_1	0 S_2	Solution
8	X_2	0	1	1/5	−1/4	2 + (9/2)t
6	X_1	1	0	−1/5	1/2	8 − 7t
	Z_j	6	8	2/5	1	64 − 6t
	$C_j - Z_j$	0	0	−2/5	−1	

To maintain the feasibility of the solution in Table 16.9, the value of each of the basic variables should be greater than or equal to zero as shown below:

$$2 + \frac{9}{2}t \geq 0 \qquad \text{or} \qquad t \geq -\frac{4}{9}$$

and

$$8 - 7t \geq 0 \quad \text{or} \quad t \leq \frac{8}{7}$$

with respect to basic variables X_2 and X_1, respectively. Therefore, the effective interval of the parameter t over which the feasibility is unaffected is:

$$-\frac{4}{9} \leq t \leq \frac{8}{7}$$

If t is greater than 8/7, the feasibility of the solution corresponding to row X_1 will be affected. The infeasibility of row X_1 can be removed using dual simplex method. The variable X_1 leaves the basis. The variable S_1 is selected as the entering variable based on the calculations in Table 16.10.

Table 16.10 Calculations for Selection of Entering Variable

	Non-basic variable	
	S_1	S_2
$-(C_j - Z_j)$	2/5	1
X_1 row	−1/5	1/2
Absolute ratio	2*	−

The results after performing pivot operations are shown in Table 16.11.

Table 16.11 Results after Pivot Operations when $t > 8/7$

CB_i	C_j	6	8	0	0	
	Basic variable	X_1	X_2	S_1	S_2	Solution
8	X_2	1	1	0	1/4	$10 - (5/2)t$*
0	S_1	−5	0	1	−5/2	$-40 + 35t$
	Z_j	8	8	0	2	$80 - 20t$
	$C_j - Z_j$	−2	0	0	−2	

To maintain the feasibility of the solution in Table 16.11, the value of each of the basic variables should be greater than or equal to zero. That is,

$$10 - \frac{5}{2}t \geq 0 \quad \text{or} \quad t \leq 4$$

and

$$-40 + 35t \geq 0 \quad \text{or} \quad t \geq \frac{8}{7}$$

with respect to basic variables X_2 and S_1. Therefore, the second effective interval of the parameter t over which the feasibility is unaffected is shown below.

$$\frac{8}{7} \leq t \leq 4$$

If t is greater than 4, the feasibility of the solution corresponding to row X_2 will be affected. The infeasibility of row X_2 can be removed using dual simplex method. The variable X_2 leaves the basis. The entering variable is selected based on the calculations in Table 16.12.

Table 16.12 Calculations for Selection of Entering Variable

	Non-basic variable	
	X_1	S_2
$-(C_j - Z_j)$	2	2
X_2 row	1	1/4
Absolute ratio	–	–

In Table 16.12, none of the denominators is negative. Hence, the solution will be infeasible for all the values of t beyond 4.

Now, consider Table 16.9. In this table, the feasibility of row X_2 will be affected if the value of t is less than $-4/9$. This infeasibility can be removed using the dual simplex method.

The basic variable X_2 is the leaving variable. The variable S_2 is selected as the entering variable based on the calculations in Table 16.13.

Table 16.13 Calculations for Selection of Entering Variable

	Non-basic variable	
	S_1	S_2
$-(C_j - Z_j)$	2/5	1
X_2 row	1/5	–1/4
Absolute ratio	–	4*

The results after performing pivot operations on Table 16.9 are shown in Table 16.14.

Table 16.14 Results after Pivot Operations when $t < -4/9$

CB_i	C_j	6	8	0	0	
	Basic variable	X_1	X_2	S_1	S_2	Solution
0	S_2	0	–4	–4/5	1	$-8 - 18t$
6	X_1	1	2	1/5	0	$12 + 2t$
	Z_j	6	12	6/5	0	$72 + 12t$
	$C_j - Z_j$	0	–4	–6/5	0	

To maintain the feasibility of the solution in Table 16.14, the value of each of the basic variables should be greater than or equal to zero. That is,

$$-8 - 18t \geq 0 \qquad \text{or} \qquad t \leq -\frac{4}{9}$$

and

$$12 + 2t \geq 0 \quad \text{or} \quad t \geq -6$$

with respect to basic variables S_2 and X_1. Therefore, the third effective interval of the parameter t over which the feasibility is unaffected is

$$-6 \leq t \leq -\frac{4}{9}$$

If t is lesser than -6, the feasibility of the row X_1 will be affected. This infeasibility is removed by using dual simplex method. The leaving variable is X_1. The entering variable is determined based on calculations in Table 16.15.

Table 16.15 Calculations for Selection of Entering Variable

	Non-basic variable	
	X_2	S_1
$-(C_j - Z_j)$	4	6/5
X_1 row	2	1/5
Absolute ratio	–	–

Since all the denominators of the ratios in Table 16.15 are positive, the solution will be infeasible for all values of t which are lesser than -6. So, the critical values of t in ascending order are -6, $-4/9$, $8/7$ and 4. The different intervals and the corresponding solutions are summarized in Table 16.16.

Table 16.16 Summary of Results

Interval of t	Critical value	Optimal solution				
		Z	X_1	X_2	S_1	S_2
$-4/9 \leq t \leq 8/7$	$t_1 = 8/7$	$64 - 6t$	$8 - 7t$	$2 + (9/2)t$	0	0
$8/7 \leq t \leq 4$	$t_2 = 4$	$80 - 20t$	0	$10 - (5/2)t$	$-40 + 35t$	0
$t > 4$	Infeasible					
$-4/9 \leq t \leq 8/7$	$t_3 = -4/9$	$64 - 6t$	$8 - 7t$	$2 + (9/2)t$	0	0
$-6 \leq t \leq -4/9$	$t_4 = -6$	$72 + 12t$	$12 + 2t$	0	0	$-8 - 18t$
$t < -6$	Infeasible					

16.4 INTRODUCTION TO CHANGES IN RESOURCE REQUIREMENTS VECTOR(S), P_j

The technological coefficients may vary as a function of some parameter, viz. labour efficiency index. The product-mix problem shown in Section 16.2 is presented below along with a parameter t affecting the technological coefficients.

$$\text{Maximize } Z = C_1 X_1 + C_2 X_2 + \cdots + C_j X_j + \cdots + C_n X_n$$

subject to

$$(a_{11} + a'_{11}t)X_1 + (a_{12} + a'_{12}t)X_2 + \cdots + (a_{1n} + a'_{1n}t)X_n \le b_1$$

$$(a_{21} + a'_{21}t)X_1 + (a_{22} + a'_{22}t)X_2 + \cdots + (a_{2n} + a'_{2n}t)X_n \le b_2$$

$$\vdots \qquad \vdots \qquad \qquad \vdots \qquad \vdots$$

$$(a_{i1} + a'_{i1}t)X_1 + (a_{i2} + a'_{i2}t)X_2 + \cdots + (a_{in} + a'_{in}t)X_n \le b_i$$

$$\vdots \qquad \vdots \qquad \qquad \vdots \qquad \vdots$$

$$(a_{m1} + a'_{m1}t)X_1 + (a_{m2} + a'_{m2}t)X_2 + \cdots + (a_{mn} + a'_{mn}t)X_n \le b_m$$

$$X_j \ge 0, \qquad j \ge 1, 2, 3,..., n$$

Under such situation, the optimal table of the non-parametric problem with 0 value for the parameter t is to be obtained first. Then the criterion row values of the above optimal table are to be updated using the following formula to get the starting table for the analysis of the parametric problem.

$$C_j - Z_j(t) = C_j - [C_B]\,[B^{-1}]\,[P_j]$$

where, $[C_B]$ is the objective function coefficients matrix of the basic variables in the optimal table. The size of this $[C_B]$ matrix is $1 \times m$. $[B^{-1}]$ is the technological coefficient matrix in the optimal table of the non-parametric problem with respect to the basic variables of the initial table. The size of $[B^{-1}]$ is $m \times m$. $[P_j]$ is the resource requirements vector (part of technological coefficients matrix) of the non-basic variable in the column j of the initial table. The size of $[P_j]$ is $m \times 1$.

As long as the values of the non-basic variables in the criterion row are less than or equal to 0 in the case of maximization problems (greater than or equal to 0 in the case of minimization problems), the optimality of the solution will not be affected.

By applying this condition to all the columns corresponding to the non-basic variables, one can determine the effective interval of the parameter t. If we carry out further iterations based on a value of the parameter t beyond (before) its critical value, any one of the parameterized non-basic variable will become as a basic variable. Beyond this stage, the problem is not amenable for further analysis and hence this type of parametric analysis introduces difficulty in identifying the critical values of the parameter. Based on these reasons, the details of the parametric analysis applied to the resource requirement vectors are not presented in this book.

QUESTIONS

1. What is parametric programming? What are the reasons for changes in the parameters? Explain them in brief.

2. What are the types of analysis under parametric programming? Explain them in brief.

3. Consider the parametric linear programming problem.

$$\text{Maximize } Z = (t - 1)X_1 + X_2$$

subject to

$$X_1 + 2X_2 \le 10$$

$$2X_1 + X_2 \le 11$$

$$X_1 - 2X_2 \le 3$$

$$X_1 \text{ and } X_2 \ge 0$$

Perform a complete parametric analysis. Identify the critical values of the parameter t and all optimal solutions.

4. Solve the following parametric cost problem.

$$\text{Minimize } Z = (2 + t)X_1 + (1 + 4t)X_2$$

subject to

$$3X_1 + X_2 \geq 3$$
$$4X_1 + 3X_2 \geq 6$$
$$X_1 + 2X_2 \leq 3$$

$X_1, X_2 \geq 0$ and t is a non-negative parameter.

5. Solve the following problem.

$$\text{Minimize } Z = 4X_1 + X_2$$

subject to

$$3X_1 + X_2 = 3 + 3t$$
$$4X_1 + 3X_2 \geq 6 + 2t$$
$$X_1 + 2X_2 \leq 3 + 4t$$

$X_1, X_2 \geq 0$ and t is non-negative parameter.

6. Solve the following problem.

$$\text{Maximize } Z = 3X_1 + 2X_2 + \cdot 5X_3$$

subject to

$$X_1 + 2X_2 + X_3 \leq 430 + 500t$$
$$3X_1 + 2X_3 \leq 460 + 100t$$
$$X_1 + 4X_2 \leq 420 - 200t$$

$X_1, X_2, X_3 \geq 0$ and t is a non-negative parameter.

7. Perform a complete parametric analysis of the following problem.

$$\text{Minimize } Z = tX_1 - X_2$$

subject to

$$3X_1 - X_2 \geq 5$$
$$2X_1 + X_2 \leq 3$$

$X_1, X_2 \geq 0$ and t is a parameter.

NONLINEAR PROGRAMMING

17

17.1 INTRODUCTION

Nonlinear programming is an extension of linear programming. In many real-life problems, the objective function may be nonlinear but the set of constraints may be linear or nonlinear. Such problems are called as nonlinear problems. A set of nonlinear programming problems is as follows:

- Nonlinear programming problem of general nature
- Quadratic programming problems
- Separable programming problems

Each of these problems requires different solution procedure. In this chapter, the procedures for the above topics are demonstrated with examples:

17.2 LAGRANGEAN METHOD

Let us consider the following form of nonlinear programming:

$$\text{Maximize or minimize } Z = f(X_1, X_2, ..., X_j, ..., X_n)$$

subject to

$$G_i(X_1, X_2, ..., X_j, ..., X_n) = b_i, \quad i = 1, 2, ..., m$$

$$X_j \geq 0, \quad j = 1, 2, 3, ..., n$$

A modified form of the above model is shown below:

$$\text{Maximize or minimize } Z = f(X_1, X_2, ..., X_j, ..., X_n)$$

subject to

$$g_i(X_1, X_2, ..., X_j, ..., X_n) = 0, \quad i = 1, 2, ..., m$$

$$X_j \geq 0, \quad j = 1, 2, 3, ..., n$$

where

$$g_i(X_1, X_2, ..., X_j, ..., X_n) = G_i(X_1, X_2, ..., X_j, ..., X_n) - b_i$$

This model consists of n variables and m constraints. The objective function is nonlinear and the constraints are in linear form.

Let, L be the Lagrangean function and ϕ_i be the Lagrangean multiplier of the ith constraint, then

$$L = f(X_1, X_2, ..., X_j, ..., X_n) - \sum_{i=1}^{m} \phi_i[g_i(X_1, X_2, ..., X_j, ..., X_n)]$$

Steps of Lagrangean method

Step 1: This step forms the Lagrangean function as:

$$L = f(X_1, X_2,..., X_j,..., X_n) - \sum_{i=1}^{m} \phi_i[g_i(X_1, X_2,..., X_j,..., X_n)]$$

Step 2: Here the first partial derivative of L with respect to X_j is obtained, where j varies from 1 to n, and also with respect to ϕ_i, where i varies from 1 to m. Then equate each of them to 0 as shown below.

$$\frac{\delta L}{\delta X_j} = 0, \qquad j = 1, 2,..., n$$

$$\frac{\delta L}{\delta \phi_i} = 0, \qquad i = 1, 2,..., m$$

So, the system consists of $n + m$ unknown variables with $n + m$ simultaneous equations of the first order partial derivatives.

Step 3: In this step, the solution to the above system of equations is found out.

Step 4: Then bordered Hessian square matrix $[H^B]$ of size $n + m$ is formed as shown in Table 17.1.

Table 17.1 General Form of Bordered Hessian Matrix

	1	2	⋯	i	⋯	m	1	2	⋯	j	⋯	n
1	0	0	⋯	0	⋯	0						
2	0	0	⋯	0	⋯	0	Coefficients of					
⋮	⋮	⋮		⋮		⋮	$\dfrac{\delta L}{\delta \phi_i}, \quad i = 1, 2,..., m$					
i	0	0	⋯	0	⋯	0						
⋮	⋮	⋮		⋮		⋮	written row-wise					
m	0	0	⋯	0	⋯	0						
1												
2	Coefficients of						$\dfrac{\delta L}{\delta X_i} \dfrac{\delta L}{\delta X_j}$					
⋮	$\dfrac{\delta L}{\delta \phi_i}, \quad i = 1, 2,..., m$						For					
j												
⋮	written column-wise						$i = 1, 2,..., n$ and $j = 1, 2,..., n$					
n												

Step 5: The guidelines for testing the stationary point $(X_1^*, X_2^*, X_3^*,..., X_j^*,..., X_n^*)$ for maximization/minimization type objective function are presented below.

Maximization type objective function.   The stationary point will give the maximum objective function value if the sign of each of the last $(n - m)$ principal minor determinants of the bordered Hessian matrix is same as that of $(-1)^{m+1}$, ending with the $(2m + 1)$th principal minor determinant.

Minimization type objective function.   The stationary point will give the minimum objective function value if the sign of each of the last $(n - m)$ principal minor determinants of the bordered Hessian matrix is same as that of $(-1)^m$, ending with the $(2m + 1)$th principal minor determinant.

As an example, the last three principal minor determinants of the bordered Hessian matrix which is shown in the Figure 17.1(a) are presented from Figure 17.1(b) to Figure 17.1(d).

$$\begin{vmatrix} 0 & 35 & 2 \\ 35 & 5 & 2 \\ 2 & 0 & 6 \end{vmatrix}$$

(a) Bordered Hessian matrix

$$\begin{vmatrix} 0 \end{vmatrix}$$

(b) First principal minor determinant

$$\begin{vmatrix} 0 & 35 \\ 35 & 5 \end{vmatrix}$$

(c) Second principal minor determinant

$$\begin{vmatrix} 0 & 35 & 2 \\ 35 & 5 & 2 \\ 2 & 0 & 6 \end{vmatrix}$$

(d) Third principal minor determinant

Figure 17.1 Sample of bordered Hessian matrix.

Example 17.1 Solve the following nonlinear programming using Lagrangean method:

$$\text{Maximize } Z = 4X_1 - 0.02X_1^2 + X_2 - 0.02X_2^2$$

subject to

$$X_1 + 2X_2 = 120$$
$$X_1 \text{ and } X_2 \geq 0$$

Solution The general form of the given nonlinear programming problem is presented below:

$$\text{Maximize } Z = 4X_1 - 0.02X_1^2 + X_2 - 0.02X_2^2$$

subject to

$$X_1 + 2X_2 - 120 = 0$$
$$X_1, X_2 \geq 0$$

Here, the number of variables, $n = 2$ and, the number of constraints, $m = 1$. Then

$$L = 4X_1 - 0.02X_1^2 + X_2 - 0.02X_2^2 - \phi(X_1 + 2X_2 - 120)$$

The first order partial derivatives with respect to X_1, X_2 and ϕ are:

$$\frac{\delta L}{\delta X_1} = 4 - 0.04X_1 - \phi = 0, \quad \text{or} \quad 0.04X_1 + \phi = 4 \tag{17.1}$$

$$\frac{\delta L}{\delta X_2} = 1 - 0.04X_2 - 2\phi = 0, \quad \text{or} \quad 0.04X_2 + 2\phi = 1 \tag{17.2}$$

$$\frac{\delta L}{\delta \phi} = -(X_1 + 2X_2 - 120) = 0, \quad \text{or} \quad X_1 + 2X_2 = 120 \tag{17.3}$$

The solution of the above system of equations is:

$$(X_1^*, X_2^*, \phi^*) = (94, 13, 0.24) \quad \text{and} \quad Z^* = 208.9$$

The second order partial derivatives are summarized in Table 17.2.

Table 17.2   Second Order Partial Derivatives

$$\frac{\delta L}{\delta X_1}\frac{\delta L}{\delta X_1} = -0.04 \qquad \frac{\delta L}{\delta X_1}\frac{\delta L}{\delta X_2} = 0$$

$$\frac{\delta L}{\delta X_2}\frac{\delta L}{\delta X_1} = 0 \qquad \frac{\delta L}{\delta X_2}\frac{\delta L}{\delta X_2} = -0.04$$

The bordered Hessian matrix of the given problem is shown in Table 17.3.

Table 17.3   Bordered Hessian Matrix of Example 17.1

0	1	2
1	−0.04	0
2	0	−0.04

The value of $n - m$ is 1 and the corresponding last one principal minor determinant is same as shown in Table 17.3. The value of the determinant in Table 17.3 is 0.12. The sign of this value is same as that of $(-1)^{1+1}$. Hence, the solution (X_1^*, X_2^*, ϕ^*) corresponds to the maximum objective function value. Therefore, the optimal results are presented below:

$$(X_1^*, X_2^*) = (94, 13) \quad \text{and} \quad Z(\text{maximum}) = 208.9.$$

Example 17.2   Solve the following nonlinear programming using Lagrangean method:

$$\text{Minimize } Z = 2X_1^2 - 3X_2^2 + 18X_2$$

subject to

$$2X_1 + X_2 = 8$$
$$X_1 \text{ and } X_2 \geq 0$$

Solution   The general form of the given nonlinear programming problem is:

$$\text{Minimize } Z = 2X_1^2 - 3X_2^2 + 18X_2$$

subject to

$$2X_1 + X_2 - 8 = 0$$
$$X_1 \text{ and } X_2 \geq 0$$

Here, the number of variables, $n = 2$ and the number of constraints, $m = 1$. Then

$$L = 2X_1^2 - 3X_2^2 + 18X_2 - \phi(2X_1 + X_2 - 8)$$

The first order partial derivatives with respect to X_1, X_2 and ϕ are:

$$\frac{\delta L}{\delta X_1} = 4X_1 - 2\phi = 0 \tag{17.4}$$

$$\frac{\delta L}{\delta X_2} = -6X_2 + 18 - \phi = 0, \quad \text{or} \quad 6X_2 + \phi = 18 \tag{17.5}$$

$$\frac{\delta L}{\delta \phi} = -(2X_1 + X_2 - 8) = 0, \quad \text{or} \quad 2X_1 + X_2 = 8 \tag{17.6}$$

The solution of the above system of equations is:

$$(X_1^*, X_2^*, \phi^*) = (3, 2, 6) \quad \text{and} \quad Z^* = 42$$

The second order partial derivatives of the given problem are summarized in Table 17.4.

Table 17.4 Second Order Partial Derivatives

$$\frac{\delta L}{\delta X_1} \frac{\delta L}{\delta X_1} = 4 \qquad \frac{\delta L}{\delta X_1} \frac{\delta L}{\delta X_2} = 0$$

$$\frac{\delta L}{\delta X_2} \frac{\delta L}{\delta X_1} = 0 \qquad \frac{\delta L}{\delta X_2} \frac{\delta L}{\delta X_2} = -6$$

The bordered Hessian matrix of the given problem is shown in Table 17.5.

Table 17.5 Bordered Hessian Matrix of Example 17.2

0	2	1
2	4	0
1	0	-6

The value of $n - m$ is 1 and the corresponding last one principal minor determinant is same as the bordered Hessian matrix as shown in the Table 17.5. The value of this principal minor determinant is -28 which is negative. The sign of this value is same as that of $(-1)^1$. Hence, the solution (X_1^*, X_2^*, ϕ^*) corresponds to the minimum objective function value. The optimal results are:

$$(X_1^*, X_2^*) = (3, 2) \quad \text{and} \quad Z(\text{minimum}) = 42$$

Example 17.3 Solve the following nonlinear programming problem using Lagrangean method:

$$\text{Maximize } Z = X_1^2 + 2X_2^2 + X_3^2$$

subject to

$$2X_1 + X_2 + 2X_3 = 30$$
$$X_1, X_2 \text{ and } X_3 \geq 0$$

Solution The general form of the given nonlinear programming problem is presented below:

$$\text{Maximize } Z = X_1^2 + 2X_2^2 + X_3^2$$

subject to

$$2X_1 + X_2 + 2X_3 - 30 = 0$$
$$X_1 \text{ and } X_2 \geq 0$$

Here, the number of variables, $n = 3$ and the number of constraints, $m = 1$. Then

$$L = X_1^2 + 2X_2^2 + X_3^2 - \phi(2X_1 + X_2 + 2X_3 - 30)$$

The first partial derivatives with respect to X_1, X_2, X_3 and ϕ are:

$$\frac{\delta L}{\delta X_1} = 2X_1 - 2\phi = 0 \tag{17.7}$$

$$\frac{\delta L}{\delta X_2} = 4X_2 - \phi = 0 \qquad (17.8)$$

$$\frac{\delta L}{\delta X_3} = 2X_3 - 2\phi = 0 \qquad (17.9)$$

$$\frac{\delta L}{\delta \phi} = -(2X_1 + X_2 + 2X_3 - 30) = 0 \quad \text{or} \quad 2X_1 + X_2 + 2X_3 = 30 \qquad (17.10)$$

The solution of the above system of equations is:

$$(X_1^*, X_2^*, X_3^*, \phi^*) = \left(\frac{120}{17}, \frac{30}{17}, \frac{120}{17}, \frac{120}{17} \right) \quad \text{and} \quad Z^* = 105.88$$

The second order partial derivatives are summarized in Table 17.6.

Table 17.6 Second Order Partial Derivatives

$\dfrac{\delta L}{\delta X_1} \dfrac{\delta L}{\delta X_1} = 2$	$\dfrac{\delta L}{\delta X_1} \dfrac{\delta L}{\delta X_2} = 0$	$\dfrac{\delta L}{\delta X_1} \dfrac{\delta L}{\delta X_3} = 0$
$\dfrac{\delta L}{\delta X_2} \dfrac{\delta L}{\delta X_1} = 0$	$\dfrac{\delta L}{\delta X_2} \dfrac{\delta L}{\delta X_2} = 4$	$\dfrac{\delta L}{\delta X_2} \dfrac{\delta L}{\delta X_3} = 0$
$\dfrac{\delta L}{\delta X_3} \dfrac{\delta L}{\delta X_1} = 0$	$\dfrac{\delta L}{\delta X_3} \dfrac{\delta L}{\delta X_2} = 0$	$\dfrac{\delta L}{\delta X_3} \dfrac{\delta L}{\delta X_3} = 2$

The bordered Hessian matrix of the given problem is shown in Table 17.7.

Table 17.7 Bordered Hessian Matrix of Example 17.3

0	2	1	2
2	2	0	0
1	0	4	0
2	0	0	2

The value of $n - m$ is 2 and the corresponding last 2 principal minor determinants of the bordered Hessian matrix are shown in Tables 17.8 and 17.9, respectively.

Table 17.8 Third Principal Minor Determinant

0	2	1
2	2	0
1	0	4

Value = 14

Table 17.9 Fourth (last) Principal Minor Determinant

0	2	1	2
2	2	0	0
1	0	4	0
2	0	0	2

Value = 4

The sign of the value of each of the last 2 principal minor determinants is positive which is same as that of $(-1)^{1+1}$. Hence, the solution $(X_1^*, X_2^*, X_3^*, \phi^*)$ corresponds to the maximum objective function value. Therefore, the optimal results are:

$$(X_1^*, X_2^*, X_3^*) = \left(\frac{120}{17}, \frac{30}{17}, \frac{120}{17}\right) \quad \text{and} \quad Z(\text{maximum}) = 105.88$$

Example 17.4 Solve the following nonlinear programming problem using Lagrangean method:

$$\text{Minimize } Z = X_1^2 + 2X_2^2 + 1.5X_3^2$$

subject to

$$2X_1 + 2X_2 + 3X_3 = 30$$
$$3X_1 - 4X_2 + 4X_3 = 25$$
$$X_1, X_2, X_3 \geq 0$$

    Solution The general form of the given nonlinear programming problem is as presented below:

$$\text{Minimize } Z = X_1^2 + 2X_2^2 + 1.5X_3^2$$

subject to

$$2X_1 + 2X_2 + 3X_3 - 30 = 0$$
$$3X_1 - 4X_2 + 4X_3 - 25 = 0$$
$$X_1, X_2, X_3 \geq 0$$

    Here, the number of variables, $n = 3$ and the number of constraints, $m = 2$. Then, the Lagrangean function, L is:

$$L = X_1^2 + 2X_2^2 + 1.5X_3^2 - \phi_1 (2X_1 + 2X_2 + 3X_3 - 30) - \phi_2 (3X_1 - 4X_2 + 4X_3 - 25)$$

The first order partial derivatives with respect to X_1, X_2, X_3, ϕ_1 and ϕ_2 are:

$$\delta L/\delta X_1 = 2X_1 - 2\phi_1 - 3\phi_2 = 0 \tag{17.11}$$

$$\delta L/\delta X_2 = 4X_2 - 2\phi_1 + 4\phi_2 = 0 \tag{17.12}$$

$$\delta L/\delta X_3 = 3X_3 - 3\phi_1 - 4\phi_2 = 0 \tag{17.13}$$

$$\delta L/\delta \varphi_1 = -2X_1 - 2X_2 - 3X_3 + 30 = 0 \text{ or } 2X_1 + 2X_2 + 3X_3 = 30 \tag{17.14}$$

$$\delta L/\delta \varphi_2 = -3X_1 + 4X_2 - 4X_3 + 25 = 0 \text{ or } 3X_1 - 4X_2 + 4X_3 = 25 \tag{17.15}$$

The solution to the above system of equations is:

$$X_1 = 5, \; X_2 = 2.5, \; X_3 = 5, \; \phi_1 = 5, \; \phi_2 = 0 \text{ and } Z^* = 75$$

The second order partial derivatives are summarized in Table 17.10.

Table 17.10 Second Order Partial Derivatives of Example 17.4

$\delta L/\delta X_1 \, \delta L/\delta X_1 = 2$	$\delta L/\delta X_1 \, \delta L/\delta X_2 = 0$	$\delta L/\delta X_1 \, \delta L/\delta X_3 = 0$
$\delta L/\delta X_2 \, \delta L/\delta X_1 = 0$	$\delta L/\delta X_2 \, \delta L/\delta X_2 = 4$	$\delta L/\delta X_2 \, \delta L/\delta X_3 = 0$
$\delta L/\delta X_3 \, \delta L/\delta X_1 = 0$	$\delta L/\delta X_3 \, \delta L/\delta X_2 = 0$	$\delta L/\delta X_3 \, \delta L/\delta X_3 = 3$

The bordered Hessian matrix of the given problem is shown in Table 17.11.

Table 17.11   Bordered Hessian Matrix of Example 17.4

0	0	2	2	3
0	0	3	−4	4
2	3	2	0	0
2	−4	0	4	0
3	4	0	0	3

The value of n–m is 1 and the corresponding last one principal minor determinant is same as the bordered Hessian matrix as shown in Table 17.11. The value of the principal minor determinant is 1304 which is positive. The sign of this value is same as that of $(-1)^2$ [i.e. $(-1)^m$]. Hence, the solution $(X_1, X_2, X_3, \phi_1, \phi_2)$ corresponds to the minimum objective function value. Therefore, the optimal results are:

$$X_1, X_2, X_3 = (5, 2.5, 5) \text{ and } Z(\text{minimum}) = 75$$

17.3   KUHN–TUCKER CONDITIONS

Consider the following general form of nonlinear programming problem which is having a maximization objective function with all less than or equal to type constraints.

$$\text{Maximize } Z = f(X_1, X_2, ..., X_j, ..., X_n)$$

subject to

$$G_i(X_1, X_2, ..., X_j, ..., X_n) \leq b_i, \qquad i = 1, 2, ..., m$$

$$X_j \geq 0, \qquad j = 1, 2, 3, ..., n$$

A modified form of the above model is:

$$\text{Maximize } Z = f(X_1, X_2, ..., X_j, ..., X_n)$$

subject to

$$g_i(X_1, X_2, ..., X_j, ..., X_n) \leq 0, \qquad i = 1, 2, ..., m$$

$$X_j \geq 0, \qquad j = 1, 2, 3, ..., n$$

where

$$g_i(X_1, X_2, ..., X_j, ..., X_n) = G_i(X_1, X_2, ..., X_j, ..., X_n) - b_i$$

Again, the above model is modified as:

$$\text{Maximize } Z = f(X_1, X_2, ..., X_j, ..., X_n)$$

subject to

$$g_i(X_1, X_2, ..., X_j, ..., X_n) + S_i^2 = 0, \qquad i = 1, 2, ..., m$$

$$X_j \geq 0, \qquad j = 1, 2, 3, ..., n$$

where S_i^2 is a complementary slack of the ith constraint. This model consists of $n + m$ variables and

m constraints. Let L be the Lagrangean function, and ϕ_i be the Lagrangean multiplier of the ith constraint. Then

$$L[(X_1, X_2,..., X_n), (S_1, S_2,..., S_m), (\phi_1, \phi_2,..., \phi_m)]$$

$$= f(X_1, X_2,..., X_j,..., X_n) - \sum_{i=1}^{m} \phi_i[g_i(X_1, X_2,..., X_j,..., X_n) + S_i^2]$$

For the above maximization problem with concave objective function and with all less than or equal to type constraints (convex type constraints), the value of ϕ_i should be greater than or equal to 0. To maintain this relation of ϕ_i, Kuhn–Tucker has established the following necessary conditions:

(a) $\phi_i \geq 0, \ i = 1, 2, 3,..., m$

(b) $\dfrac{\delta L}{\delta X_j} = 0, \quad j = 1, 2, 3,..., n$ (first partial derivatives)

(c) $\phi_i g_i(X_1, X_2,..., X_n) = 0, \quad i = 1, 2, 3,..., m$

(d) $g_i(X_1, X_2,..., X_n) \leq 0, \quad i \leq 1, 2, 3,..., m$

For a minimization problem with convex objective function and with all greater than or equal to type constraints (concave type constraints), the value of ϕ_i should be greater than or equal to 0. If L is concave in the case of maximization problem and convex in the case of minimization problem, the different possibilities of ϕ_i are presented below:

(i) For maximization objective function and with '$\leq$' type constraints, $\phi_i \geq 0, i = 1, 2, 3,..., m$.

(ii) For maximization objective function and with '$\geq$' type constraints, $\phi_i \leq 0, i = 1, 2, 3,..., m$.

(iii) For maximization objective function and with '$=$' type constraints, ϕ_i is unrestricted in sign for $i = 1, 2, 3,..., m$.

(iv) For minimization objective function and with '$\leq$' type constraints, $\phi_i \leq 0, i = 1, 2, 3,..., m$.

(v) For minimization objective function and with '$\geq$' type constraints, $\phi_i \geq 0, i = 1, 2, 3,..., m$.

(vi) For minimization objective function and with '$=$' type constraints, ϕ_i is unrestricted in sign for $i = 1, 2, 3,..., m$.

Example 17.5 Solve the following nonlinear programming problem using Kuhn–Tucker conditions:

$$\text{Maximize } Z = 3X_1^2 + 14X_1X_2 - 8X_2^2$$

subject to

$$3X_1 + 6X_2 \leq 72$$
$$X_1 \text{ and } X_2 \geq 0$$

Solution The given problem is modified as:

$$\text{Maximize } Z = 3X_1^2 + 14X_1X_2 - 8X_2^2$$

subject to

$$3X_1 + 6X_2 - 72 \leq 0$$
$$X_1 \text{ and } X_2 \geq 0$$

Then, we have the Lagrangean function:

$$L = 3X_1^2 + 14X_1X_2 - 8X_2^2 - \phi(3X_1 + 6X_2 - 72)$$

The four sets of Kuhn–Tucker conditions are as given below:

(a) $\phi \geq 0$ (17.16)

(b) $\dfrac{\delta L}{\delta X_1} = 6X_1 + 14X_2 - 3\phi = 0,$ or $6X_1 + 14X_2 = 3\phi$ (17.17)

$\dfrac{\delta L}{\delta X_2} = 14X_1 - 16X_2 - 6\phi = 0,$ or $14X_1 - 16X_2 = 6\phi$ (17.18)

(c) $\phi(3X_1 + 6X_2 - 72) = 0$ (17.19)

(d) $3X_1 + 6X_2 - 72 \leq 0$ (17.20)

From Eqs. (17.17) and (17.18), we get

$$-2X_1 + 44X_2 = 0 \tag{17.21}$$

From Eq. (17.19), if ϕ is equated to 0, X_1 and X_2 should be equal to 0, which is not true. Therefore,

$$3X_1 + 6X_2 - 72 = 0, \quad \text{or} \quad 3X_1 + 6X_2 = 72 \tag{17.22}$$

By solving Eqs. (17.21) and (17.22), we get

$$X_1^* = 22, \qquad X_2^* = 1 \qquad Z(\text{maximum}) = 1752$$

Example 17.6   Solve the following nonlinear programming problem using Kuhn-Tucker conditions.

$$\text{Maximize } Z = X_1^2 + X_1X_2 - 2X_2^2$$

subject to

$$4X_1 + 2X_2 \leq 24$$
$$5X_1 + 10X_2 \leq 30$$
$$X_1, X_2 \geq 0$$

Solution   The given problem is modified as:

$$\text{Maximize } Z = X_1^2 + X_1X_2 - 2X_2^2$$

subject to

$$4X_1 + 2X_2 - 24 \leq 0$$
$$5X_1 + 10X_2 - 30 \leq 0$$
$$X_1, X_2 \geq 0$$

The Lagrangean function of this model is:

$$L = X_1^2 + X_1X_2 - 2X_2^2 - \phi_1(4X_1 + 2X_2 - 24) - \phi_2(5X_1 + 10X_2 - 30) \tag{17.23}$$

The four sets of Kuhn-Tucker conditions are as given below:

(a) $\phi_1 \geq 0$ (17.24)

$\phi_2 \geq 0$ (17.25)

(b) $\delta L/\delta X_1 = 2X_1 + X_2 - 4\phi_1 - 5\phi_2 = 0$ (17.26)

$\delta L/\delta X_2 = X_1 - 4X_2 - 2\phi_1 - 10\phi_2 = 0$ (17.27)

(c) $\phi_1(4X_1 + 2X_2 - 24) = 0$ (17.28)

$\phi_2(5X_1 + 10X_2 - 30) = 0$ (17.29)

(d) $4X_1 + 2X_2 - 24 \leq 0$ (17.30)

$5X_1 + 10X_2 - 30 \leq 0$ (17.31)

From equation (17.28), if ϕ_1 is equated to 0, X_1 and X_2 must be equal to 0, which is not true. Therefore,

$$4X_1 + 2X_2 - 24 = 0 \tag{17.32}$$

From equation (17.29), if ϕ_2 is equated to 0, X_2 and X_3 must be equal to 0, which is not true. Therefore,

$$5X_1 + 10X_2 - 30 = 0 \tag{17.33}$$

Now, solving equations (17.26), (17.27), (17.32) and (17.33) gives the following solution:

$$X_1^* = 6, \ X_2^* = 0, \ \phi_1^* = 3, \ \phi_2^* = 0, \text{ and}$$
$$Z^*(\text{maximum}) = 36$$

Example 17.7   Solve the following nonlinear programming problem using Kuhn-Tucker conditions:

$$\text{Maximize } Z = X_1^2 + X_1 X_2 - 2X_2^2$$

subject to

$$4X_1 + 2X_2 \leq 24$$
$$X_1, X_2 \geq 0$$

Solution.   The given problem is modified as:

$$\text{Maximize } Z = X_1^2 + X_1 X_2 - 2X_2^2$$

subject to

$$4X_1 + 2X_2 - 24 \leq 0$$
$$X_1, X_2 \geq 0$$

The Lagrangean function of this model is:

$$L = X_1^2 + X_1X_2 - 2X_2^2 - \phi(4X_1 + 2X_2 - 24)$$

The four sets of Kuhn-Tucker conditions are as given below:

(a) $\phi \geq 0$ $\hfill (17.34)$

(b) $\delta L/\delta X_1 = 2X_1 + X_2 - 4\phi = 0$ $\hfill (17.35)$

    $\delta L/\delta X_2 = X_1 - 4X_2 - 2\phi = 0$ $\hfill (17.36)$

(c) $\phi(4X_1 + 2X_2 - 24) = 0$ $\hfill (17.37)$

(d) $4X_1 + 2X_2 - 24 \leq 0$ $\hfill (17.38)$

From equation (17.37), if ϕ is equated to 0, X_1 and X_2 must be equal to 0, which is not true. Therefore,

$$4X_1 + 2X_2 - 24 = 0 \qquad \text{or} \qquad 4X_1 + 2X_2 = 24 \tag{17.39}$$

Now, solving equations (17.35), (17.36) and (17.39) gives the following solution.

$$X_1^* = 6, \ X_2^* = 0, \ \phi^* = 3 \text{ and } Z^*(\text{maximum}) = 36$$

17.4   QUADRATIC PROGRAMMING

Let us consider the following form of nonlinear programming problem in which the highest order of the polynomial of the objective function is restricted to at most 2.

$$\text{Maximize } Z = c_1 X_1 + c_2 X_2 + \cdots + c_n X_n$$
$$+ c_{n+1} X_1^2 + c_{n+2} X_2^2 + \cdots + c_{2n} X_n^2$$
$$+ c_{2n+1} X_1 X_2 + c_{2n+2} X_1 X_3 + \cdots + c_{2n+n_{c_2}} X_{n-1} X_n$$

subject to

$$a_{i1} X_1 + a_{i2} X_2 + \cdots + a_{in} X_n \le b_i, \qquad i = 1, 2,..., m$$
$$X_1, X_2,..., X_n \ge 0$$

The above quadratic programming model may be represented in short, follows:

$$\text{Maximize } Z = CX + X^T DX$$

subject to

$$AX \le B, \qquad i = 1, 2, 3,..., m$$
$$X \ge 0$$

where

$$C = [c_1, c_2,..., c_n]$$

$$X = \begin{bmatrix} X_1 \\ X_2 \\ \vdots \\ X_n \end{bmatrix} \qquad D = \begin{bmatrix} d_{11} & d_{12} & \cdots & d_{1n} \\ d_{21} & d_{22} & \cdots & d_{2n} \\ \vdots & \vdots & & \vdots \\ d_{n1} & d_{n2} & \cdots & d_{nn} \end{bmatrix}$$

$$A = \begin{bmatrix} a_{11} & a_{12} & \cdots & a_{1n} \\ a_{21} & a_{22} & \cdots & a_{2n} \\ \vdots & \vdots & & \vdots \\ a_{i1} & a_{i1} & \cdots & a_{in} \\ \vdots & \vdots & & \vdots \\ a_{m1} & a_{m2} & \cdots & a_{mn} \end{bmatrix} \qquad B = \begin{bmatrix} b_1 \\ b_2 \\ \vdots \\ b_i \\ \vdots \\ b_m \end{bmatrix}$$

In matrix [D], d_{jj} represents the coefficient of the term X_j^2 in the objective function and d_{ij} (when i is not equal to j) represents the coefficient of the term $X_i X_j$ in the objective function.

The above model is further modified by balancing the constraints as shown below.

$$\text{Maximize } Z = CX + X^T DX$$

Subject to

$$AX + S_i^2 = B, \quad i = 1, 2, 3,..., m \qquad \text{(Type I constraints)}$$
$$X_j \ge 0, \quad j = 1, 2, 3,..., n; \qquad -X_j \le 0, \quad j = 1, 2, 3,..., n; \qquad \text{(Type II constraints)}$$

Let ϕ_i be the Lagrangean multiplier of the ith constraint in Type I set of constraints, μ_j be the Lagrangean multiplier associated with the jth constraint in Type II set of constraints, and S_i^2 be the complementary slack of the constraint i of Type I. Application of Kuhn–Tucker conditions to this model results into the following system of constraints.

$$
\left[\begin{array}{c|c|c|c}
-2D & A^T & -I & 0 \\
\hline
A & 0 & 0 & I
\end{array}\right]
\begin{bmatrix} X \\ Q \\ M \\ S \end{bmatrix}
=
\begin{bmatrix} C^T \\ B \end{bmatrix}
$$

$$
\mu_j X_j = 0, \qquad j = 1, 2, 3,..., n
$$
$$
\phi_i S_i = 0, \qquad i = 1, 2, 3,..., m
$$

where, I is an identity matrix, and

$$
Q = \begin{bmatrix} \phi_1 \\ \phi_2 \\ \vdots \\ \phi_i \\ \vdots \\ \phi_m \end{bmatrix}
\qquad
M = \begin{bmatrix} \mu_1 \\ \mu_2 \\ \vdots \\ \mu_j \\ \vdots \\ \mu_n \end{bmatrix}
\qquad
S = \begin{bmatrix} S_1 \\ S_2 \\ \vdots \\ S_i \\ \vdots \\ S_m \end{bmatrix},
$$

where S is a set of slack variables.

Wolfe has developed a method to solve the above system of constraints. The steps of Wolfe's method are listed as follows:

Procedure to solve quadratic programming problem

Step 1: The system of constraints is to be formed first.

Step 2: Next, add R_j as the artificial variable for the jth constraint of the first n constraints, since each of these first n constraints has no basic variable in it.

Step 3: A minimization type objective function is then formed by summing the artificial variables.

Step 4: Solve the model consisting of the above objective function and system of constraints using the two-phase simplex method.

Example 17.8 Solve the following quadratic programming problem using Wolfe's method:

$$
\text{Maximize } Z = 6X_1 + 3X_2 - 2X_1^2 - 3X_2^2 - 4X_1X_2
$$

subject to

$$
X_1 + X_2 \leq 1
$$
$$
2X_1 + 3X_2 \leq 4
$$
$$
X_1 \text{ and } X_2 \geq 0
$$

Solution The given problem is written as follows:

$$
\text{Maximize } Z = c_1X_1 + c_2X_2 + c_3X_1^2 + c_4X_2^2 + c_5X_1X_2
$$
$$
= 6X_1 + 3X_2 - 2X_1^2 - 3X_2^2 - 4X_1X_2
$$

ubject to

$$
X_1 + X_2 \leq 1 \tag{17.40}
$$
$$
2X_1 + 3X_2 \leq 4 \tag{17.41}
$$
$$
- X_1 \leq 0 \tag{17.42}
$$
$$
- X_2 \leq 0 \tag{17.43}
$$

All the matrices that are required to generate the system of constraints except Matrix D can be obtained from the given problem easily. The guideline for obtaining the Matrix D are as follows:

$$D = \begin{bmatrix} c_3 & \dfrac{c_5}{2} \\ \dfrac{c_5}{2} & c_4 \end{bmatrix} = \begin{bmatrix} -2 & -2 \\ -2 & -3 \end{bmatrix}$$

The model of the given problem is:

$$\text{Maximize } Z = \begin{bmatrix} 6 & 3 \end{bmatrix} \begin{bmatrix} X_1 \\ X_2 \end{bmatrix} + \begin{bmatrix} X_1 & X_2 \end{bmatrix} \begin{bmatrix} -2 & -2 \\ -2 & -3 \end{bmatrix} \begin{bmatrix} X_1 \\ X_2 \end{bmatrix}$$

subject to

$$\begin{bmatrix} 1 & 1 \\ 2 & 3 \end{bmatrix} \begin{bmatrix} X_1 \\ X_2 \end{bmatrix} \leq \begin{bmatrix} 1 \\ 4 \end{bmatrix}, \qquad \text{where } -X_1 \leq 0, \quad -X_2 \leq 0$$

Let ϕ_1 and ϕ_2 are the Lagrangean multipliers associated with Constraint 1 and Constraint 2 of the system of constraints, respectively. μ_1 and μ_2 are the Lagrangean multiplier associated with Constraints 3 and 4 of the system of constraints, respectively. Application of Kuhn–Tucker conditions to this model results into the following system of constraints.

$$\left[\begin{array}{cc|cc|cc|cc} 4 & 4 & 1 & 2 & -1 & 0 & 0 & 0 \\ 4 & 6 & 1 & 3 & 0 & -1 & 0 & 0 \\ \hline 1 & 1 & 0 & 0 & 0 & 0 & 1 & 0 \\ 2 & 3 & 0 & 0 & 0 & 0 & 0 & 1 \end{array}\right] \begin{bmatrix} X_1 \\ X_2 \\ \phi_1 \\ \phi_2 \\ \mu_1 \\ \mu_2 \\ S_1 \\ S_2 \end{bmatrix} = \begin{bmatrix} 6 \\ 3 \\ 1 \\ 4 \end{bmatrix}$$

$$\mu_1 X_1 = \mu_2 X_2 = 0$$
$$\phi_1 S_1 = \phi_2 S_2 = 0$$

The above system of equations is written as follows:

$$4X_1 + 4X_2 + \phi_1 + 2\phi_2 - \mu_1 = 6$$
$$4X_1 + 6X_2 + \phi_1 + 3\phi_2 - \mu_2 = 3$$
$$X_1 + X_2 + S_1 \qquad\qquad = 1$$
$$2X_1 + 3X_2 + S_2 \qquad\qquad = 4$$

$$\mu_1 X_1 = \mu_2 X_2 = 0$$
$$\phi_1 S_1 = \phi_2 S_2 = 0$$

The model for Phase I of the two-phase simplex method is presented below with a minimization objective function irrespective of the type of the objective function in the original problem.

$$\text{Minimize } Z = R_1 + R_2$$

subject to

$$4X_1 + 4X_2 + \phi_1 + 2\phi_2 - \mu_1 + R_1 = 6 \tag{17.44}$$

$$4X_1 + 6X_2 + \phi_1 + 3\phi_2 - \mu_2 + R_2 = 3 \tag{17.45}$$

$$X_1 + X_2 + S_1 = 1 \tag{17.46}$$

$$2X_1 + 3X_2 + S_2 = 4 \tag{17.47}$$

$$\mu_1 X_1 = 0 \tag{17.48}$$

$$\mu_2 X_2 = 0 \tag{17.49}$$

$$\phi_1 S_1 = 0 \tag{17.50}$$

$$\phi_2 S_2 = 0 \tag{17.51}$$

$$X_1, X_2, R_1, R_2, \phi_1, \phi_2, \mu_1, \mu_2, S_1 \text{ and } S_2 \geq 0$$

Note: The artificial variables R_1, R_2 are included in the constraints (17.44) and (17.45), respectively to have a basic variable in each of them. Table 17.12 represents the initial table of Phase I of the two-phase simplex method.

Table 17.12 Initial Table

CB_i	C_j Basic variable	0 X_1	0 X_2	0 ϕ_1	0 ϕ_2	0 μ_1	0 μ_2	1 R_1	1 R_2	0 S_1	0 S_2	Solution
1	R_1	4	4	1	2	−1	0	1	0	0	0	6
1	R_2	4	$\boxed{6}$	1	3	0	−1	0	1	0	0	3*
0	S_1	1	1	0	0	0	0	0	0	1	0	1
0	S_2	2	3	0	0	0	0	0	0	0	1	4
	Z_j	8	10	2	5	−1	−1	1	1	0	0	9
	$C_j - Z_j$	−8	−10*	−2	−5	1	1	0	0	0	0	

In Table 17.12, the non-basic variable X_2 has the maximum negative criterion value. Since, μ_2 is not in the basis, X_2 is the entering variable. [In this iteration, both μ_2 and X_2 should not be present in the basis as per the constraint (17.49)]. The corresponding leaving variable is R_2. As per this combination of entering variable and leaving variable, pivot operations are shown in Table 17.13.

Table 17.13 Iteration 1

CB_i	C_j Basic variable	0 X_1	0 X_2	0 ϕ_1	0 ϕ_2	0 μ_1	0 μ_2	1 R_1	1 R_2	0 S_1	0 S_2	Solution
1	R_1	4/3	0	1/3	0	−1	2/3	1	−2/3	0	0	4
0	X_2	$\boxed{2/3}$	1	1/6	1/2	0	−1/6	0	1/6	0	0	1/2*
0	S_1	1/3	0	−1/6	−1/2	0	1/6	0	−1/6	1	0	1/2
0	S_2	0	0	−1/2	−3/2	0	1/2	0	−1/2	0	1	5/2
	Z_j	4/3	0	1/3	0	−1	2/3	1	−2/3	0	0	4
	$C_j - Z_j$	−4/3*	0	−1/3	0	1	−2/3	0	5/3	0	0	

In Table 17.13, the non-basic variable X_1 has the maximum negative criterion value. Since μ_1 is not in the basis, X_1 is the entering variable. [In this iteration, as per the constraint (17.48), both μ_1 and X_1 should not be present in the basis. The corresponding leaving variable is X_2. As per this combination of entering variable and leaving variable, pivot operations are shown in Table 17.14.

Table 17.14 Iteration 2

CB_i	Basic variable	C_j 0 X_1	0 X_2	0 ϕ_1	0 ϕ_2	0 μ_1	0 μ_2	1 R_1	1 R_2	0 S_1	0 S_2	Solution
1	R_1	0	−2	0	−1	−1	1	1	−1	0	0	3
0	X_1	1	3/2	1/4	3/4	0	−1/4	0	1/4	0	0	3/4
0	S_1	0	−1/2	−1/4	−3/4	0	1/4	0	−1/4	1	0	1/4*
0	S_2	0	0	−1/2	−3/2	0	1/2	0	−1/2	0	1	5/2
	Z_j	0	−2	0	−1	−1	1	1	−1	0	0	3
	$C_j - Z_j$	0	2	0	1	1	−1*	0	2	0	0	

In Table 17.14, the non-basic variable μ_2 has the maximum negative criterion value. In this iteration, as per constraint (17.49), both X_2 and μ_2 should not be present in the basis. Since X_2 is not in the basis, μ_2 is the entering variable. The corresponding leaving variable is S_1. As per this combination of entering variable and leaving variable, pivot operations are shown in Table 17.15.

Table 17.15 Iteration 3

CB_i	Basic variable	C_j 0 X_1	0 X_2	0 ϕ_1	0 ϕ_2	0 μ_1	0 μ_2	1 R_1	1 R_2	0 S_1	0 S_2	Solution
1	R_1	0	0	1	2	−1	0	1	0	−4	0	2*
0	X_1	1	1	0	0	0	0	0	0	1	0	1
0	μ_2	0	−2	−1	−3	0	1	0	−1	4	0	1
0	S_2	0	1	0	0	0	0	0	0	−2	1	2
	Z_j	0	0	1	2	−1	0	1	0	−4	0	2
	$C_j - Z_j$	0	0	−1*	−2	1	0	0	1	4	0	

In Table 17.15, the non-basic variable ϕ_2 has the maximum negative criterion value. Since S_2 is in the basis, ϕ_2 cannot enter the basis. So, the non-basic variable ϕ_1 which has the next highest negative criterion row value is considered as the entering variable. Since S_1 is not in the basis, ϕ_1 is the entering variable. The corresponding leaving variable is R_1. As per this combination of entering variable and leaving variable, pivot operations are shown in Table 17.16.

Table 17.16 Iteration 4

CB_i	Basic variable	X_1	X_2	ϕ_1	ϕ_2	μ_1	μ_2	R_1	R_2	S_1	S_2	Solution
	C_j	0	0	0	0	0	0	1	1	0	0	
0	ϕ_1	0	0	1	2	−1	0	1	0	−4	0	2
0	X_1	1	1	0	0	0	0	0	0	1	0	1
0	μ_2	0	−2	0	−1	−1	1	1	−1	0	0	3
0	S_2	0	1	0	0	0	0	0	0	−2	1	2
	Z_j	0	0	0	0	0	0	0	0	0	0	0
	$C_j - Z_j$	0	0	0	0	0	0	1	1	0	0	

Since the values in the criterion row of Table 17.16 are all nonnegative, the optimality is reached. The optimal solution of the problem is: $X_1^* = 1$, $X_2^* = 0$ and the corresponding maximum value of the objective function is 4.

Example 17.9   Solve the following quadratic programming problem using Wolfe's method:

$$\text{Maximize } Z = 40X_1 + 6X_2 - 4X_1X_2 - 2X_1^2 - 8X_2^2$$

subject to

$$6X_1 + 2X_2 \leq 36$$

$$X_1, X_2 \geq 0$$

Solution   The given problem is written as follows:

$$\text{Maximize } Z = c_1X_1 + c_2X_2 + c_3X_1^2 + c_4X_2^2 + c_5X_1X_2$$
$$= 40X_1 + 6X_2 - 2X_1^2 - 8X_2^2 - 4X_1X_2$$

subject to

$$6X_1 + 2X_2 \leq 36 \tag{17.52}$$

$$-X_1 \leq 0 \tag{17.53}$$

$$-X_2 \leq 0 \tag{17.54}$$

From the above model, the matrices that are required to generate the system of constraints except matrix D can be easily written. The matrix D is written as given below:

$$D = \begin{pmatrix} c_3 & c_5/2 \\ c_5/2 & c_4 \end{pmatrix} = \begin{pmatrix} -2 & -2 \\ -2 & -8 \end{pmatrix}$$

Therefore, the model of the given problem is rewritten as:

$$\text{Maximize } Z = \begin{bmatrix} 40 & 6 \end{bmatrix}\begin{pmatrix} X_1 \\ X_2 \end{pmatrix} + \begin{bmatrix} X_1 & X_2 \end{bmatrix}\begin{pmatrix} -2 & -2 \\ -2 & -8 \end{pmatrix}\begin{pmatrix} X_1 \\ X_2 \end{pmatrix}$$

subject to

$$\begin{bmatrix} 6 & 2 \end{bmatrix}\begin{pmatrix} X_1 \\ X_2 \end{pmatrix} \leq \begin{bmatrix} 36 \end{bmatrix}, \text{ where } -X_1 \leq 0 \text{ and } -X_2 \leq 0$$

Let, ϕ_1 be the Lagrangean multiplier associated with the constraint 1 of the system of constraints. μ_1 and μ_2 be the Lagrangean multipliers associated with the constraint 2 and constraint 3 of the system of constraints. S_1 is the slack variable of the constraint in the given model. Application of Kuhn-Tucker conditions to this model results into the following system of constraints.

$$\left|\begin{array}{cc|c|cc|c} 4 & 4 & 6 & -1 & 0 & 0 \\ 4 & 16 & 2 & 0 & -1 & 0 \\ \hline 6 & 2 & 0 & 0 & 0 & 1 \end{array}\right| \begin{pmatrix} X_1 \\ X_2 \\ \phi_1 \\ \mu_1 \\ \mu_2 \\ S_1 \end{pmatrix} = \begin{pmatrix} 40 \\ 6 \\ 36 \end{pmatrix}$$

$$\mu_1 X_1 = \mu_2 X_2 = 0$$
$$\phi_1 S_1 = 0$$

The above system of equations is simplified into the following:

$$4X_1 + 4X_2 + 6\phi_1 - \mu_1 = 40$$
$$4X_1 + 16X_2 + 2\phi_1 - \mu_2 = 6$$
$$6X_1 + 2X_2 + S_1 = 36$$
$$\mu_1 X_1 = \mu_2 X_2 = 0$$
$$\dot{\phi}_1 S_1 = 0$$

The model for Phase 1 of the two-phase simplex method irrespective of the nature of the objective function is presented below:

$$\text{Minimize } Z = R_1 + R_2$$

subject to

$$4X_1 + 4X_2 + 6\phi_1 - \mu_1 + R_1 = 40 \tag{17.55}$$
$$4X_1 + 16X_2 + 2\phi_1 - \mu_2 + R_2 = 6 \tag{17.56}$$
$$6X_1 + 2X_2 + S_1 = 36 \tag{17.57}$$
$$\mu_1 X_1 = 0 \tag{17.58}$$
$$\mu_2 X_2 = 0 \tag{17.59}$$
$$\dot{\phi}_1 S_1 = 0 \tag{17.60}$$

$X_1, X_2, R_1, R_2, \mu_1, \mu_2, \phi_1, S_1 \geq 0$, where, R_1 and R_2 are the artificial variables introduced in the constraints (17.55) and (17.56), respectively to serve as basic variables in the respective constraints.

The different iterations of the Phase 1 of the two-phase simplex method are shown in Table 17.17. In the first iteration of this table, entering variable is X_2 and the corresponding leaving variable is R_2. In the second iteration, the possibility that ϕ_1 becoming as the entering variable is prevented because of the presence of S_1 in the basis at this stage [i.e. $\phi_1 S_1$ must be zero]. Hence, in this iteration, X_1 is the entering variable and the corresponding leaving variable is X_2. Similarly, in the third iteration, the possibility that ϕ_1 becoming as the entering variable is prevented because of the presence of S_1 in the basis at this stage [i.e. $\phi_1 S_1$ must be zero]. Hence, in this iteration, μ_2 is the entering variable and the corresponding leaving variable is S_1. In the fourth iteration, the entering variable is ϕ_1 and leaving variable is R_1. From the final iteration of this table, the results are: $X_1 = 6$ and $X_2 = 0$, and the corresponding maximum objective function value, Z(optimum) is 168.

Table 17.17 Iterations of Two-Phase Simplex Method for Example 17.9

CB_i	C_j → Basic variable	0 X_1	0 X_2	0 ϕ_1	0 μ_1	0 μ_2	1 R_1	1 R_2	0 S_1	Solution	Ratio
1	R_1	4	4	6	−1	0	1	0	0	40	10
1	R_2	4	16	2	0	−1	0	1	0	6	3/8*
0	S_1	6	2	0	0	0	0	0	1	36	18
	Z_j	8	20	8	−1	−1	1	1	0	46	
	$C_j - Z_j$	−8	−20**	−8	1	1	0	0	0		
1	R_1	3	0	11/2	−1	1/4	1	−1/4	0	77/2	77/6
0	X_2	1/4	1	1/8	0	−1/16	0	1/16	0	3/8	3/2*
0	S_1	11/2	0	−1/4	0	1/8	0	−1/8	1	141/4	141/22
	Z_j	3	0	11/2	−1	1/4	1	−1/4	0	77/2	
	$C_j - Z_j$	−3**	0	−11/2	1	−1/4	0	5/4	0		
1	R_1	0	−12	4	−1	1	1	−1	0	34	34
0	X_1	1	4	1/2	0	−1/4	0	1/4	0	3/2	−
0	S_1	0	−22	−3	0	3/2	0	−3/2	1	27	18*
	Z_j	0	−12	4	−1	1	1	−1	0	34	
	$C_j - Z_j$	0	12	−4	1	−1**	0	2	0		
1	R_1	0	8/3	6	−1	0	1	0	−2/3	16	8/3*
0	X_1	1	1/3	0	0	0	0	0	1/6	6	−
0	μ_2	0	−44/3	−2	0	1	0	−1	2/3	18	−
	Z_j	0	8/3	6	−1	0	1	0	−2/3	16	
	$C_j - Z_j$	0	−8/3	−6**	1	0	0	1	2/3		
0	ϕ_1	0	4/9	1	−1/6	0	1/6	0	−1/9	8/3	
0	X_1	1	1/3	0	0	0	0	0	1/6	6	
0	μ_2	0	−124/9	0	−1/3	1	1/3	−1	4/9	70/3	
	Z_j	0	0	0	0	0	0	0	0	0	
	$C_j - Z_j$	0	0	0	0	0	0	0	0		

Note: *Key row, **Key column

17.5 SEPARABLE PROGRAMMING

Let us consider the following nonlinear programming problem:

$$\text{Maximize } Z = f(X_1, X_2, X_3,..., X_j,..., X_n)$$

subject to

$$g_i(X_1, X_2, X_3,..., X_j,..., X_n) \le b_i, \ i = 1, 2, 3,..., m$$
$$X_j \ge 0, \qquad j = 1, 2, 3,..., n$$

In the model, the terms of the objective function as well as the terms of the constraints may be nonlinear. *Separable programming* is a technique to deal with such problems. In separable programming, the nonlinear terms of the objective function and the constraints are converted into linear form by using piecewise linear approximation to each of the terms.

Consider a specific problem:

$$\text{Maximize } Z = 5X_1 + 3X_2^4$$

subject to

$$4X_1 + X_2^2 \le 20$$
$$X_1 \text{ and } X_2 \ge 0$$

The objective function can be rewritten as:

$$\text{Maximize } Z = f_1(X_1) + f_2(X_2)$$

subject to

$$g_{11}(X_1) + g_{12}(X_2) \le 20$$
$$X_j \ge 0, \qquad j = 1 \text{ and } 2$$

where,

$$f_1(X_1) = 5X_1$$
$$f_2(X_2) = 3X_2^4$$
$$g_{11}(X_1) = 4X_1$$
$$g_{12}(X_2) = X_2^2$$

Based on the above guidelines, a generalized model of separable programming is

$$\text{Maximize } Z = f_1(X_1) + f_2(X_2) + \cdots + f_j(X_j) + \cdots + f_n(X_n)$$

subject to

$$g_{11}(X_1) + g_{12}(X_2) + \cdots + g_{1j}(X_j) + \cdots + g_{1n}(X_n) \le b_1$$
$$g_{21}(X_1) + g_{22}(X_2) + \cdots + g_{2j}(X_j) + \cdots + g_{2n}(X_n) \le b_2$$
$$\vdots \qquad \vdots \qquad \qquad \vdots \qquad \qquad \vdots \qquad \vdots$$
$$g_{i1}(X_1) + g_{i2}(X_2) + \cdots + g_{ij}(X_j) + \cdots + g_{in}(X_n) \le b_i$$
$$\vdots \qquad \vdots \qquad \qquad \vdots \qquad \qquad \vdots \qquad \vdots$$
$$g_{m1}(X_1) + g_{m2}(X_2) + \cdots + g_{mj}(X_j) + \cdots + g_{mn}(X_n) \le b_m$$
$$X_j \ge 0, \qquad j = 1, 2, 3,..., n$$

Since in the above example, the term of the objective function as well as the term of the constraint with respect to the variable X_1 is in linear form, piecewise linear approximation is not needed for this variable. But, the term of the objective function and the term of the constraint with respect to the variable X_2 are nonlinear. So, piecewise linearization is applied to these terms as shown.

Let K be the number of break points of the variable X_2 starting from 0. The value of K can be estimated from the given constraint by setting the variable X_1 to 0. (A detailed guideline to determine the value of K is given in the solved problem.) Based on these informations, the variable X_2 is defined as:

$$X_2 = t_{21}a_{21} + t_{22}a_{22} + \cdots + t_{2k}a_{2k} + \cdots + t_{2K}a_{2K}$$

where a_{2k} is the kth break-point value of X_2, t_{2k} is the weight of the kth break-point value of X_2 and

$$t_{21} + t_{22} + \cdots + t_{2k} + \cdots + t_{2K} = 1$$

Using the guidelines, $f_2(X_2)$ and $g_{12}(X_2)$ can be derived using the following formulae.

$$f_2(X_2) = t_{21}\, f_2(a_{21}) + t_{22}\, f_2(a_{22}) + \cdots + t_{2k}\, f_2(a_{2k}) + \cdots + t_{2K}\, f_2(a_{2K})$$

$$g_{12}(X_2) = t_{21}\, g_{12}(a_{21}) + t_{22}\, g_{12}(a_{22}) + \cdots + t_{2k}\, g_{12}(a_{2k}) + \cdots + t_{2K}\, g_{12}(a_{2K})$$

Now, these formulae are to be substituted in the original model in place of $f_2(X_2)$ and $g_{12}(X_2)$ to get a revised linear model. Then, one can apply the conventional simplex algorithm with restricted basis method to solve this problem.

Restricted basis method

While carrying out the iterations of the simplex method applied to any separable programming problem, the current basis should not contain any of the t_{ik} type variables, or one of its immediate neighbouring weights (preceding weight $t_{i(k-1)}$ or succeeding weight $t_{i(k+1)}$, if any of these exists) must be present in the existing basis.

If t_{ik} is to be treated as an entering variable, any one of the above two conditions should be satisfied; otherwise, the non-basic variable which has the next best criterion value should be tried as an entering variable.

Example 17.10 Solve the following nonlinear programming problem using the restricted basis method of separable programming:

$$\text{Maximize } Z = 2X_1^3 + \frac{5}{2}\, X_2$$

subject to

$$2X_1^2 + 3X_2 \leq 16$$
$$X_1,\, X_2 \geq 0$$

Solution   Let,

$$f_1(X_1) = 2X_1^3, \qquad f_2(X_2) = \frac{5}{2}X_2, \qquad g_{11}(X_1) = 2X_1^2, \qquad g_{12}(X_2) = 3X_2$$

Based on the above equations, the revised model is:

$$\text{Maximize } Z = f_1(X_1) + f_2(X_2)$$

subject to

$$g_{11}(X_1) + g_{12}(X_2) \leq 16$$
$$X_1 \text{ and } X_2 \geq 0$$

In this model, $f_2(X_2)$ and $g_{12}(X_2)$ are already in linear form. But, the terms $f_1(X_1)$ and $g_{11}(X_1)$ are in nonlinear form. Hence, these terms require piecewise linearization.

Determination of number of break points (K) of X_1. From the given constraint, one can verify the fact that the maximum permitted value of X_1 by assuming zero for X_2 is $8^{1/2}$ which is approximately 3. So, the number of break points is assumed as 4 (i.e. maximum permitted value of X_1 plus one, $3 + 1$) and the corresponding integer break point values of X_1 starting from 0 are 0, 1, 2 and 3.

The expression for X_1 in terms of its break-point values (a_{1k}, $k = 1, 2, 3, 4$) and their weightages (t_{1k}, $k = 1, 2, 3, 4$) is

$$X_1 = t_{11}a_{11} + t_{12}a_{12} + t_{13}a_{13} + t_{14}a_{14}$$

The evaluations of the nonlinear terms for different break-point values are summarized in Table 17.18.

Table 17.18 Evaluations of $f_1(a_{1k})$ and $g_{11}(a_{1k})$

Break point (k)	Break point value (a_{1k})	$f_1(a_{1k})$	$g_{11}(a_{1k})$
1	0	0	0
2	1	2	2
3	2	16	8
4	3	54	18

Therefore,

$$f_1(X_1) = t_{11}\,f_1(a_{11}) + t_{12}\,f_1(a_{12}) + t_{13}\,f_1(a_{13}) + t_{14}\,f_1(a_{14}) = 2t_{12} + 16t_{13} + 54t_{14}$$

and

$$g_{11}(X_1) = t_{11}\,g_{11}(a_{11}) + t_{12}\,g_{12}(a_{12}) + t_{13}\,g_{13}(a_{13}) + t_{14}\,g_{14}(a_{14}) = 2t_{12} + 8t_{13} + 18t_{14}$$

Here, t_{11}, t_{12}, t_{13} and t_{14} are the weights of the break-point values (a_{11}, a_{12}, a_{13} and a_{14}), respectively. The sum of all these weights must be equal to 1. The modified model after substituting these formulae is:

$$\text{Maximize } Z = 2t_{12} + 16t_{13} + 54t_{14} + \frac{5}{2}X_2$$

subject to

$$2t_{12} + 8t_{13} + 18t_{14} + 3X_2 \leq 16$$
$$t_{11} + t_{12} + t_{13} + t_{14} = 1$$
$$t_{1k} \geq 0,\ k = 1, 2, 3, 4 \text{ and } X_2 \geq 0$$

The canonical form of the above model is:

$$\text{Maximize } Z = 2t_{12} + 16t_{13} + 54t_{14} + \frac{5}{2}X_2$$

subject to

$$2t_{12} + 8t_{13} + 18t_{14} + 3X_2 + S_1 = 16 \tag{17.61}$$
$$t_{11} + t_{12} + t_{13} + t_{14} = 1 \tag{17.62}$$
$$t_{1k} \geq 0,\ k = 1, 2, 3, 4 \quad \text{(Set 1)}$$
$$X_2,\ S_1 \geq 0 \quad \text{(Set 2)}$$

where S_1 is the slack variable of the first constraint.

In the above model, t_{11} is treated as the basic variable of the constraint (17.62) because it has unit coefficient in that constraint and zero coefficient in the constraint (17.61). This means that there is no need to introduce an artificial variable in the second constraint. The initial table of this linear programming problem is shown in Table 17.19.

Table 17.19 Initial Table

| CB_i | C_j | 0 | 2 | 16 | 54 | 5/2 | 0 | |
	Basic variable	t_{11}	t_{12}	t_{13}	t_{14}	X_2	S_1	Solution
0	S_1	0	2	8	18	3	1	16
0	t_{11}	1	1	$\boxed{1}$	1	0	0	1*
	Z_j	0	0	0	0	0	0	0
	$C_j - Z_j$	0	2	16*	54	5/2	0	

Here, t_{14} has the highest criterion value. As per feasibility condition, S_1 should leave the basis. Under this circumstance, t_{14} can enter the basis provided its neighbour t_{13} presents in the current basis. Since, t_{13} is not in the current basis, t_{14} cannot enter the current basis. Then, the non-basic variable t_{13} which has the next maximum criterion value is tried as an entering variable. If the introduction of t_{13} into the basis is compulsory, t_{11} (which is not a neighbour of t_{13}) must be dropped from the basis. With respect to the variable t_{13}, based on the feasibility condition, t_{11} is the leaving variable which helps t_{13} to enter the basis. So, t_{13} is treated as the entering variable. The results of pivot operations with respect to this combination of entering variable and leaving variable are summarized in Table 17.20.

Table 17.20 Iteration 1

| CB_i | C_j | 0 | 2 | 16 | 54 | 5/2 | 0 | |
	Basic variable	t_{11}	t_{12}	t_{13}	t_{14}	X_2	S_1	Solution
0	S_1	−8	−6	0	$\boxed{10}$	3	1	8*
16	t_{13}	1	1	1	1	0	0	1
	Z_j	16	16	16	16	0	0	16
	$C_j - Z_j$	−16	−14	0	38*	5/2	0	

In Table 17.20, t_{14} has the highest criterion value for which the corresponding leaving variable is S_1. Since, t_{14} has its neighbour t_{13} in the current basis, the variable t_{14} is treated as the entering variable at this stage. The results of pivot operations with respect to this combination of entering variable and leaving variable are summarized in Table 17.21.

In Table 17.21, t_{11} has the maximum criterion row value. So, it can enter the basis provided it has its neighbour t_{12} in the current basis. Since, t_{12} is absent in the current basis, t_{11} cannot enter the basis. The nonbasic variable t_{12} has the next highest criterion row value and the corresponding leaving variable is t_{13}. The introduction of t_{12} into the basis will remove its neighbour t_{13} from the basis. So, t_{12} cannot act as the entering variable in this iteration. Since all other criterion row values are non-positive, the optimality is reached. The corresponding results are $t_{13}^* = 1/5$, $t_{14}^* = 4/5$, and all other variables are zero, $Z(\text{maximum}) = 232/5$.

Table 17.21 Iteration 2

CB_i	Basic variable	C_j	0	2	16	54	5/2	0	
			t_{11}	t_{12}	t_{13}	t_{14}	X_2	S_1	Solution
54	t_{14}		−4/5	−3/5	0	1	3/10	1/10	4/5
16	t_{13}		9/5	8/5	1	0	−3/10	−1/10	1/5
	Z_j		−72/5	−34/5	16	54	57/5	19/5	232/5
	$C_j - Z_j$		72/5	44/5	0	0	−89/10	−19/5	

$$X_1^* = t_{11}a_{11} + t_{12}a_{12} + t_{13}a_{13} + t_{14}a_{14} = 0 \times 0 + 0 \times 1 + \frac{1}{5} \times 2 + \frac{4}{5} \times 3 = 2.8$$

and

$$X_2^* = 0$$

The value of the objective function by substituting the value of X_1^* and X_2^* in it is 43.904. So, the final results are as presented below. It should be noted that the solution need not be optimal.

$$X_1^* = 2.8, \quad X_2^* = 0, \quad Z = 43.904$$

The optimal results of this problem using a computer program for complete enumeration technique are given below:

$$X_1 = 2.8284, \quad X_2 = 0.0001, \quad Z(\text{optimum}) = 45.25378$$

It is clear that the solution using the separable programming problem is almost closer to the optimal problem.

17.6 CHANCE-CONSTRAINED PROGRAMMING OR STOCHASTIC PROGRAMMING

Consider the following mathematical model wherein a chance is associated with each constraint.

$$\text{Maximize } Z = \sum_{j=1}^{n} C_j X_j$$

subject to

$$P\left(\sum_{j=1}^{n} a_{ij} X_j \leq b_i \right) = (1 - \alpha_i), \qquad i = 1, 2, 3,..., m$$

$$X_j \geq 0, \qquad j = 1, 2, 3,..., n$$

where

α_i = Significance level of the constraint i. (This means that the probability of the constraint i becoming untrue is α_i.)

$1 - \alpha_i$ = Level of confidence of the constraint i. [This means that the probability of the constraint i becoming true is $(1 - \alpha_i)$.]

a_{ij} = Random variable following, say, normal distribution with a mean of $\mu 1_{ij}$ and a variance of $\sigma 1_{ij}^2$, representing the coefficient of the variable X_j in the constraint i.

b_i = Random variable following say, normal distribution with a mean of $\mu2_i$ and a variance of $\sigma2_{ij}^2$, representing the right-hand side value in the constraint i.

Therefore,

$$\sum_{j=1}^{n} \mu1_{ij} X_j + K_{\alpha i}\left(\sum_{j=1}^{n} \sigma1_{ij}^2 X_j^2\right)^{1/2} \leq \mu2_i + K_{\alpha i}\sigma2_i, \qquad i = 1, 2,..., m$$

$K_{\alpha i}$ is the value of Z (standard normal statistic) from normal table for the significance level of α_i. Now, the revised deterministic model is:

$$\text{Maximize } Z = \sum_{j=1}^{n} C_j X_j$$

subject to

$$\sum_{j=1}^{n} \mu1_{ij} X_j + K_{\alpha i}\left(\sum_{j=1}^{n} \sigma1_{ij}^2 X_j^2\right)^{1/2} \geq \mu2_i + K_{\alpha i}\sigma2_i, \qquad i = 1, 2,..., m$$

$$X_j \geq 0, \qquad j = 1, 2, 3,..., n$$

This model can be solved using separable programming.

Example 17.11 Develop a deterministic version of the following stochastic model:

$$\text{Maximize } Z = X_1 + 2X_2 + 5X_3$$

subject to

$$P(10X_1 + 15X_2 + 2X_3 \leq b_1) = 0.80 \qquad (17.63)$$

$$P(4X_1 + a_2X_2 + a_3X_3 \leq 20) = 0.90 \qquad (17.64)$$

$$X_1, X_2, X_3 \geq 0$$

In this model b_1, a_2 and a_3 are random variables following normal distribution. The mean and variance of each of these random variables are summarized in Table 17.22.

Table 17.22 Mean and Variance of Random Variables

Random variable	Mean, μ	Variance, σ^2
b_1	40	25
a_2	6	4
a_3	3	2

Solution In the constraint (17.63), $1 - \alpha_1$ is equal to 0.8. Therefore, the value of α_1 is equal to 0.2. Similarly, in the second constraint, $1 - \alpha_2$ is equal to 0.9. Therefore, the value of α_2 is equal to 0.1. The corresponding Z values of the normal table are 0.84 and 1.28, respectively, which means,

$$K_{\alpha 1} = 0.84, \qquad K_{\alpha 2} = 1.28$$

Based on these guidelines, the constraints (17.63) is:

$$10X_1 + 15X_2 + 2X_3 \leq \mu_1 + K_{\alpha 1}\sigma_1$$

Substitution of μ_1, $K_{\alpha 1}$ and σ_1 in the above constraint yields the following form:

$$10X_1 + 15X_2 + 2X_3 \leq 40 + 0.84(25)^{1/2}$$

or

$$10X_1 + 15X_2 + 2X_3 \leq 44.2 \tag{17.65}$$

Similarly, the general format of the constraint (17.64) is:

$$4X_1 + \mu_{22}X_2 + \mu_{23}X_3 + K_{\alpha 2}(\sigma_{22}^2 X_2^2 + \sigma_{23}^2 X_3^2)^{1/2} \leq 20$$

or

$$4X_1 + 6X_2 + 3X_3 + 1.28(4X_2^2 + 2X_3^2)^{1/2} \leq 20$$

In this constraint, make the following substitution:

$$4X_2^2 + 2X_3^2 = Y^2$$

Therefore, the modified final form of the constraint is:

$$4X_1 + 6X_2 + 3X_3 + 1.28Y \leq 20 \tag{17.66}$$

Based on the formula for Y^2 which is given earlier [under the constraint (17.64)], an *additional constraint* is generated as follows:

$$4X_2^2 + 2X_3^2 - Y^2 = 0 \tag{17.67}$$

Based on the foregoing discussions, the final *deterministic model* of the given problem is as follows:

$$\text{Maximize } Z = X_1 + 2X_2 + 5X_3$$

subject to

$$10X_1 + 15X_2 + 2X_3 \leq 44.2 \qquad \text{[same as (17.65)]}$$
$$4X_1 + 6X_2 + 3X_3 + 1.28Y \leq 20 \qquad \text{[same as (17.66)]}$$
$$4X_2^2 + 2X_3^2 - Y^2 = 0 \qquad \text{[same as (17.67)]}$$
$$X_1, X_2 \text{ and } Y \geq 0$$

The above model can be solved using the separable programming which is left as an exercise.

Example 17.12   A company manufactures two products, P_1 and P_2. The details of machine-hour requirements, profit per unit and the maximum available machine hours of the two machine types are summarized in Table 17.23. Each non-deterministic data follows normal distribution and its mean and variance are given in the same table. Formulate a chance constrained programming model for this situation to find the production volume of each product such that the total profit is maximized.

Table 17.23   Data of Example 17.12

Machine Type	Product		Maximum hours available	Probability
	P_1	P_2		
1	8	a_{12}	b_1	0.95
		$\mu_{12} = 6$	$\mu_1 = 96$	
		$\sigma_{12}^2 = 25$	$\sigma_1^2 = 64$	
2	a_{21}	12	100	0.90
	$\mu_{21} = 4$			
	$\sigma_{21}^2 = 4$			
Profit/Unit	Rs. 50	Rs. 65		

Solution Let, X_1 and X_2 be the production volume of P_1 and P_2, respectively. The model of this problem is shown below:

$$\text{Maximize } Z = 50X_1 + 65X_2$$

subject to

$$P(8X_1 + a_{12}X_2 \leq b_1) \geq 0.95$$
$$P(a_{21}X_1 + 12X_2 \leq 100) \geq 0.90$$
$$X_1, X_2 \geq 0$$

In the first constraint, $1 - \alpha_1 = 0.95$. Therefore, $\alpha_1 = 0.05$ and $K_{\alpha 1} = 1.64$.
In the second constraint, $1 - \alpha_2 = 0.90$. Therefore, $\alpha_2 = 0.10$ and $K_{\alpha 2} = 1.28$.

The first constraint is rewritten as:

$$P(8X_1 + a_{12}X_2 - b_1 \leq 0) \geq 0.95$$

The general form of the first constraint is:

$$8X_1 + \mu_{12}\, X_2 - \mu_1 + K_{\alpha 1}\, (\sigma_{12}^2 X_2^2 + \sigma_1^2)^{0.5} \leq 0$$

From the above general constraint, the specific constraint for the first constraint is shown below:

$$8X_1 + 6X_2 - 96 + 1.64\,(25X_2^2 + 64\,)^{0.5} \leq 0$$

By substituting $Y^2 = 25\,X_2^2 + 64$ in the above constraint, we get,

$$8X_1 + 6X_2 - 96 + 1.64Y \leq 0$$
$$\text{i.e. } 8X_1 + 6X_2 + 1.64Y \leq 96 \tag{17.68}$$

The general form of the second constraint is:

$$\mu_{21}X_1 + K_{\alpha 2}\,(\sigma_{21}^2 X_1^2)^{0.5} + 12X_2 \leq 100$$

From the above general constraint, the specific constraint for the second constraint is shown below.

$$4X_1 + 1.28\,(4X_1^2\,)^{0.5} + 12X_2 \leq 100$$
$$6.56X_1 + 12X_2 \leq 100 \tag{17.69}$$

From the equation $Y^2 = 25\,X_2^2 + 64$, we get the following constraint.

$$Y^2 - 25X_2^2 = 64 \tag{17.70}$$

The final model is as presented below.

$$\text{Maximize } Z = 50X_1 + 65X_2$$

subject to

$$8X_1 + 6X_2 + 1.64Y \leq 96 \qquad \text{same as (17.68)}$$
$$6.56X_1 + 12X_2 \leq 100 \qquad \text{same as (17.69)}$$
$$Y^2 - 25\,X_2^2 = 64 \qquad \text{same as (17.70)}$$
$$X_1, X_2,\, Y \geq 0$$

This model can be solved using separable programming which is left as an exercise.

QUESTIONS

1. List different types of nonlinear programming problems. Also, explain their application areas.

2. How will you construct the Lagrangean function for a model with maximization type objective function and with all 'less than or equal to' type constraints?

3. What is bordered Hessian matrix? Give an example.

4. Explain the steps of Lagrangean method.

5. State and explain Kuhn–Tucker conditions.

6. Solve the following nonlinear programming problem using Lagrangean multipliers method.

$$\text{Minimize } Z = 4X_1^2 + 2X_2^2 + X_3^2 - 4X_1X_2$$

subject to

$$X_1 + X_2 + X_3 = 15$$
$$2X_1 - X_2 + 2X_3 = 20$$
$$X_1, X_2 \text{ and } X_3 \geq 0$$

7. Solve the following nonlinear programming problem using Lagrangean multipliers method.

$$\text{Minimize } Z = X_1^2 + X_2^2 + X_3^2$$

subject to

$$X_1 + X_2 + 3X_3 = 2$$
$$5X_1 + 2X_2 + X_3 = 5$$
$$X_1, X_2 \text{ and } X_3 \geq 0$$

8. Solve the following nonlinear programming problem using Kuhn–Tucker conditions.

$$\text{Maximize } Z = 8X_1 + 10X_2 - X_1^2 - X_2^2$$

subject to

$$3X_1 + 2X_2 \leq 6$$
$$X_1 \text{ and } X_2 \geq 0$$

9. Explain the steps of Wolfe's method for solving quadratic programming problem.

10. Solve the following quadratic programming problem using Wolfe's method.

$$\text{Maximize } Z = 50X_1 + 15X_2 - 6X_1X_2 - 0.5X_1^2 - 10X_2^2$$

subject to

$$4X_1 + X_2 \leq 50$$
$$X_1 + X_2 \leq 40$$
$$X_1 \text{ and } X_2 \geq 0$$

11. What is separable programming? How will you determine the number of break points in the separable programming?

12. Solve the following nonlinear programming problem using separable programming:

$$\text{Maximize } Z = 4X_1^2 + 2X_2$$

subject to

$$3X_1^2 + 4X_2 \leq 24$$
$$X_1 \text{ and } X_2 \geq 0.$$

13. What is stochastic programming? Give some application areas of stochastic programming.

14. Develop a deterministic version of the following stochastic model:

$$\text{Maximize } Z = 2X_1 + 4X_2$$

subject to

$$P(9X_1 + 8X_2 \leq b_1) = 0.80$$

$$P(a_1X_1 + a_2X_2 \leq 25) = 0.90$$

$$X_1 \text{ and } X_2 \geq 0$$

In this model b_1, a_1 and a_2 are random variables following normal distribution. The mean and variance of each of these random variables are summarized in the following table:

Random variable	Mean, μ	Variance, σ^2
b_1	60	9
a_1	25	4
a_2	16	4

15. Develop a deterministic version of the following stochastic model:

$$\text{Maximize } Z = 2X_1 + 4X_2 + 6X_3$$

subject to

$$P(8X_1 + 6X_2 + 4X_3 \leq b_1) = 0.85$$

$$P(a_1X_1 + 4X_2 + a_3X_3 \leq 30) = 0.95$$

$$X_1, X_2 \text{ and } X_3 \geq 0$$

In this model b_1, a_1 and a_3 are random variables following normal distribution. The mean and variance of each of these random variables are summarized in the following table:

Random variable	Mean, μ	Variance, σ^2
b_1	50	16
a_1	16	4
a_3	25	9

Area Under Standard Normal Distribution from its Mean

Z	.00	.01	.02	.03	.04	.05	.06	.07	.08	.09
.0	.0000	.0040	.0080	.0120	.0160	.0199	.0239	.0279	.0319	.0359
.1	.0398	.0438	.0478	.0517	.0557	.0596	.0636	.0675	.0714	.0753
.2	.0793	.0832	.0871	.0910	.0948	.0987	.1026	.1064	.1103	.1141
.3	.1179	.1217	.1255	.1293	.1331	.1368	.1406	.1443	.1480	.1517
.4	.1554	.1591	.1628	.1664	.1700	.1736	.1772	.1808	.1844	.1879
.5	.1915	.1950	.1985	.2019	.2054	.2088	.2123	.2157	.2190	.2224
.6	.2257	.2291	.2324	.2357	.2389	.2422	.2454	.2486	.2517	.2549
.7	.2580	.2611	.2642	.2673	.2703	.2734	.2764	.2794	.2823	.2852
.8	.2881	.2910	.2939	.2967	.2995	.3023	.3051	.3078	.3106	.3133
.9	.3159	.3186	.3212	.3238	.3264	.3289	.3315	.3340	.3365	.3389
1.0	.3413	.3438	.3461	.3485	.3508	.3531	.3554	.3577	.3599	.3621
1.1	.3643	.3665	.3686	.3708	.3729	.3749	.3770	.3790	.3810	.3830
1.2	.3849	.3869	.3888	.3907	.3925	.3944	.3962	.3980	.3997	.4015
1.3	.4032	.4049	.4066	.4082	.4099	.4115	.4131	.4147	.4162	.4177
1.4	.4192	.4207	.4222	.4236	.4251	.4265	.4279	.4292	.4306	.4319
1.5	.4332	.4345	.4357	.4370	.4382	.4394	.4406	.4418	.4429	.4441
1.6	.4452	.4463	.4474	.4484	.4495	.4505	.4515	.4525	.4535	.4545
1.7	.4554	.4564	.4573	.4582	.4591	.4599	.4608	.4616	.4625	.4633
1.8	.4641	.4649	.4656	.4664	.4671	.4678	.4686	.4693	.4699	.4706
1.9	.4713	.4719	.4726	.4732	.4738	.4744	.4750	.4756	.4761	.4767
2.0	.4772	.4778	.4783	.4788	.4793	.4798	.4803	.4808	.4812	.4817
2.1	.4821	.4826	.4830	.4834	.4838	.4842	.4846	.4850	.4854	.4857
2.2	.4861	.4864	.4868	.4871	.4875	.4878	.4881	.4884	.4887	.4890
2.3	.4893	.4896	.4898	.4901	.4904	.4906	.4909	.4911	.4913	.4916
2.4	.4918	.4920	.4922	.4925	.4927	.4929	.4931	.4932	.4934	.4936
2.5	.4938	.4940	.4941	.4943	.4945	.4946	.4948	.4949	.4951	.4952
2.6	.4953	.4955	.4956	.4957	.4959	.4960	.4961	.4962	.4963	.4964
2.7	.4965	.4966	.4967	.4968	.4969	.4970	.4971	.4972	.4973	.4974
2.8	.4974	.4975	.4976	.4977	.4977	.4978	.4979	.4979	.4980	.4981
2.9	.4981	.4982	.4982	.4983	.4984	.4984	.4985	.4985	.4986	.4986
3.0	.4987	.4987	.4987	.4988	.4988	.4989	.4989	.4989	.4990	.4990

SUGGESTED FURTHER READING

Baker, K.R., 1972, *Introduction to Sequencing and Scheduling*, John Wiley and Sons, New York.

Budnick, F.S., McLeavey, D. and Mojena, R., 1988, *Principles of Operations Research for Management*, 2nd ed., Richard D. Irwin Inc., Illinois.

Conway, R.W., Maxwell, W.L. and Miller, L.W., 1967, *Theory of Scheduling*, Addison-Wesley, Reading, Mass.

Curry, G.L. and Skeith, R.W., 1969, A dynamic programming algorithm for facility location and allocation, *AIIE Transactions*, 1, 133–138.

Francis, R.L., et al., 1974, *Facility Layout and Location*, Prentice-Hall, New Jersey.

Gopalakrishnan, P. and Sundaresan, M., 1979, *Materials Management: An Integrated Approach*, Prentice-Hall of India, New Delhi.

Hamdy, A. Taha, 1997, *Operations Research—An Introduction*, 6th ed., Prentice-Hall of India, New Delhi.

Panneerselvam, R., 1990, A Heuristic algorithm for total covering problem, *Industrial Engineering Journal*, Vol. 19, **2**, 1–10. (Received the best article award in traditional areas of Industrial Engineering for the year 1990.)

Panneerselvam, R., 1991, Modelling the single machine scheduling problem to minimize total tardiness and weighted total tardiness, *International Journal of Management and Systems*, Vol. 7, **1**, 37–48.

Panneerselvam, R., 1997, Strategies for productivity improvement, *Pondicherry Catalyst*, Vol. 1, **1**, 51–60.

Panneerselvam, R., 2005, *Production and Operations Management*, 2nd ed., Prentice-Hall of India, New Delhi.

Panneerselvam, R. and Balasubramanian, K.N., 1991, Single machine scheduling model to minimize the sum of the weighted number of early and tardy jobs, *Industrial Engineering Journal*, Vol. 20, **7**, 6–9.

Panneerselvam, R. and Rajkumar, G., 1991, Modelling the transportation problem with quantity discounts, *International Journal of Management and Systems*, Vol. 7, **3**, 197–210.

Panneerselvam, R., et al., 1990, Group technology machine-component cell formation, *Industrial Engineering Journal*, **9**, 22–32. (Also a corrigendum. *Industrial Engineering Journal*, **19**, 21)

Panneerselvam, R., Balasubramanian, K.N. and Thiagarajan, M.T., 1990, Models for warehouse location problem, *International Journal of Management and Systems*, **6**, 1–8.

Panneerselvam, R., et al., 1996, Heuristic for partial covering problem, *Industrial Engineering Journal*, Vol. 25, **10**, 11–16.

Plossl, G.W. and Wight, O.W., 1967, *Production and Inventory Control*, Prentice Hall, New Jersey.

Rajkumar, G. and Panneerselvam, R., 1992, Quantitative tools for public systems design and operations, *Indian Economic Panorama*, Vol. 2, **3**, 23–27.

Shah, A.M., 1996, Environmental factors for strategy formulation, *Productivity*, Vol. 36, **4**, 594–599.

Toregas, C., Swain, R., Revelle, C. and Bergman, L., 1971, The location of emergency service facilities, *Operations Research*, **19**, 1363–1373.

ANSWERS TO QUESTIONS

Chapter 2

2.4 Maximize $Z = 1000X_1 + 800X_2$, subject to: $2X_1 + X_2 \leq 40$, $2X_1 + 3X_2 \leq 80$, where X_1 and $X_2 \geq 0$.

Solution: $X_1 = 10$, $X_2 = 20$ and $Z = 26{,}000$.

2.5 Model: Maximize $Z = 300X_1 + 200X_2 + 400X_3$, subject to: $4X_1 + 3X_2 + 5X_3 \leq 2000$, $2X_1 + 2X_2 + 4X_3 \leq 2500$, where $X_1 \leq 100$, $X_2 \leq 200$, $X_3 \leq 50$ and $X_1 \geq 50$; and X_1, X_2 and $X_3 \geq 0$.

Solution: $X_1 = 100$, $X_2 = 200$, $X_3 = 50$ and $Z = 90{,}000$.

2.7 $X_1 = 0$, $X_2 = 15$ and $Z = 1200$.

2.8 $X_1 = 6$, $X_2 = 12$ and $Z = 240$.

2.9 $X_1 = 40$, $X_2 = 0$ and $Z = 2400$.

2.10 $X_1 = 8$, $X_2 = 0$ and $Z = 360$.

2.11 The solution is unbounded.

2.12 $X_1 = 5$, $X_2 = 0$, $X_3 = 4$ and $Z = 35$.

2.13 $X_1 = 6$, $X_2 = 0$, $X_3 = 8$ and $Z = 86$.

2.14 $X_1 = 8$, $X_2 = 0$ and $Z = 48$. The model has alternate optima at all points on the line joining the points $(2.4, 8.4)$ and $(8, 0)$.

2.15 The solution is infeasible.

2.16 $X_1 = 3$, $X_2 = 2$, $X_3 = 0$ and $Z = 7$.

2.17 $X_1 = 0.25$, $X_2 = 1.25$ and $Z = 10$.

2.18 Maximize $Y = 7Y_1 + 4Y_2 - 10Y_3 + 3Y_4 + 2Y_5$, subject to:

$3Y_1 + 6Y_2 - 7Y_3 + Y_4 + 4Y_5 \leq 3$, $5Y_1 + Y_2 + 2Y_3 - 2Y_4 + 7Y_5 \leq -2$, $4Y_1 + 3Y_2 + Y_3 + 5Y_4 - 2Y_5 \leq 4$, where Y_1, Y_2, Y_3, Y_4, $Y_5 \geq 0$.

2.19 Minimize $Y = 20Y_1 + 10Y_2 + 25Y_3$, subject to: $4Y_1 + 5Y_2 + 6Y_3 \geq 5$, $7Y_1 + 2Y_2 + 8Y_3 \geq 6$, where Y_1, Y_2, Y_3 are unrestricted in sign,

2.20 $X_1 = 4$, $X_2 = 2$ and $Z = 10$.

2.21 $X_1 = 1$, $X_2 = 1$, $X_3 = 0$ and $Z = 5$. Solution after change in RHS: $X_1 = 2$, $X_2 = 0$, $X_3 = 1$ and $Z = 1$.

2.22 $X_1 = 2$, $X_2 = 0$, $X_3 = 1$ and $Z = 5$. Solution after change in objective coefficients: $X_1 = 2.2$, $X_2 = 0.4$, $X_3 = 0$ and $Z = 6.4$.

2.23 $X_1 = 0$, $X_2 = 4$ and $Z = 20$. There is no change in the solution after introducing a new variable.

2.24 $X_1 = 0$, $X_2 = 15$ and $Z = 1200$. There is no change in the solution after introducing a new constraint.

Chapter 3

3.5 $X_{11} = 20$, $X_{21} = 20$, $X_{22} = 6$, $X_{23} = 4$, $X_{33} = 4$, $X_{34} = 11$, $X_{44} = 7$, $X_{45} = 6$, Total cost $= 878$.

3.6   (a) $X_{11} = 2$, $X_{12} = 3$, $X_{22} = 6$, $X_{23} = 2$, $X_{33} = 7$, $X_{43} = 9$, $X_{4\text{Dummy}} = 5$. Total cost $= 112$.

       (b) $X_{1\text{Dummy}} = 5$, $X_{23} = 8$, $X_{32} = 7$, $X_{41} = 2$, $X_{42} = 2$, $X_{43} = 10$. Total cost $= 70$.

3.7 $X_{\text{F1W1}} = 5$, $X_{\text{F1W4}} = 2$, $X_{\text{F2W3}} = 7$, $X_{\text{F2W4}} = 2$, $X_{\text{F3W2}} = 8$, $X_{\text{F3W4}} = 10$. Total cost $= 734$.

3.8   (a) $X_{\text{AA1}} = 3$, $X_{\text{AB1}} = 1$, $X_{\text{BB1}} = 2$, $X_{\text{BC1}} = 4$, $X_{\text{BD1}} = 2$, $X_{\text{CD1}} = 3$, $X_{\text{CE1}} = 6$. Total cost $= 153$.

       (b) $X_{\text{AD1}} = 4$, $X_{\text{BB1}} = 2$, $X_{\text{BE1}} = 6$, $X_{\text{CA1}} = 3$, $X_{\text{CB1}} = 1$, $X_{\text{CC1}} = 4$, $X_{\text{CD1}} = 1$. Total cost $= 68$.

3.9 $X_{\text{F1S1}} = 1500$, $X_{\text{F1S4}} = 4000$, $X_{\text{F1S6}} = 1500$, $X_{\text{F2S6}} = 2000$, $X_{\text{F2S7}} = 2000$, $X_{\text{F3S2}} = 2000$, $X_{\text{F3S3}} = 4500$, $X_{\text{F3S5}} = 2500$, $X_{\text{F3S7}} = 1000$. Total cost $= 63{,}000$.

3.10   (a) $X_{\text{XB}} = 76$, $X_{\text{YA}} = 21$, $X_{\text{YC}} = 41$, $X_{\text{ZA}} = 51$, $X_{\text{ZB}} = 26$.

       (b)   Total cost $= 2424$.

       (c)   Solution is not unique: $X_{\text{XB}} = 76$, $X_{\text{YB}} = 21$, $X_{\text{YC}} = 41$, $X_{\text{ZA}} = 72$, $X_{\text{ZB}} = 5$.

3.11 $X_{\text{1D}} = 200$, $X_{\text{2B}} = 15$, $X_{\text{2E}} = 160$, $X_{\text{3B}} = 75$, $X_{\text{3C}} = 45$, $X_{\text{3D}} = 30$, $X_{\text{4A}} = 110$, $X_{\text{4C}} = 75$, $X_{\text{4F}} = 140$. Total cost $= 7{,}285$.

3.12 $X_{\text{AI}} = 50$, $X_{\text{AII}} = 100$, $X_{\text{AIV}} = 150$, $X_{\text{BI}} = 200$, $X_{\text{CI}} = 150$, $X_{\text{CIII}} = 350$, $X_{\text{DummyII}} = 150$. Total profit $= 30{,}700$.

3.14 $X_{\text{S1S2}} = 200$, $X_{\text{S2D1}} = 450$, $X_{\text{S3D2}} = 200$, $X_{\text{S4D2}} = 450$, $X_{\text{D2D1}} = 100$. Total cost 7950.

3.15 $X_{\text{S1D2}} = 400$, $X_{\text{S1D3}} = 400$, $X_{\text{S2D1}} = 500$, $X_{\text{S2D3}} = 200$. Total cost $= 9900$.

3.16 $X_{\text{AW1}} = 400$, $X_{\text{BW2}} = 500$, $X_{\text{CW2}} = 600$, $X_{\text{W1Z}} = 400$, $X_{\text{W2X}} = 300$, $X_{\text{W2Y}} = 700$, $X_{\text{W2Z}} = 100$. Total cost $= 72{,}000$.

Chapter 4

4.5   Minimize $Z = 5X_{11} + 6X_{12} + 8X_{13} + 6X_{14} + 4X_{15} + 4X_{21} + 8X_{22} + 7X_{23} + 7X_{24} + 5X_{25} + 7X_{31} + 7X_{32} + 4X_{33} + 5X_{34} + 4X_{35} + 6X_{41} + 5X_{42} + 6X_{43} + 7X_{44} + 5X_{45} + 4X_{51} + 7X_{52} + 8X_{53} + 6X_{54} + 8X_{55}$

Subject to

$$\sum_{j=1}^{5} X_{ij} = 1, \quad i = 1, 2, 3, 4, 5 \quad \text{and} \quad \sum_{i=1}^{5} X_{ij} = 1, \quad j = 1, 2, 3, 4, 5$$

$$X_{ij} = 0 \text{ or } 1, \quad i = 1, 2, 3, 4, 5, \quad j = 1, 2, 3, 4, 5$$

4.6 $X_{15} = 1$, $X_{21} = 1$, $X_{32} = 1$, $X_{43} = 1$. $X_{54} = 1$. Total time $= 71$ hrs.

4.7 $X_{11} = 1$, $X_{26} = 1$, $X_{34} = 1$, $X_{43} = 1$, $X_{52} = 1$, $X_{65} = 1$. Total time $= 118$ hrs.

4.8 $X_{12} = 1$, $X_{23} = 1$, $X_{31} = 1$, $X_{44} = 1$. Total annual sales $=$ Rs. 39 crores.

4.9   Pairing of flights: 204–101 (base city Y), 102–203 (base city X), 202–103 (base city Y) and 104–201 (base city X). Total minimum stay-over time of the crews in non-base cities $= 22.5$ hrs.

4.10 $X_{12} = 1$, $X_{21} = 1$, $X_{34} = 1$, $X_{43} = 1$. Total time $= 33$ hrs.

Chapter 5

5.6 (b) Shortest path is 1–2–7–9 and the corresponding total distance is 24.

    (c) Same as in (b).

5.7 (b)

$$D^5 \qquad\qquad\qquad P^5$$

	1	2	3	4	5
1	0	3	7	9	15
2	3	0	4	6	12
3	7	4	0	2	8
4	9	6	2	0	6
5	15	12	8	6	0

	1	2	3	4	5
1	0	1	2	3	3
2	2	0	2	3	3
3	2	3	0	3	3
4	2	3	4	0	4
5	2	3	5	5	0

    (c) (i) Shortest path from 1 to 5: 1–2–3–5 and distance = 15.

        (ii) Shortest path from 2 to 5: 2–3–5 and distance = 12.

5.8 (b) Arcs of minimum spanning tree: 1–2, 1–3, 2–5, 3–4, 3–6, 3–7, 6–8, 6–9, 6–10.

    Total distance = 46.

5.9 Same as in 5.8.

5.11 Maximal flow = 95. Flow details in arcs: 1–2 = 60, 1–3 = 35, 2–3 = 25, 2–4 = 10, 2–5 = 25, 3–4 = 30, 3–5 = 30, 4–6 = 40 and 5–6 = 55.

Chapter 6

6.2 Maximize $Z = 75{,}000X_1 + 1{,}00{,}000X_2 + 80{,}000X_3$, subject to: $2000X_1 + 1500X_2 + 1000X_3 \le 38{,}000$, $1000X_1 + 1500X_2 + 2000X_3 \le 33{,}000$, where X_1, X_2 and $X_3 \ge 0$ and integers.

6.5 $X_1 = 17$, $X_2 = 0$, $X_3 = 7$ and $Z = 206$.

6.6 $X_1 = 4$, $X_2 = 13$ and $Z = 110$.

6.7 $X_1 = 1$, $X_2 = 1$ and $Z = 7$.

6.8 $X_1 = 4$, $X_2 = 1$ and $Z = 32$.

6.9 $X_1 = 6$, $X_2 = 0$ and $Z = 64$.

6.10 $Y_1 = Y_2 = Y_3 = 1$, $Y_4 = Y_5 = 0$ and $Z = 105$ thousands of rupees.

Chapter 7

7.6 EOQ = 849 units, No. of orders = 28.27 and $t^* = 0.035$ year.

7.7 EOQ = 1200 units, $t^* = 0.0334$ year.

7.8 (a) 1929 units (b) 1867 units (c) 62 units (d) 0.0536 year (e) 0.0519 year (f) 0.0017 year.

7.9 $Q^* = 5543$ units, $Q^*_1 = 2310$ units, $Q^*_2 = 462$ units, $t^* = 0.231$ year, $t^*_1 = 0.0963$ year, $t^*_2 = 0.0963$ year, $t^*_3 = 0.01925$ year and $t^*_4 = 0.01925$ year.

7.10 (a) 656 units (b) 2156 units.

7.11 When ordering cost per order = Rs. 600, EOQ = 3000 units.

    When ordering cost per order = Rs. 300, EOQ = 3000 units.

7.12 (a) 3547 units (b) 7355 units.

7.14 $\mu = 3.08404$, $Q_1 = 720.2772$ units = 720 units (approx.)

$Q_2 = 519.8152$ units = 520 units (approx.)

7.15

Item number (i)	EOQ in units (Q_i)	Interval multiple (n_i)
1	1683.1080	1
2	224.4143	1
3	2244.1430	1

7.16 $Q_1 = 1162.873$ units, $Q_2 = 830.6233$ units, $Q_3 = 1328.997$ units, $Q_4 = 1495.122$ units and Total cost = Rs. 3994.81.

7.17

Item number (i)	EOQ in units (Q_i)	Interval multiple (n_i)
1	387.1050	1
2	172.0467	1
3	258.0700	1

7.18 39 packets.

7.19 406 meals.

7.20 810.5 kg.

Chapter 8

8.2 Plant 1: Alternative 2; Plant 2: Alternative 3; Plant 3: Alternative 1. Total return = Rs. 32 crores.

8.3 Component 1: Alternative 2, Component 2: Alternative 2, Component 3: Alternative 1. Or, Component 1: Alternative 3, Component 2: Alternative 1, Component 3: Alternative 1. Total reliability = 0.6864.

8.4 Shortest path: 1–4–7–9–11. Total distance = 19.

8.5 Optimal cargo mix: No. of units of item 1, item 2 and item 3 are 0, 3 and 1, respectively. Optimal total return = Rs. 3000.

8.6 The optimal sequence of jobs: $4 - 1 - 3 - 2 - 5$ and total tardiness = 41.

8.7 $Z = \left(\dfrac{k}{n}\right)^n$, $p_1 = p_2 = \cdots = p_n = \dfrac{k}{n}$ units.

8.8 $Z = 625$, $p_1 = p_2 = p_3 = p_4 = 5$ units.

8.9 $X_1 = 30$, $X_2 = 0$ and $Z = 900$.

8.10 $X_1 = 35$, $X_2 = 0$ and $Z = 1050$.

Chapter 9

9.4 (a) 0.33 (b) 0.0134 (c) 0.0027 (d) $L_s = 2.03$ customers, $L_q = 1.36$ customers, $W_s = 0.0667$ hr and $W_q = 0.0447$ hr.

9.5 $L_q = 6.125$ vehicles and $P_0 = 32\%$. One more gate is not justified.

9.6 (a) $P_0 = 0.263$ and $p_3 = 0.1031$.

(b) $L_q = 0.02559$ vehicle, $L_s = 1.356$ vehicles, $W_q = 0.00107$ hr and $W_s = 0.05663$ hr.

9.7 $C = 3$. Total cost = Rs. 251.

9.8 $L_q = 2.38$ cars, $L_s = 3.3133$ cars, $W_q = 0.116$ hr and $W_s = 0.1615$ hr.

9.9 $P_0 = 0.1998$, $L_q = 1.0371$ ships, $L_s = 2.9671$ ships, $W_q = 0.04478$ week and $W_s = 0.1281$ week.

9.10 $P_0 = 0.0000004103$, $L_q = 2.8634$ machines, $L_s = 5.863$ machines, $W_q = 0.4772$ hr and $W_s = 0.9772$ hr.

9.11 $P_0 = 9.4018 \times 10^{-7}$, $L_q = 6.5$ machines, $L_s = 6.5$ machines, $W_q = 3.25$ hrs and $W_s = 3.25$ hrs.

9.17 GPSS program:

```
1    GENERATE     20, 2
2    QUEUE        CUSQ
3    SEIZE        CLERK
4    DEPART       CUSQ
5    ADVANCE      25, 7
6    RELEASE      CLERK
7    TERMINATE
8    GENERATE     480
9    TERMINATE    1
     START 1
```

9.18 GPSS program:

```
1    LANE    STORAGE      6
2            GENERATE     150, 30
3            QUEUE        VEHQU
4            ENTER        LANE
5            DEPART       VEHQU
6            ADVANCE      20, 10
7            LEAVE        LANE
8            TERMINATE
9            GENERATE     72000
10           TERMINATE    1
             START 1
```

9.19 GPSS program segment:

```
1    ITIME FUNCTION R4, D6
0.19, 1/0.44, 2/0.79, 3/0.89, 4/0.97, 5/0.99, 6
```

9.20 GPSS program:

```
        1   ITIME FUNCTION R2, D6
0.04, 1/0.19, 2/0.39, 3/0.69, 4/0.89, 5/0.99, 6
        2   MACHN   STORAGE      5
        3           GENERATE     ITIME
        4           QUEUE        JOBQ
        5           ENTER        MACHN
        6           DEPART       JOBQ
        7           ADVANCE      6, 2
        8           LEAVE        MACHN
        9           TERMINATE
       10           GENERATE     600
       11           TERMINATE    1
                    START        1
```

Chapter 10

10.9 (b) B–G–M–Q

(c)

Activity	A	C	D	E	F	H	I	J	K	L	N	O	P
Total float	13	4	13	13	8	4	6	13	13	8	4	6	8
Free float	0	0	0	0	0	0	0	5	5	0	4	6	8

10.10 Critical path: $A–C–D–E–F–G$, Project completion time = 35 days.

10.11 (b)

Activity	A	B	C	D	E	F	G	H
Expected duration	5.17	7.17	5.50	5.17	6.17	8	6	3
Variance	0.695	0.25	0.695	0.695	0.695	1	1	1.78

(c) Critical path: B–D–F–H. Project completion time = 23.34 weeks

(d) 0.9999

(e) 25.8 weeks.

10.12 (b)

Activity	A	B	C	D	E	F	G	H	I	J	K	L	M	N
Expected duration	5	3	6	4	2	5	3	4	3	7	3	5	3	7
Variance	1	1.8	4	1	0.1	1.8	1.8	0	1	0.1	1	0	1.8	0.1

(c) Critical paths: A–F–L–N and C–H–L–N. Project completion time = 22 weeks

(d) 0.9999

(e) 24.73 weeks.

10.13

Crashed activity	1–2	2–3	3–4	4–5
Crashed duration (weeks)	4	8	4	5

Critical paths: 1–2–3–4–5 and 1–2–3–5. Project completion time = 21 weeks and total cost = Rs. 10,300.

10.14

Crashed activity	1–2	1–4	4–6	5–7
Crashed duration (weeks)	5	7	5	1

Critical paths: 1–2–5–7 and 1–4–6–7. Project completion time = 16 weeks and total cost = Rs. 12,350.

10.15 Maximum manpower requirement = 19 and the project completion time = 30 weeks.

10.16 Maximum manpower requirement = 24 and the project completion time = 23 weeks.

10.17 Minimum project completion time as per resource allocation = 30 months.

Chapter 11

11.5 Alternative 2 is the best alternative.

11.6 Alternative 2 is the best alternative.

11.7 (a) 300 kg daily (b) 300 kg daily.

11.8 75 units

11.9 Based on *savage minimax regret* criterion: 225 units. Based on *Hurwicz* criterion: 225 units.

11.10 The company should drill the said property for oil (expected rupee consequences = Rs. 77,200).

11.11 Mr. Ramesh should develop the new drug (expected present value of the net profit = Rs. 1,24,200).

Chapter 12

12.2 $A(1, 0, 0)$, $B(1, 0, 0)$ and $V = 40$.

12.3 (a) $A(0, 1, 0)$, $B(0, 1, 0)$ and $V = 50$. (b) $A(0, 1, 0)$, $B(0, 0, 1)$ and $V = 40$.

12.4 (a)

Player B

		25p	50p	100p
	25p	25	25	25
Player A	50p	50	50	100
	100p	100	50	100

(b) $A(0, 1, 0)$, $B(0, 1, 0)$ and $V = 50$

12.5 $A\left(\dfrac{3}{4}, \dfrac{1}{4}\right)$, $B\left(\dfrac{1}{2}, \dfrac{1}{2}\right)$ and $V = 9$.

12.6 $A\left(\dfrac{7}{17}, \dfrac{10}{17}\right)$, $B\left(\dfrac{9}{17}, \dfrac{8}{17}\right)$ and $V = \dfrac{216}{17}$.

12.7 $A\left(0, \dfrac{7}{10}, \dfrac{3}{10}\right)$, $B\left(0, \dfrac{2}{5}, \dfrac{3}{5}\right)$ and $V = 6.2$.

12.8 $A\left(\dfrac{1}{5}, 0, \dfrac{4}{5}, 0\right)$, $B\left(\dfrac{1}{5}, \dfrac{4}{5}, 0, 0\right)$ and $V = 5.6$.

12.9 $A(1, 0, 0)$, $B(0, 0, 0, 0, 1)$ and $V = 6$.

12.10 $A\left(\dfrac{13}{21}, \dfrac{8}{21}\right)$, $B\left(0, \dfrac{10}{21}, 0, \dfrac{11}{21}, 0\right)$, $V = -\dfrac{46}{21}$.

12.11 $A(1, 0)$, $B(0, 0, 1, 0, 0)$ and $V = 4$.

12.12 $A(0, 1, 0, 0)$, $B(0, 1)$ and $V = 4$.

12.14 $A(0, 0.36, 0.64)$, $B(0.728, 0.272, 0)$ and $V = 3.09877$.

12.15 $A(0.2, 0, 0.8)$, $B(0, 0.867, 0.133)$ and $V = 34$.

Chapter 13

13.4   (a)   Economic life = 6 years and corresponding annual equivalent cost = Rs. 542.

         (b)   Economic life = 8 years and corresponding annual equivalent cost = Rs. 638.56.

13.5   Economic life = 13 years and corresponding annual equivalent cost = Rs. 34,510.

13.6   (a)   Economic life = 4 years and corresponding annual equivalent cost = Rs. 7100.

         (b)   Economic life = 5 years and corresponding annual equivalent cost = Rs. 9171.

13.7   For Machine A, economic life = 8 years and corresponding annual equivalent cost = Rs. 2780. For, Machine B, economic life = 10 years and corresponding annual equivalent cost = Rs. 2873. Hence, the manufacturer should purchase Machine A.

13.8   Follow group replacement and replace transistors once in 3 weeks.

13.9   Follow group replacement and replace resistors once in 2 months.

Chapter 14

14.9   (a)   Sequence: 5–1–8–7–3–4–2–6

         (b)   Sequence: 5–1–3–7–8–2–4–6

         (c)   Sequence: 5–8–1–6–7–4–3–2

14.10   Sequence to minimize mean tardiness: 2–4–5–1–3.

         Sequence to minimize maximum lateness: 2–4–1–5–3.

14.11   Sequence with respect to optimal makespan: 1–5–3–6–2–4 and the optimal makespan = 61.

14.12   Sequence with respect to optimal makespan: 4–1–2–3–5 (or) 4–2–1–3–5 and optimal makespan = 92.

14.13   Sequence with respect to optimal makespan: 1–2–3–4 and optimal makespan = 71.

14.14   Optimal makespan = 37 hrs.

Chapter 15

15.4 Let X_1 = Number of hours allotted to the Team X and X_2 = Number of hours allotted to the Team Y.

$$\text{Minimize } Z = P_1 d_1^- + P_2 d_4^+ + P_3 d_5^+ + P_4(6d_2^- + 10d_3^-)$$

subject to

$$6X_1 + 10X_2 + d_1^- - d_1^+ = 825$$
$$X_1 + d_2^- - d_2^+ = 50$$
$$X_2 + d_3^- - d_3^+ = 50$$
$$d_2^+ + d_4^- - d_4^+ = 10$$
$$d_3^+ + d_5^- - d_5^+ = 15$$
$$X_1, X_2, d_1^-, d_2^-, d_3^-, d_4^-, d_5^-, d_1^+, d_2^+, d_3^+, d_4^+, d_5^+ \geq 0$$

Answer: $X_1 = 50$, $X_2 = 52.5$, $d_3^+ = 2.5$, $d_4^- = 10$, $d_5^- = 12.5$, all other variables are zero and total number of units = **825**.

15.5 Let X_1 = Production volume of product X, and X_2 = Production volume of Product Y.

$$\text{Minimize } Z = P_1 d_1^- + P_2 d_2^+ + P_3 d_1^+$$

subject to

$$2X_1 + 2X_2 + d_1^- - d_1^+ = 20$$
$$d_1^+ + d_2^- - d_2^+ = 4$$
$$X_1, X_2, d_1^-, d_2^-, d_1^+, d_2^+ \geq 0$$

Answer: $X_1 = 10$, $X_2 = 0$, $d_1^- = 4$ and all other variables are zero.

15.6 Let X_1 = Production volume of product A, and X_2 = Production volume of Product B.

$$\text{Minimize } Z = P_1 d_1^- + P_1 d_2^- + P_1 d_3^-$$

subject to

$$3X_1 + 2X_2 + d_1^- - d_1^+ = 1700$$
$$2X_1 + X_2 \leq 1400$$
$$X_1 + 3X_2 + d_2^- - d_2^+ = 1600$$
$$600X_1 + 800X_2 + d_3^- - d_3^+ = 1,00,000$$
$$X_1, X_2, d_1^-, d_2^-, d_3^-, d_1^+, d_2^+, d_3^+ \geq 0$$

Results: $X_1 = 271.45$, $X_2 = 442.85$, $d_3^+ = 4,17,150$, monthly profit = Rs. 5,17,150.

15.7 Let X_1 = Production volume of product A, X_2 = Production volume of Product B and X_3 = production volume of Product C.

$$\text{Minimize } Z = P_1 d_1^- + P_1 d_2^-$$

subject to

$$X_1 + 3X_2 + 4X_3 \leq 1800$$
$$2X_1 + 2X_2 + 3X_3 + d_1^- - d_1^+ = 1500$$
$$450X_1 + 550X_2 + 675X_3 + d_2^- - d_2^+ = 1,20,000$$
$$X_1, X_2, X_3, d_1^-, d_2^-, d_1^+, d_2^+ \geq 0$$

Results: $X_1 = 120$, $X_3 = 420$, $d_2^+ = 2,17,500$, monthly profit = Rs. 3,37,500.

Chapter 16

16.3   When $0 \le t \le \dfrac{3}{2}$, $X_1 = 0$ and $X_2 = 5$.

When $\dfrac{3}{2} \le t \le 3$, $X_1 = 4$ and $X_2 = 3$.

When $t > 3$, $X_1 = 5$ and $X_2 = 1$.

16.6   When $0 \le t \le \dfrac{1}{55}$, $X_1 = 0$, $X_2 = 100 + 225t$, $X_3 = 230 + 50t$, $Z = 1350 + 700t$.

When $\dfrac{1}{55} \le t \le 2.1$, $X_1 = 0$, $X_2 = 105 - 50t$, $X_3 = 230 + 50t$, $Z = 1360 + 150t$.

When $t > 2.1$, solution is infeasible.

16.7   When $-2 \le t \le 3$, $X_1 = \dfrac{8}{5}$, $X_2 = -\dfrac{1}{5}$, $Z = \dfrac{1}{5} + \dfrac{8}{5}t$.

When $t = 3$, multiple solution exists.

Chapter 17

17.6 $X_1 = \dfrac{11}{3}$, $X_2 = \dfrac{10}{3}$, $X_3 = 8$ and $Z = 91.111$.

17.7 $X_1 = 0.8043$, $X_2 = 0.3478$, $X_3 = 0.2826$ and $Z = 0.8477261$.

17.8 $X_1 = \dfrac{4}{5}$, $X_2 = \dfrac{9}{5}$ and $Z = 20.52$.

17.10 $X_1 = 12.5$, $X_2 = 0$ and $Z = 546.875$.

17.12 $X_1 = 2.8$, $X_2 = 0$ and $Z = 32$.

17.14   Maximize $Z = 2X_1 + 4X_2$

subject to:
$$9X_1 + 8X_2 \le 62.52$$
$$25X_1 + 16X_2 + 1.28Y \le 25$$
$$4X_1^2 + 4X_2^2 - Y^2 = 0$$
$$X_1, X_2, Y \ge 0$$

17.15   Maximize $Z = 2X_1 + 4X_2 + 6X_3$

subject to:
$$8X_1 + 6X_2 + 4X_3 \le 54.16$$
$$16X_1 + 4X_2 + 25X_3 + 1.64Y \le 30$$
$$4X_1^2 + 9X_3^2 - Y^2 = 0$$
$$X_1, X_2, X_3, Y \ge 0$$

INDEX

Additivity, 12
ADVANCE block, 342–343
Algorithm
    cutting-plane, 201–209
    Dijkstra's, 163–171
    to determine mixed strategies, 428–429
    Floyd's, 171–174
    Fractional (pure), 202–203
    Johnson's, 517–521
    Kruskal's, 181–185
    for $2 \times n$ games, 440
    for least cost cell method, 77–78
    for $m \times 2$ games, 440–441
    for northwest corner cell method, 77
    PRIM, 175–181
    for two–phase method, 38
    for Vogel's approximation method, 78
All Quantity Discount Scheme (AQDS), 115–116
    models for, 117–118
Alternate optima, 44–45
Alternate optimum solution, 12
Alternative, 275
Artificial variable, 30
Assignment problem, 127–153
    types of, 130

Backward recursive function, 275–276
Balanced transportation problem, 74
Balking, 299
Basic of interest formulas, 473–475
Basic variable, 26
Big M Method, 30–34
Bordered–Hessian square matrix, 563
Branch-and-bound
    algorithm for integer programming, 209–219
    method for flow shop problem, 521–532
    technique for assignment problem, 147–151
    technique for single-machine scheduling, 502–512
Bulk arrival, 299

Capital budgeting, 276–278
Canonical form, 25
Cargo–loading problem, 17, 284–286

Chance-constrained programming, 585–588
Chance point, 417
Completion time, 495
Computer
    era, 1, 2
    simulation, 326–327
Constraints
    adding new, 61–63
    changes in right-hand side constraints of, 57–58
Continuous distribution, guidelines to handle, 348–350
Costs trade-off, 230–231
Covering problem, 199–201
Crash
    cost, 375
    limit, 375
    time, 375
Crashing of project network, 375–390
    with cost trade-off, 377–390
Critical path, 359
    method (CPM), 359–368
Current best lower bound, 210
Cutting-plane algorithm, 201–209

Decision
    deterministic, 409
    point, 417
    theory, 409–421
    tree, 417–421
    types of, 409
    under certainty, 409
    under risk, 409–410
    under uncertainty, 411–416
    variable, 10
Degeneracy in LP, 45–46
Degenerate solution, 12–13
DEPART block, 339–340
Determination of economic life of an asset, 472–485
Deterministic situation, 429
Dijkstra's algorithm, 163–171
Divisibility, 12
Dominance property,
    for columns, 431
    for rows, 431
    in single machine scheduling, 504

Duality, 47–56
Dual simplex method, 34–38
Due date, 495
    earliest, 498–499
Dynamic programming, 275–295
    application areas of, 276–295

Economic life, 472–485
Empirical queueing models, 300–325
Entering variable, 26
Equal payment series capital recovery factor, 474–475
Expected value criterion, 410–411

Fathomed node, 209
Feasible
    region, 22–25
    solution, 12
Fixed-charge problem, 197–199
Fixed order quantity system (Q system), 250–251
Flow shop scheduling, 494, 515–532
Flow time, 495
    maximum, 496
    mean, 495
Floyd's algorithm, 171–174
Free float, 362

Game
    with mixed strategies, 428–430
    with pure strategies, 426–428
    theory, 424–467
Gantt chart, 365–368
General Purpose Simulation System, (GPSS), 337–352
Goal programming, 538–547
Gomory's cut, 34
Graphical method, 21–25
    for $2 \times n$ and $m \times 2$ games, 440–452
Group replacement policy, 485

Hungerian's method, 130–131
Hurwicz criterion, 416

Identity simulation, 326
Implicit enumeration algorithm, zero-one, 219–228
Incremental Quantity Discount Scheme (IQDS), 115–116
    models for, 118–119
Individual replacement policy, 485
Infeasible solution, 12

Initial basic solution, 77
Integer programming, 191–228
    formulations, 197–201
Integer programming algorithms, need for, 197
Inventory
    control, 230–272
    decisions, 230
    model for multi-item joint replenishment, 258–264
    model for multi-item joint replenishment with space constraint, 265–269
    model for multi–item with carrying cost constraint, 256–258
    model for perishable items, 269–272
    system, operation of, 245–247
Inventory model, Implementation of purchase, 250–254

Job shop
    problem, 532–535
    scheduling, 494
    two jobs and m machines, 533–535
Johnson's algorithm, 517–519
    extension of, 519–521
Joykeying, 299

Kendall notations, 300–301
Key
    column, 26
    element, 26
    row, 26
Kruskal's algorithm, 181–185
Kuhn-Tucker conditions, 569–570

Laboratory simulation, 326
Lagrangean method, 562–569
Laplace criterion, 412
Lateness, 495
Latest completion time, 362
Leaving variable, 27
Linearity, 12
Linear programming, 9
    approach for game theory, 453–467
    assumptions in, 12
    methods, 25–40
    modelling of maximal flow problem, 186–188
    parametric, 549–560
    special cases of, 40–47
Linear programming model,
    concepts of, 9–13
    of game theory, guidelines for, 453–459
    properties of, 12–13

Linear programming model, development of
    with respect to player A, 453–456
    with respect to player B, 456–459
Lower bound, 209

Maintenance, types of, 471–472
Manpower scheduling problem, 14–15
Manufacturing model
    without shortages, 236–238
    without shortages for multi-item joint replenishment, 262–264
    with shortages, 241–245
Maximal flow
    algorithm, 188–193
    problem, 186–193
Maximin
    criterion, 413–414
    principle, 425
Measure of performance, 495–496
Minimax
    criterion, 414
    principle, 425
Minimum spanning tree problem, 174–185
Mixed strategy, 428
Model
    to minimize total tardiness, 499–502
    for weighted number of early and tardy jobs, 512–515
Modelling,
    parallel processors under scheduling, 200–201
    steps of, 3
Multi-item
    joint replenishment, EOQ model for, 258–264
    joint replenishment without shortages, manufacturing model for, 262–264
    joint replenishment without shortages, purchase model for, 259–262
    joint replenishment with space constraint, EOQ for purchase model, 265–269
    with carrying cost constraint, purchase model of inventory for, 256–258
Multiple item with storage limitation, 254–256

Network
    construction, guidelines for, 359
    crashing, guidelines for, 376–377
    techniques, 154–193
Non-negativity, 12
    constraints, 11
Nonlinear programming, 562–588
Normal
    cost, 375
    time, 375

Objective function coefficients, 10
    changes in, 59–61
Operation of inventory system, 245–247
Operations research, 1, 6–7
    during the World War II, 1, 2
    scope of, 4–6
    topics of, 3–4
Optimal solution, 12

Parallel server queueing system, 332–337
Parametric linear programming, 549–560
Parametric programming
    changes in objective coefficients, 550–554
    changes in resource requirements, 559–560
    changes in RHS constants, 554–559
Path model, shortest, 154–174
Payoff matrix, 425
Periodic review system (P system), 251–252
PERT, 368–375
Pivot element, 26
Players, 424
Post-World War II developments in OR, 1, 2
Pre-World War II development in OR, 1, 2
PRIM algorithm, 175–181
Probabilistic model, 269–272, 485–491
Probabilistic situation, 424
Processing time, 495
    shortest (SPT), 496–497
Production scheduling, 494–535
Productivity improvement tool, 7–8
Project
    management, 355–404
    scheduling with constrained resources, 390–404
Project management
    phases of, 358
    techniques of, 356
Purchase inventory model, implementation of, 250–254
Purchase model
    for multi-item joint replenishment without shortages, 259–262
    for multi-item joint replenishment with space constraint, 265–269
    for multi-item with carrying cost constraint, 256–258
    without shortages, 231–235
    with shortages, 238–241
Pure strategy, 426

Q system, 250–251
Quadratic programming problem, 572–580
    procedure to solve, 574
Quantity discount, 247–250

Quasi-identity simulation, 326
QUEUE block, 339–340
Queueing models, empirical, 300–325
Queueing system, application areas of, 298
Queueing theory, 298–352

Ready time, 495
Recursive function
    backward, 275–276
    best alternative, 275
    forward, 276
RELEASE block, 341–342
Reliability improvement problem, 278–281
Reneging, 299
Reorder level, 245–247
Replacement and maintenance analysis, 471–491
Replacement problem, types of, 472
Research direction to design superior cut, 207
Resource
    allocation, 398–404
    availability, 11
    levelling, 390–397
Restricted basis method, 582–585

Saddle point, 426
Safety stock, 245–246
Sampling probability distribution, 347–350
Savage minimax regret criterion, 414–415
SEIZE block, 341–342
Sensitivity analysis, 56–65
Separable programming, 581–585
Set of constraints, 11
Silver's algorithm, 260
Simplex method, 25–30
    dual, 34–38
    feasibility condition of, 26, 36
    for goal programming, 541–547
    optimality condition of, 26, 36
Simplex table, identification of special cases from,
    46–47
Simulation, 326–352
    need for, 325–326
    steps of, 327
    types of, 326–327
    using high–level languages, 327–337
Single-machine scheduling, 494
Single-server queueing system, 328–332
Slack variable, 25
Space with finite solution, 43–44
SPT, 496

Stage, 275
Stage-Coach problem (*see also* shortest-path
    problem), 281–284
START block, 340–341
Start time, earliest, 361
State variable, 275
Stochastic programming, 585–588
STORAGE block, 343–345
Strategy, 424
Subdividing problem, optimal, 290–292
Systematic method, 154–163

Tardiness, 495
    maximum, 496
    mean, 495
    minimizing total, 286–290
    model to minimize total, 499–502
Tardy jobs, number of, 512
Technological coefficient, 10–11
Time chart (*see* Gantt chart)
Tools for decision support, 6
Total float, 362
Transportation problem, 71–121
    balanced, 74
    mathematical model, 72–73
    methods to solve, 76–79
    types of, 74–76
    unbalanced, 74–76
Transshipment model, 106–115
Two-person zero-sum game, 426
Two-phase method, 38–40

Uncertainty models, 1, 2
Uncertainty situation, 424
Unconditional transfer block, 350–351
Upper bound, 209

Value of the game, 426
Variable, adding a new, 63–65

Weighted shortest processing time (WSPT) rule,
    497–498
Wolfe's method, 574

Zero-one
    implicit enumeration algorithm, 219–228
    programming model for assignment problem, 128